Springer Series in Advanced Microelectronics

Volume 57

More information about this series at http://www.springer.com/series/4076

The Springer Series in Advanced Microelectronics provides systematic information on all the topics relevant for the design, processing, and manufacturing of microelectronic devices. The books, each prepared by leading researchers or engineers in their fields, cover the basic and advanced aspects of topics such as wafer processing, materials, device design, device technologies, circuit design, VLSI implementation, and subsystem technology. The series forms a bridge between physics and engineering and the volumes will appeal to practicing engineers as well as research scientists.

Yan Li • Deepak Goyal
Editors

3D Microelectronic Packaging

From Fundamentals to Applications

Springer

Editors
Yan Li
Intel Corporation
Chandler, AZ, USA

Deepak Goyal
Intel Corporation
Chandler, AZ, USA

ISSN 1437-0387 ISSN 2197-6643 (electronic)
Springer Series in Advanced Microelectronics
ISBN 978-3-319-83086-5 ISBN 978-3-319-44586-1 (eBook)
DOI 10.1007/978-3-319-44586-1

Printed on acid-free paper

This Springer imprint is published by Springer Nature
The registered company is Springer International Publishing AG
The registered company address is: Gewerbestrasse 11, 6330 Cham, Switzerland

Contents

Contributors

Liangbiao Chen Department of Mechanical Engineering, Lamar University, Beaumont, TX, USA

Paul Conway The Wolfson School of Mechanical, Electrical and Manufacturing Engineering, Loughborough University, Loughborough, Leicestershire, UK

Indranath Dutta School of Mechanical and Materials Engineering, Washington State University, Pullman, WA, USA

Xuejun Fan Department of Mechanical Engineering, Lamar University, Beaumont, TX, USA

Amlan Ganguly Computer Engineering Department, Rochester Institute of Technology, Rochester, NY, USA

Deepak Goyal Intel Corporation, Chandler, AZ, USA

Ran He The University of Tokyo, Tokyo, Japan

Zhiheng Huang School of Materials Science and Engineering, Sun Yat-sen University, Guangzhou, China

Tengfei Jiang Department of Material Science and Engineering, University of Central Florida, Orlando, FL, USA

Satish G. Kandlikar Mechanical Engineering Department, Rochester Institute of Technology, Rochester, NY, USA

Praveen Kumar Department of Materials Engineering, Indian Institute of Science, Bangalore, India

Sangil Lee, Ph.D. Invensas Corporation, San Jose, CA, USA

Menglu Li Department of Materials Science and Engineering, University of California, Los Angeles (UCLA), Los Angeles, CA, USA

Yan Li Intel Corporation, Chandler, AZ, USA

Kwang-Lung Lin Department of Materials Science and Engineering, National Cheng Kung University, Tainan, Taiwan, R.O.C.

Pilin Liu Intel Corporation, Chandler, AZ, USA

Yingxia Liu Department of Materials Science and Engineering, University of California, Los Angeles (UCLA), Los Angeles, CA, USA

Ravi Mahajan High Density Interconnect Pathfinding, Assembly Technology, Intel Corporation, Chandler, AZ, USA

Antonio La Manna IMEC Consortium, Leuven, Belgium

Erasenthiran Poonjolai Intel Corporation, Chandler, AZ, USA

Bob Sankman High Density Interconnect Pathfinding, Assembly Technology, Intel Corporation, Chandler, AZ, USA

Hualiang Shi Intel Corporation, Chandler, AZ, USA

Songhua Shi Medtronic, Tempe, AZ, USA

Tadatomo Suga The University of Tokyo, Tokyo, Japan

Peter Tortorici Medtronic, Tempe, AZ, USA

K. N. Tu Department of Materials Science and Engineering, University of California, Los Angeles (UCLA), Los Angeles, CA, USA

George Vakanas Intel Corporation, Chandler, AZ, USA

Paul Vianco Sandia National Laboratories, Albuquerque, NM, USA

Yaodong Wang Department of Materials Science and Engineering, University of California, Los Angeles (UCLA), Los Angeles, CA, USA

Luhua Xu Intel Corporation, Chandler, AZ, USA

About the Authors

Deepak Goyal graduated (Ph.D.) from State University of New York, Stony Brook, and joined Intel as a Failure Analysis Engineer. He is currently the Technical Leader of the Assembly and Test Technology Development (ATTD) Failure Analysis Labs at Intel, supporting the materials and failure analysis of all assembly technology development at Intel and developing the next generation of analytical tools and techniques. He is an expert in the failure analysis of packages and has taught Professional Development courses on Package FA/FI and failure mechanisms at the ECTC from 2003 to present. He has won two Intel Achievement Awards and around 25 Division Recognition Awards at Intel. He has around 50 publications and has been awarded three patents and has ten patents filed. He was the chair of the Package and Interconnect Failure Analysis Forum sponsored by the International Sematech. He is a senior member of the IEEE. He has coauthored a chapter titled "Applications of terahertz technology in the semiconductor industry" in the *Handbook of Terahertz Technology for Imaging, Sensing and Communications*, 2013. He has also coauthored and edited two chapters for a book titled *Characterization and Metrology for Nanoelectronics and Nanostructures*, which will be published in November 2016.

Yan Li is currently a senior staff packaging engineer in the Assembly Test and Technology Development Failure Analysis Lab of Intel Corporation located in Chandler, Arizona. She joined Intel in 2006, and her work focuses on the quality and reliability of electronic packages, fundamental understanding of failure modes and failure mechanisms of electronic packages, and developing new tools and techniques for fault isolation and failure analysis of electronic packages. Yan received her Ph.D. in materials science and engineering from Northwestern University in 2006 on studies related to the synthesis and characterization of boron-based nanowires and nanotubes. She is actively involved in professional associations, such as Minerals, Metals and Materials Society (TMS), American Society for Metals (ASM) International, and Electronic Device Failure Analysis Society (EDFAS), serving as conference organizers and session chairs for more than 5 years.

Chapter 1
Introduction to 3D Microelectronic Packaging

Yan Li and Deepak Goyal

1.1 Introduction

Microelectronic packaging is the bridge between the Integrated Circuit (IC) and the electronic system, which incorporates all technologies used between them [1]. Advanced 3D microelectronic packaging technology is the industry trend to meet portable electronics demand of ultra-thin, ultra-light, high performance with low power consumption. It also opens up a new dimension for the semiconductor industry to maintain Moore's law with a much lower cost [1–3].

A wide variety of real products assembled by the advanced 3D packaging technology have been unveiled in the recent years. For example, the Apple A7 inside the iPhone 5S, introduced in September 2013, is a 3D package with Package on Package (POP) configuration [4]. As displayed in Fig. 1.1, the wire bond Elpida (now Micron) memory (low power double data rate type-3 (LPDDR3) mobile random access memory (RAM)) package is stacked on top of the Apple A7 flip-chip package to achieve better performance with smaller form factor. In early 2014, SK Hynix announced its high bandwidth memory (HBM) products having higher bandwidth, less power consumption, and substantially small form factor, achieved by stacking up to eight DRAM dice interconnected through Through Silicon Vias (TSV) and micro-bumps [5]. In July of 2015, AMD introduced the AMD Radeon™ Fury graphics cards, the first Graphic Processing Unit (GPU) to implement HBM by TSVs and micro-bumps [6]. Figure 1.2 shows a top-down and schematic cross-sectional view of the advanced 3D package. The big GPU die is integrated into the Si interposer along with four HBM memory stacks by micro-bumps and TSVs to ensure faster and shorter connection between chips [6]. These real products bring

Y. Li (✉) • D. Goyal
Intel Corporation, Chandler, AZ, USA
e-mail: yan.a.li@intel.com

Y. Li, D. Goyal (eds.), *3D Microelectronic Packaging*, Springer Series in Advanced Microelectronics 57, DOI 10.1007/978-3-319-44586-1_1

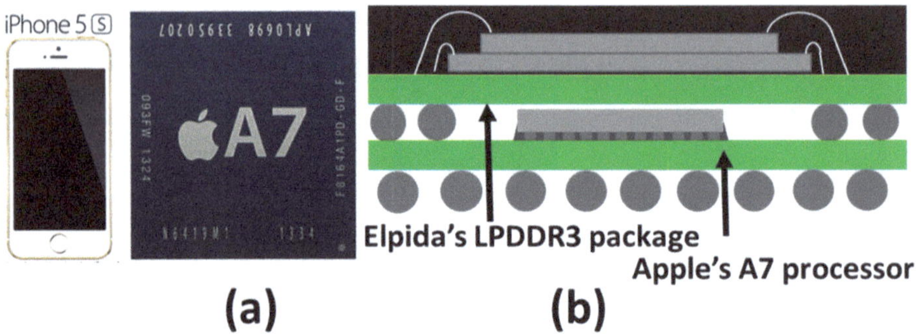

Fig. 1.1 POP inside iPhone 5s. (**a**) Top view of the Apple A7 package, (**b**) schematic of the cross-sectional view (not in scale), (adapted from Ref. [4])

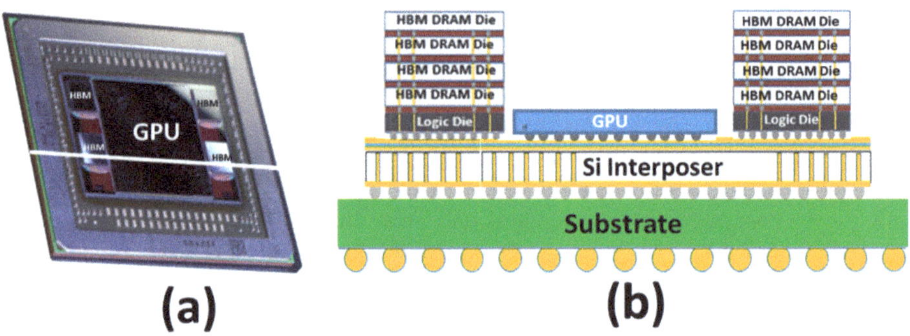

Fig. 1.2 Top view (**a**) and schematic cross-sectional view (not in scale), (**b**) of the AMD Radeon™ Fury. The big GPU die is integrated into the Si interposer along with four HBM memory stacks by micro-bumps and TSVs (adapted from Ref. [6])

the 3D packaging techniques from paper to reality and indicate the extensive applications of 3D packaging technology to microelectronics.

3D packaging technology involves multiple disciplines, for example, materials science, mechanical engineering, physics, chemistry, and electrical engineering. A technical book, which could provide a comprehensive scope of 3D microelectronic packaging technology is desirable for graduate students and professionals in both academic and industry area. Current available books on 3D integration typically focus on processing of wafers, especially TSV fabrication, and do not cover other key elements in 3D packaging. This book is proposed to fill in the gap. It presents a thorough extent of 3D packaging, covering the fundamentals of interconnects, bonding process, advanced packaging materials, thermal management, thermal mechanical modeling, architecture design, quality and reliability, and failure analysis of 3D packages, which are critical for the success of advanced 3D packaging.

This chapter provides detailed illustration of motivations as well as various architectures of 3D packaging. Challenges in 3D packaging, including fabrication, assembly, cost, design, modeling, thermal management, material, substrate, quality, reliability, and failure analysis, are reviewed with brief introduction to the chapters addressing these challenges.

1.2 Why 3D Packaging

1.2.1 Moore's Law

Since Intel introduced the world's first single-chip microprocessor, the Intel 4004, in 1971, an exponential growth of ICs has been observed following Moore's law in terms of transistor number per chip [7]. As illustrated in Fig. 1.3, the number of transistors per Si chip doubles approximately every 18 months, resulting in a straight line on a log scale [7, 8]. In 1990, the bipolar transistor technology switched to CMOS in order to reduce thermal power, circuit size, and manufacturing costs, at the same time increase the operating speed and energy efficiency [3]. In the early 2000s, multi-core processors were developed to address the challenging thermal power issue in conventional single-core processors [3]. Since multi-core processors require enormous cache capacity and memory bandwidth to achieve the designed performance, 3D packaging becomes one of the viable solutions to provide the required cache and bandwidth with a relatively low cost [3].

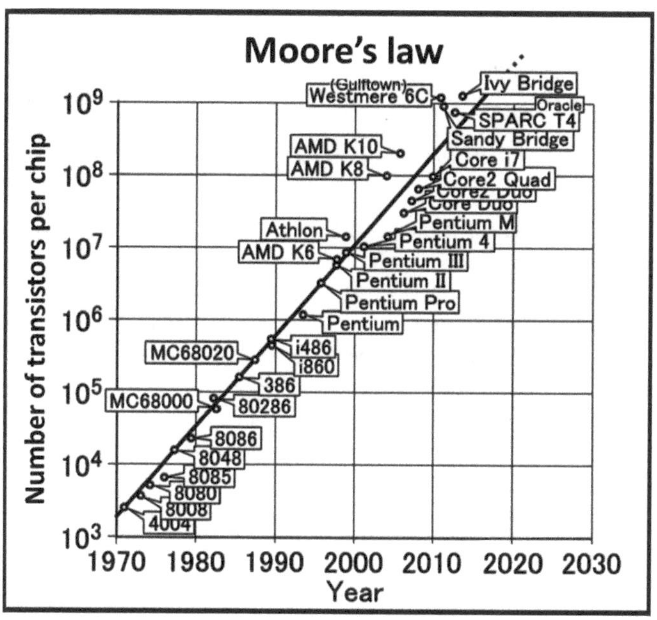

Fig. 1.3 Moore's law predicts the exponential growth of ICs since 1970s (adapted from Ref. [8])

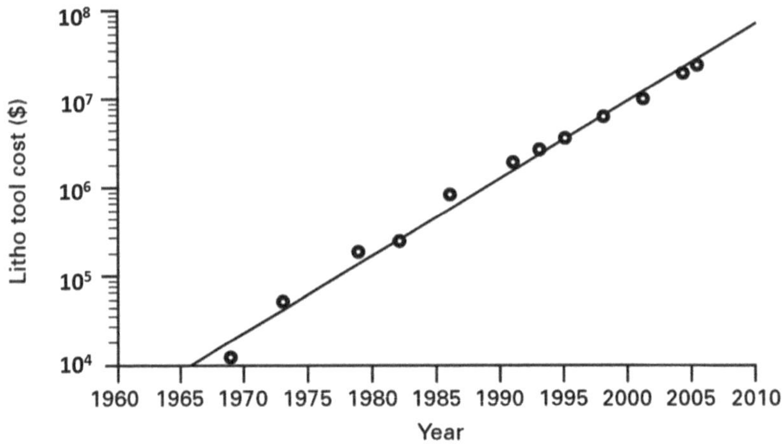

Fig. 1.4 The exponential growth of lithography equipment cost since 1970s (adapted from Ref. [8])

The conventional method to maintain Moore's law is to decrease the dimensions of components by lithography, which is becoming more and more sophisticated and expensive [8]. Figure 1.4 illustrates the exponential growth of lithography equipment cost since 1970s, which presents an economic challenge as the capital cost rises faster than semiconductor industry revenue [8]. 3D integration technology, which has been recognized as an enabling technology for future low cost ICs, provides the third dimension to extend Moore's law to ever higher density, more functionality, better performance with lower cost [3].

1.2.2 Small Form Factor Requires 3D Packaging

Market demands of small form factor microelectronics head to 3D packages, which are ultra-light, ultra-thin, and with small chip footprint. Si chips in 3D packages are typically 50–100 μm thick, about 90 % thinner compared with those in conventional packages. Substrate core thickness of 3D packaging is about 0–100 μm, more than 90 % thinner than that of traditional packaging. High density interconnects in 3D packaging are on the order of 5–20 μm in diameter, more than 90 % smaller than those in 2D packaging. Thus, tremendous reduction in size and weight could be achieved by replacing conventional packaging with 3D technology [2].

Small form factor requires small chip footprint, which is defined as the printed circuit board area occupied by the Si chip, as illustrated in Fig. 1.5 [2]. By stacking multiple dice on top of each other using 3D packaging techniques, the chip footprint could be reduced dramatically. Figure 1.5 schematically demonstrates the difference between conventional 2D packages and 3D packages

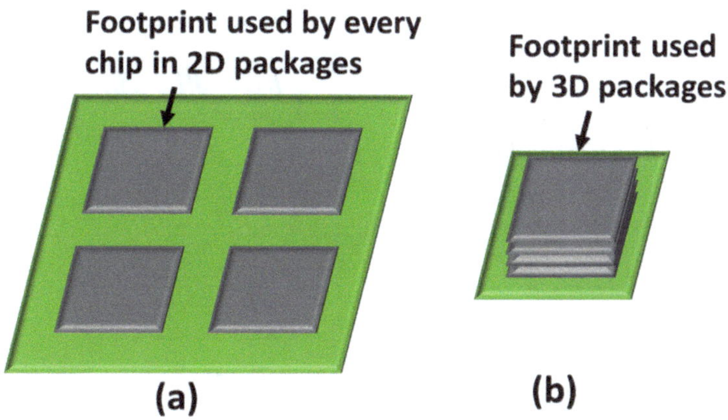

Fig. 1.5 (**a**) Schematic illustration of the footprint difference between conventional 2D packages (**a**) and 3D packages (**b**) (adapted from Ref. [2])

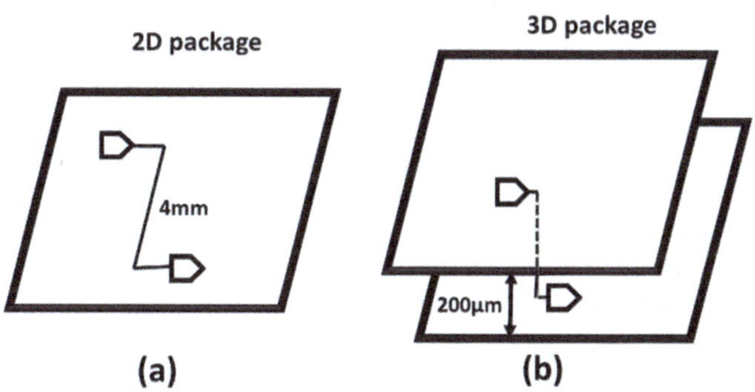

Fig. 1.6 Schematic comparison between the wiring length in 2D packages (**a**) and 3D packages (**b**) (adapted from Ref. [2])

1.2.3 Improved System Performance with Reduced Power

Interconnect length in 3D packages can be significantly reduced compared with conventional 2D packaging [2]. Figure 1.6a shows a typical interconnect length of 4 mm in a 2D package. The interconnect length could be reduced to 200 μm in a 3D package as demonstrated in Fig. 1.6b. Additionally, 3D packaging can also greatly improve the interconnect usability and accessibility [2]. Figure 1.7 shows a comparison between 2D and 3D packaging in terms of the accessibility and usability of interconnection. In contrast to eight neighbors to the center element in the case of 2D packaging technology, the utilization of a 3D packaging configuration provides access to 116 neighbors within an equal interconnect length to a center element in the stack.

The significant decrease in interconnect length as well as dramatic improvement in interconnect usability and accessibility result in much less delay in 3D devices, which is primarily limited by the time taken for the signal to travel along the

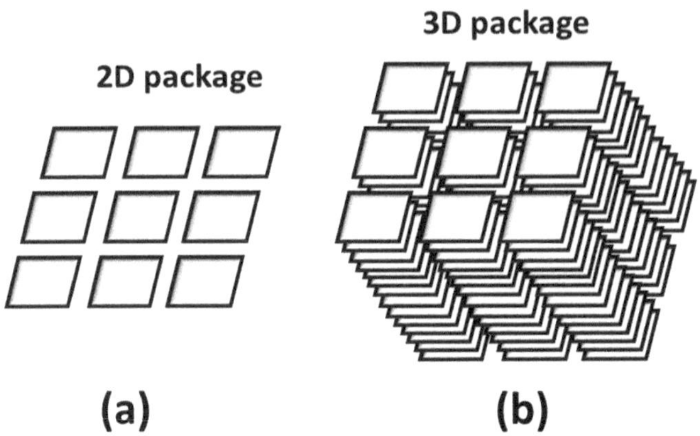

Fig. 1.7 Schematic comparison between 2D packages (**a**) and 3D packages (**b**) in terms of the interconnect accessibility and usability (adapted from Ref. [2])

interconnects [2]. Low latency and wide buses in 3D microelectronic systems lead to significant improvement to the system bandwidth [2]. Noise in well-designed 3D microelectronic systems, including reflection noise, crosstalk noise, simultaneous switching noise, and electromagnetic interference, can be reduced as a result of the reduction of interconnection length [2]. Additionally, as the parasitic capacitance in microelectronic packages is proportional to the interconnection length, the total power consumption in 3D packages is also reduced because of the reduced parasitic capacitance [2]. The power saving achieved by 3D technology enables 3D devices to perform at a faster rate or transition per second (frequency) with less power consumption. The overall system performance is greatly improved by applying 3D packaging technology [2].

1.3 3D Microelectronic Packaging Architectures

The various 3D packaging architectures could be divided into the following three categories: die-to-die 3D integration, package-to-package 3D integration, and heterogeneous 3D integration combining both package and die stacking [3, 9, 10]. Chapter 2 discusses different 3D packaging architectures in detail along with assembly and test flows.

1.3.1 Die-to-Die 3D Integration

Die-to-die 3D integration is enabled by through silicon via (TSV) interconnections and thinned die-to-die bonding [3]. As illustrated in Fig. 1.8, two memory dice are stacked on top of a logic die with TSVs and micro-bumps. First Level Interconnect

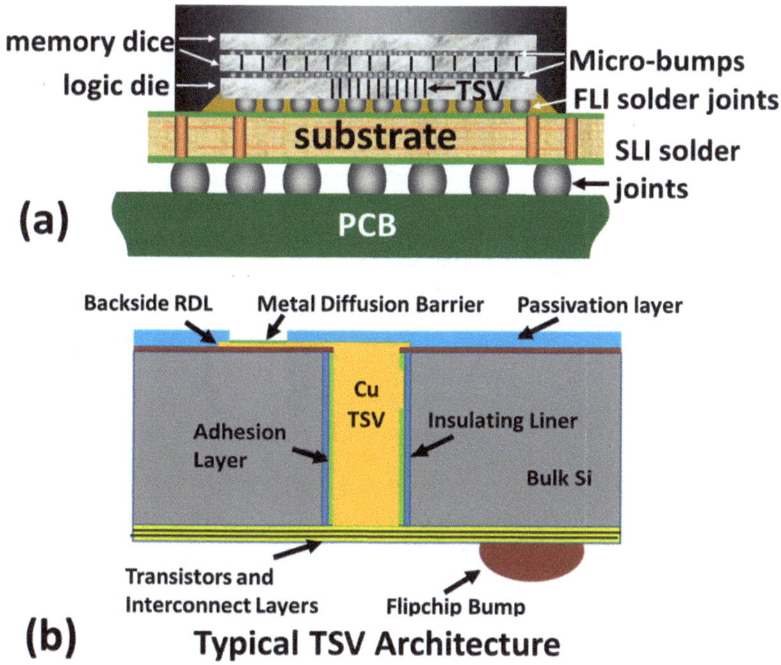

Fig. 1.8 Schematic illustration of die-to-die 3D integration enabled by TSV and thinned die-to-die bonding (adapted from Ref. [10])

(FLI) solder joints connects the logic die with the substrate, while Second Level Interconnect (SLI) solder joints provides the connection between this 3D package to the Printed Circuit Board (PCB) [10]. The TSVs are formed by laser drilling or deep reactive ion etching, followed by liner deposition and copper fill. There are three typical manufacturing processes for TSVs, Via First, Via Middle, and Via Last [11–13]. The detailed process flow of each process as well as the Pros and Cons of each process are discussed in Chaps. 2 and 3.

Die-to-die bonding is implemented by either Thermal Compression Bonding (TCB) process for solder micro-bumps or other alternative bonding process, for example, Cu-Cu bonding. Conventional solder mass reflow process in 2D packages, which includes flux dispensing, die attaching, and solder reflow in ovens, is not able to assemble advanced 3D packages having much thinner dice and organic packages, along with much smaller and denser interconnects. As the extent of warpage from both dice and substrates at the reflow temperature overcomes solder surface tension, leading to die misalignment, and results in die tilting, solder joint non-contact opens, and solder bump bridging [14]. TCB bonding process is developed to replace the conventional solder mass reflow process for solder-based micro-bump assembly in 3D packages. As illustrated schematically in Fig. 1.9, the substrate with pre-applied flux is kept flat on hot pedestal under vacuum to eliminate the substrate warpage. The die is picked up by the bond head, secured and kept flat with vacuum

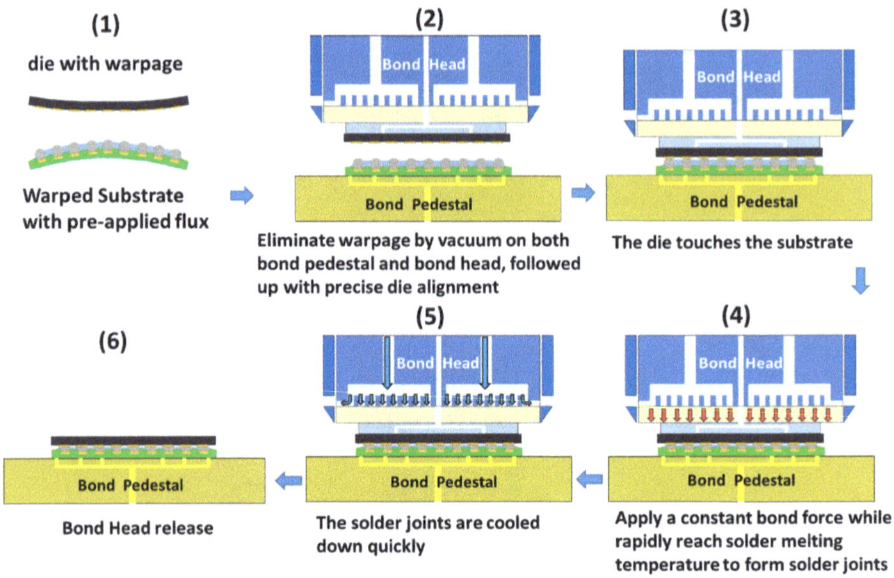

Fig. 1.9 Schematic illustration of a typical thermal compression bonding process (adapted from Ref. [14])

to remove any incoming die warpage. After the die is precisely aligned to the substrate, the bond head with die touches the substrate. A constant bond force is then applied on the die through the bond head, while the die is heated up rapidly beyond the solder melting temperature, with a ramping rate higher than 100 °C/s. As soon as the solder joints melt, the die is moved further down to ensure all the solder joints are at the same height. The die is held in this position long enough so that the solder joint forms between the die and the substrate. While the solder is still in the molten state, the bond head with die could retract upwards to control the solder joint height. Subsequently, the solder joints are cooled abruptly below the solidus temperature, with a cooling rate of more than 50 °C/s, followed up with die release from the bond head [14]. Unlike the traditional solder mass reflow process, with up to 10 min of process time for units in batches, the TCB bonding process assemble units one by one with about a couple of seconds per unit [14]. Additionally, thermal ramping rates during both the heating and cooling cycles are much higher than the conventional method. These higher rates result in solder grain size and orientation differences that can affect mechanical properties as detailed in Chap. 7.

Solder-based micro-bumps are more compliant, thus could compensate bump height variations, lack of co-planarity, and misalignment issues during high volume manufacture. However, the TCB process peak temperature needs to be higher than the melting point of solder material, typically in the range of 250–300 °C, which brings more assembly and reliability challenges. Additionally, solder bridging risk gets much higher as bump pitch shrinks from more than 100 μm to less than 40 μm. Alternative bonding process, like direct Cu to Cu bonding, which could address interconnects with

Fig. 1.10 Schematic illustration of package-to-package 3D integration (adapted from Ref. [3])

less than 5 μm bump pitch, and assemble at relatively lower temperature is very promising. Various types of alternative bonding process is reviewed in Chap. 6 as well as Pros and Cons comparing with solder-based TCB process.

1.3.2 Package-to-Package 3D Integration

System in Package (SIP) and Package on Package (POP) are typical configurations of package-to-package 3D integration, which is enabled by stacking packages through wire bonding or flip-chip bonding [3]. Comparing to die-to-die stacking, package-to-package stacking technique has a shorter development cycle, thus help bring products to market faster with a low price. As displayed in Fig. 1.10, a wire bonding package is stacked on top of the other wire bonding package by flip-chip bonding. The two packages are then stacked on a flip-chip package to form a POP. The conventional solder mass reflow process could still be used in package-to-package stacking if the package warpages are within control, and the interconnect size and density is comparable with traditional 2D packages. However, market demands require packages to be ultra-thin, which limit the number of packages that could be stacked together. Additionally, solder joints and package materials in SIP and POP need to go through multiple cycles of reflow, bringing process and reliability challenges, like solder joint open, delamination between multiple layers in packages, and moisture control between each reflow process. Chapter 13 discusses in detail the processing and reliability of stacked packaging technique, as well as the Pros and Cons comparing with die stacking.

1.3.3 Heterogeneous 3D Integration

Depending on product needs, complex 3D packages can have the combination of both die stacking and package stacking [10]. As shown in Fig. 1.11, a 3D Dynamic Random Access Memory (DRAM) package formed by stacking four memory dice

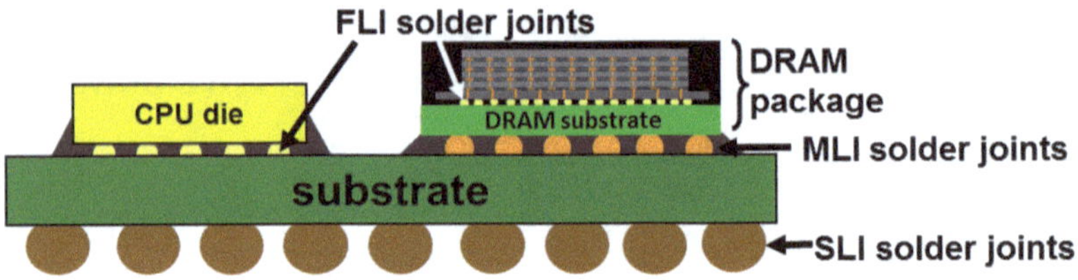

Fig. 1.11 Schematic illustration of 3D packaging architectures having the combination of both die stacking and package stacking (adapted from Ref. [10])

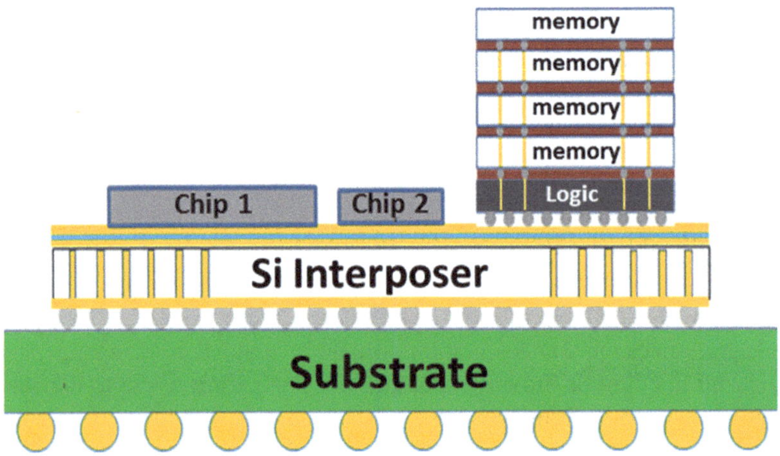

Fig. 1.12 Schematic illustration of Silicon interposer technology providing connection between chips and stacked dice through TSVs (adapted from Ref. [10])

on top of the logic die through TSVs and micro-bumps is integrated along with a flip CPU chip by package stacking. FLIs between the CPU chip and the 3D package, MLIs between the DRAM package and the 3D package, as well as interconnects in the substrate provide connection between the CPU chip and the DRAM package [10]. Smaller and denser interconnects between chips and packages are highly desired for better performance, higher bandwidth, and lower power consumption. Figure 1.12 demonstrates the Silicon interposer technology, which could provide better connection between chips and stacked dice through TSVs [10, 15]. Embedded Multi-Die Interconnect Bridge (EMIB) technology is an alternative approach to provide localized high density interconnects between dice without TSVs [16]. As illustrated in Fig. 1.13, the link between dice is provided by fine Cu interconnects in Si bridges embedded in the organic substrate and confined denser FLIs between Si bridges and chips. Comparing with Si interposer technology, EMIB technology is able to provide similar performance with a much lower cost, thus opens up new opportunities for heterogeneous 3D packaging [16].

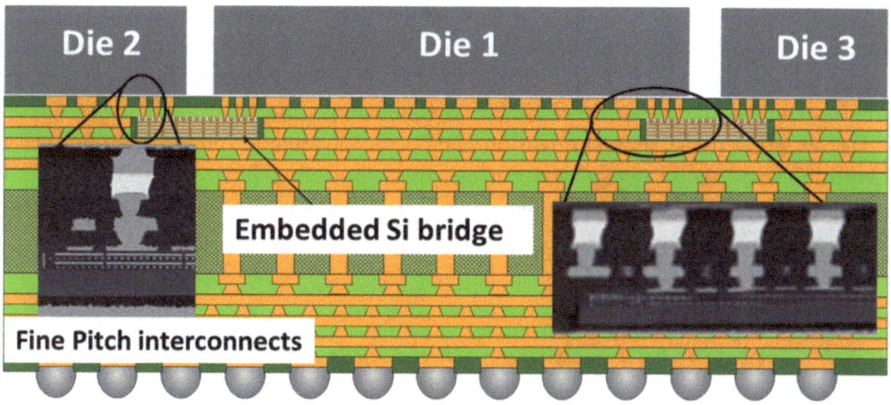

Fig. 1.13 Schematic illustration of Embedded Multi-Die Interconnect Bridge (EMIB) technology providing localized high density interconnects between chips without TSVs (adapted from Ref. [16])

1.4 3D Microelectronic Packaging Challenges

1.4.1 Assembly Process, Yield, Test, and Cost Challenges

3D packaging involves more challenging assembly steps than conventional packaging, such as TSV wafer fabrication and die singulation process (reviewed in Chap. 5), TCB of micro-bumps (discussed in Chap. 7), and multiple solder reflow process for POP (refer to Chap. 13). The complicated process results in yield, test, and cost challenges [2], which could be addressed by redundancy or fault-tolerant designs, throughput time (TPT) improvement of assembly process, and minimize process steps based on product quality and reliability requirement [2], [3].

1.4.2 Thermal Management, Package Design, and Modeling Challenges

The 3D integration of heat generation components in close vicinity increases the heat flux density as well as complexity of coolant routing, thus leads to big challenges to the thermal management of 3D packaging [2, 3]. Chapter 10 presents the fundamentals of heat transfer along with advanced guidelines helpful to address the issue.

Due to the increased system complexity, designing 3D packages could be very challenging, but could be addressed by designing and developing design software [2]. In addition to increased system complexity, 3D packaging involves multilevels of solder joints, underfill, and molding compounds. Thermal stress due to the mismatch in coefficients of thermal expansion (CTE) and hygroscopic stress caused by excessive moisture absorption are often combined together, which complicates the stress modeling in 3D packaging [17, 18]. For interconnects with a couple of micrometer in diameter, like TSVs, it is found that microstructure, anisotropy of material properties, recrystallization, and time-dependent phase morphological evolution need to be considered during stress modeling [19]. Chapters 4 and 10 provide through discussions on the modeling of thermal mechanical and moisture stresses in 3D packaging.

1.4.3 Material and Substrate Challenges

3D packages typically have smaller interconnect size, tighter bump pitch, and reduced chip gap, which brings challenges to underfill, chip attachment, and deflux process. Additionally, the TCB process widely used in the chip attachment of 3D packages is very different from the conventional mass reflow process, as the whole bonding cycle completes in a few seconds rather than over 10 min [14]. Major modification of traditional underfill and flux material is essential to prevent underfill process, flux residue, and interconnect integrity-induced yield loss, such as underfill voids, delamination because of flux residue, solder bump bridging, or non-wets. Chapter 7 reviews the material challenges and provides guidelines for epoxy and flux material selection.

To enable the highly integrated 3D packaging, both substrate and PCB need to fulfill the much higher signal and power density requirements. Smaller substrate vias, through holes (TH), and traces, along with much tighter pitches are desired. Furthermore embedded components, including both Si chip and packages, integrating into substrates or PCB is one of the approaches to achieve the product miniaturization with higher performance and lower power consumption goal. These results in significant challenges in substrate warpage control and flawless fabrication process to enable much finer interconnect size and pitch. Chapter 11 reviews the substrate material and fabrication technology evolution, along with general recommendations in selecting and applying appropriate material and process technologies for 3D packaging.

1.4.4 Quality, Reliability, and Failure Analysis Challenges

Complex 3D packages have multiscale interconnects ranging from a few to 1000 micrometers. For instance, TSVs and micro-bumps are about a couple of micron in diameters, while the SLI connecting packages to the PCB could be up to 1000 μm in diameter. During 3D integration, interconnects need to go through multiple solder reflow process, defects generated during fabrication and assembly, CTE mismatch between different materials, and microstructure evolution in interconnects could lead to new quality and reliability issues. Additionally, the extended application of 3D packaging in products requiring much higher reliability, for instance, Advanced Driver Assistance Systems (ADAS), avionics, and high-end servers brings extra challenges to the quality and reliability of 3D packages. Chapters 3 and 4 review the quality and reliability of TSVs. Chapter 9 reviews the electro migration concerns in interconnects of 3D packages. Focusing on the reliability of multilevel solder joints in stacked packages, Chap. 13 explains the different reliability requirements between consumer electronics and high-reliability electronics. It also provides detailed discussions of use conditions, the roles of encapsulants and underfills, reliability tests and modeling to address complex reliability concerns in 3D packages. Chapter 14 provides an overview of quality and reliability of 3D packaging and demonstrates with case studies, along with field performance prediction.

Failure analysis is critical for the technology development of 3D packaging, as in-depth root cause analysis of failures provides solution paths for resolving quality and reliability issues. Due to the complexity of 3D packages, Fault Isolation (FI) and Failure analysis (FA) become very challenging. First of all, multiple failures could exist in one unit post-reliability tests, flawless failure analysis on each failure requires nondestructive high-resolution techniques, including fault isolation, imaging, and material analysis. Additionally, each electrical failure in a 3D package could come from various dice, assembly layers or interconnects, high-resolution fault isolation techniques, which could provide 3D information of defects are highly desired. After the identification of defects, physical failure analysis needs to be performed for the root cause study. However interconnects in 3D packages, like TSVs, have small diameters (2–10 μm) and long lengths (40–200 μm), artifact free cross-sectional techniques with short throughput time is critical for characterizing small defects in a relatively large cross-sectional plane. Chapter 15 reviews advanced high-resolution nondestructive FI and FA techniques, such as Electro Optic Terahertz Pulse Reflectometry (EOTPR), 3D X-ray Computed Tomography (CT), Lock-in Thermography (LIT), and acoustic microscopy. The applications of novel physical sample preparation techniques and various material analysis methods in 3D packaging failure analysis are also discussed. It also provides guidelines for building up efficient 3D packaging FI-FA flow and performing in-depth root cause studies, along with case study demonstration.

1.4.5 Summary

3D packaging has provided a new dimension for semiconductor industry to maintain Moore's law with much lower cost and has been adopted as an effective approach to provide portable microelectronics with better performance, smaller power consumption, and less cost. There has been numerous novel technological innovations invented for the development of advanced 3D packaging in the recent years. However, due to the much smaller and denser interconnects, complicated assembly process, unique TSV and micro-bump TCB process, and the extended application in high-reliability products, there are tremendous challenges in highly integrated 3D packaging.

Chapters in this book are written by experts from both academia and semiconductor industry. Chapter 2 provides an insightful overview of 3D packaging architecture and assembly process design. Chapters 3, 4, and 5 focus on the fundamentals of TSV processing, reliability, and mechanical properties. Chapters 6 and 7 discuss the fundamentals of thermal compression bonding of micro-bumps, process materials, direct Cu to Cu bonding, and other alternative interconnects in 3D packaging. Chapters 8 and 9 provide fundamentals of solder alloys and electro migration in interconnects of 3D packages. Chapter 10 presents a thorough review of thermal management in 3D packaging. Chapter 11 displays in great details the fundamentals of substrate materials and manufacture process. Chapter 12 covers the thermal mechanical and moisture modeling in 3D packaging. Chapters 13, 14, and 15 illustrate a comprehensive overview of quality, reliability, fault isolation, and failure analysis of advanced 3D packages. Readers could obtain all-around knowledge about 3D packaging, including the fundamentals, developing areas, technique gaps, and guidelines for future research and development.

Acknowledgement The editors would like to thank John Elmer from Lawrence Livermore National Laboratory for his critical review of this chapter.

References

1. I. Szendiuch, Radioengineering. **20**(1), 214 (2011)
2. S.F. Al-sarawi, D. Abbott, P.D. Franzon, IEEE Trans. Comp. Packag. Manufact. Technol. **21** (1), 2 (1998)
3. J.Q. Lu, Proc. IEEE **97**(1), 18 (2009)
4. J.H. Lau, Chip Scale Rev. **18**(1), 32 (2014)
5. L. Li, P. Chia, P. Ton, M. Nagar, S. Patil, J. Xue, J. DeLaCruz, M. Voicu, J. Hellings, B. Isaacson, M. Coor, R. Havens, in *Conference Proceedings from the 66th Electronic Components and Technology Conference. (ECTC)*, Las Vegas, 31 May–3 June 2016, p. 1445
6. C. Lee, C. Hung, C. Cheung, P. Yang, C. Kao, D. Chen, M. Shih, C. C. Chien, Y. Hsiao, L. Chen, M. Su, M. Alfano, J. Siegel, J. Din, B. Black, in *Conference Proceedings from the 66th Electronic Components and Technology Conference. (ECTC)*, Las Vegas, 31 May–3 June 2016, p. 1439
7. R.R. Tummala, *Fundamentals of Microsystems Packaging* (McGraw-Hill, New York, 2001), pp. 4–41
8. D.C. Brock, *Understanding Moore's Law: Four Decades of Innovation* (Chemical Heritage Foundation, Philadelphia, 2006), pp. 67–84
9. Y. Li, P.K. Muthur Srinath, D. Goyal, J. Electron. Mater. **45**(1), 116 (2016)
10. Zhiyong Ma and David G. Seiler, "Metrology and Diagnostic Techniques for Nanoelectronics", Pan Stanford Publishing, (2016), pp. 1089-1119. (To be published).
11. T.M. Bauer, S.L. Shinde, J.E. Massad, D.L. Hetherington, in *Conference Proceedings from the 59th Electronic Components and Technology Conference. (ECTC)*, San Diego, 26–29 May 2009, p. 1165

12. G. Pares, N. Bresson, S. Minoret, V. Lapras, P. Brianceau, J.F. Lugand, R. Anciant, N. Sillon, in *Conference Proceedings from the 11th Electronics Packaging Technology Conference (EPTC)*, Singapore, 9–11 Dec. 2009, p. 772
13. K-W Lee, H. Hashimoto, M. Onishi, Y. Sato, M. Murugesan, J.-C. Bea, T. Fukushima, T. Tananka, M. Koyanagi, in *Conference Proceedings from the 64th Electronic Components and Technology Conference (ECTC)*, Orlando, 27–30 May 2014, p. 304
14. A. Eitan, K. Hung, in *Conference Proceedings from the 65th Electronic Components and Technology Conference (ECTC)*, San Diego, 26–29 May 2015, p. 460
15. K. Zoschke, M. Wegner, M. Wilke, N. Jürgensen, C. Lopper, I. Kuna, V. Glaw, J. Röderl, O. Wünschl, M.J. Wolf, O. Ehrmann, H. Reichl, in *Conference Proceedings from the 60th Electronic Components and Technology Conference (ECTC)*, Las Vegas, 1–4 June 2010, p. 1385
16. R. Mahajan, R. Sankman, N. Patel, D. Kim, K. Aygun, Z. Qian, Y. Mekonnen, I. Salama, S. Sharan, D. Iyengar, D. Mallik, in *Conference Proceedings from the 66th Electronic Components and Technology Conference (ECTC)*, Las Vegas, 31 May–3 June 2016, p. 558
17. C.-H. Liu, J.-L. Tsai, C. Hung-Hsien, C.-L. Lu, S.-C. Chen, in *Conference Proceedings from the 64th Electronic Components and Technology Conference (ECTC)*, Orlando, 27–30 May 2014, p. 1628
18. C.-H. Liu, Y.-H. Liao, W.-T. Chen, C.-L. Lu, S.-C. Chen, in *Conference Proceedings from the 65th Electronic Components and Technology Conference (ECTC)*, San Diego, 26–29 May 2015, p. 1502
19. N. Nabiollahi, N. Moelans, M. Gonzalez, J.D. Messemaeker, C.J. Wilson, K. Croes, E. Beyne, I.D. Wolf, Microelectron. Reliab. **55**(5), 765–770 (2015)

Chapter 2
3D Packaging Architectures and Assembly Process Design

Ravi Mahajan and Bob Sankman

Acronyms

2D	Two dimensional
3D	Three dimensional
BEOL	Back end of line
BI	Burn-In
CMP	Chemical mechanical polishing
D2D	Die-to-die
D2W	Die-to-wafer
ECD	Electro-chemical deposition
ECG	Deleted in chapter
EMIB	Embedded multi-die interconnect bridge
FEOL	Front end of line
IP	Intellectual property
KGD	Known good die
KOZ	Keep out zone
MCM	Multi chip module
MCP	Multi chip package
MEOL	Middle end of line
MPM	Multi package module
PECVD	Plasma enhanced chemical vapor deposition
PVD	Plasma vapor deposition
Rx	Receiver
SBS	Side by side

R. Mahajan (✉) • B. Sankman
High Density Interconnect Pathfinding, Assembly Technology, Intel Corporation,
MS CH5-157, 5000 West Chandler Blvd., Chandler, AZ 85226, USA
e-mail: ravi.v.mahajan@intel.com; bob.sankman@intel.com

Y. Li, D. Goyal (eds.), *3D Microelectronic Packaging*, Springer Series in Advanced
Microelectronics 57, DOI 10.1007/978-3-319-44586-1_2

SIP	System in package
SOC	System on chip
TDP	Thermal design power
TIM	Thermal interface material
TSV	Through silicon via
Tx	Transmitter
W2W	Wafer-to-wafer

2.1 Introduction

Increasing transistor density enabled by Moore's Law [1, 2] has led to increasingly powerful and pervasive computer systems that enable multiple personal and business applications (see Fig. 2.1 for an estimate of the overall size and growth trends of the semiconductor market). The trend in increased compute capabilities has been fueled by increased wired and wireless connectivity [3] which has led to powerful interconnected computer networks. These computers and associated networks utilize a myriad of computing and communication functions that are implemented within digital circuits (e.g., Microprocessors, Field Programmable Gate Arrays), memory circuits (e.g., SRAM, DRAM), and analog circuits (e.g., power supplies, clocks, RF front end modules, amplifiers, SERDES, USB, PCIe, DDR). Different computing and communication functions can be integrated on a monolithic silicon chip (typically referred to as System On Chip or SOC integration) or on a package (typically referred to as System In Package or SIP integration[1]). Integration on chip has the advantages of improved signal transmission fidelity due to reduced interconnect lengths, lower system power due to efficient on-chip connections between IP blocks, and overall reduced silicon resulting from Moore's Law scaling. Thus, on-chip integration is usually preferred when:

(a) The integrated functions can easily be implemented on the same silicon fabrication process. For example, digital logic and SRAM can be built using compatible silicon processes. Conversely, high performance digital logic and DRAM are rarely fabricated on similar silicon manufacturing processes, and hence are not commonly included in the same chip.

---

[1] As a general definition, SIP refers to the on-package integration of multiple heterogeneous and/or homogenous ICs each of which may be in the form of unpackaged die, individually packaged die, or packaged modules. iNEMI (International Electronics Manufacturing Initiative) defines SIP as: *System in Package is characterized by any combination of more than one active electronic component of different functionality plus optionally passives and other devices like MEMS or optical components assembled preferred into a single standard package that provides multiple functions associated with a system or subsystem.* SIP is considered to be subset of the broader concept of System On Package (SOP) [4] where an entire computer system is built on a package.

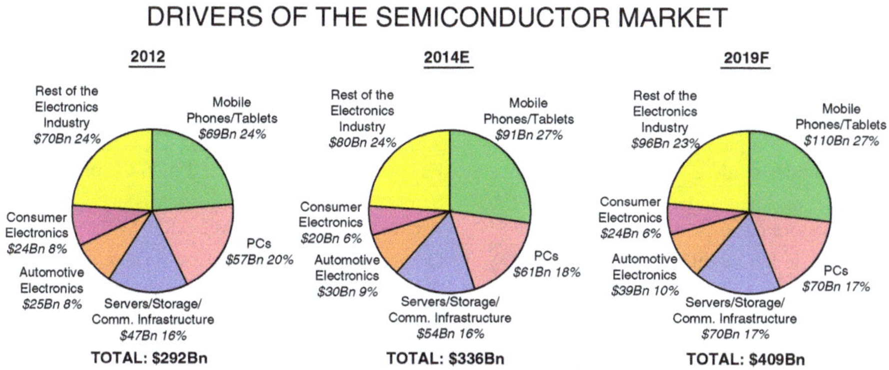

	CAAGR 2012-2014	CAAGR 2014-2019
Mobile Phones/Tablets	15%	3.9%
PCs	3.5%	2.8%
Server/Storage/Comm. Infra.	7.2%	5.3%
Automotive Electronics	9.5%	5.4%
Consumer Electronics	-8.7%	3.7%
Rest of the Electronics Indus.	6.9%	3.7%
TOTAL	7.3%	4.0%

N215.267bp-semimkt12-19

Fig. 2.1 Overall size and growth trends of the semiconductor market (Source: Prismark Partners LLC)

(b) The IP[2] blocks desired for the SOC are available in the same silicon fabrication process and the resulting chip meets the cost target required to make the product economically viable.

SIP is preferred for the integration of heterogeneous functions [5] (i.e., functions manufactured using disparate semiconductor technologies) and to help bring products to market quicker when technical and/or business reasons prevent timely SOC integration. As seen in Table 2.1 and Fig. 2.2, there continues to be significant interest in SIP configurations. SIP architectures can be broadly classed in three categories:

(a) Planar configurations where two or more die or packages are placed side by side and connected to each other through lateral interconnects in a multilayer substrate.
(b) Stacked configurations where two or more die, or packages, are stacked one on top of the others and connected through a combination of both lateral and vertical interconnects. The value of implementing stacking in a product

[2] An IP (Intellectual Property) block is reusable circuit block that performs a certain specialized functions and serves as a building block for constructing the SOC.

Table 2.1 Volume forecast of different SIP configurations (Source: Prismark Partners LLC)

SiP/MCP forecast			
Product/package type volume (Bn units)	2014	2019 F	Leading suppliers/players
Stacked die in package and memory card	8.3	10.5	Samsung, Micron, SKHynix, Toshiba, SanDisk PTI, ASE, SPIL, Amkor, STATS ChipPAC
Stacked package on package: bottom package only	0.95	1.2	Samsung, Apple, Qualcomm, MediaTek Amkor, STATS ChipPAC, ASE, SPIL
PA centric RF module	4.5	5.9	Qorvo, Skyworks, Anadigics, Avago, Amkor, ASE, Inari, HEG, JCET. Unisem, ShunSin
Connectivity module (bluetooth/WLAN)	0.6	0.9	Murata, Taiyo Yuden, Samsung, ACSIP, ALPS, USI
Graphics/CPU or ASIC MCP	0.2	0.2	Intel, AMD, Nvidia, Xilinx, Altera
Leadframe module (power SiP)	3.2	4.7	NXP, STMicro, TI, Freescale, Toshiba, Infineon/IR, Renesas, ON Semi
MEMS and controller	5 4	8.2	ST, Analog, Bosch, Freescale, Knowles, InvenSense, Denso
Camera module	3.7	5.3	LG Innotek, SEMCO, Hon Hai, Lite-on, Toshiba, Sunny Optical, Sharp, Cowell
Fingerprint sensor	0.35	1.5	Apple, Synaptics, Fingerprint Cards, Cypress/ IDEX, Silead, Goodix, NEXT Biometrics, Qualcomm
Total	*27.2*	*38.4*	

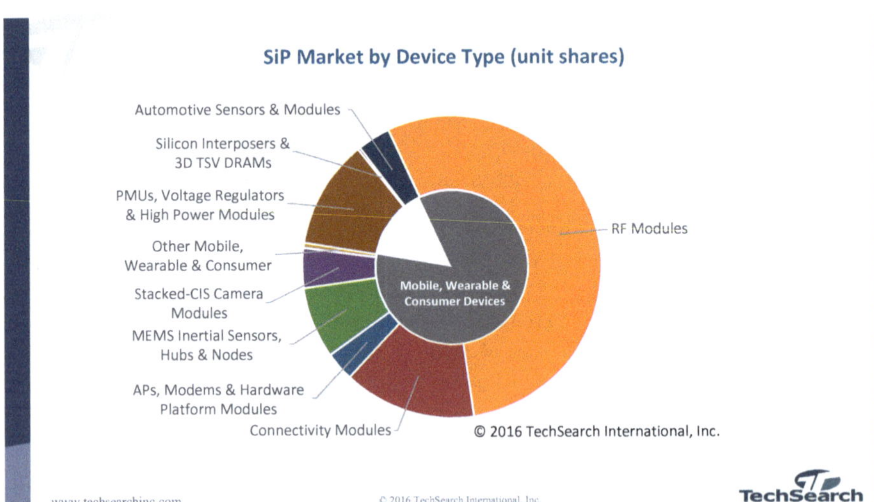

Fig. 2.2 2015 SIP Market by device type (Source: TechSearch International)

design, and thus leveraging the vertical dimension in a package, has been discussed in detail in [6].

(c) Hybrid configurations that combine both the planar and stacked configurations.

There are a number of innovative ways to construct SIP architectures using planar and stacked structures. Some of the more well-known SIP architectures are shown in Fig. 2.3. Figure 2.3 illustrates how different architectures have evolved to meet specific needs for different markets. Figure 2.3 is not intended to be a complete list of all the possible architectures. It is clear from the configurations shown in Fig. 2.3a–f that a number of SIP combinations are possible, offering feature, size, configuration, and cost flexibility.

One class of SIP configurations that has driven the most significant technology changes is through silicon vias (TSV)-based SIP. This chapter focuses on 3D stacks that are enabled by TSV technology. TSV-based SIPs have been the subject of considerable research for more than two decades and numerous papers have explored applications, architecture and design opportunities, as well as process, material, and equipment complexities [11–14]. Since it is difficult to comprehensively discuss all these aspects in a general overview, this chapter will attempt to provide a broad perspective of the architectural and process opportunities and complexities. The process of TSV formation has been previously discussed in depth in Chap. 3 and will not be repeated here, except for a brief reference in Sect. 2.3.

The most commonly used interconnect between stacked die for currently available products with TSVs is solder based (Fig. 2.4a, b) with interconnect pitches as low as 40 μm. Solder-based interconnects have an advantage of being compliant and hence they are more tolerant to misalignment and lack of coplanarity between bonded surfaces during assembly[3]. However as the joints become increasingly small, with decreasing interconnect pitches projected below 40 μm for future 3D stacks, the available solder volume will be reduced and a greater proportion of the solder joint will become intermetallic compounds [15], thus decreasing its compliance. Additionally, with shrinking interconnect pitch, there is an increasing risk of solder bridging during the assembly process since the joints are closer to each other. Various research groups have suggested the need for alternate interconnects; the most common among these are Cu-Cu bonding [17–21], a subject covered in detail in Chap. 6. In 2016, there are two types of widely available products with TSVs. These are DRAM memory stacks [7, 22, 23] and image sensors [24–26].

[3] In most applications, Thermo-Compression Bonding (TCB) is used to create the fine pitch interconnect typically needed between two stacked die because of its superior alignment capability over reflow based flip-chip bonding [16].

Fig. 2.3 (**a**) Multi Chip Module (MCP)—Two or more die are attached in a planar configuration to a package substrate (red line in schematic drawing indicates that the two or more die are electrically connected using lateral on-package interconnects). Image shows the Intel Iris Pro processor with a CPU (larger of the two die) and a DRAM chip. (**b**) Multi Package Module (MPM)—One or more die are packaged before being attached to the final package. Image above shows the Intel's Knights Landing Processor which uses the MC-DRAM Memory Module. The MC-DRAM memory module (a memory stack that conforms to the Hybrid Memory Cube specification [6]) has four stacked memory die connected using TSVs, placed on top of a logic chip). (**c**) Embedded SIP modules—Embedded SIP modules have solder-less die-to-package interconnects and are usually built using reconstituted wafer level or panel level processes [7, 8]. (**d**) Package-on-Package (POP) SIP module—Two packages are stacked on top of each other and connected using peripheral interconnects. The peripheral interconnects can be made using solder balls, Cu posts, or solder-coated Cu balls. Example of an MCeP® (Molded Core embedded Package) is shown above. MCeP® is the registered trademark of SHINKO Electric Industries CO., Ltd. (**e**) EMIB (Embedded Multi-Die Interconnect Bridge) [9]—EMIB-based SIPs use embedded silicon for localized high-density interconnects. Figure above shows a Stratix-10 part from Altera [5]. (**f**) Silicon interposer-based SIP Module—A silicon interposer is used to for fine feature interconnects between the different chips or chip stacks. The interposer has TSVs to connect the chips to the package substrate. See [10] for a product implementation. (**g**) 3D TSV-Stacked module—Multiple die are stacked using TSV-based interconnects. Image above shows a HMC memory stack from Micron. (**h**) SIP Variants that include combinations of wirebond and flip-chip interconnects. The Possum die is from AMKOR

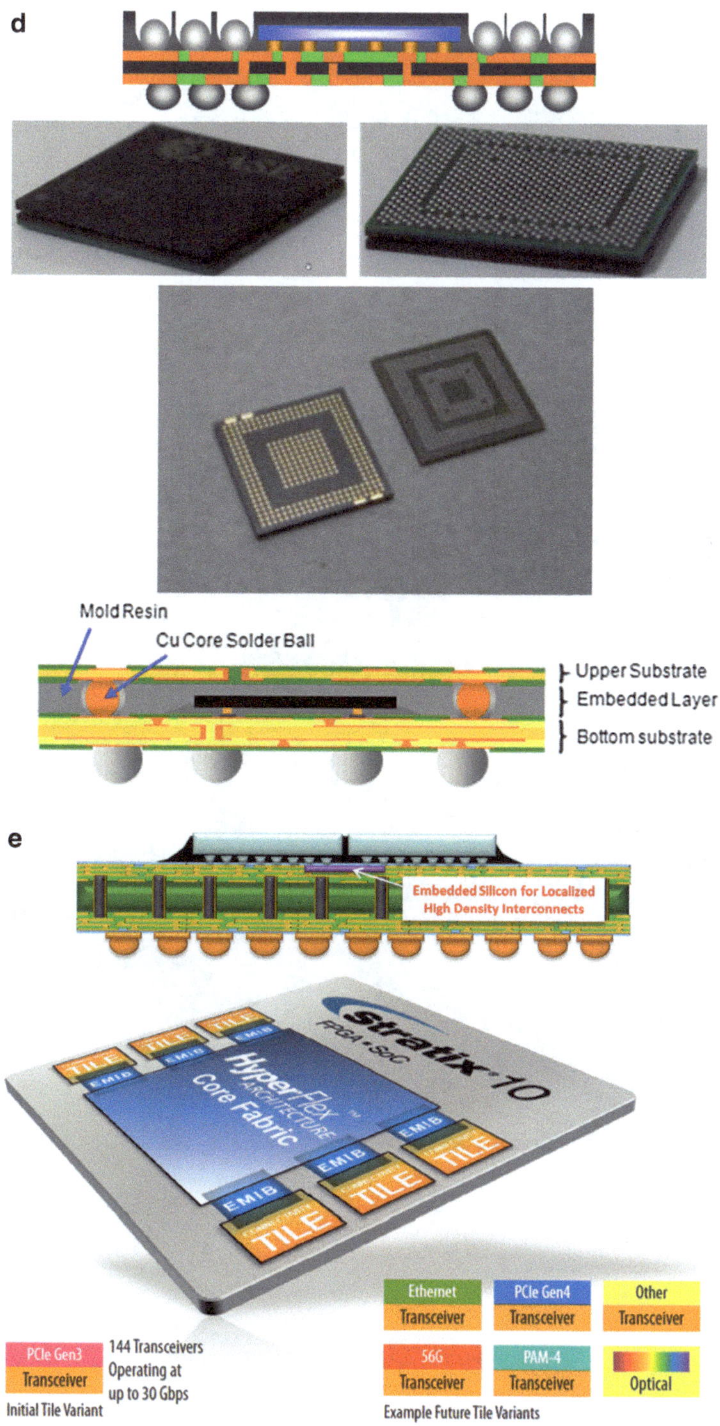

Fig. 2.3 (continued)

f

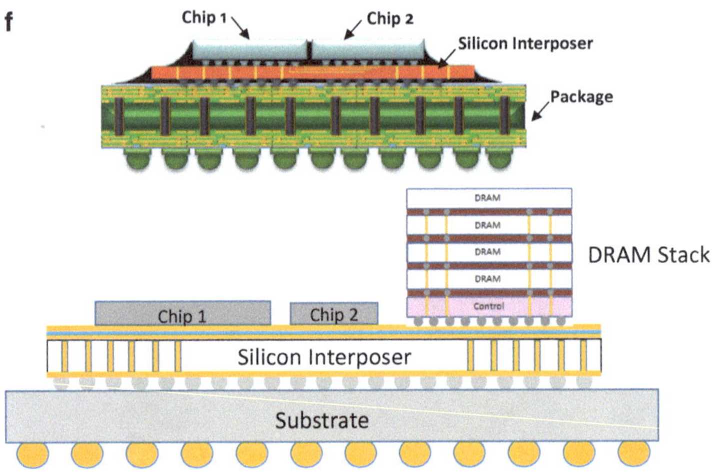

g

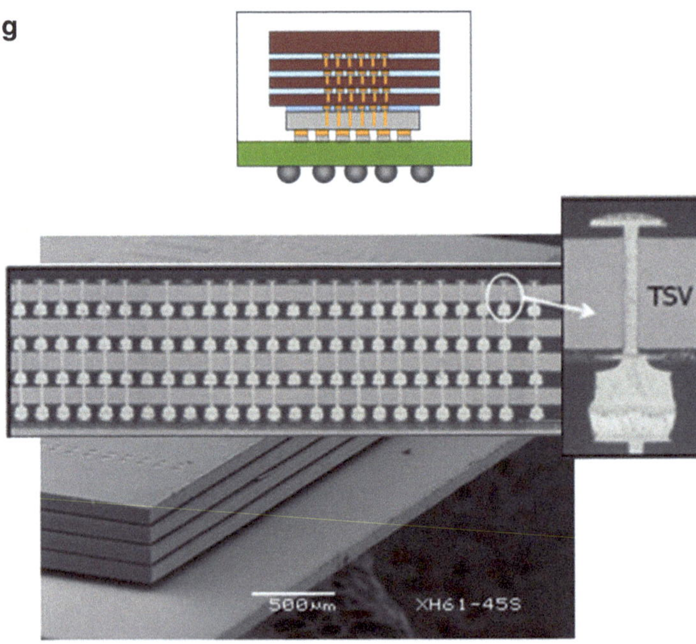

Fig. 2.3 (continued)

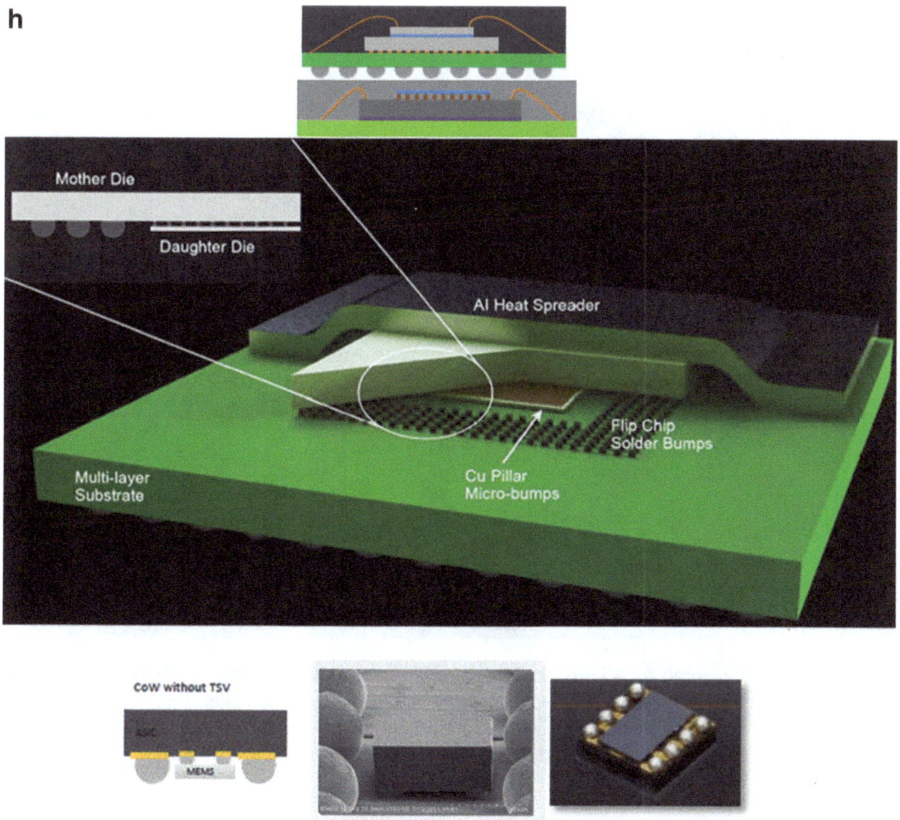

Fig. 2.3 (continued)

2.2 3D TSV-Based Architectures: Advantages and Limitations

In the past few decades, there has been considerable interest in 3D TSV-based SIP architectures [27–30] because of their compact construction and electrical advantages. The advantages include reduced signal latency, and lower signaling power dissipation due to reduced interconnect lengths. Figure 2.5 shows the energy consumption and dissipation during a $0 \rightarrow 1$ and a $1 \rightarrow 0$ bit transition based on a simple capacitance model. Assuming the same Tx and Rx capacitance, energy draw, and dissipation is directly proportional to the capacitance of the interconnect, and this is where TSVs have an advantage over traditional planar wire connections. Since TSVs are short length interconnects, they can exhibit a significantly lower (typically 4–20 times lower [31]) interconnect capacitance compared to a side-by-side or planar interconnect. In Fig. 2.6a, a memory is connected to the processor

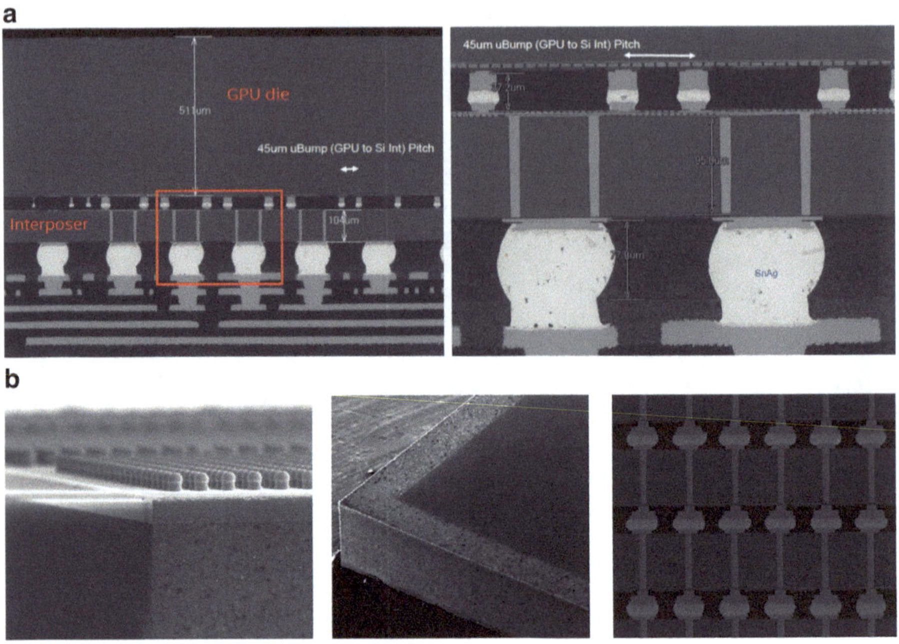

Fig. 2.4 (**a**) Solder joints @ 45 μm pitch at the GPU to silicon interposer (AMD Fury [10]). (**b**) HBM Stacks. Leftmost image shows solder bumps on the HBM stack, center image shows the top view of a HBM package with side mold. Image on the right shows the *M*emory–*M*emory *I*nterconnect (MMI) and the TSVs in the memory die. The MMI pitch is 55 μm

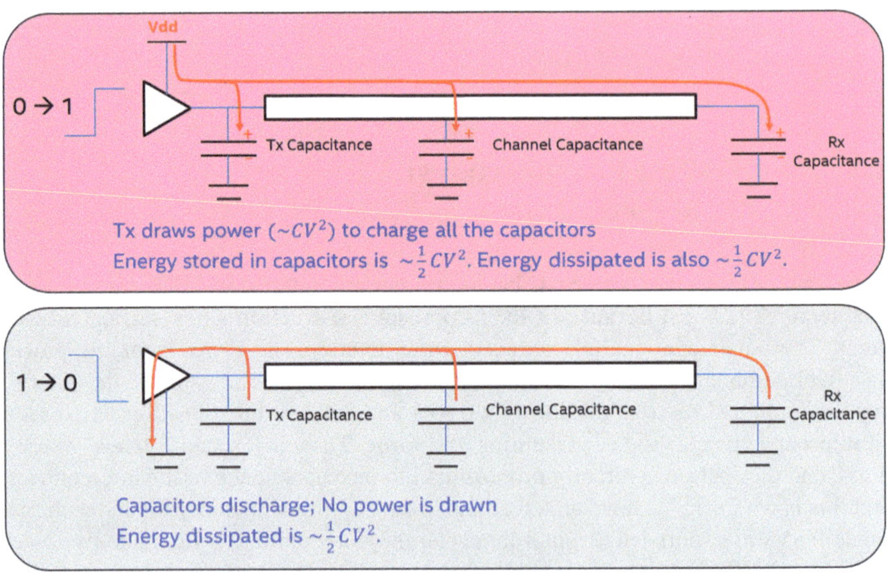

Fig. 2.5 Simple capacitance model describing the energy consumption in a Tx (Transmitter)—Rx (Receiver) interconnect link

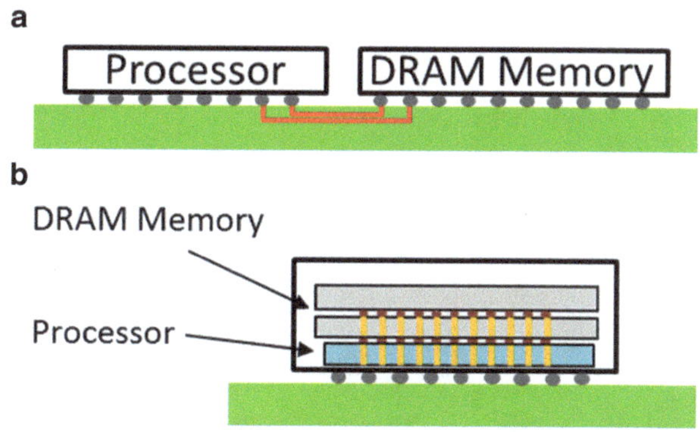

Fig. 2.6 (**a**) Schematic illustrating a side-by-side Processor–DRAM interconnect. (**b**) Schematic illustrating a side-by-side Processor–DRAM interconnect

through a planar on-package interconnect and in Fig. 2.6b the connection uses TSVs. A comprehensive calculation [32] shows that the interconnect power efficiency (mW/Gbps) dropped from 15.65 mW/Gbps (for a bounding case where a DDR memory in a DIMM slot was accessed by the CPU) to 0.55 mW/Gbps when the CPU accessed a WIO[4] memory through TSVs). This is more than a 28× improvement[5]. This power efficiency, multiplied by the number of wires in the interface, combined with reduced data latency due to reduced interconnect lengths is what has motivated designers to consider TSV-based architectures [33]. The power efficiency advantage is appealing across the application spectrum from handheld devices where longer usage times between battery recharging can be enabled to servers where lower energy costs make 3D TSV stacks attractive.

While energy efficiency and latency are the significant advantages for 3D TSV-based architectures, a key limitation is that the maximum *T*hermal *D*esign *P*ower (TDP) can be significantly lower than with comparable 2D configurations. This is due to the thermal resistances in a 3D stack which are in series and hence additive (Fig. 2.7). Consider the case where a DRAM memory is stacked on top of a logic processor[6]. The Processor–DRAM interface, the bulk DRAM silicon, and in the case of multiple DRAM die, the DRAM-DRAM interfaces all add thermal resistances to the heat flow. This compares to a planar case where heat from the processor has to only flow through the silicon of a single chip and a *T*hermal *I*nterface *M*aterial (TIM) before reaching a Heat Spreader or the base of a heatsink. Thus, the thermal impact of stacking is that it reduces the total Thermal Design Power (TDP) available for the system designer to utilize for the processor. This is

[4] WIO, i.e., Wide IO is a JEDEC standard memory, where the memory die are connected by TSVs [22].

[5] Power efficiency quoted for the ESD case. See [*Ref Hazkazemi paper*] for a detailed review of power and performance differences between LPDDR and Wide-IO.

[6] This is architecturally a more likely scenario for CPU-DRAM 3D stacks, since the CPU typically needs more bump interconnects than a DRAM and power delivery to a DRAM would require fewer TSVs compared to the converse case, where power is delivered to the CPU through the DRAM.

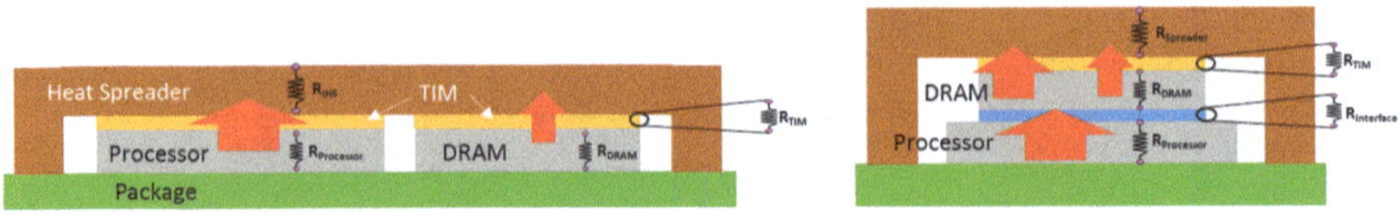

Fig. 2.7 Schematic showing heat paths in a 2D vs. 3D stack

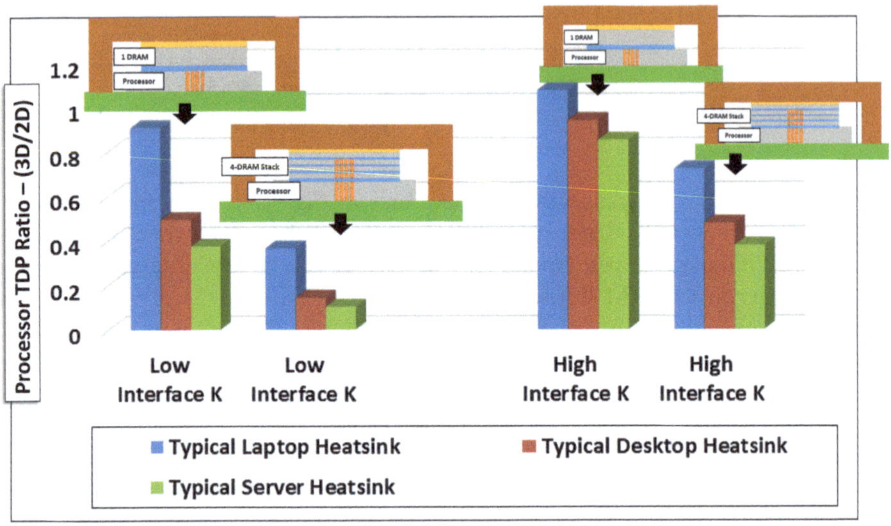

Fig. 2.8 TDP Impact of 3D stacking

quantitatively shown in Fig. 2.8 using a simple case study comparing TDP between the 3D case (where 1 and 4 DRAMs are stacked on top of a processor) and the 2D case where 1 DRAM or a 4 DRAM stack is placed side-by-side to the processor. In this case study, the following assumptions are made:

1. The processor and DRAM silicon are both 100 μm thick.
2. Processor power is uniform across the processor die and is time independent.
3. Junction temperature limits for both the processor and DRAM are 105 °C.
4. Three different system cooling[7] solutions are considered, i.e.,
   (a) High thermal resistance cooling solutions which depend mainly on conduction (Typically encountered in laptop environments where active airflow is not possible)
   (b) Medium thermal resistance cooling solutions which use both conductive and convective heat transfer (Typically encountered in desktop environments that use cost-performance optimized cooling solutions)

[7] System cooling refers to the cooling solution attached to the SIP.

(c) Low thermal resistance cooling solutions which use both conductive and convective heat transfer (Typically encountered in server environments that use performance optimized solutions to manage high TDP envelopes).

5. The Processor–DRAM and DRAM–DRAM interfaces, comprising metallic interconnects and protective underfill, are key interfaces that limit heat transport since the effective thermal conductivity (K) across that interface tends to be in the approximate range (0.3–3.0 W/m °C), which is significantly lower than the thermal conductivity of silicon (110 W/m °C) or Cu (390 W/m °C). Two different interface conditions are considered for the analysis including a low interface with $K = 0.3$ W/m °C and high interface with $K = 3.0$ W/m °C

In cases where the system thermal resistance is high relative to the package resistances, the impact of changes in package resistance has a lesser influence on the overall TDP. In these cases, a system designer can increase memory capacity and still take advantage of improved memory capacity and power savings. As the performance of the system thermal cooling solution improves, increasing package resistance (with increasing number of stacks) will have a greater influence on the product's TDP capability. In this scenario, the TDP degradation and increasing the number of stacks is more significant. It can be seen from Fig. 2.8 that since the bulk of the heat transfer in the package is conduction based. For this case, improvements in the effective thermal conductivity of the interfaces between the stacked silicon chips are critical to improving the overall TDP capability. The effective thermal conductivity of the interface can also be improved by increasing the number of micro-bumps between the die, and by increasing the effective thermal conductivity of the underfill, or polymeric encapsulant, used to increase the reliability of the micro-bumps. Additionally, there is a need to improve the quality of system thermal solutions. Chapter 10 summarizes the various thermal solution strategies developed to address the thermal management problem in 3D stacking.

It should also be noted that the assumption of uniform power distribution in the processor is simplistic. In most cases, the processor power distribution is nonuniform. The presence of the additional thermal interfaces from the DRAM stack will exacerbate the hot spots compared to a side-by-side configuration. Design approaches that minimize the thermal burden on the package have been shown to be capable of achieving thermal equivalence with 2D configurations [34].

Another key consideration is the impact of TSVs on the stress state in silicon. Cu and W are typical materials used for TSVs and polysilicon TSVs have also been mentioned in literature [35]. TSVs are typically filled using *E*lectro-*C*hemical *D*eposition (ECD) at temperatures greater than 200 °C[8]. The ECD temperatures are significantly higher than the typical operation temperature of a processor or a memory device which tend to be in the (90 °C–110 °C) range. At the deposition temperature, the TSV is in an equilibrium (i.e., stress-free) state with the

[8] 200 °C quoted as a typical lower temperature bound. A number of FEOL and MEOL processes have deposition temperatures significantly higher than 200 °C.

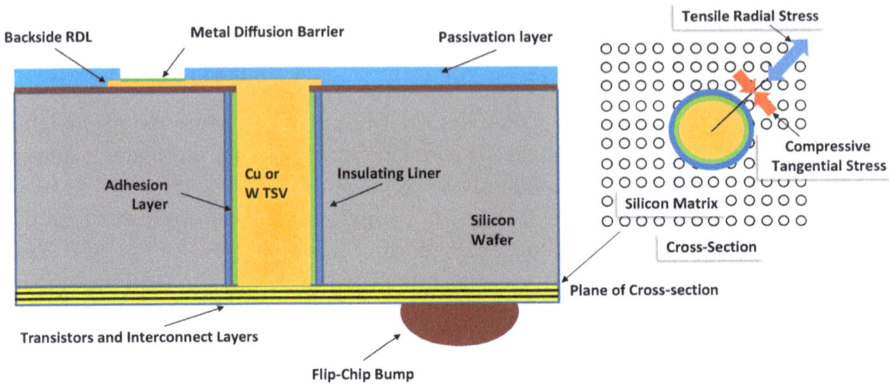

Fig. 2.9 Stresses in silicon due to TSV integration

surrounding silicon, however at operating temperatures, the TSV induces radially tensile and tangentially compressive stresses in the surrounding silicon (see Fig. 2.9 for a schematic illustration) resulting from the CTE differential between the Cu/W fill and the surrounding silicon. These stresses have an impact on the electron and hole mobility in the transistors and hence an impact on transistor performance and reliability [36–38]. Keep out zones (KOZ), i.e., regions in the silicon surrounding the TSVs where transistors cannot be placed, are specified for silicon designers so that performance and reliability impact is reduced to an acceptable level. The consequence is that there is a silicon area increase due to both the TSV and its associated KOZ.

In summary, the power efficiency advantage due to short TSV interconnects must be balanced against the disadvantages in TDP reduction and increased chip area due to the TSV integration. A designer must ensure that the disadvantages don't adversely affect the overall performance or value of the product being designed.

2.3 Methods of Fabrication and Other TSV Attributes

The structure and location of a TSV with reference to the transistors and back-end interconnect stack in a typical wafer is shown schematically in Fig. 2.10. TSVs are typically created using a process with the following steps:

- A photoresist is coated on the silicon wafer and lithographic exposure is used to define the TSV locations.
- The silicon is etched, typically using the Bosch process [39], which uses multiple sequences of *etch and coat* to create the via holes.
- The photoresist is then stripped and the wafer surface is cleaned.

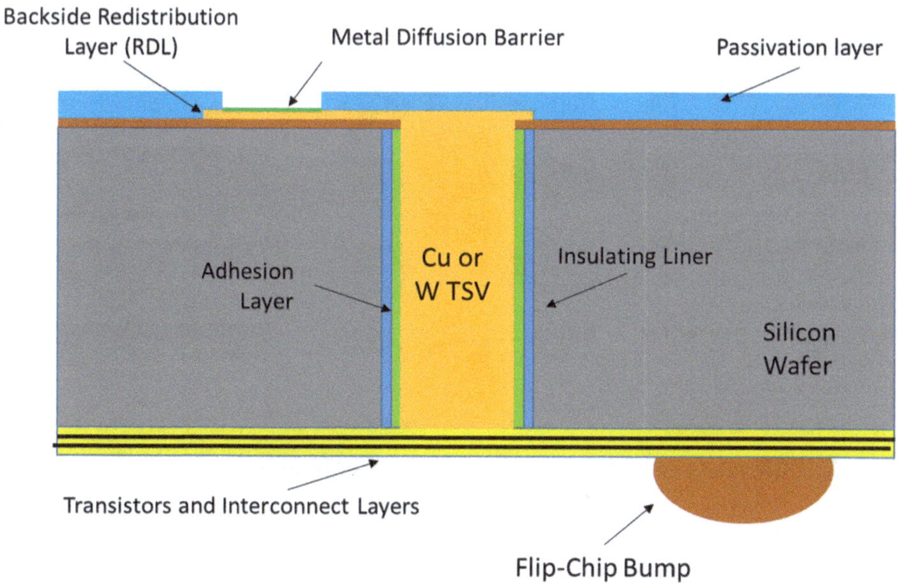

Fig. 2.10 Schematic showing the structure of a TSV in the silicon back-end interconnect stack

- The inner walls of the TSV are coated with a dielectric liner (typically SiO_2 though Si_3N_4 has also been reported [40] using thermal oxidation or PECVD (Plasma Enhanced Chemical Vapor Deposition). The liner electrically isolates the TSV from the bulk silicon.
- Next an adhesion layer, typically Ti or Ta, is added on the inner lining of the TSV using PVD (Physical Vapor Deposition).
- A seed layer is deposited in the via hole next and the vias are ready for deposition of the TSV metal.
- Electro-Chemical Deposition (ECD) is used to fill the via holes with Cu or W.
- Vias are annealed to stabilize the TSV microstructure and to relieve stresses within the TSV.
- Finally, Chemical Mechanical Polish (CMP) is used to planarize the wafer.

TSVs are typically manufactured using three different processes (Fig. 2.11)

1. *Via First Process (Also referred to as the Front End Of Line (FEOL) Via Process)*: In this process, the TSVs are created before the transistors [40, 41]. Key steps in this process are
   (a) The wafer surface is patterned and TSVs are etched and filled on the transistor side (front or active side) of the wafer.
   (b) After the TSV formation, the wafer is planarized and readied for transistor creation.

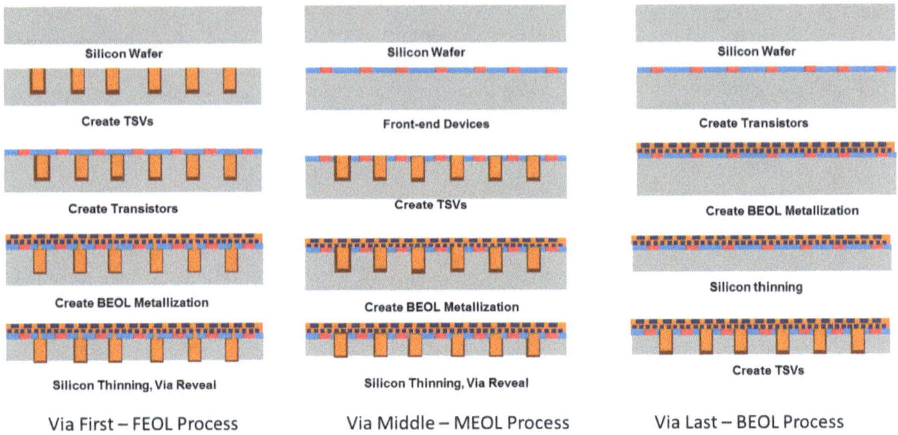

Fig. 2.11 High-level process flow for the three processes of creating TSVs

(c) Transistor creation is followed by the Back End of Line (BEOL) process which creates the multiple metal/insulator layers on the silicon wafer.
(d) Finally, the wafer backside is thinned to reveal the vias and creates a Redistribution Layer (RDL) (Fig. 2.12).

An important advantage of the Via First approach is that since the vias are created before transistor creation, only known good wafers with defect-free TSVs can be used for subsequent steps. Additionally, some authors [40] have claimed that the Via First process allows for a fewer design constraints and a higher density of vias compared to the Via Middle and Via Last processes[9]. A key limitation of the Via First approach is that all the TSV materials and processes need to be compatible with CMOS processes and temperatures (~1200 °C). Polysilicon vias are mostly used in the Via First process [41]. Bauer et al. [40] discuss a process where high density vias are created in a Via First process using a silicon as a sacrificial material during the initial via formation step. The silicon is replaced with W after the transistor formation processes are completed.

2. *Via Middle (Also referred to as Mid End of Line (MEOL) Via Process*: In this process, TSV formation is after front-end device creation and before the creation of all the back-end metal layers [41–47]. Key steps in this process are as follows:

(a) Front-end devices are fabricated including transistors and several lower metal layers.
(b) Next, TSVs are created from the active device side typically at temperatures in the 400 °C–450 °C.
(c) The TSV is then plated with Cu or W.

[9] While this statement seems intuitively feasible, the authors are not aware of an authoritative study that establishes the design advantages of the Via First process over the Via Middle or Via Last options.

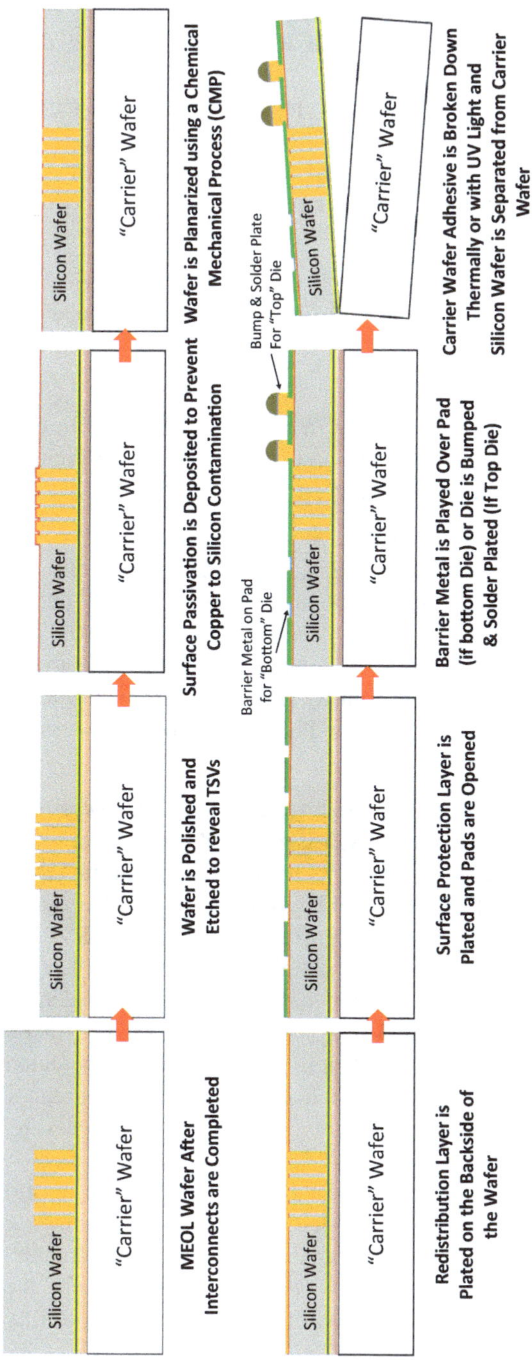

Fig. 2.12 Bumping process flow post TSV formation, shown for the MEOL/FEOL TSV process

(d) The wafer is then annealed at ~400 °C to relieve stresses in the TSV and to stabilize the TSV metal structure.
(e) Next, the wafer is planarized with a typical damascene process and readied for the remaining BEOL metallization.
(f) After BEOL metallization is complete, the wafer is mounted on a carrier and thinned to reveal the TSVs on the inactive (backside) of the wafer. Via reveal is usually (optionally) followed by creation of a Re-Distribution Layer (RDL) and backside bump formation processes that is a critical structure for die stacking (See Fig. 2.12).

An advantage of the Via Middle process is that subsequent chip fabrication processing temperatures are lower than for the Via First process, and hence the thermal stresses induced in silicon due to the CTE mismatch between the silicon and TSV metal are lower.

3. *Via Last (Also referred to as Back End Of Line (BEOL) Via Process*: In this process, TSVs are created after the transistors and interconnects are created [41, 48]. Key steps in this process are as follows:

(a) The transistors and the entire BEOL process is first completed.
(b) The wafer is then temporarily attached to a carrier using an adhesive (usually referred to as a wafer bond/debond adhesive) on the active (front side).
(c) The wafer is thinned and the TSVs are formed using a similar process to the Via Mid described earlier. This is followed by manufacturing of the RDL and backside bumping. The backside RDL and bumping process are essentially the same as described in Fig. 2.12.

In the Via Last process, TSV processing temperatures are typically lower than 200 °C. This is important since the fully formed wafer is mounted on a carrier using a low temperature adhesive. An advantage of this approach is that the thermal stress issues are lower than with the Via First and Via Middle process options. However, there are two significant concerns with the Via Last process. First since TSVs are created after the BEOL, the TSVs need to land at the right metal layer in the chip metal wiring. Special landing surfaces, also called catch cups, are included in silicon metal layers to precisely locate the TSVs. Special care needs to be taken that the TSVs land on catch cups and don't accidentally punch through the catch cups. Integration of catch cups requires careful design and restricts the placement of the TSVs. Secondly since the TSVs are formed after complete wafer processing, yield loss due to TSV formation could result in loss of valuable silicon.

Some of the key design attributes of TSVs are their diameter, pitch, aspect ratio (i.e., the ratio of TSV diameter to its depth), electrical characteristics (including resistivity, inductance, and frequency dependent capacitance), and stress-related KOZs. TSV aspect ratios are influenced by a few key parameters such as the ability to get good insulation coverage as a function of via depth and the ability

Table 2.2 Some recent TSV dimensions reported in the literature

Reference	TSV diameter (μm)	Silicon thickness (μm)	Aspect ratio (diameter:depth)	TSV material	Process
[40]	2	45	1:22.5	W	FEOL
[41]	5	150	1:30	Poly-silicon	FEOL
[42]	3	100	1:33.3	W	MEOL
[48]	10	50	1:5	Cu	BEOL
[43]	6	55	1:9	Cu	MEOL
[44]	10	50	1:5	Cu	MEOL
[49]	5	50	1:10	Cu	MEOL/BEOL
[50]	20	50	1:2.5	Cu	BEOL
[51]	2	30	1:15	Cu	Not specified
[46]	3	50	1:17	Cu	MEOL

to get void-free via filling. Table 2.2 shows some of the published data on TSV diameters and depth. Electrical and stress characteristics are determined by the TSV materials choices (including via metal, liner, and adhesion materials) and process choices (deposition and anneal temperatures).

2.4 Assembly Process Flows

3D stacks can be assembled using three different approaches. They are commonly referred to as Wafer-to-Wafer (W2W) attach [52, 53], Die-to-Wafer (D2W) attach [53–56], or Die-to-Die (D2D) attach.

(a) *Wafer-to-Wafer Attach (W2W)*

In this process, entire wafers are aligned and then bonded, followed by the singulation of individual die stacks. Two W2W flows are schematically illustrated in Fig. 2.13.

- Back-to-Face Flow: where the inactive side (back) of the top wafer is bonded to the active (front) side of the bottom wafer. This approach can enable multiple wafers with TSVs to be bonded one after the other, while retaining the back-to-face connectivity between the die.
- Face-to-Face Flow: where the active side of the top wafer is attached to the active side of the bottom wafer. This approach is less valued for stacking more than 2 wafers.

A significant advantage of a W2W process is that both bonding surfaces are extremely flat, and thus there can be excellent wafer-to-wafer alignment (<3 μm across the wafer) and hence very fine interconnect pitches can be achieved. In the case of solder-based interconnects, pitch scaling in W2W attach is limited by solder

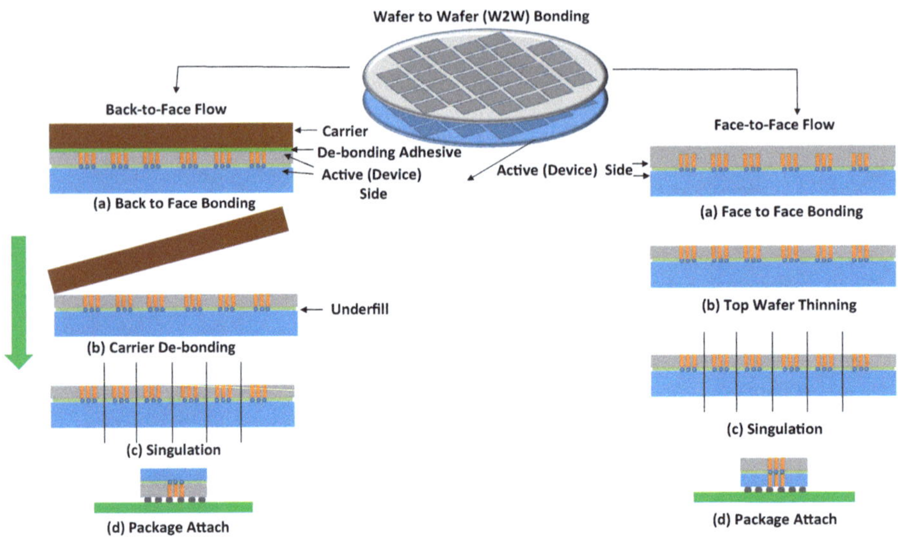

Fig. 2.13 Schematic describing the key steps in a W2W attach process

bridging between interconnects and the implications of intermetallic compound formation. Another advantage of the W2W process is that it is a batch process and with high manufacturing throughput. Key limitations of the W2W process include: (1) all the die have to be of the same size; (2) the process precludes the ability to bond known good die together; (3) yield loss due to misalignment during this process can be significantly expensive, especially in the case where multiple wafers are stacked.

(b) *Die-to-Wafer Attach (D2W)*

In this process, individual die (with or without TSVs) are bonded on a base wafer incorporating TSVs mounted on a carrier (Fig. 2.14). This process has similar advantages of alignment as the W2W process, and it is not limited by the requirement for the die to be the same size. The top stacked die used in successive stacking steps can be the same size or smaller than the corresponding die below it. If only known good die are stacked, better yields than the W2W process can be expected. Additionally, top die with different functionality than the bottom die can be stacked, allowing for increased heterogeneous integration. Unlike the W2W flow, D2W is a sequential process which will have considerably slower process throughput times.

(c) *Die-to-Die Attach (D2D)*

In this process flow, bottom die is first assembled to a package substrate and subsequently other die or die stacks are then stacked on the die connected to an assembled package (Fig. 2.15). This process resolves the die size limitations of

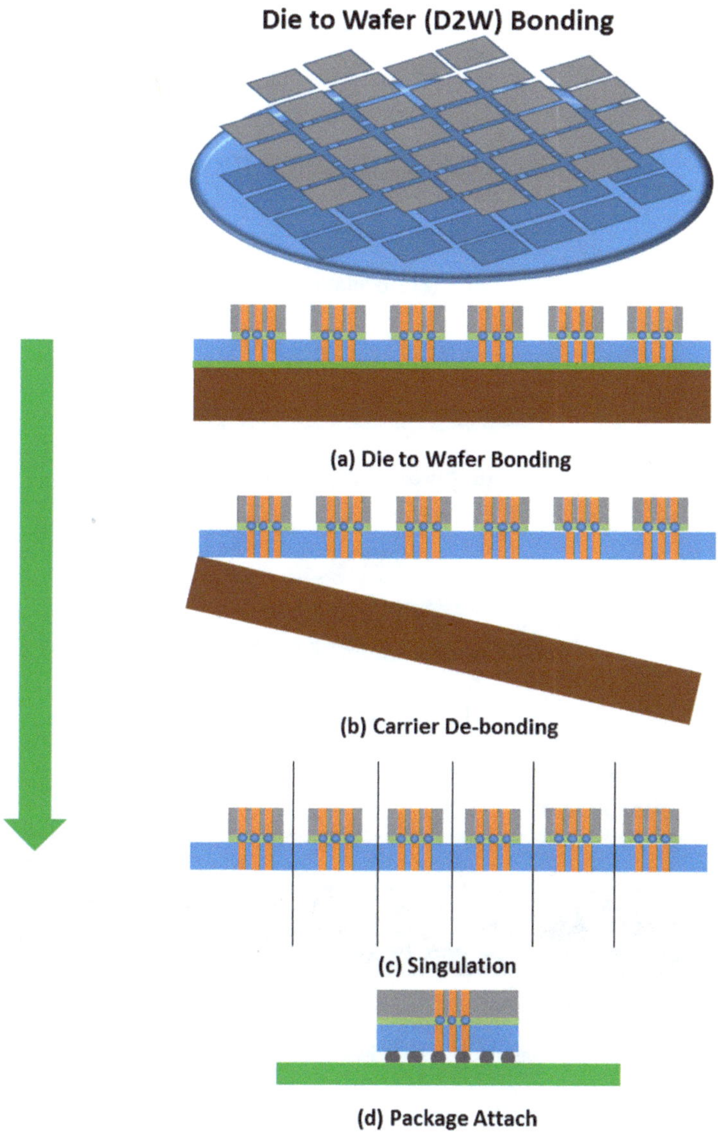

Fig. 2.14 Schematic describing the key steps in a D2W attach process. (**a**) Singulation; (**b** and **c**) die to package attach; (**d**) WIO Memory on an application processor

the W2W and D2W flows through careful alignment methods. In this way, die larger than the bottom die can be stacked on top. Since the bottom die can be fully tested prior to committing the top die, this process has best chance among the three process of creating known good stacks. However, a key disadvantage of this process is that since the bottom die is fully assembled to a package substrate

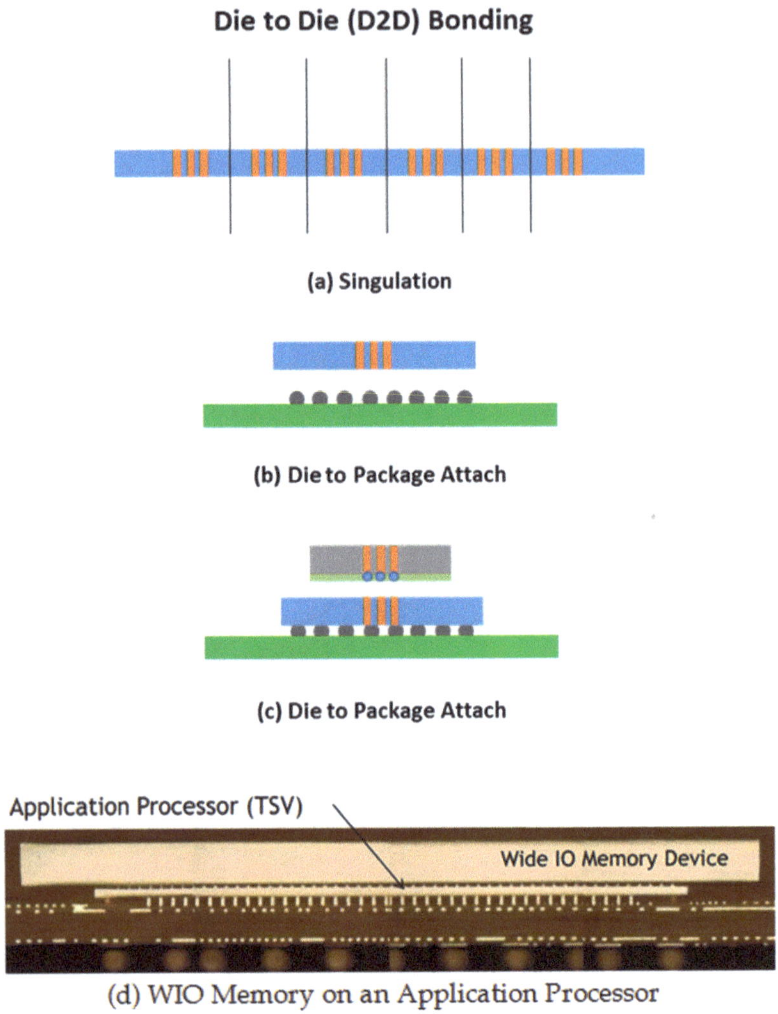

Fig. 2.15 Schematic describing the key steps in a D2D attach process

(typically to an organic package), it can become quite warped because of the CTE mismatch with the package. Hence during the attach of the top die to the bottom package, the die to package yield can be compromised due to the alignment challenges that are created. In general, the 3D stacking process is an exercise in precise tolerance and process control and requires very good characterization of the surfaces being assembled as a function of temperature. Some of this is schematically illustrated in Fig. 2.16.

An important consideration in all three attach processes is the ability to successfully underfill the die–die interconnects which is especially challenging at very fine

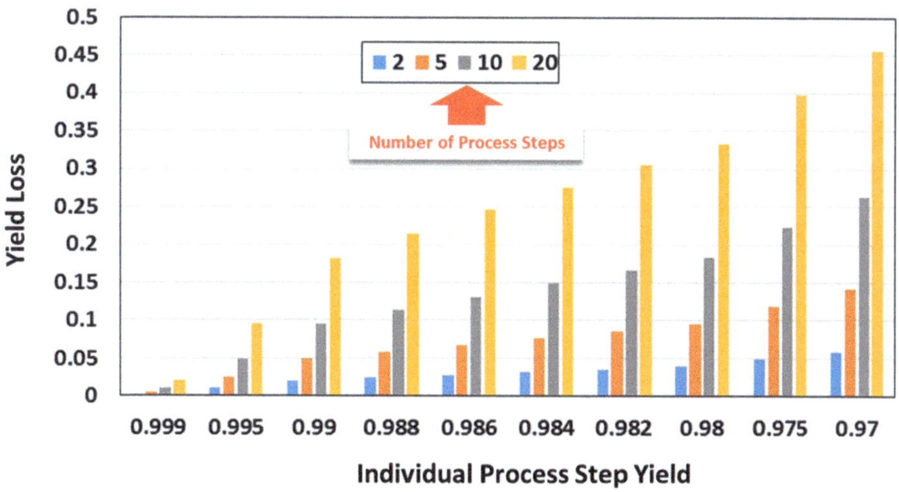

Fig. 2.16 Yield loss in SIPs as a function of individual process step yields

pitches. Underfill is needed to improve the reliability of the die-to-die and die-topackage interconnects. Since the die–die interconnect pitches are significantly lower (typically $\leq$ 55 μm) compared to typical die–package interconnect pitches (typically $\geq$ 100 μm) the gaps or empty space between bumps become smaller for die–die interconnects. This creates another disadvantage of the D2W process. Since it is a sequential stacking process, a dispense underfill process is possible, but underfill bleed-out and unintentional cure in adjacent unattached chip area is a concern. Dispense of underfill after all the die are attached is a possibility that needs to be carefully planned ahead to ensure there is a sufficient gap between the die for adequate underfill flow between the solder-based bump interconnects. Pre-applied film underfills are a better choice for D2W stacking, however if filled underfills are used to deliver the improved thermal performance (Sect. 2.2) filler entrapment in the solder joints during chip attach becomes an important concern.

2.5 Manufacturing Yields and the Role of Test

One of the key goals in manufacturing multi-die modules is to maximize manufacturing yields. Minimizing the number of good die scrapped in modules that fail when tested, has a large impact on average product cost. This proposition can be quantified using a simple example. Consider a module that has n number of die, each of which costs a, and is attached to a single package substrate that costs b. For simplicity, assume that each die is attached to the package in a single process step (i.e., the entire assembly process is a single integrated step), hence the overall SIP assembly process has n steps—one step for each die. If the yield in each

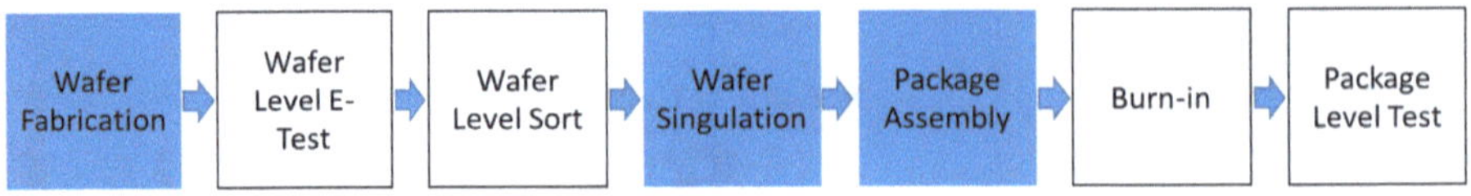

Fig. 2.17 Key steps in a package assembly process

assembly step is z, then the yield after n steps is z^n; the yield loss is $(1 - z^n)$. The yield loss is plotted as a function of number of steps and individual process step yield in Fig. 2.16[10]. The cost due to scrapped modules that fail due to a flaw in the assembly process is $(na + b) \times (1 - z^n)$. There are two primary ways of reducing the cost of scrapped units.

(a) *Increasing individual step yield*
Increasing yield of each individual step in the assembly process is the main focus of packaging technology development. Process, materials, and design parameters for each assembly step, and the impact to upstream and downstream steps of the process are carefully studied and optimized for maximal yield.

(b) *Ensuring each key component "Known Good"*
If only known good components are assembled, the chances that the overall module will perform as intended is significantly increased. Components and subassemblies need to be fully tested prior to assembly to ensure that they are *known good*[11] and cause no yield loss when modules are tested. Designing an efficient test flow that maximizes quantity of known good components while minimizing any added test cost is a key focus area.

Before discussing the challenges involved in testing SIP modules with 3D TSV stacks, it is necessary to understand the different steps involved. A high-level assembly process flow for a single component is illustrated in Fig. 2.17 to show the typical points in the flow where the wafer, die, and assembled package are tested to check for manufacturing quality or performance. Key steps involved in testing products include:

(a) *E-Test (or Electrical-Test)*—This test step is a process characterization step used in the wafer fab to check the manufacturing quality and device parametric values. E-Test can be used after an intermediate process step to check the quality of that step and/or at the end of wafer processing to assess overall quality. Note that this test step does not influence the SiP Module yield, but can affect Module costs indirectly as lower wafer yields will translate into higher die cost.

---

[10] It should be pointed out that this model while illustrative is simplistic in a number of ways. In real life situations, the number of process steps is higher than n; individual process step yields vary and are not always independent of each other.

[11] In industry parlance, the acronym KGD (Known Good Die) is used to describe the need for working pretested die.

(b) *Wafer Level Sort*—This is a wafer level test where each individual die is probed and functionally tested to see if it is good enough for assembly[12]. Contact is accomplished using a probe card that has individual probes to make electrical contact with the interconnect pads on the die. Typically, a probe card can contact several die simultaneously with each one individually tested. The test results are stored in a database that tracks each wafer, and die indexed by its position on the wafer. For SiP Modules, Sort is expected to deliver Known Good Die to the assembly process.
(c) *Burn-In*—The function of Burn-In is to accelerate the failure of latent defects (such as fab and assembly process-induced defects or design marginalities) in an assembled SiP unit to become detectable at a Test step. The Burn-In step helps remove Modules that would have contributed to early customer failures typically called "Infant Mortality." By removing these units, product failures observed by customers are limited to "wearout" mechanisms which occur far in the future. Acceleration of latent failures is accomplished by subjecting the units to higher voltages and temperatures then they would experience in the field, and at long stress times, usually in the range of many minutes to many hours[13]. The goal of Burn-In is to aid in identifying process and design factors responsible for latent failures. As manufacturing processes mature, Burn-In times typically are reduced.
(d) *Package Level Functional Test*—This step is the most comprehensive of all test steps and is used to ensure that the assembled units are defect free, meet performance requirements, and to grade the performance level of the assembled units. Since this is the final and most comprehensive of the Test steps, the level of Test coverage (i.e., how much of the product functionality is evaluated) and the precise control of test parameters are critically monitored.

Now that the key test steps have been described, specific challenges related to the testing of 3D TSV stacks can be discussed with reference to Fig. 2.18 which shows a two chip stack although the conclusions can be extended to higher number stacks. Chips 1 and 2 will be tested at Sort before the wafer thinning and TSV reveal process steps for both the FEOL and MEOL flows, and before TSV formation in the BEOL flow. Only the active side of the die are probed at Sort, and the die backside is ignored. In the W2W and D2W flows, each chip is similarly sorted before TSV and RDL formation. The chips go through the assembly process and package stacks are created. After the entire die stack is mounted on a package, it is functionally tested.

[12] Determining viability for assembly requires a careful optimization of cost (i.e., cost of probing die on a wafer needs to be balanced against the cost of package waste and need for additional test steps later in the flow) and test coverage (while checking a greater degree of die functionality before packaging is financially viable, it can also require more sophisticated Sort technology).

[13] One of the key considerations in manufacturing costs is the time it takes to test units. The greater the amount of time, the lower the throughput and hence higher the costs. The goal in E-Test, Sort, and Package Level Test steps is to focus on test time minimization without impacting quality. Burn-In processes, on the other hand, are designed to run longer so that latent defects are screened out.

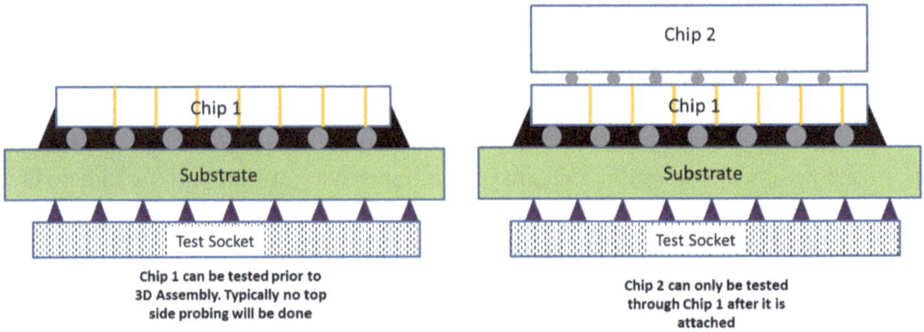

Fig. 2.18 Test considerations for a 3D stack with TSVs

The advantage of the D2D flow is that Chip 1 can be tested for full functionality before Chip 2 is attached. Thus, only known good packaged Chip 1 die will be used for assembly. Full functionality of Chip 2 can only be verified through connectivity through Chip 1. This constraint would not exist if the two die were assembled side-by-side in a package. Additionally, since the heat path for Chip 1 is impeded by Chip 2, precise temperature control of Chip 1 during full functional test is a more challenging than in the corresponding side-by-side case. Both of these constraints imply that there are more challenges with stacked die testing vs. side-by-side, and the potential for higher compound yield loss is greater too.

2.6 Challenges with 3D TSV Architectures

It should be clear from the discussion so far that TSV-based architectures, while attractive from a performance and power efficiency perspective, can be thermally limited. Increasing the thermal envelope of 3D stacks is an important challenge. Considerable research in enhancing conduction and convection modes of heat transfer are described in Chap. 10. 3D TSV stacks break existing paradigms and require new design tools that accommodate the area and stress impacts on transistor performance while taking advantage of the newly available vertical TSV interconnects.

From a manufacturing perspective, TSV manufacturing processes create greater overlap between back-end silicon fabrication and assembly technologies. Additionally, it requires considerable investment in new equipment such as deep via etchers, high aspect ratio Cu plating, ultra-thin wafer handling, and stacked die assembly. Stacked die assembly drives the need for new materials such as fluxes that work in narrow chip gaps with minimal residue, and high thermal conductivity underfills that are compatible with different assembly flows. 3D stacks drive new Test paradigms including fine pitch test and a very detailed understanding of KGD and KGSD.

2.7 Summary

A broad overview of the value and importance of SIP packages has been provided along with motivation for TSV-based 3D architectures. TSV-based 3D stacking has generated considerable interest in the past two decades and the considerable research and development effort in architecture and manufacturing has resulted in a detailed understanding of multiple aspects of the technology.

Acknowledgments The authors would like to acknowledge Prismark Partners LLC, TechSearch International Inc, Shinko Electric Industries Co., Ltd, Amkor Technology®, ASE Group, SK Hynix, for their generous permission to use their pictures. Thanks are also due to Dr. Zhiguo Qian (Intel Corporation) for his help on the section on IO power dissipation, Upendra Sheth (Intel Corporation) for his help in compiling information on different MCP technologies, Dr. Arnab Choudhury (Intel Corporation) for help with thermal analysis and Prasad Ramanathan (Intel Corporation) for help with images. Guidance from Chris Nelson (Intel Corporation) on test processes is also gratefully acknowledged. Finally, thanks are due to Dheeraj Reddy (Intel Corporation) for a thorough review of this chapter.

References

1. G. Moore, Cramming more components onto integrated circuits. Electronics **38**(8), 114 (1965)
2. R. Thakur, 50 years of Moore's law. Solid State Technology **58**(4), 41 (2015)
3. ISSCC 2016 Trends, http://isscc.org/doc/2016/ISSCC2016_TechTrends.pdf
4. R.R. Tummala, *System on Package: Miniaturization of the Entire System* (McGraw-Hill, New York, 2008)
5. https://www.altera.com/products/fpga/stratix-series/stratix-10/overview.html
6. W.R. Davis, J. Wilson, J. Xu, L. Luo, H. Hua, A. Sule, C.A. Mineo, M.B. Steer, P.D. Franzon, Demystifying 3D ICs: The pros and cons of going vertical. IEEE Des. Test Comput. **22**(6), 498–510 (2005)
7. http://www.hybridmemorycube.org/
8. C.C. Liu, S.-M. Chen, F.-W. Kuo, H.-N. Chen, E.-H. Yeh, C.-C. Hsieh, L.-H. Huang, M.-Y. Chiu, J. Yeh, T.-S. Lin, T.-J. Yeh, S.-Y. Hou, J.-P. Hung, J.-C. Lin, C.-P. Jou, C.-T. Wang, S.-P. Jeng, D.C.H. Yu, High performance integrated fan-out wafer level packaging (InFO-WLP): Technology and system integration, in *IEEE IEDM*, 2012, pp. 323–326
9. R. Mahajan, R. Sankman, N. Patel, D.-W. Kim, K. Aygun, Z. Qian, Y. Mekonnen, I. Salama, S. Sharan, D. Iyengar, D. Mallik, Embedded multi-die interconnect bridge (EMIB)—a high density, high bandwidth packaging interconnect. Paper presented at the 66th electronic components and technology conference, Las Vegas, NV, pp. 557–565, June 2016
10. http://www.anandtech.com/show/9390/the-amd-radeon-r9-fury-x-review
11. J.D. Meindl, Interconnect opportunities for gigascale integration, in *IEEE Micro 2003*, pp. 28–35
12. J.D. Meindl, Beyond Moore's law: The interconnect era, in *Computing in Science and Engineer, 2003 IEEE*, 2003, pp. 20–24
13. R.S. Patti, Three-dimensional integrated circuits and the future of system-on-chip designs. Proc. IEEE **94**(6), 1214–1224 (2006)
14. F. Mukta, S.S. Iyer, 3D integraion review. Sci. China Inf. Sci. **54**(5), 1012–1025 (2011)
15. G.G. Vakanas, O. Minho, B. Dimcic, K. Vanstreels, B. Vandecasteele, I. De Preter, J. Derakhshandeh, K. Rebibis, M. Kajihara, I. De Wolf, Formation, processing and

characterization of Co-Sn intermetallic compounds for potential integration in 3D interconnects. Microelectron. Eng. **140**, 72–80 (2015)
16. A. Eitan, K.-Y. Hung, Thermo-compression bonding for fine-pitch copper-pillar flip-chip interconnect—tool features as enablers of unique technology, in *Proceedings IEEE 65th Electronic Components and Technology Conference (ECTC)*, pp. 460–464, May 2015
17. A. Klumpp, R. Merkel, P. Ramm, J. Weber, R. Weiland, Vertical system integration by using inter-chip vias and solid-liquid interdiffusion bonding. Jpn. J. Appl. Phy. **43**(7A), L829–L830 (2004)
18. P. Batra, S. Skordas, D. LaTulipe, K. Winstel, C. Kothandaraman, B. Himmel, G. Maier, B. He, D.W. Gamage, J. Golz, W. Lin, T. Vo, D. Priyadarshini, A. Hubbard, K. Cauffman, B. Peethala, J. Barth, T. Kirihata, T. Graves-Abe, N. Robson, S. Iyer, Three-dimensional wafer stacking using Cu TSV integrated with 45nm high performance SOI-CMOS embedded DRAM technology. J. Low Power Electron. Appl. **4**, 77–89 (2014). doi:10.3390/jlpea4020077
19. K. Takahashi, M. Umemoto, N. Tanaka, K. Tanida, Y. Nemoto, Y. Tomita, M. Tago, M. Bonkohara, Ultra-high-density Interconnection technology of three-dimensional packaging. Microelectron. Reliab. **43**, 1267–1279 (2003)
20. C.S. Tan, G.Y. Chong, High throughput Cu-Cu bonding by non-thermo-compression method. Paper presented at the 63rd electronic components and technology conference, Las Vegas, NV, pp. 1158–1164, May 2013
21. P. Guegen, C. Ventosa, L. Di Cioccio, H. Moriceau, F. Grossi, M. Rivoire, P. Leduc, L. Clavelier, Physics of direct bonding: applications to 3D heterogeneous or monolithic integration. Microelectron. Eng. **87**, 477–484 (2010)
22. https://www.jedec.org/standards-documents/docs/jesd229-2
23. https://www.jedec.org/standards-documents/docs/jesd235a
24. S. Lhostis, A. Farcy, E. Deloffre, F. Lorut, S. Mermoz, Y. Henrion, L. Bethier, F. Bailly, D. Scevola, F. Gyuader, F. Gigon, C. Besset, S. Pellssier, L. Gay, N. Hetellier, M. Arnoux, A.-L. Le Berrigo, S. Moreau, V. Balan, F. Fournel, A. Jouce, S. Cheramy, B. Rebhan, G. Maier, L. Chitu, Reliable 300mm wafer level hybrid bonding for 3D stacked CMOS image sensors. Paper presented at the 66th electronic components and technology conference, Las Vegas, NV, pp. 869–876, June 2016
25. V.C. Venezia, C. Shih, W.Z. Yang, B. Zhang, H. Rhodes, Stack chip technology: a new direction for CMOS imagers. Paper presented at the IISW conference 2015
26. R. Fontaine, The state-of-the-art of mainstream CMOS image sensors. Paper presented at the IISW conference 2015
27. B. Black, M. Annavaram, N. Brekelbaum, J. DeVale, L. Jiang, G.H. Loh, D. McCauley, P. Morrow, D.W. Nelson, D. Pantuso, P. Reed, J. Rupley, S. Shankar, J. Shen, C. Webb, Die stacking (3D) microarchitecture. Paper presented at the 39th annual IEEE/ACM international symposium on microarchitecture (MICRO'06) 2006
28. J.M. Stern, V.H. Ozguz, 3D system architectures, in *Intelligent Integrated Microsystems*, ed. by R.A. Athale, J.C. Wolper, Proc. of SPIE **6232** 6232K (2006). doi:10.1117/12.667381
29. P. Jacob, O. Erdogan, A. Zia, P.M. Belemjian, R.P. Kraft, J.F. McDonald, Predicting the performance of a 3D processor-memory chip stack. IEEE Des. Test Comput. **22**, 540–547 (2005)
30. P. Franzon, E. Rotenberg, J. Tuck, W.R. Davis, H. Zhou, J. Schabel, Z. Zhang, J.B. Dwiel, E. Forbes, J. Huh, S. Lipa, Computing in 3D. Presented at the 2015 custom integrated circuits conference (CICC), 2015 IEEE, pp. 1–6, 28–30 Sept. 2015
31. K. Chandrashekar, W. Weis, B. Akesson, N. When, K. Goossens, System and circuit level power modeling of energy-efficient 3D-stacked wide I/O DRAMs. Presented at the 2013 DATE conference, pp. 236–241
32. M.A. Karim, P.D. Franzon, A. Kumar, Power comparison of 2D, 3D and 2.5D interconnect solutions and power optimization of interposer interconnects, in *ECTC 2013*, pp. 860–866
33. M.H. Hajkazemi, M.K. Tavana, H. Homayoun, Wide I/O or LPDDR? exploration and analysis of performance, power and temperature trade-offs of emerging DRAM technologies in

embedded MPSoCs. Paper presented at the 33rd IEEE international conference on computer design, 2015, pp. 70–77
34. M. Saeidi, K. Samadi, A. Mittal, R. Mittal, Thermal implications of mobile 3D-ICs. Presented at the 2014 3D systems integration conference (3DIC) in Kinsdale. doi:10.1109/3DIC.2014.7152160, pp. 1–7
35. M. Koyanagi, H. Kurino, K.W. Lee, K. Sakuma, N. Miyakawa, H. Itani, Future system-on-silicon LSI chips. IEEE Micro. **18**(4), 17–22 (1998)
36. A. Mercha, G. Van der Plas, V. Moroz, I. De Wolf, P. Asimakopoulos, N. Minas, S. Domae, D. Perry, M. Choi, A. Redolifi, C. Okoro, Y. Yang, J. Van Olmen, S. Thangaraju, D. Sabuncuoglu Tezcan, P. Soussan, J.H. Cho, A. Yakovlev, P. Marchal, Y. Travaly, E. Beyne, S. Biesemans, B. Swinnen, Comprehensive analysis of the impact of single and arrays of through silicon vias induced stress on high-K/metal gate CMOS performance, in *IEDM*, pp. 2.2.1–2.2.4, 2010
37. W. Guo, G. Van der Plas, A. Ivankovic, V. Cherman, G. Eneman, B. De Wachter, M. Togo, A. Redolfi, S. Kubicek, Y. Civale, T. Chiarella, B. Vandevelde, K. Croes, I. De Wolf, I. Debusschere, A. Mercha, A. Thean, G. Beyer, B. Swinnen, E. Beyne. Impact of through silicon via induced mechanical stress on fully depleted bulk finFET technology, in *IEDM*, pp. 18.4.1–18.4.4, 2012
38. T. Kauerauf, A. Branka, K. Croes, A. Redolfi, Y. Civale, C. Torregiani, G. Groeseneken, E. Beyne, Effect of TSV presence on FEOL yield and reliability, pp. 5C.6.1–5C.6.4, 2013
39. M. Tanaka, M. Sekine, I. Sakai, Y. Kusuda, T. Nonaka, O. Ysuji, K. Kondo, TSV processes, in *Three dimensional integration of semiconductors*, pp. 43–96 (2015). http://rd.springer.com/chapter/10.1007/978-3-18675-7_3/fulltext.html
40. T.M. Bauer, S.L. Shinde, J.E. Massad, D.L. Hetherington, Front end of line integration of high density, electrically isolated, metallized through silicon vias, in *Proceedings of the 58th Electronic Components and Technology Conference (ECTC)*, pp. 1165–1169, May 2009
41. M. Puech, J.M. Thevenoud, J.M. Gruffat, N. Launay, N. Arnal, P. Godinat, Fabrication of 3D packaging TSV using DRIE, design, test, integration and packaging of MEMS/MOEMS, 2008, in *MEMS/MOEMS 2008. Symposium on*, 2008, pp. 109–114. doi:10.1109/DTIP.2008.4752963
42. G. Pares, N. Bresson, S. Minoret, V. Lapras, P. Brianceau, J.F. Lugand, R. Anciant, N. Sillon, Through silicon via technology using tungsten metallization, in *2011 I.E. International Conference on IC Design & Technology*, 2011, pp. 1–4. doi:10.1109/ICICDT.2011.5783204
43. R. Agarwal, D. Hiner, S. Kannan, K. Lee, D. Kim, J. Paek, S. Kang, Y. Song, S. Dej, D. Smith, S. Thangaraju, J. Paul, TSV integration on 20nm logic: 3D Assembly and reliability results, in *Proceedings of the 64th Electronic Components and Technology Conference (ECTC)*, pp. 590–595, May 2014
44. D.J. Na, K.O. Aung, W.K. Choi, T. Kida, T. Ochiai, T. Hashimoto, M. Kimura, K. Kata, S.W. Yoon, A.C.B. Yong, TSV MEOL (Mid End of Line) and packaging technology of mobile 3D-IC stacking, in *Proceedings of the 64th Electronic Components and Technology Conference (ECTC)*, pp. 596–600, May 2014
45. N. Kumar, S. Ramaswami, J. Dukovic, J. Tseng, R. Ding, N. Rajagopalan, B. Eaton, R. Mishra, R. Yalamanchili, Z. Wang, S. Xia, K. Sapre, J. Hua, A. Chan, G. Mori, B. Linke, Robust TSV via-middle and via-reveal process integration accomplished through characterization and management of sources of variation, in *2012 I.E. 62nd Electronic Components and Technology Conference*, pp. 787–793. doi:10.1109/ECTC.2012.6248922
46. E. Beyne, Reliable via-middle copper through-silicon via technology for 3-D integration. IEEE Trans. Compon. Packag. Manuf. Technol. **6**(7), 983–992 (2016). doi:10.1109/TCPMT.2015.2495166
47. S.W. Yoon, D.J. Na, K.T. Kang, W.K. Choi, C.B. Yong, Y.C. Kim, P.C. Marimuthu, TSV MEOL (Mid-End-Of-Line) and its assembly/packaging technology for 3D/2.5D solutions, in *ICEP-IAAC 2012 Proceedings*, pp. 1–5
48. K.-W. Lee, H. Hashimoto, M. Onishi, Y. Sato, M. Murugesan, J.-C. Bea, T. Fukushima, T. Tananka, M. Koyanagi, A resilient 3D stacked multicore processor fabricated using

die-level 3D integration and backside TSV technologies. in *Proceedings of the 64th Electronic Components and Technology Conference (ECTC)*, pp. 304–308, May 2014
49. M.-J. Tsai, Overview of ITRI's TSV technology, in *7th Annual SEMATECH Symposium Japan*, June 2011, http://www.sematech.org/meetings/archives/symposia/9237/Session%205%203D%20interconnect/1%20MJ_Tsai_ITRI.pdf
50. H. Ikeda, Heterogeneous 3D stacking technology developments in ASET, in *CPMT Symposium Japan, 2012 2nd IEEE*, 2012, pp. 1–4. doi:10.1109/ICSJ.2012.6523453
51. H.B. Chang, H.Y. Chen, P.C. Kuo, C.H. Chien, E.B. Liao, T.C. Lin, T.S. Wei, Y.C. Lin, Y.H. Chen, K.F. Yang, H.A. Teng, W.C. Tsai, Y.C. Tseng, S.Y. Chen, C.C. Hsieh, M.F. Chen, Y.H. Liu, T.J. Wu, S.Y. Hou, W.C. Chiou, S.P. Jeng, C.H. Yu, High-aspect ratio through silicon via (TSV) technology, in *2012 Symposium on VLSI Technology Digest of Technical Papers*, pp. 173–174
52. P. Lindner, V. Dragoi, T. Glinsner, C. Schaefer, R. Islam, 3D interconnect through aligned wafer level bonding, in *Proceedings of the 52nd Electronic Components and Technology Conference (ECTC)*, pp. 1439–1443, May 2002
53. T. Ohba, Wafer level three-dimensional integration (#DI) using bumpless TSV interconnects for tera-scale generation, in *2013 IEEE*, 2013, pp. 1–4
54. Q. Chen, D. Zhang, Z. Wang, L. Liu, J.J.-Q. Lu, Chip-to-wafer (C2W) 3D integration with well-controlled template alignment and wafer-level bonding, in *Proceedings of the 61st Electronic Components and Technology Conference (ECTC)*, pp. 1–6, May 2011
55. W.K. Choi, C.S. Premchandran, L. Xie, S.C. Ong, J.H. He, G.J. Yap, A. Yu, A novel die to wafer (D2W) collective bonding method for MEMs and electronics a heterogeneous 3D integration, in *Proceedings of the 60th Electronic Components and Technology Conference (ECTC)*, pp. 829–833, May 2010
56. K. Sakuma, P.S. Andy, C.K. Tsang, S.L. Wright, B. Dang, C.S. Patel, B.C. Webb, J. Maria, E.J. Sprogis, S.K. Kang, R.J. Polastre, R.R. Horton, J.U. Knockerbocker, 3D chip-stacking technology with through-silicon vias and low-volume lead-free interconnections. IBM J. Res. Dev. **52**(6), 611–622 (2008)

Chapter 3
Materials and Processing of TSV

Praveen Kumar, Indranath Dutta, Zhiheng Huang, and Paul Conway

3.1 Introduction

With the evolution of electronics towards increasingly energy-efficient, mobile, and high performance platforms, the requirements for packages with smaller form-factor, lighter weight and lower power, along with higher bandwidth and memory capacity, have been accelerating. This has led to increasing use of 3-dimensional (3-D) packaging of microelectronic devices, which can result in substantial reduction of energy consumption per bit of information processed, due to a dramatic reduction of package parasitics. Of the different types of stacked-die packages, area-array dies interconnected with through-silicon vias (TSVs) have emerged as probably the best 3D-package architecture. In a typical 3-D package, multiple chips, each performing a well-defined and independent task, are stacked in a vertical column. The vertically stacked chips are then connected through solder microbumps, and communicate with each other through metal-filled TSVs, which

P. Kumar
Department of Materials Engineering, Indian Institute of Science, Bangalore 560012, India
e-mail: praveenk@materials.iisc.ernet.in

I. Dutta (✉)
School of Mechanical and Materials Engineering, Washington State University, Pullman, WA 99164, USA
e-mail: idutta@wsu.edu

Z. Huang
School of Materials Science and Engineering, Sun Yat-sen University, 135 West Xingang Road, Guangzhou 510275, China
e-mail: hzh29@mail.sysu.edu.cn

P. Conway
The Wolfson School of Mechanical, Electrical and Manufacturing Engineering, Loughborough University, Loughborough, Leicestershire LE11 3TU, UK
e-mail: p.p.conway@lboro.ac.uk

Y. Li, D. Goyal (eds.), *3D Microelectronic Packaging*, Springer Series in Advanced Microelectronics 57, DOI 10.1007/978-3-319-44586-1_3

are fabricated in the thinned silicon dies. This circumvents the need for horizontal integration and reduces communication bus-lengths, thereby improving performance while simultaneously reducing the energy footprint of the package.

In this chapter, we focus on the materials and critical steps associated with fabricating TSVs and the flow process to fabricate TSV-enabled 3-D integration of silicon dies. In the following chapter, we will introduce the microstructural and reliability issues of TSVs.

3.2 Overview of TSV Materials and Processes

A TSV is composed of a conductor, passing through the Si substrate of the stacked dies. The conductor is electrically insulated from the substrate by a dielectric layer, i.e., the TSV liner, and interconnects the metal wires of the stacked dies. The TSV liner also determines the TSV parasitic capacitance. In order to avoid diffusion of metal from the TSV into the Si substrate, a barrier layer is used between the liner and the TSV metal [1, 2]. The geometry of the TSV conductor may vary depending on the 3-D stacking technology. The area crossed by current may have different shapes, e.g., square, rectangular, circular, elliptical, and polygonal [3, 4]. In addition, the lateral surface of the conductor can be cylindrical or conical [2].

The 3-D interconnection technology based on TSVs basically consists of three main process modules: (1) The TSV module itself, (2) wafer thinning and back side processing, and (3) the die or wafer stacking process (permanent bonding and/or temporary bonding) [1]. The sequence of these process modules may vary and different process flows may be characterized by the following four key differentiating characteristics [1]:

1. The order of the TSV process with respect to the device wafer fabrication process. So we have via-first, via-middle, and via-last processes, cf. Sect. 3.4.1.
2. The order of the TSV processing and 3D bonding.
3. The order of wafer thinning and 3D bonding.
4. The method of 3D bonding. We have wafer-to-wafer (W2W) bonding, die-to-wafer (D2W) bonding and die-to-die (D2D) bonding.

In addition to these four main characteristics, three secondary characteristics are identified [1]:

1. Face-to-face (F2F) or back-to-face (B2F) bonding (the face or top surface of the wafer being the side with the active devices and back-end interconnect layers), cf. Sect. 3.4.2.
2. In case of a via-last approach, the TSV may be realized starting from the wafer front side or from the wafer back side. Starting from the front side of the wafer requires etching through the back end of line (BEOL) stack.
3. Removal of the carrier-wafer before or after bonding (i.e., temporary bonding and permanent bonding).

Table 3.1 provides an overview of the TSV materials and associated processes.

Table 3.1 An overview of TSV materials and processes[a]

	Materials	Processes	Comments
Filler	Cu	ECD[b]	Well-established process with high conductivity, compatible with metal BEOL process.
	W	CVD[c]	Lower conductivity than Cu, but with better coefficient of thermal expansion (CTE) match with Si and better filling to small via.
	Poly-Si	LPCVD[d]	CVD process compatible with FEOL process, but with low conductivity and high contact resistance.
	Solder alloy	LMF[e] [6, 7], IMS[f] [8]	Reduced principal stress compared to Cu TSV.
	Polymer	Spin-on [9], NanoFill [10]	Polymer filling process, but lower conductivity than metal.
	CNT[g] [11, 12]	CVD/PECVD[h]	CNT via is promising but materials and processing issues remain.
Barrier	Ta, TaN	CVD	Common for Cu TSV.
	TiN	CVD	Common for W TSV.
	NiB, etc.	ELD[i]	Used for special process (e.g., for electrografting deposition of insulator, diffusion barrier, and copper seed layers).
Liner	SiO_2	CVD	Both SiO_2 and SiN_x used as common insulators, often combined.
	SiN_x	CVD	
	Polymer	Spin-on	Polymer can serve as a stress buffer.

[a] This is an updated version of the original table from [5] [b]Electrochemical deposition
[c] Chemical vapor deposition
[d] Low-pressure CVD
[e] Liquid metal filling
[f] Injection molded soldering
[g] Carbon nanotube
[h] Plasma-enhanced CVD
[i] Electroless deposition

3.3 Fabrication of TSV and TSV Assembly

Fabrication of a TSV structure (or TSV assembly) comprises four main steps: (1) etching of Si, where a hole or via in Si wafer is created, (2) filling, where the via created in the previous step is sequentially filled with a dielectric layer, a diffusion barrier and/or adhesion layer, a seed layer, and a filler material, (3) planarization and thinning of the structure after the filling step, followed by (4) deposition of a redistribution layer with embedded circuitry to re-route electrical signals to the chip stacked on top. Steps (1) through (3) are schematically illustrated in Fig. 3.1. Most of the processes utilized during these three steps were originally developed for fabrication of micro-electro-mechanical systems (MEMS) and integrated circuits (ICs). Hence, these processes and the corresponding technologies are not strictly new; however, these processes, as it will be explained below, have been

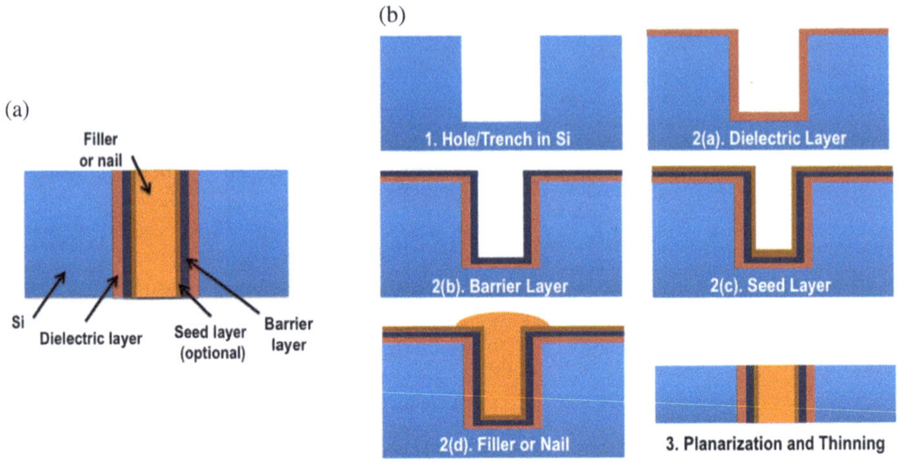

Fig. 3.1 Schematic illustration of (**a**) a TSV assembly or TSV structure comprising of TSV, dielectric layer, barrier layer, seed layer, and the filler, and (**b**) three major steps required for fabricating a TSV assembly

significantly modified and tailored for catering to the requirements of creating TSV structures. The most unique feature of a TSV is that it is a high aspect ratio (HAR) structure having very small cross-sectional area. Thus, the unique challenges associated with fabricating TSV and TSV-assembly stem primarily from drilling a HAR hole in Si wafer and then sequentially conformally filling it with a myriad of materials belonging to different classes (e.g., metal, ceramics, etc.). Below, we describe the main steps of fabricating a TSV assembly, and also highlight associated challenges and a few methods of overcoming them.

3.3.1 Creating a Via or Trench in Si Wafer

A TSV usually has a diameter and a height in the range of 1–10 μm and 10–150 μm, respectively [13–15]. Hence, the aspect ratios of TSVs are often in the range of 1–50. Generally, TSVs with small diameters are preferred for increasing the layout efficiencies[1] and minimizing keep-out zone (KOZ).[2] In addition, the inner walls of the holes should be slightly tapered (85–88° [16]) and smooth for conformal and void-free material filling [16–18]. The smooth wall of TSV also reduces stress concentrations; this is particularly useful as the TSV assembly usually has high residual stresses and it is often exposed to large thermal stresses during fabrication

[1] Layout efficiency is understood as the number of conductors per unit area.

[2] KOZ is the region where functional properties of Si are significantly affected by the stress field of the TSV.

of transistors, etc., as well as during service [19, 20]. Therefore, fabrication of TSV requires drilling small sized, slightly tapered, smooth holes of HAR in Si wafer. For this purpose, laser drilling, powder blast micromachining (PBM), anisotropic wet etching, and plasma etch have been explored. However, as explained below, the plasma-etching-based Bosch-process type deep reactive ion etching (DRIE) method is the most widely used commercial method for creating holes or trenches in Si for fabricating TSVs [21].

3.3.1.1 Laser Drilling

Laser drilling for fabricating TSVs is a process, where a hole or trench in Si is created by focusing a high-powered laser beam at a desired location on a Si wafer. Due to the excessive heating, Si below the laser beam melts and vaporizes, thus creating a vertical trench with small diameter [22]. The typical diameter of the laser beam used in this process is 10–20 μm [22, 23]. To confine the spread of the heat-affected zone and remove the machined debris, a jet of coolant (e.g., deionized water) may also be splashed onto Si wafer near the laser spot. As shown in Fig. 3.2,

(a)

(b)

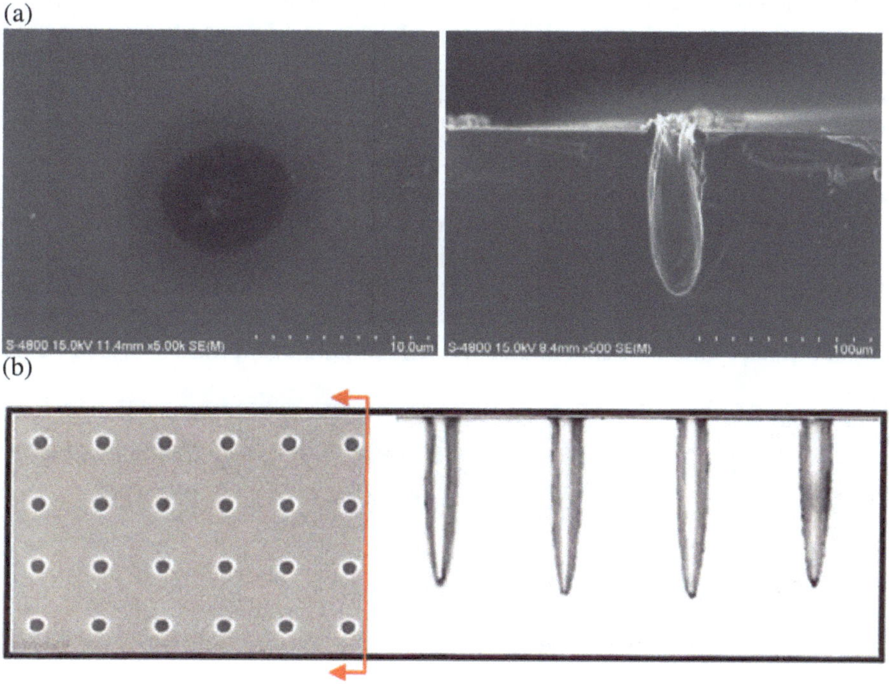

Fig. 3.2 A few examples of blind holes drilled by laser: (**a**) scanning electron micrograph of the *top* and the *profile view* of a blind hole in Si having a diameter of ~ 8 μm [24], and (**b**) a regular array of holes in Si of ~ 10 μm diameter and ~ 70 μm depth [22]. Note the tapered nature of drilled holes

this process has been successfully employed to drill blind holes of HAR and having diameters in range of 10–80 μm in Si [25]. As multiple laser beams can be simultaneously used, this process can be used to simultaneously drill multiple holes (see Fig. 3.2b) and hence may also have very high throughput ($>$2000 holes/s [23]).

Nevertheless, this process involves heating, although locally but excessively, of Si wafer that can damage Si wafer near the hole. This is particularly a problem if the Si wafer is single crystalline, as such heating may generate defects and also, induce polycrystallization, thus leading to a deterioration of functional properties of Si wafer [23]. In addition, excessive heating may also affect front-end and back-end structures, if they are already fabricated on the wafer. A laser pulse of very short durations (i.e., ultra-fast lasers) may be used to minimize the extent of heat-affected zone and residual stresses (or strains) in Si [23]. Due to this inherent existence of heat-affected zone, this process has so far found only limited application in creating TSVs.

3.3.1.2 Powder Blast Micromachining

PBM or abrasive jet machining (AJM) is a drilling method that works on the mechanical removal of material from precise locations. In this process, very fine hard powders are bombarded through high-pressure air nozzle on the brittle Si wafer. As the hard powders moving at very high speed ($\sim$100 m/s [26]) collide with Si, it abrasively removes bulk Si, thus creating a trench or hole in Si [26]. Al_2O_3 powders of sizes in the range of 10–20 μm can be used for creating trenches in Si wafer [26, 27]. In this process, a mask of a ductile material, such as electroplated Cu, polymer photoresists, etc., can be placed on top of Si wafer to protect the region from getting abrasively removed. Usage of such a mask decreases the aspect ratio of the trench [27]. As shown in Fig. 3.3, PBM has been successfully used to drill holes in Si. However, this process has not been able to drill holes of very HAR (usually aspect ratio is less than 5 [26]). Furthermore, this process imposes large

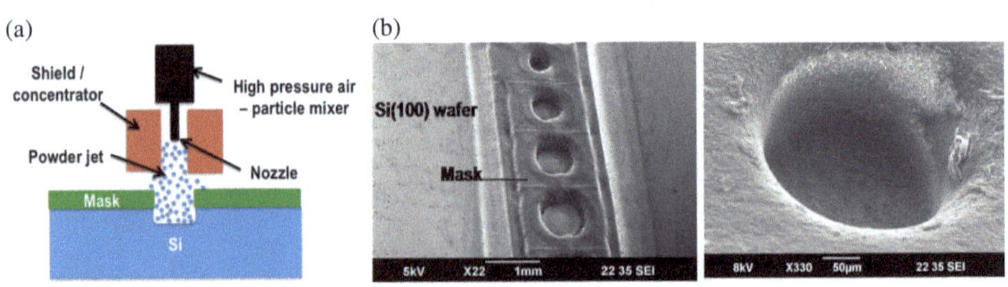

Fig. 3.3 (**a**) A schematic representation of the setup for powder blast micromachining. (**b**) A few examples of blind holes of different diameters created in Si wafer using high-resolution powder blast micromachining [27]. The mask used in (**b**) was SU-8

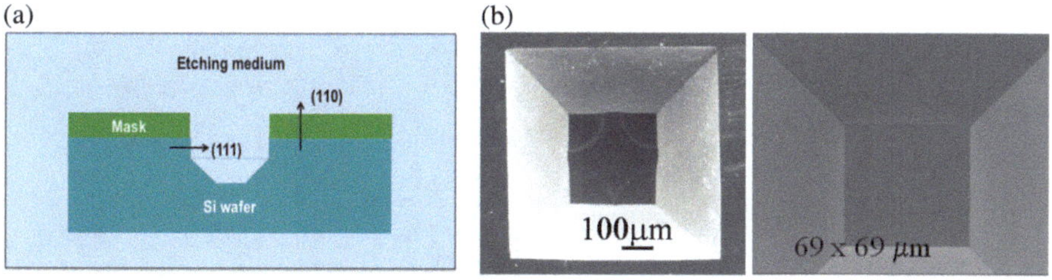

Fig. 3.4 (**a**) Schematic illustration of the workings of anisotropic wet etching of Si wafer. (**b**) A trench prepared in (1 0 0)-oriented Si wafer using etching with 9 M KOH at 60–65 °C [28]. In (**b**), a hard mask of SiO_2 on *top* of Si wafer was used

mechanical stresses on Si wafers, resulting in significant residual stresses, chipping of the corners, generation of defects, and polycrystallization. Therefore, this process is still not popular for producing TSVs.

3.3.1.3 Wet Etching

A widely used method for fabricating trenches in Si wafer is anisotropic wet etching of Si using hydroxides, e.g., KOH, NaOH, CeOH, RbOH, NH_4OH, $(CH_3)_4NOH$ (i.e., tetra methyl ammonium hydroxide or TMAH), etc. Reaction of hydroxides with Si produces a water soluble Si-oxide complex (i.e., $SiO_2(OH)_2^{2-}$), which is then flushed out, leaving a trench in Si. Since KOH etches the {110} family of planes almost 600 times faster than {1 1 1} planes, proper masking and exposing Si wafers of appropriate orientations to KOH can lead to anisotropic etching along one direction, resulting in HAR trenches (see Fig. 3.4a). Furthermore, since the kinetics of oxidation of {1 1 0}-planes is significantly slower as compared to {1 1 1}-planes, addition of an oxidizing agent in KOH may further suppress etching rate of {1 1 1}-planes as compared to {1 1 0}-planes [21]. This may further enhance the anisotropic etching, resulting in holes with an aspect ratio as high as 600 [29]. Similarly, ethylene diamine pyrocatechol (EDP) and TMAH etch 1 0 0-planes ~ 30 times faster than {1 1 1}-planes, and can, thus also be used to etch holes in {1 0 0}-oriented single crystals of Si [30]. In the context of wet chemical etching, amine gallate can also be used to anisotropically etch {1 0 0}-planes as compared to {1 1 1}-planes—the latter planes are etched at rates that are 50–100 times slower than {1 0 0}-planes. As shown in Fig. 3.4b, anisotropic wet etching of Si can be used to fabricate HAR TSVs. Although this process is easy to use and can produce structures with very HAR, this is limited mostly to produce rectangular grooves and is not popularly used for creating TSV structures with vertical walls [21].

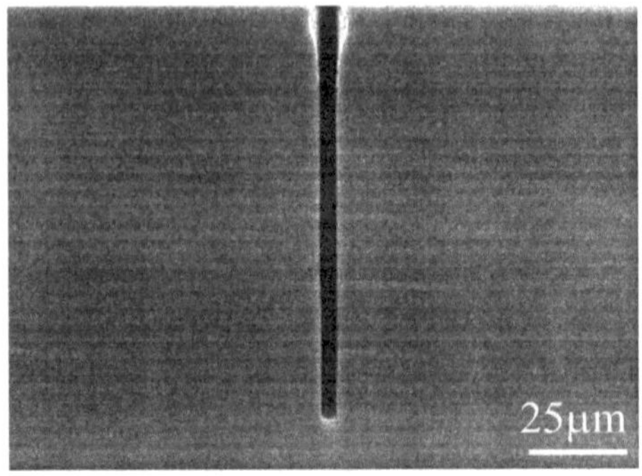

Fig. 3.5 A high aspect ratio blind hole etched in Si wafer by cryogenic plasma etching process [31]

3.3.1.4 Plasma-Based Methods

The most common method for creating holes or trenches in Si for fabricating TSVs with almost vertical walls is based on anisotropic etching of Si using fast moving and colliding plasma; this method is termed as 'deep reactive ion etching' or DRIE [21]. There are two main DRIE processes suitable for etching vertical holes in Si: (1) cryogenic DRIE and (2) Bosch process. In cryogenic plasma etching, the Si wafer is cooled to a very low temperature, often below −110 °C, using liquid nitrogen or liquid helium cells [21] and then unidirectionally bombarded with etchant plasma. Due to the low temperatures, the chemical etching of Si due to the reactive ions (which may be isotropic) is reduced dramatically and the Si wafer is primarily etched mechanically at a particular location in the vertical direction due to the bombardment of the highly energetic plasma [21]. Thus, as shown in Fig. 3.5, cryogenic DRIE produces holes or trenches of very HAR with almost vertical walls [28]. The major issues with the cryogenic process are (1) the need of a hard mask which does not etch away under plasma-bombardment and (2) cracking of Si, especially near the corners, due to the extremely low temperatures [21]. Due to these reasons, this process is not widely used commercially.

The Bosch process, which is another DRIE process used for etching HAR holes or trenches in Si, falls in the general category of time-multiplexed alternating process [21]. As the name suggests, in a time-multiplexed alternation process, multiple processes occur in a synchronous fashion. For example, in the Bosch process, etching and passivation processes are performed separately in a synchronous fashion, where etching is followed by passivation, and vice versa [32, 33]. Such an alternating, synchronous process becomes effective in making very deep trenches or holes in Si, as it is often very difficult to achieve anisotropic etching from sheer bombardment of the plasma. The passivation layer is applied especially on the sidewalls of the hole, so that the diameter of the hole does not increase as the depth of the hole is increased. With the help of schematic shown in Fig. 3.6, we describe different steps of this process below:

1. The top layer of the Si wafer is patterned using standard photolithography (see Fig. 3.6a). The plasma assisted etching process, as described in (b), does not etch

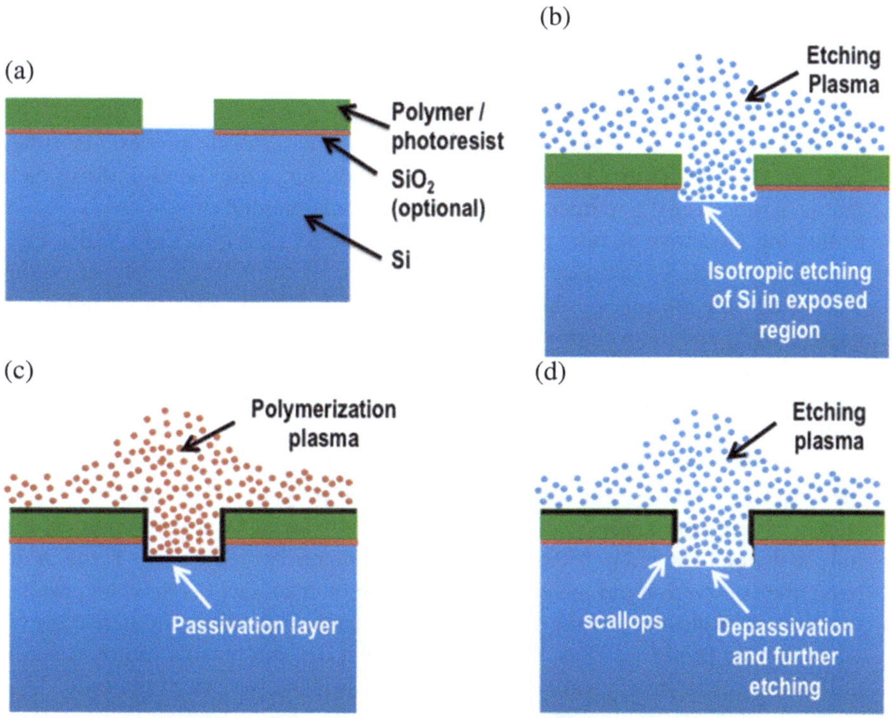

Fig. 3.6 Schematic illustration of different stages of Bosch process for etching holes or trenches in Si: (**a**) placement of a polymer mask on Si surface, (**b**) flow of etchant plasma on exposed surface of Si and subsequent etching of Si, (**c**) flowing passivation gases and subsequent formation of passivation layer on walls of Si, and (**d**) erosion of polymer layer from the *bottom* of the Si hole (depassivation) and subsequent further etching of Si at the *bottom* of the hole

the polymeric photoresist coated over the Si wafer. A SiO_2 layer may or may not be grown on top of Si wafer.

2. The Si wafer is etched by highly energetic plasma of a suitable gas, e.g., SF_6, CF_4, NF_3, Cl_2, etc. (see Fig. 3.6b) [21]. The gas creates a plasma consisting of electrons and neutral species as well as positively charged species. The positively charged species, which accelerate under an applied electric field, impinge on the Si surface normal to its motion path at high speed, removes any passivation layer or debris from there, and then reacts with Si [21]. Interestingly, due to the high momentum of the plasma, it also removes the formed chemical compound from the end of the hole in Si. The most common etching gas (and also the gas originally suggested in the Bosch process) is sulfur hexafluoride, SF_6 [32]. The following chemical reaction occurs between SF_6 and Si:

$$Si + 2SF_6(\mathrm{g}) \rightarrow SiF_4(\mathrm{g}) + 2SF_4(\mathrm{g}) \qquad T > 100^\circ\mathrm{C} \tag{3.1}$$

$$2Si + 2SF_6(g) \rightarrow 2SiF_4(g) + S_2F_4(g) \qquad T < -30°C \qquad (3.2)$$

Since all reaction products in above equations are gaseous, i.e., volatile, Si atoms are gradually removed from the wafer as the above process continues. The above reactions are highly exothermic and hence Bosch process is often performed at low temperatures (< 5 °C). In addition, a good and uniform etch rate is achieved if the temperature is uniformly maintained.

3. Following the above etching process, another gas, for example, C_4F_8, C_4F_6, etc., is passed over Si for depositing a thin layer of passivation polymer layer on the walls and the bottom of the hole (see Fig. 3.6c) [33]. The most common passivation gas is octa-fluoro-cyclo-butane, C_4F_8, which breaks down and produces the repeating unit -[CF_2]- of a Teflon type polymer in presence of the plasma created in the etching process. The polymer then gets deposited on the walls of Si hole [21].
4. In the next sequence of etching (i.e., step 1), the fast moving plasma impinges on the bottom of the hole and abrasively erodes the polymer layer from there; this step is called *depassivation* (see Fig. 3.6d). This exposes Si at the bottom of the hole to the etchant and hence the depth of the hole further increases. On the other hand, the passivation of the vertical sidewalls remains intact (as they are parallel to the flow of plasma and do not collide head-on with plasma) and hence the Si walls remain protected from the etchants. However, the plasma removes some polymer layer from sidewalls also, especially near the bottom of the hole. This leads to sideways etching or undercutting of the Si walls, leading to a scallopy structure (see Fig. 3.7). Minimization of the scallopy features or surface roughness is a challenge in Bosch process, giving rise to several innovative technologies, as will be explained below.

Using the Bosch process, holes in Si of various diameters and aspect ratios have been successfully prepared (see Fig. 3.7 for a few examples). It should be noted that the polymer removal rate from the bottom reduces with the increase in depth of the hole (as the velocity of the plasma is reduced). Hence, a strategy of gradually increasing accelerating voltage with the depth of the Si hole is often used for attaining very HAR holes [35]. It is also important to know that a one-step etching-passivation process can also be used in Bosch process. However, the etch-rate in the one-step process is very low as compared to the two-step cyclic Bosch process, and hence the latter is more widely used. In the cyclic Bosch process, the two steps of etching and passivation may last for a few seconds each. In a cyclic etch-passivation step, the depth of the Si-hole may increase by 0.5–5 μm. The original Bosch process involving the SF_6/C_4F_8 sequence results in an etch rate of 10 μm/min, which can be further improved by manipulating the gas chemistry [21]. A cycle with longer time not only increases the etching rate, but also increases the roughness of the Si sidewall [21].

As mentioned above and shown in Fig. 3.7a, the Si holes produced by the Bosch process are generally very scallopy. It has been reported that power to pressure ratio is an important process parameter affecting the surface roughness, where a higher

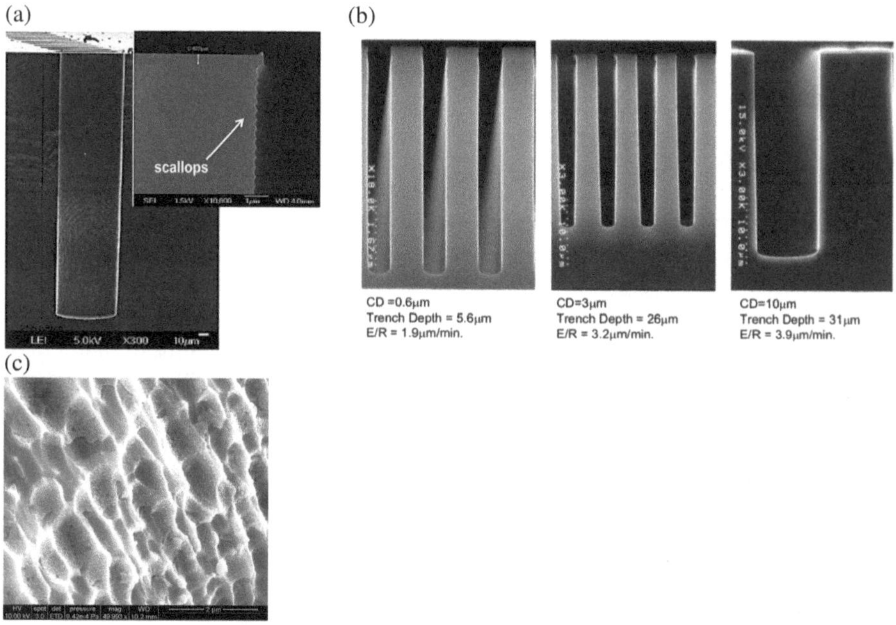

Fig. 3.7 A few representative Si blind holes prepared by Bosch process using: (**a**) SF_6/C_4F_8 [17] and (**b**) $SF_6/HBr/O_2$ [34] gas chemistries. The inset in (**a**) shows a magnified view of the wall of the hole, clearly revealing formation of scallops. A higher magnification image of the rough, scallopy structure of the sidewall is shown in (**c**). The marker bars in inset of (a) and (c) are 1 μm and 2 μm, respectively.

ratio produces smoother walls [36]. When the Bosch process is well controlled, the maximum distance between the hill and trough is less than 50 nm. Furthermore, it has been reported that adding postetch wet-etching process using KOH and isopropyl alcohol can reduce the surface roughness to 6 nm [37]. It has also been reported that adding new steps in the above time multiplex alternate process by including flow of either oxygen plasma or oxygen and argon plasmas [38, 39] can also reduce the surface roughness as well as increase the overall depth of the Si holes. Another strategy for reducing surface roughness is to perform a single step anisotropic etching at the beginning of the etching process, followed by the usual time multiplexing alternating method [21]. This method stems from the observation that the roughness of Si walls created by the Bosch process decreases with an increase in the depth of the hole [21].

3.3.2 Sequential Filling of Si Via

Once a blind hole is etched in Si, it is conformally filled with different classes of materials. Of all of the material filling processes, the two most critical criteria are:

(1) step coverage and (2) void-free filling. For ensuring step coverage, it is desired to have some taper in the Si via [16, 17]. Often a taper of 2–5° (i.e., TSV-sidewall inclination of 85–88° relative to the Si-surface) is optimum for conformal coating [16]. Below, with help of Fig. 3.1b, we describe the filling processes in detail:

1. Dielectric layer: A thin layer of low-dielectric, i.e., a dielectric with high breakdown voltage, is placed over the walls of the Si hole to establish electrical isolation between the current-carrying TSV-filler and the rest of the die. The most common dielectric materials are SiO_2, silicon nitrides, silicon oxynitride, tetra-ethyl ortho-silicate (TEOS), etc. Most of these dielectrics are deposited by plasma-enhanced chemical vapor deposition (PE-CVD) technique, whereas SiO_2 can also be grown using dry etching process. It is important to maintain a low thermal budget during this process, especially when the TSVs are prepared after fabricating transistors, etc. Polymers are another class of dielectric materials, which can be coated at low temperatures and are suitable for low power applications. A common polymer dielectric material is parylene or poly (p-xylene) [40]. Another advantage of using a soft and compliant polymer dielectric layer is that during thermal cycling, it can accommodate the differential thermal strains produced between Si and the metal filler (e.g., Cu) during thermal excursions, and hence minimize stresses produced in the TSV assembly [41]. As shown in Fig. 3.8a, one of the ways to coat polymers on Si is by first etching an annular trench concentric to the final via and then filling it by spin-on polymer, followed by curing of the polymer [41]. Following coating of the polymer in the Si annular trench, the solid Si inside the trench is etched, resulting in a blind hole with walls coated with polymers (see Fig. 3.8a). As shown in Fig. 3.8b, this method allows conformal coating of Si walls with the dielectric layer without voids. The thicknesses of oxide-based and polymer-based dielectrics are 50–200 nm and 1–5 μm, respectively [41].
2. Barrier and adhesion layers: Following the placement of an appropriate dielectric layer on Si walls, a thin barrier layer is placed on the dielectric layer for inhibiting diffusion of the conducting filler material into the dielectric layer and,

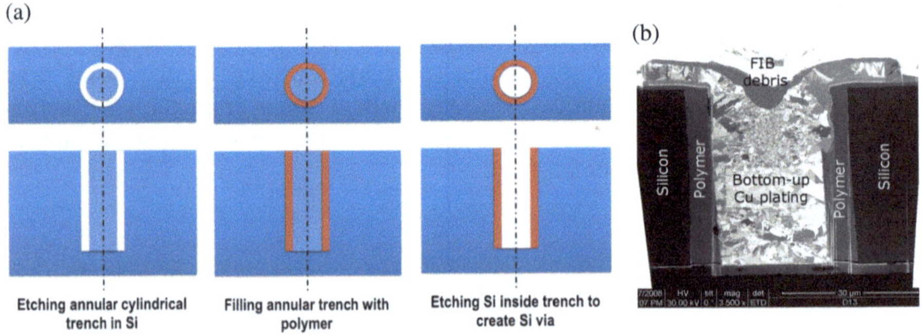

Fig. 3.8 (**a**) Schematic illustration of polymer filling strategy. (**b**) Scanning electron micrograph of a Cu filled TSV with polymer (Epoxy 8023-10) dielectric layer [41]

eventually, into Si. Selection of the barrier layer is specific to the filler material. For example, Ta, TaN, TaN/Ta bilayer, Re, Ti, and TiN may be used as the barrier layer for Cu [42]. An additional adhesion layer may be deposited on the barrier layer for improving attachment of the seed layer on top of the barrier layer [43, 44]. A thin layer of TiW, Ti, and Ta may be used as adhesion layer for Cu [44, 45]. It is interesting to note that a few materials, such as TaN, Ti, and Mn_4N, may act as both the barrier and the adhesion layers [44, 45]. In the particular case of TaN, often a thin layer of Ta is placed on top of it to further improve the adhesion. The barrier layer can be deposited using ionized pressure vapor deposition (PVD), atomic layer deposition (ALD), and chemical vapor deposition (CVD) processes. ALD usually gives the most conformal and continuous coating with minimum thickness of the barrier layer. The usual thickness of the barrier layers is a few tens of nanometers.

3. Seed layer: Since the resistivity of the barrier layer is very high, it cannot be used directly as a substrate (i.e., the cathode) for electroplating metal filler. Thus, a thin seed layer of the filler metal is deposited on the barrier layer for enabling electroplating of the metal filler. Hence, the seed layer may not be needed if either a paste of conducting material is squeeze-filled into the hole, or a moderately conducting barrier layer (e.g., Re), etc., is used. Often PVD is used for depositing the seed layer of metal filler. However, it is challenging to achieve conformal and continuous coating of uniform thickness of HAR holes through PVD. This leads to porous filling of material in the hole, resulting in a well-known phenomenon of bottom voiding [46]. CVD may produce a better conformal coating of the Cu seed layer; however, the adhesion of the Cu CVD layer and the barrier layer remains an issue [46]. Thus, electrochemical seed-layer enhancement (SLE) technologies based on wet coatings may be used for repairing the seed layer by filling the discontinuities in the PVD seed layer [46]. This process is also an electroplating process (which, as will be discussed below, is used to fill the hole by a metallic filler) and hence the same setup as of standard electroplating can be used for this process also. However, SLE uses different bath chemistry so that metal plating can be take place on both seed layer and the barrier layer, without corroding the existing seed layer [46]. Figure 3.9 shows a pair of scanning electron micrographs showing the effect of SLE process on a PVD-grown seed layer. As shown in Fig. 3.9, the usual thickness of seed layers is 10–40 nm. Since the grain size, grain shape, and grain orientation of the metal filler depend on the microstructure of the seed layer [47], it is a very critical step and is a subject of constant innovation.
4. Filler: Once the Si hole has been filled with the dielectric layer, barrier layer, and seed layer, it is then filled with the conducting filler, which occupies most of the volume of an etched hole. Figure 3.10 shows a few micrographs of a Cu filled TSV, also revealing different layers. The most common method for depositing filler material is electroplating. This is due to the fact that metals, especially Cu, are the most widely used fillers for TSV assemblies. Besides electroplating, squeezing-in a conducting paste (called paste printing) and CVD can also be

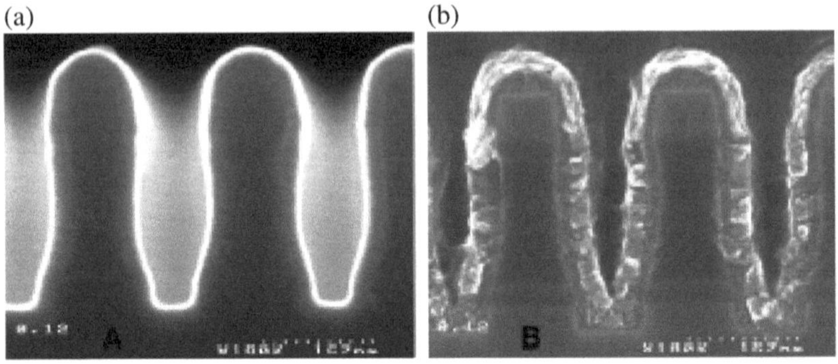

Fig. 3.9 Scanning electron micrographs showing the effect of SLE process on seed layer of Cu deposited in trenches with 0.15 μm openings: (**a**) 15-nm thick seed layer deposited by PVD process, and (**b**) an additional 40-nm thick layer deposited through SLE process [46]

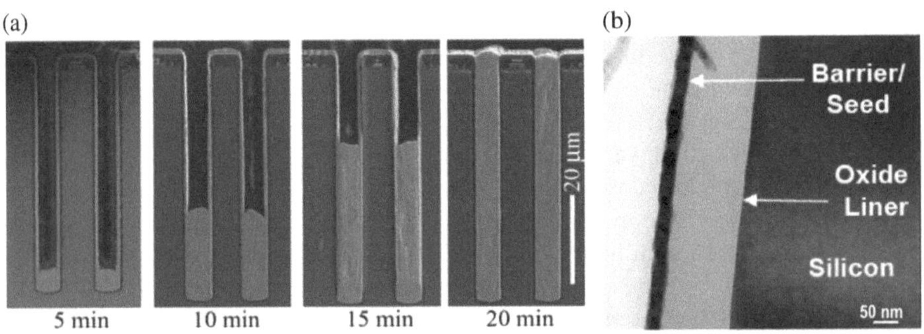

Fig. 3.10 Micrographs showing cross-sectional view of Cu filled TSVs: (**a**) time lapse images at low magnification showing the fill front during bottom-up electroplating [48], and (**b**) at high magnification TEM micrograph showing different material layers [49]. Cu electroplating in (**a**) was performed at an applied potential of 0.600 V using electrolytic bath comprising of 1.0 mol/L $CuSO_4$, 0.5 mol/L H_2SO_4, 0.001 mol/L NaCl, and 10 μmol/L poloxamine (Tetronic 701) [48]

used to fill the hole [50]. A filling process should result in a void-free and stress-free filler material; however, these two critical requirements are not easily fulfilled. Hence, in practice, the goal becomes their minimization. Cu, W, and other metals are electroplated, whereas heavily doped poly-silicon is filled by low-pressure CVD [51], and metal-polymer (e.g., Ag/polypyrrole) composites are often paste printed [50]. One of the common commercial electrolytes for Cu electroplating is an acidic bath comprising 0.88 M $CuSO_4$, 0.54 M H_2SO_4, and 60 ppm Cl^- (NaCl) [51]. Since most often the strategy for filling the via is bottom-up (i.e., directional and not isotropic), a suppressor, an accelerator, and a leveler are also added to the electrolytic bath for efficient via filling [52]. Suppressors interact with chloride ions to inhibit Cu deposition. An accelerator is adsorbed on the electrode surface, where it gradually replaces the suppressor, thus allowing electroplating of Cu to take place. Furthermore, coverage of an accelerator moves upwards, thus enabling Cu filling to take place in a directional fashion from the bottom to the top [52]. Levelers are used to improve filling performance, decrease the surface roughness and prevent Cu deposition at the opening of the via [52]. Most of the suppressors are polymers (e.g., polyethylene glycol (PEG), etc.) whereas common accelerators and levelers for Cu plating are 3-mercapto-1-propanesulfonate (MPS), (3-sulfopropyl) disulfide (SPS), 3,3-thiobis (1-propanesulfonate) (TBPS), etc., and Janus Green B (JGB), Diazine Black, Alcian Blue, etc., respectively [52]. As mentioned above, Cu is the most widely used filler for its very high electrical conductivity and compatibility with the back-end structures, although in some applications, W fillers are also used [53]. On the other hand, poly-silicon is compatible with the front-end structures (i.e., transistors) and its thermal expansion coefficient is also similar to that of Si.

3.3.3 Planarization and Die-Thinning

As shown in Figs. 3.1b and 3.11a, after filling the metal in the Si via, there is some excess metal or paste that is leftover on top of the wafer; this feature is called overburden. There is a need for extra filling so that voids inside the filler can be minimized. Also, due to the Cu-pumping phenomenon [56] (which will be discussed in next chapter), Cu metal can protrude relative to the Si. In addition, if the metal is not filled with proper overburden, the filler may shrink inside the Si via, leaving a filler gap near its mouth. Therefore, there is a need for planarization of Si wafer on the side with the TSV-mouth or opening (i.e., from the side where TSV was created and filled). In addition, the Si wafer may also need to be thinned down

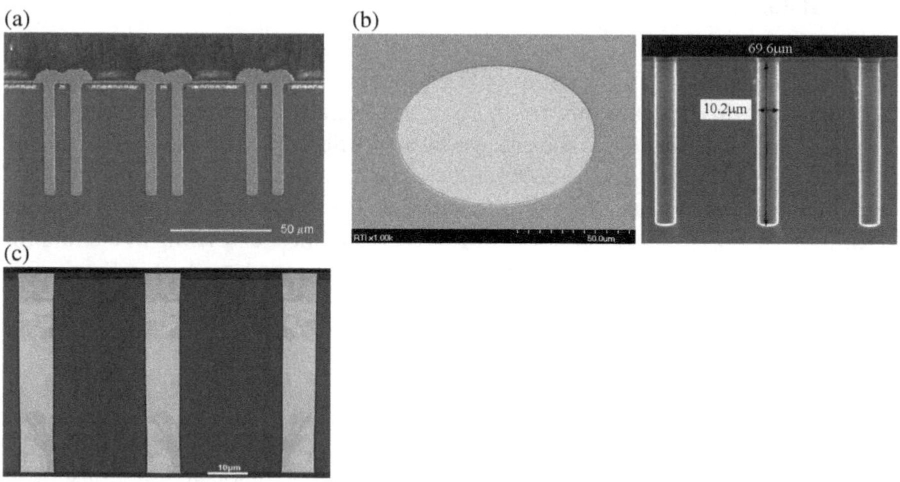

Fig. 3.11 The sequence of events showing cross-section of a TSV after (**a**) metal filling [48] (**b**) planarization of Si wafer on the surface having via mouth or opening (*left* and *right* picture shows *top* and *profile* views, respectively) [redrawn from [54]], and (**c**) thinning of Si wafer from the back-side [55]

to the desired thickness. It is convenient to thin down the Si wafer from the back side, where TSV processing was not performed, primarily to avoid polishing the ductile metal and brittle Si at the same time. Fig. 3.11 shows the process flow of these two processes and their impact on the TSV assembly.

As mentioned above, Cu pumping, which manifests itself in protrusion of Cu relative to Si [56–58], may take place during high temperature processing of front-end and back-end structures, and deposition of redistribution layers (RDL). Since FEOL and BEOL processes, as well as soldering and die stacking steps typically, require heating the device to temperatures in the range of 250–450 °C, the as-deposited Cu TSV structures are often annealed at temperatures moderately higher than that used in subsequent fabrication steps, after filling process and before chemical-mechanical planarization (CMP) [58]. Such pre-CMP annealing reduces the adverse effects of Cu-pumping during subsequent service, and also improves the adhesion of the via-metal to the via-walls, and stabilizes the microstructure of the metal filler [59].

Mechanical grinding and chemo-mechanical polishing (CMP) are subsequently used to planarize the Si-die or wafer at the via-opening side, as well as to thin the Si-wafer. Following planarization and thinning, the Si surfaces possess a smooth finish (typically better than 1 μm), as well as small residual stresses in TSV. For performing mechanical grinding and CMP, the Si wafer containing TSV assembly can be temporarily bonded with a glass sheet or another thick wafer using epoxy, etc.

3.4 Flow Process for Fabricating TSVs and Integration of Dies

A 3-D architectured microelectronic package comprising TSVs can be fabricated using one of 9 different algorithms: three of these relate to the sequence in which a TSV assembly is fabricated in the process flow (see Sect. 3.4.1), whereas for each of these three possibilities, there are three possible configurations in which the TSV containing die can be integrated with other dies (see Sect. 3.4.2). In any of these algorithms, the basic steps for fabricating TSV assemblies, consisting of etching Si, filling, and planarization and thinning, remain the same as described in Sect. 3.3.

3.4.1 Sequence of Flow Process

Based on the sequence of the process flow, TSVs can be fabricated first, last or in the middle, relative to the front end of line (FEOL) processes and the fabrication of BEOL structures. Accordingly these are called via-first, via-last, and via-middle sequences, respectively [13]. As the names of these sequences suggest and as shown

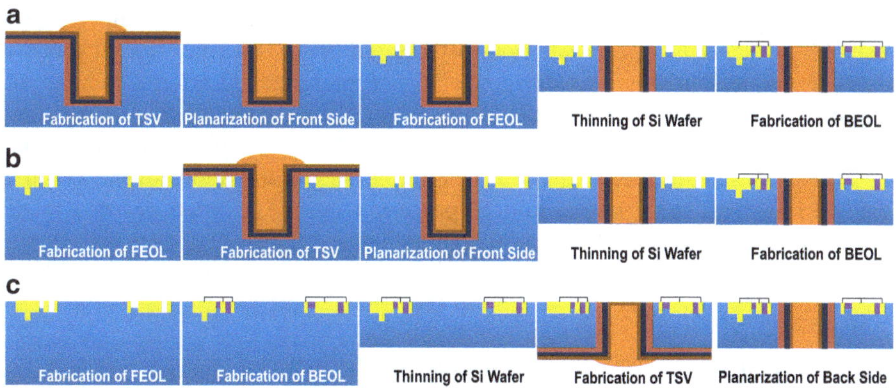

Fig. 3.12 Flow-process sequence for fabricating TSV assembly with respect to the fabrication of FEOL and BEOL structures: (**a**) via-first, (**b**) via-middle and (**c**) via-last. It should be noted that some of the sequences may interchange depending on the process optimization; for example, BEOL structures may be formed before thinning the Si wafer in the via-first process sequence, etc.

in Fig. 3.12, in the via-first and the via-last process sequences, the TSV structures are fabricated before FEOL and after BEOL processes, respectively, whereas the TSV structures are fabricated after FEOL and before BEOL in the via-middle process. In addition and as shown in Fig. 3.12, the Si wafer is etched from the top (where the FEOL and BEOL structures reside) and thinned from the bottom, after fabrication of TSV structures in the via-first and via-middle process sequences. In contrast, in the via-last process, the Si wafer is thinned before fabrication of TSV structures, following which the via is etched from the bottom of the wafer [33]. Therefore, maintaining proper alignment between the TSV and the BEOL is challenging in the via-last process.

Another important aspect that affects the selection of the via-processing sequence is the thermal budget during FEOL and BEOL processes. FEOL processing involves oxidation and etching of Si and hence it is often conducted at high temperatures ($>400\,^{\circ}$C). BEOL processes, which comprise placing dielectrics, metal interconnects, etc., are generally conducted at relatively low temperatures ($<400\,^{\circ}$C), although dielectric deposition temperatures can sometimes reach 425–450 °C. Hence, selecting the sequence in which TSVs should be fabricated is important, and has to be often optimized on a case-by-case basis. For example, fabricating TSV assemblies in the via-first sequence is the easiest of all process sequences, as it allows working on a pristine Si wafer with no other structure. However, since there may be considerable difference between the CTE of the metal filler and Si, the FEOL processing may lead to the build up of significant thermal stresses in the structure, resulting in metal pumping, interfacial sliding, fracture, and other reliability issues. In addition, certain filler metals, for example, Cu, are difficult to use in the via-first process sequence due to their very fast diffusivity into Si at high temperatures. This limits the choices for materials to be used as fillers, e.g., W and highly doped poly-silicon, which have CTEs close to that of Si. On the other hand, the via-last process sequence may avoid such thermal loading of TSV assemblies. Nevertheless, in addition to the alignment issue, the via-last process also affords less process-flexibility due to the presence of a large number of pre-formed structures (i.e., BEOL and FEOL) on the Si wafer. Hence, fabricating vias with very small diameter ($< 5\,\mu$m) and very HAR (> 50) is relatively more difficult in via-last process sequence. Due to the above limitations of the via-first and via-last processes, the via-middle process sequence is often deemed to be advantageous and suitable for several applications.

3.4.2 Integration of Dies Comprising TSVs

Once a die comprising TSV structures is fabricated, it can then be integrated with another die in the microelectronic device. As schematically shown in Fig. 3.13a–c, one of the following three strategies can be used for integration [15]: (1) face-to-face (F2F), where the front side, i.e., the side of the Si wafer which has the FEOL and BEOL structures, of both dies face each other, (2) back-to-back (B2B), where the back side of both dies face each other, and (3) face-to-back (F2B), where the

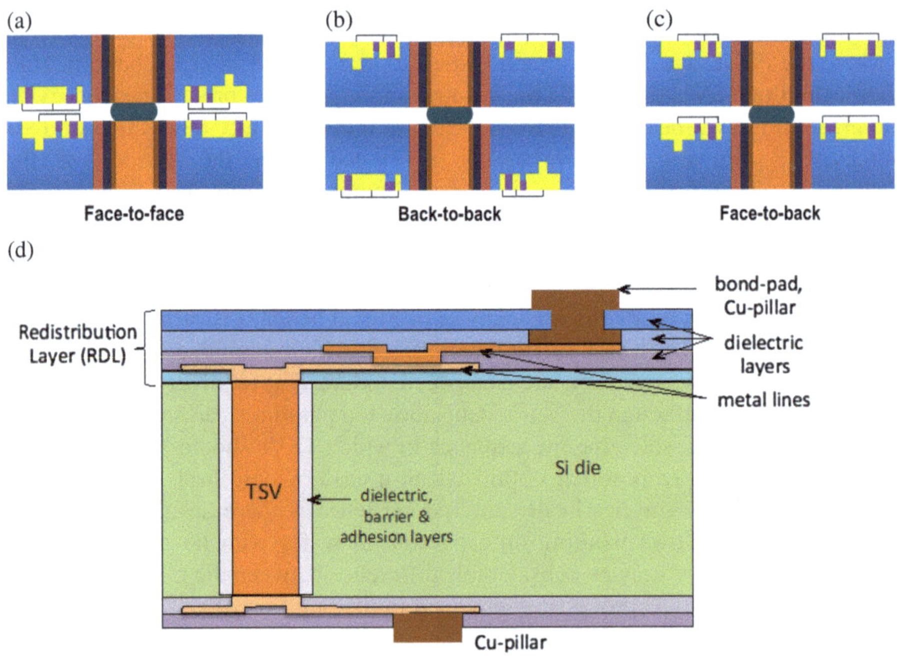

Fig. 3.13 Schematic of integration strategies via die-stacking and connecting with solder-microbumps: (**a**) face-to-face, (**b**) back-to-back and (**c**) face-to-back. (**d**) A redistribution layer (RDL) is often utilized to connect dies, allowing TSVs to be placed at different locations from the bond-pads

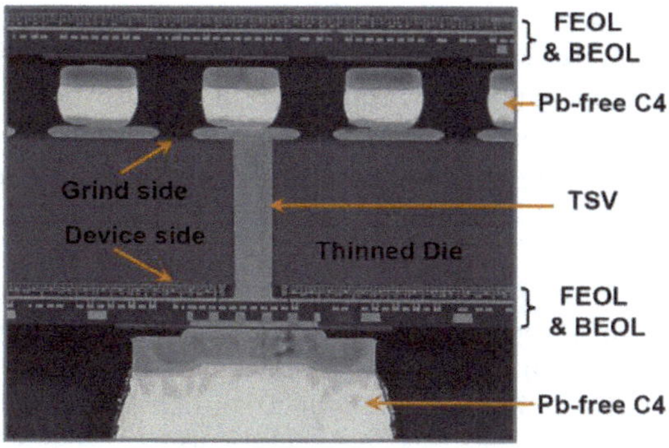

Fig. 3.14 Cross-section of a device employing face-to-back strategy of integration [61]

front side of one of the dies faces the back side of another die. In each of these integration strategies, a joint between the properly aligned vias may be formed using solder microbumps, often utilizing a bond-pad or copper pillar attached to the end of the TSV. To minimize challenges associated with perfect alignment, and to fan out and/or enlarge the inter-chip joints, a redistribution layer (RDL), comprising metal circuitry (e.g., Cu) embedded in a dielectric layer (e.g., polyimide or benzocyclobutene), is often utilized to re-route the electrical connections from the TSV to larger bond-pads, as illustrated in Fig. 3.13d. In addition to relaxing constraints associated with joint spacing and size, multi-layered RDLs can reduce TSV-insertion losses, while also improving signal transmission and reliability [60]. This strategy also obviates the need for perfect alignment of pads connected to TSVs, allowing more flexibility in soldering with very thin joints. Figure 3.14 shows cross-section of a device employing face-to-back integration technique. Two key challenges during integration of two dies are the handling of very thin wafers ($< 100\,\mu m$) and minimizing warpage during assembly, respectively. To minimize these complications, the thinned Si wafer is often temporarily attached to a glass or a thick wafer using curable epoxy, high temperature adhesives or sometimes, by using electrostatic bonding. However, for electrostatic bonding, the surfaces of the glass plate and the Si wafer need to be very smooth (< 10 nm roughness) and clean. In addition, since electrostatic bonding is performed at moderately high temperatures ($> 300\,^{\circ}C$), a glass with CTE similar to that of Si should be used.

3.5 Summary

TSVs are a critical feature of vertically stacked area-array dies, and therefore constitute a very important enabling feature of 3D packages. In this chapter, the approaches for the fabrication of TSVs in dies have been presented. Briefly, fabrication processes comprise four steps, including drilling of via-holes in silicon, sequential via-filling (with a dielectric, a diffusion barrier, an adhesion layer, a seed layer, and a metal filler), chemical–mechanical planarization and thinning of the silicon, and finally, deposition of a redistribution layer on the surface of the silicon. These processes and associated materials have been described in some detail. Processing challenges as well as methods to overcoming them are highlighted and discussed.

Acknowledgements The editors would like to thank Purushotham Kaushik Muthur Srinath and Shengquan E Ou from Intel Corporation for their critical review of this chapter. The authors (PK and ID) acknowledge financial support for some of the reported work by the National Science Foundation (DMR-0513874 and DMR-1309843), Cisco Research Council and the Semiconductor Research Corporation. PK would also like to acknowledge financial support from Department of Science and Technology, India (grant number DSTO 1164). The contributions of, and collaborations with several colleagues (Dr. Lutz Meinshausen, formerly of Washington State University, and currently at Global Foundries, Dresden, Germany; Dr. Tae-Kyu Lee, formerly of Cisco Systems, and currently at Portland State University; Dr. Ravi Mahajan of Intel Corporation; Dr. Vijay Sarihan of Freescale Semiconductor, and Professor Muhannad Bakir of Georgia Tech) are gratefully acknowledged. The assistance of current and former colleagues (Dr. Hanry Yang of Washington State University, and Dr. Zhe Huang, formerly of Washington State University, and currently at Seagate Technologies) with the literature survey is also gratefully acknowledged. The author (ZH) acknowledges financial support for his research by the Pearl River Science and Technology Nova Program of Guangzhou under grant no. 2012J2200074, the National Natural Science Foundation of China (NSFC) under grant no. 51004118 and Guangdong Natural Science Foundation under grant no. 2015A030312011.

References

1. E. Beyne, Through-silicon via technology for 3D IC, in: *Ultra-Thin Chip Technology and Applications*, ed. by J.N. Burghartz (Springer, New York, 2011)
2. M. Stucchi, G. Katti, D. Velenis, TSV characterization and modeling, in: *Three Dimensional System Integration: IC Stacking Process and Design*, ed. by A. Papanikolaou, D. Soudris, R. Radojcic (Springer, New York, 2011)
3. P.S. Andry, C.K. Tsang, B.C. Webb, E.J. Sprogis, S.L. Wright, B. Dang, D.G. Manzer, Fabrication and characterization of robust through-silicon vias for silicon-carrier applications. IBM J. Res. Dev. **52**, 571–581(2008)
4. Z. Xu, J.Q. Lu, High-speed design and broadband modeling of through-strata-vias (TSVs) in 3D integration. IEEE Trans. Compon. Pack. Manuf. Technol. **1**, 154–162 (2011)
5. J.Q. Lu, Advances in materials and processes for 3D-TSV integration. ECS Trans. **45**, 119–129 (2012)
6. A. Tsukada, R. Sato, S. Sekine, R. Kimura, K. Kishi, Y. Sato, Y. Iwata, H. Murata, Study on TSV with new filling method and alloy for advanced 3D-SiP, in: *ECTC: Electronic Components and Technology Conference*, 31 May–3 June 2011, Lake Buena Vista, FL, New York, USA (IEEE, 2011), p. 1981
7. R. Sato, A.Tsukada, Y. Sato, Y. Iwata, H. Murata, S. Sekine, R. Kimura, K. Kishi, Study on high performance and productivity of TSV's with new filling method and alloy for advanced 3D-SiP, in: *3DIC: International 3D Systems Integration Conference*, 31 Jan–2 Feb 2012, Osaka, New York, USA (IEEE, 2012), p. 1
8. A. Horibe, K. Sueoka, T. Aoki, K. Toriyama, K. Okamoto, S. Kohara, H. Mori, Y. Orii, Through silicon via process for effective multi-wafer integration, in: *ECTC: Electronic*

Components and Technology Conference, 26–29 May 2015, San Diego, CA (IEEE, 2015), p. 1808
9. M. Bouchoucha, L.L. Chapelon, P. Chausse, S. Moreau, N. Sillon, Through silicon via polymer filling for 3D-WLP applications, in: *ESTC: Electronic System-Integration Technology Conference*, 13–16 Sept 2010, Berlin (IEEE, 2010), p. 1
10. A. Peic, *Lithography Process Innovations for 2.5/3D Part 1: Alleviating TSV Stress* (2014), http://www.3dincites.com/2014/08/lithography-process-innovations-for-2-53d-part-1-alleviating-tsv-stress/. Accessed 21 June 2016
11. L. Zhu, D.W. Hess, C.P. Wong, Carbon nanotube electrical and thermal properties and applications for interconnects, in: *Integrated Interconnect Technologies for 3D Nanoelectronic Systems*, ed. by M.S. Bakir, J.D. Meindl (Artech House, Norwood, MA, 2009)
12. D. Jiang, W. Mu, S. Chen, Y. Fu, K. Jeppson, J. Liu, Vertically stacked carbon nanotube-based interconnects for through silicon via application. IEEE Electron Dev. Lett. **36**, 499–501 (2015)
13. J.U. Knickerbocker, P.S. Andry, B. Dang, et al., Three dimensional silicon integration. IBM J. Res. Dev. **52**, 553–569 (2008)
14. J.Q. Lu, 3-D hyperintegration and packaging technologies for micro-nano systems. Proc. IEEE **97**, 18–30 (2009)
15. M. Koyanagi, T. Fkushima, T. Tanaka, High-density through silicon vias for 3-D LSIs. Proc. IEEE **97**, 49–59 (2009)
16. S Spiesshoefer, Z. Rahman, G. Vangara, S. Polamreddy, S. Burkett, L. Schaper, Process integration for through-silicon vias. J. Vac. Sci. Technol. A **23**, 824–829 (2005)
17. R. Nagarajan, K. Prasad, L. Ebin, B. Narayanan, Development of dual-etch via tapering process for through-silicon interconnection. Sens. Actuators A **139**, 323–329 (2007)
18. R. Li, Y. Lamy, W.F.A. Besling, F. Roozeboom, P.M. Sarro, Continuous deep reactive ion etching of tapered via holes for three-dimensional integration. J. Micromech. Microeng. **18**, 125023 (2008)
19. C. Okoro, K. Vanstreels, R. Labie, O. Luhn, B. Vandevelde, B. Verlinden, D. Vandepitte, Influence of annealing conditions on the mechanical and microstructural behavior of electroplated Cu-TSV. J. Micromech. Microeng. **20**, 045032 (2010)
20. N. Ranganathan, L. Ebin, L. Linn, L.W.S. Vincent, O.K. Navas, V. Kripesh, N. Balasubramanian, Integration of high aspect ratio tapered silicon via for silicon carrier fabrication. IEEE Trans. Adv. Pack. **32**, 62–71 (2009)
21. B. Wu, A. Kumar, S. Pamarthy, High aspect ratio silicon etch: a review. J. Appl. Phys. **108**, 051101 (2010)
22. D. Gerke, NASA 2009 body of knowledge (BoK): through-silicon via technology. JPL Publication 09-28, Jet Propulsion Laboratory, Pasadena, CA (2009)
23. R. Landgraf, R. Rieske, A.N. Danilewsky, K.J. Wolter, Laser drilled through silicon vias: crystal defect analysis by synchrotron X-ray topography, in: *ESTC: 2nd Electronics System-Integration Technology Conference*, Greenwich, New York, USA 1–4 Sept 2008 (IEEE, 2008), p. 1023
24. Y.H. Lee, K.J. Choi, Analysis of silicon via hole drilling for wafer level chip stacking by UV laser. Int. J. Prec. Eng. Manuf. **11**, 501–507 (2010)
25. Industrial laser application note: laser drilling of through silicon vias (tsv). http://www.spectra-physics.com/applications/application-notes/laser-drilling-of-through-silicon-vias-tsv. Accessed 21 June 2016
26. A. Polyakov, T. Grob, R.A. Hovenkamp, H.J. Kettelarij, I. Eidner, M.A. de Samber, M. Bartek, J.N. Burghartz, Comparison of via-fabrication techniques for through-wafer electrical interconnect applications, in: *ECTC: Electronic Components and Technology Conference*, 1–4 June 2004 (IEEE, 2004), p. 1466
27. S.P. Lee, H.-W. Kang, S.-J. Lee, I.H. Lee, T.J. Ko, D.-W. Cho, Development of rapid mask fabrication technology for micro-abrasive jet machining. J. Mech. Sci. Technol. **22**, 2190–2196 (2008)

28. M.J. Archer, F.S. Ligler, Fabrication and characterization of silicon micro-funnels and tapered micro-channels for stochastic sensing applications. Sensors **8**, 3848–3872 (2008)
29. S.H. Kim, et al. (1 1 0) Silicon etching for high aspect ratio comb structures, in: *ETFA '97: Proc. Emerging Technologies and Factory Automation*, 9–12 Sept 1997, Los Angeles, CA (IEEE, 1997), p. 248
30. G.T.A. Kovacs, N.I. Maluf, K.E. Petersen, Bulk micromachining of silicon. Proc. IEEE **86**, 1536–1551 (1998)
31. S. Aachboun, P. Ranson, Deep anisotropic etching of silicon. J. Vac. Sci. Technol. A **17**, 2270–2273 (1999)
32. F. Laermer, A. Schilp, Method of anisotropically etching silicon. US Patent US5501893 A (1996)
33. M.J. Madou, *Fundamentals of Microfabrication: The Science of Miniaturization*, 2nd ed. (CRC Press, Boca Raton, 2002)
34. P. Nallan, A. Khan, S. Pamarthy, S.T. Hsu, A. Kumar, Advanced deep silicon etching for deep trench isolation, optical components and micro-machining applications, in: *ET Conference Proceedings* (Applied Materials Internal Publication, 2001), Santa Clara, USA
35. S.-B. Jo, M.-W. Lee, S.-G. Lee, E.-H. Lee, S.-G. Park, B.-H. O, Characterization of a modified Bosch-type process for silicon mold fabrication. J. Vac. Sci. Technol. A **23**, 905–910 (2005)
36. M.J. Walker, Comparison of Bosch and cryogenic processes for patterning high aspect ratio features in silicon. Proc. SPIE **4407**, 89–99 (2001)
37. I.R. Johnston, H. Ashraf, J.K. Bhardwaj, J. Hopkins, A.M. Hynes, G. Nicholls, S.A. McAuley, S. Hall, L. Atabo, G.R. Bogart, A. Kornblit, A.E. Novembre, Etching 200-mm diameter SCALPEL masks with the ASE process. Proc. SPIE **3997**, 184–193 (2000)
38. R. Nagarajan, L. Ebin, D. Lee, C.S. Soh, K. Prasad, N. Balasubramanian, Development of a novel deep silicon tapered via etch process for through-silicon interconnection in 3-D integrated systems, in: *ECTC: Electronic Components and Technology Conference*, 30 May–2 June 2006, San Diego, CA (IEEE, 2006), p. 383
39. J.-H. Lai, H.S. Yang, H. Chen, C.R. King, J. Zaveri, R. Ravindran, M.S. Bakir, A 'mesh' seed layer for improved through-silicon-via fabrication. J. Micromech. Microeng. **20**, 025016 (2010)
40. M. Miao, Y. Zhu, M. Ji, J. Ma, X. Sun, Y. Jin, Bottom-up filling of through silicon via (TSV) with Parylene as sidewall protection layer, in: *EPTC '09: Electronics Packaging Technology Conference*, 9–11 Dec 2009, Singapore (IEEE, 2009), p. 442
41. D.S. Tezcan, F. Duval, H. Philipsen, O. Luhn, P. Soussan, B. Swinnen, Scalable through silicon via with polymer deep trench isolation for 3D wafer level packaging, in: *ECTC: Electronic Components and Technology Conference*, 26–29 May 2009, San Diego, CA, New York, USA (IEEE, 2009), p. 1159
42. P. Shi, J. Enloe, R. van den Boom, B. Sapp, Direct copper electrodeposition on a chemical vapor-deposited Ruthenium seed layer for through-silicon vias, in: *IITC: International Interconnect Technology Conference*, 4–6 June 2012, San Jose, CA (IEEE, 2012), p. 1
43. M.J. Wolf, T. Dretschkow, B. Wunderle, N. Jurgensen, G. Engelmann, O. Ehrmann, A. Uhlig, B. Michel, H. Reichl, High aspect ratio TSV copper filling with different seed layers, in: *ECTC: Electronic Components and Technology Conference*, 27–30 May 2008, Lake Buena Vista, FL (IEEE, 2008), p. 563
44. V.S. Rao, S.W. Ho, W.S.V. Lee, H.Y. Li, E. Liao, R. Nagarajan, T.C. Chai, X. Zhang, P. Damaruganath, TSV interposer fabrication for 3D IC packaging, in: *EPTC '09: Electronics Packaging Technology Conference*, 9–11 Dec 2009, Singapore (IEEE, 2009), p. 431
45. Y. Au, Q.M. Wang, H. Li, J.S.M. Lehn, D.V. Shenai, R.G. Gordon, Vapor deposition of highly conformal copper seed layers for plating through-silicon vias (TSVs). J. Electrochem. Soc. **159**, D382–D385 (2012)
46. P.H. Haumesser, L.A. Roule, S. Maitrejean, G. Passemard, Seed enhancement: a bridging technology. Future Fab. Intl. **19**, 81–83 (2005)

47. K.J. Ganesh, A.D. Darbal, S. Rajasekhara, Effect of downscaling nano-copper interconnects on the microstructure revealed by high resolution TEM-orientation-mapping. Nanotechnology **23**, 135702 (2012)
48. T.P. Moffat, D. Josell, Extreme bottom-up superfilling of through-silicon-vias by damascene processing: suppressor disruption, positive feedback and Turing patterns. J. Electrochem. Soc. **159**, D208–D216 (2012)
49. J. Dukovic, S. Ramaswami, S. Pamarthy, R. Yalamanchili, N. Rajagopalan, K. Sapre, Z. Cao, T. Ritzdorf, Y. Wang, B. Eaton, R. Ding, M. Hernandez, M. Naik, D. Mao, J. Tseng, D. Cui, G. Mori, P. Fulmer, K. Sirajuddin, J. Hua, S. Xia, D. Erickson, R. Beica, E. Young, P. Kusler, R. Kulzer, S. Oemardani, H. Dai, X. Xu, M. Okazaki, K. Dotan, C. Yu, C. Lazik, J. Tran, L. Luo, Through-silicon-via technology for 3D integration, in: *IEEE International Memory Workshop*, 16–19 May 2010, Seoul (IEEE, 2010), p. 1
50. B. Horvath, J. Kawakita, T. Chikyow, Through silicon via filling methods with metal/polymer composite for three-dimensional LSI. Jpn. J. Appl. Phys. **53**, 06JH01 (2014)
51. P. Dixit, T. Vehmas, S. Vahanen, P. Monnoyer, K. Henttinen, Fabrication and electrical characterization of high aspect ratio poly-silicon filled through-silicon vias. J. Micromech. Microeng. **22**, 055021 (2012)
52. C.-F. Hsu, W.-P. Dow, H.-C. Chang, W.-Y. Chiu, Optimization of the copper plating process using the Taguchi experimental design method: I. Microvia filling by copper plating using dual leverlers. J. Electrochem. Soc. **162**, D525–D530 (2015)
53. G. Pares, N. Bresson, S. Minoret, V. Lapras, P. Brianceau, J.F. Lugand, R. Anciant, N. Sillon, Through silicon via technology using tungsten metallization, in: *ICICDT: International Conference IC Design & Technology*, 2–4 May 2011, Kaohsiung (IEEE, 2011), p. 1
54. R.L. Rhoades, Overview of CMP for TSV applications (2013), http://www.entrepix.com/docs/papers-and-presentations/Rhoades-CMP-for-TSV-AVS-June2013-shareable.pdf. Accessed 21 June 2016
55. S.W. Yoon, D.J. Na, K.T. Kang, W.K. Choi, C.B. Yong, Y.C. Kim, P.C. Marimuthu, TSV MEOL (Mid-End-Of-Line) and its assembly/packaging technology for 3D/2.5D solutions, in: *ICEP-IAAC: Joint Conference of International Conference on Electronics Packaging and IMAPS All Asia Conference*, 17–20 April 2012, Tokyo (2012), http://www.statschippac.com/~/media/Files/DocLibrary/whitepapers/2012/STATSChipPAC_ICEP2012_TSV_MEOL_and_Pkg.ashx. Accessed 21 June 2016
56. D. Smith, S. Singh, Y. Ramnath, M. Rabie, D. Zhang, L. England, TSV residual Cu step height analysis by white light interferometry for 3D integration, in: *ECTC: Electronic Components and Technology Conference*, 26–29 May 2015, San Diego, CA (IEEE, 2015), p. 578
57. A. Heryanto, W.N. Putra, A. Trigg, S. Gao, W.S. Kwon, F.X. Che, X.F. Ang, J. Wei, R.I. Made, C.L.Gan, K.L. Pey, Effect of copper TSV annealing on via protrusion for TSV wafer fabrication. J. Electron Mater. **41**, 2533–2542 (2012)
58. T. Jiang, S.-K. Ryu, Q. Zhao, J. Im, R. Huang, P.S. Ho, Measurement and analysis of thermal stresses in 3D integrated structures containing through-silicon-vias. Microelectron Reliab. **53**, 53–62 (2013)
59. K. Ohta, A. Hirate, Y. Miyachi, T. Shimizu, S. Shingubara, All-wet TSV filling with highly adhesive displacement plated Cu seed layer, in: *3DIC: International 3D Systems Integration Conference*, 31 Aug–2 Sept 2015, Sendai, New York, USA (IEEE, 2015), p. TS8.4.1
60. Q. Cui, X. Sun, Y. Zhu, S. Ma, J. Chen, M. Miao, Y. Jin, Design and optimization of redistribution layer (RDL) on TSV interposer for high frequency applications, in: *ICEPT-HDP: International Conference on Electronic Packaging Technology and High Density Packaging*, 8–11 Aug 2011, Shanghai, New York, USA (IEEE, 2011), p. 1
61. IBM 3D Semiconductor & Packaging Technology for Systems. http://researcher.ibm.com/researcher/view_group.php?id=4436. Accessed 21 June 2016

Chapter 4
Microstructural and Reliability Issues of TSV

Praveen Kumar, Indranath Dutta, Zhiheng Huang, and Paul Conway

4.1 Introduction

Following Chap. 2 on materials, processing, and assembly, this chapter discusses the microstructural and reliability issues of TSVs. This chapter begins with microstructural characterization and stress measurement; it then discusses detailed reliability issues associated with TSVs; finally it presents promising techniques towards atomistically informed reliability modeling of TSVs.

P. Kumar
Department of Materials Engineering, Indian Institute of Science, Bangalore 560012, India
e-mail: praveenk@materials.iisc.ernet.in

I. Dutta
School of Mechanical and Materials Engineering, Washington State University, Pullman, WA 99164, USA
e-mail: idutta@wsu.edu

Z. Huang (✉)
School of Materials Science and Engineering, Sun Yat-sen University, 135 West Xingang Road, Guangzhou 510275, China
e-mail: hzh29@mail.sysu.edu.cn

P. Conway
The Wolfson School of Mechanical, Electrical and Manufacturing Engineering, Loughborough University, Loughborough, Leicestershire LE11 3TU, UK
e-mail: p.p.conway@lboro.ac.uk

Y. Li, D. Goyal (eds.), *3D Microelectronic Packaging*, Springer Series in Advanced Microelectronics 57, DOI 10.1007/978-3-319-44586-1_4

4.2 Microstructural Characterization and Stress Measurement

4.2.1 Microstructural Characterization

The microstructure of TSV can be characterized at four levels, i.e., crystal structure, phase structure, grain structure, and defect structure. Bunge introduced a function

$$G(x) = \begin{cases} i(x), & \text{phase} \\ g(x), & \text{orientation} \\ D(x), & \text{defects, lattice strain} \end{cases} \quad (4.1)$$

and mathematical methods to quantitatively describe microstructure [8, 9]. For experimental characterization, standard metallographic techniques can also be applied to TSV structures. However, preparing TSV samples for microstructural characterization to investigate precise locations through the TSV sample structures using standard metallographic polishing is tedious and unpredictable due to the small size of TSVs and presence of soft materials (e.g., Cu, Ag paste, etc.) adjacent to hard and brittle Si. Focused ion beam (FIB) milling, which allows precise machining at small length scales, is widely used for preparing TSV samples for microstructural characterization. FIB can also be used to prepare ultra-thin samples, suitable for observation under transmission electron microscope (TEM). Machining samples using FIB is both a slow and an expensive process. Emerging Xe^+ plasma FIB systems, with optimized control over milling parameters, promise faster removal rates [36]. Another method for characterization of the microstructure of filler material is by via revealing: in this process, Si is wet-etched using hydroxides, which do not affect metal fillers, thus revealing the pillar [51, 87]. However, this is also a slow process and it does not allow observation of the Si-filler interface. Therefore, sample preparation, especially over the entire height of the TSV, for microstructural characterization is often very slow. The important microstructural features of TSV assemblies that are often characterized and quantified are the size and shape of filler grains, micro-texture, and grain orientation, the presence of twin boundaries in Cu fillers, shapes, and sizes of voids or cavities in the filler, micro-cracks, interfacial structure, etc.

Grain size, which can be measured using scanning electron microscope (SEM) and TEM, is important for the estimation of the strength of the metal filler (generally, the smaller the grain size, the greater the strength of the material). The hardness and hence strength of the TSV fillers can be measured using nano-indentation. Interestingly, as shown in Fig. 4.1, nano-indentation studies have shown that TSV fillers with smaller grain size may not always show an enhanced strength. Furthermore, as also shown in Fig. 4.1, the strength versus grain size behavior often does not follow the Hall–Petch relationship; in other words, the increase in the strength is not strictly proportional to the inverse of the square root of the grain size [59, 86]. These deviations can be attributed to the fact that TSV

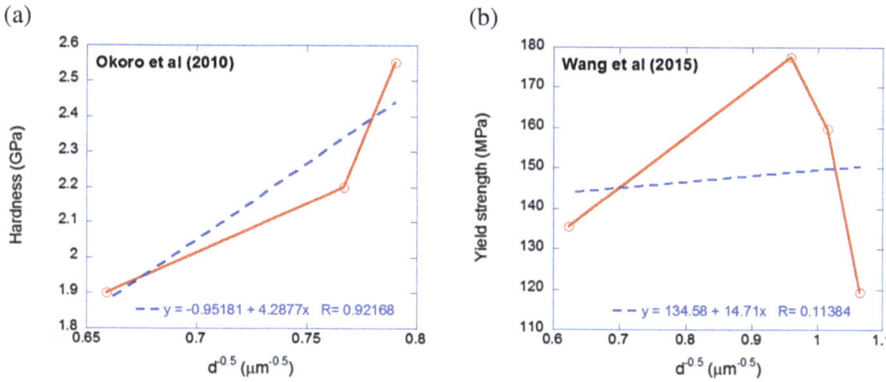

Fig. 4.1 Variation of (**a**) average hardness and (**b**) yield strength of filler material with average grain size of Cu. The *broken line* shows the best curve fit using Hall–Petch type of relationship between hardness in (**a**) and yield strength in (**b**) and the inverse of the square root of grain size. The data for (**a**) and (**b**) are taken from references [59] and [86], respectively

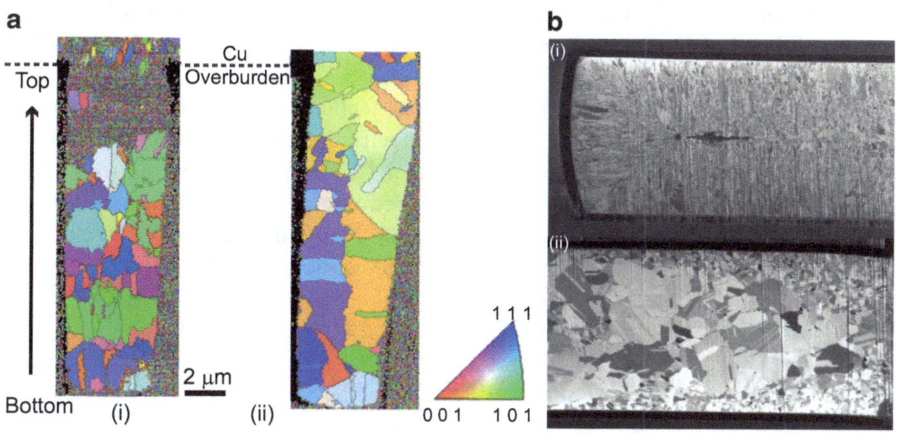

Fig. 4.2 (**a**) Electron backscatter diffraction (EBSD) patterns showing inverse pole figure (IPF) maps of cross-section of Cu fillers: (1) as-deposited and (2) after annealing at 420 °C for 20 min followed by annealing at 300 °C for additional 15 min [59]. (**b**) FIB micrographs showing grain structure of Cu filler: (1) as-deposited and (2) after annealing at 400 °C for 1 h [44]

fillers contain relatively few grains across the via-diameter, as shown in Fig. 4.2, and hence do not display generalized polycrystalline behavior. It has been shown that electrolytes with impurities can enhance the yield strength of the fillers [61]. Therefore, attention should be paid to the methods used to estimate the mechanical behavior of TSVs [19, 78]. In addition, the conventional parameter of an average grain size may not be sufficient to characterize the grain structure of TSV fillers. In the current materials genome initiative [56], new methods that consider more parameters including phase morphology and distribution have been proposed for quantification of microstructure [42, 90].

Figure 4.2 shows that the grain size of the Cu via-filler increases significantly upon annealing at elevated temperatures [59]; for example, annealing at 450 °C for 30 min leads to an increase in the grain size from 0.84 to 1.26 μm [34]. However, as shown in Fig. 4.2b, the grain growth during annealing is often inhomogeneous [44]. Furthermore, impurities in the filler material slow down the grain growth during annealing [61]. Electroplating at lower current density may lead to fillers with large grain size; for instance, plating current densities of 1 and 9 A/cm^2 resulted in Cu fillers with average grain sizes of 2.57 and 0.88 μm, respectively [86]. Annealing has also been noted to induce twinning in Cu fillers [24, 59].

The micro-texture of TSV fillers is often measured by electron back-scattered diffraction (EBSD) conducted in SEM (see Fig. 4.2a). However, if the filler grains are very small, as found in the seed layer, diffraction-scanning transmission electron microscope (D-STEM) can also be used to map the micro-texture [27]. It has been noted that although annealing increases the grain size, it does not have any other noticeable effect on the texture of the filler material [59]. The micro-texture of filler materials appears to depend on the orientation of the seed layer, electroplating conditions, e.g., current density, bath chemistry, bath temperature, etc., impurity concentration, etc. [27]. However, to date, the dependence of the crystallographic texture of the via-filler on various process and geometrical parameters has not been unambiguously determined.

The general microstructural characterization techniques, such as STEM, energy dispersive X-ray spectroscopy (EDXS), and electron probe microanalysis (EPMA), can be used to identify the chemical species in TSVs and map their concentration profile. Time-of-flight secondary ion mass spectrometry (TOF-SIMS) may also be used to study the chemical species in TSVs [61]. X-ray diffraction (XRD) can also be used to identify the phases present in the TSV structures [23].

4.2.2 Measurement of Stress State

Substantial stresses are produced in the TSV assemblies, as a result of processing and during thermal cycling. The sources and implications are discussed in Sect. 4.3. These stresses affect the overall reliability of microelectronic packages comprising these TSV structures, and therefore various approaches have been utilized to understand the nature of these stresses. Three main techniques that have been used to measure the stress state in TSVs are briefly discussed below.

4.2.2.1 Wafer Curvature Method

If a layered assembly of two or more materials having different coefficient of thermal expansion (CTE) are joined together and then heated, the assembly bends with a curvature proportional to the in-plane stresses generated in the individual components [79]. Although TSV assemblies are more complex, they also bend due

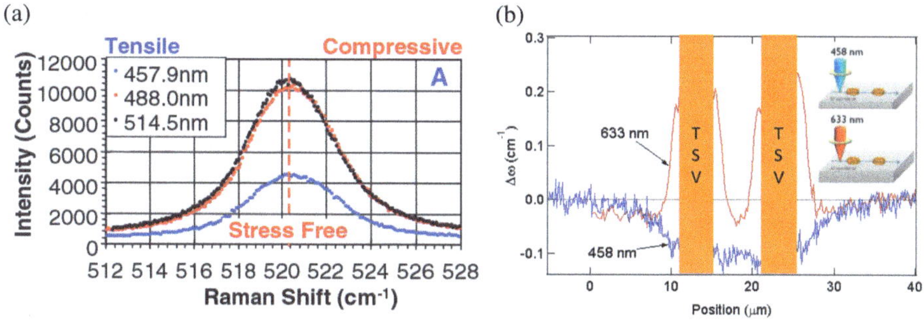

Fig. 4.3 (**a**) Representative Raman spectrum of Si wafer using lasers with wavelengths of 457.9, 488.0, and 514.5 nm, showing correlation between shift and stress [84]. (**b**) Measured stress profiles at different depths of the Si wafer near Cu-TSVs, using two different wavelengths of laser beams [17]

to generation/relief of residual stresses during thermal excursions [38, 50, 61, 72], and measurement of the curvature of the wafer can be used to determine stresses in the TSVs. However, the stress state in TSV structures is more complex than simple film-substrate systems, and as such, stress analysis based on the curvature method has usually been supplemented by finite element analysis (FEA) [72]. Although this method is meant to provide only an overall or average stress (as only one curvature is measured), FEA may be used regressively to predict the 3-D state of stress. Moreover, this method can be easily adapted for in situ measurement of stress (e.g., during thermal cycling).

4.2.2.2 Micro-Raman Spectroscopy

Since Si is Raman-active, the stress in the Si wafer can be measured using micro-Raman spectroscopy (μRS) [17, 38, 84, 91]. The most common case involving usage of μRS in backscatter mode resolves the longitudinal vibrational mode (i.e., mode 3), which can then be used to determine the sum of the two in-plane principle stresses [17, 91]. However, if a high numerical aperture (NA) (say, ≥ 0.4 [91]) is used, then it is possible to resolve all three modes of vibration and hence tensorial nature of the stresses in TSV structures [91]. As shown in Fig. 4.3a, compressive and tensile stresses in Si lead to forward and backward shifts in the Raman signal (e.g., a shift by a wavenumber of 1 cm^{-1} corresponds to a stress of $\sim$434 MPa[1]), the shift being directly proportional to the stress in Si [84]. Since a laser beam with long wavelength can penetrate deeper in Si, it can provide information from a deeper

[1] It should be noted that the exact value of the stress for a wavenumber shift may depend on the materials properties used in secular equation (i.e., set of equations in the reference axes which may be different than that of materials crystallographic reference system). It can lie in the range of 430–520 MPa [17].

depth than a beam with a longer wavelength, and thus, depth-sensitive information can be obtained. Nevertheless, the penetration depth of μRS in the backscatter configuration is still very small and hence this method can be used to obtain stress information from near surface regions only. As shown in Fig. 4.3b, the stress in Si around a TSV strongly depends on the distance from the via-end: at very near the surface (as measured using short wavelength laser), the stress becomes less tensile away from the via end, whereas at slightly below depth (as measured using long wavelength laser), the stress transitions from being compressive to tensile as one moves away from the via end. Since the stress state in the Si wafer is generally triaxial, it is challenging to measure all stress-components via Raman spectroscopy. Nevertheless, by combining with FEA, μRS has been utilized to yield good estimates of the three-dimensional stress state in the wafer [17, 84, 91]. It should be noted that metals are not Raman-active and therefore the stress state of Cu cannot be measured using this technique and has to be derived indirectly from measurements on Si.

4.2.2.3 X-Ray Diffraction-Based Techniques

To directly measure the localized strain, and hence the stress-field in Cu TSVs, X-ray diffraction, used in conjunction with a cross-section TEM or a synchrotron radiation source, may be used to determine stress-induced changes in unit cell parameter from the equilibrium [62, 77, 85]. This can also be used to measure stress in Si non-destructively with very high resolution. Although XRD typically yields only average stress values, by focusing high energy X-rays in a synchrotron to a small beam size and rastering it on the sample, full strain and stress tensors from submicrometer sample volumes in the TSV structures may be obtained [7, 48, 77, 83]. Figure 4.4 shows an example of spatially resolved stress variation in a TSV assembly, determined using high energy X-rays.

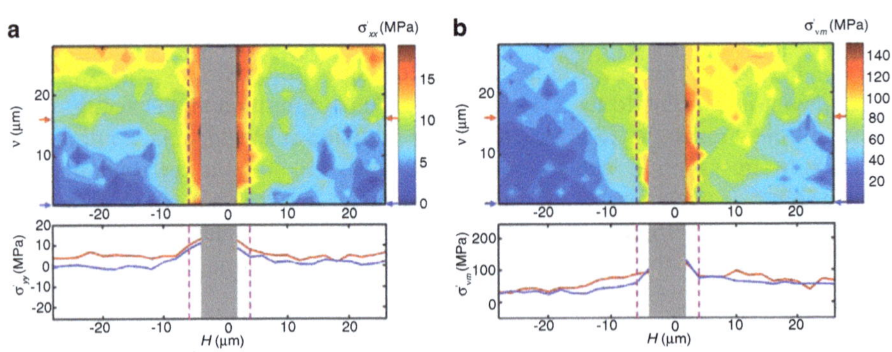

Fig. 4.4 Stress distribution on a plane as determined using synchrotron X-ray micro-diffraction: (**a**) the deviatoric σ_{xx} and (**b**) von-Mises stress [77]. The *dotted vertical lines* represent the effective size of the Cu TSV whereas the *grey shaded region* represent "non-indexable" region where the diffracted intensities from Cu are comparable with that from Si

4.2.2.4 Stress Metrology Challenges

Although the synchrotron X-ray micro-diffraction is by far the most powerful technique for stress analysis, it is the most expensive and requires access to a synchrotron radiation source [85]. None of these aforementioned techniques can directly measure stress in Si or Cu. In Raman spectroscopy phonon frequency shifts are converted into stress values [85]; in diffraction-based techniques strain tensors are mathematically converted to stress tensor using linear elastic properties of bulk single crystalline samples [85]. Although μRS can provide stress profiles in the depth direction, the penetration depth is within a submicrometer range. Therefore, the complete picture of the stress distribution in TSVs is still unavailable. However, this information is critical to elucidate the root cause for the metal pumping phenomenon as discussed in Sect. 4.3.1.3. The modeling techniques discussed in Sect. 4.4 may provide a solution towards this direction.

4.3 Reliability Issues Associated with TSVs

Reliability complications in TSVs typically arise either from stress-related or electrical sources, and are often convoluted with microstructural effects. In the following, we discuss the cause and effects of each of these reliability issues, along with potential approaches to circumvent or minimize these.

4.3.1 Stresses in TSVs

4.3.1.1 Origin and Effect of Stresses

Stresses in TSVs, and therefore in the neighboring Si, arise from two sources: (1) growth stresses, that arise as a result of via-filling by electroplating, and (2) - thermo-mechanical stresses that arise due to thermal expansion mismatch between Cu in the via and the surrounding Si.

Growth stresses arise when the electrodeposited Cu grows radially inwards on the Cu-seed layer from the via side-walls, and abut itself, often resulting in a seam along the TSV axis. Such a seam may leave a thin void-line along the TSV axis (Fig. 4.5), which can grow during post-deposition annealing, and result in mechanical and electrical performance loss. Electroplated copper undergoes significant grain growth by self-annealing at room temperature, which results in volume shrinkage of the Cu TSVs by elimination of grain boundaries [31]. This may allow any existing void to grow during both self-annealing and during the pre-CMP (chemical-mechanical polishing) annealing step. During annealing, hydrostatic stress gradients generated around pre-existing defects in the Cu may

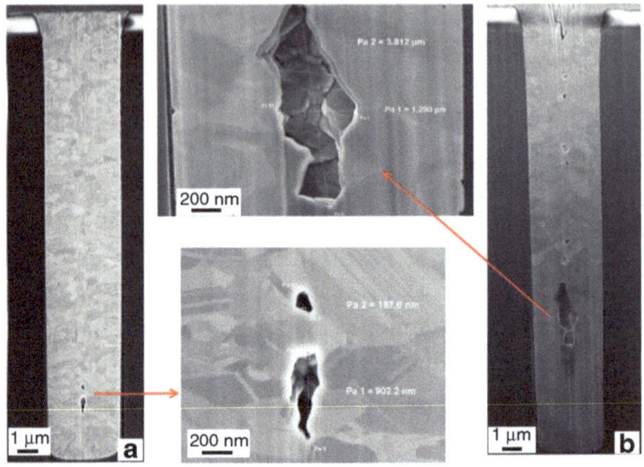

Fig. 4.5 SEM images of TSVs showing centerline void after annealing at 150 °C (**a**) and 300 °C (**b**). [43]

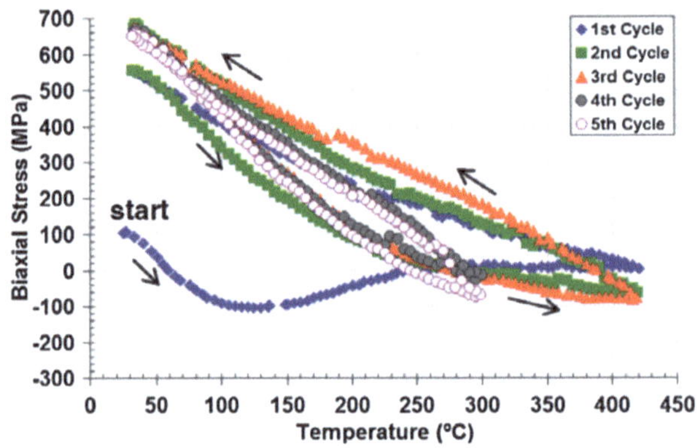

Fig. 4.6 Evolution of equi-biaxial stress in a blanket Cu film deposited on Si during thermal cycling [16]

be relieved by diffusion of lattice vacancies towards the existing void near the axis, causing the void to grow [43].

Thermo-mechanical stresses arise because of differential thermal expansion or contraction of Cu and Si during heating or cooling of the device as a result of their large difference in the coefficient of thermal expansion ($\alpha_{Si} = 2.8 \times 10^{-6}$/K, $\alpha_{Cu} = 17 \times 10^{-6}$/K). Thermal cycling resulting from fluctuating Joule heating occurs continually during the service life of electronic devices, and 3D devices are no exception. Typically, electroplated Cu deposited on Si is under residual tension to begin with (at ambient temperature). During heating, the tension is first elastically relieved, following which a compressive stress builds up. As temperature increases, the yield strength and creep resistance of Cu decrease, resulting in stress relief by plastic yielding and creep, such that at the highest temperature, there is little stress remaining (Fig. 4.6). During subsequent cooling, a tensile stress builds up, resulting in stresses in the neighborhood of 200–500 MPa (biaxial for thin films and hydrostatic for TSVs). Further cycling repeats the same behavior, although because of substantial plasticity during the first cycle, the yielding and stress-relaxation start at a higher temperature, and are therefore lower. This is shown for a Cu film on Si, measured by wafer curvature testing, in Fig. 4.6 [16]. The details of this behavior depend on the temperature range and rate of thermal cycling, but the overall behavior is qualitatively replicated in TSVs. In contrast to thin films, however, the TSV stress state is triaxial, and the TSV is typically in triaxial (i.e., hydrostatic) tension at ambient temperature.

As noted earlier, the stress state in Cu-TSVs may be studied either by measuring the stress in Si via Raman spectroscopy, or that in Si or Cu by X-ray micro-diffraction. Depth-sensitive Raman measurements, using various excitation wavelengths of an Ar^+ laser, show that Si surrounded by TSVs has a compression hydrostatic stress at ambient temperature, with the stress varying significantly along the length of the TSV [17, 84]. The hydrostatic stress in Si becomes less negative nearer the ends of the TSV. Assuming that the radial and circumferential stresses (σ_{rr}, $\sigma_{\theta\,\theta}$) in the Si surrounding the TSVs are either weakly or not dependent on the axial position relative to the TSVs, one may infer that at ambient temperature, the axial stress σ_{zz} in Si is negative (i.e., σ_{zz} is positive in Cu) near the middle of the TSV, and becomes ~ 0 at the surface. It was also noted that the stress variation along the depth is larger for smaller TSV diameters (i.e., larger TSV aspect ratios). Furthermore, the stress in Si surrounded by an array of TSVs is typically larger than that outside of TSV-arrays, and these stresses change from tensile immediately after Cu electrodeposition (where the stress is due to growth and self-annealing), to increasingly compressive when annealed at increasing temperatures [52]. Raman peak-shift measurements near the surface of the chip next to TSV-ends show that the biaxial hydrostatic stress state (i.e., $[\sigma_r + \sigma_\theta]/3$, since $\sigma_z \sim 0$) is negative at the TSV-chip interface, and rises sharply and becomes positive mid-way between TSVs (Fig. 4.7). After annealing, the hydrostatic stress in Si becomes even more compressive at the interface due to Cu volume expansion, and more tensile away from the interface [38]. Conversely, in addition to increasing the grain size, annealing lowers the compressive stresses in the Cu.

Thermal cycling also leads to accumulation of hydrostatic stress in Cu TSV or filler [62]. In addition, TSVs with high impurity levels appear to lead to high residual stresses [61]. It has been reported that large tensile radial stresses exist at the Si-Cu interface [17]; this may make these sites prone to failure in TSV

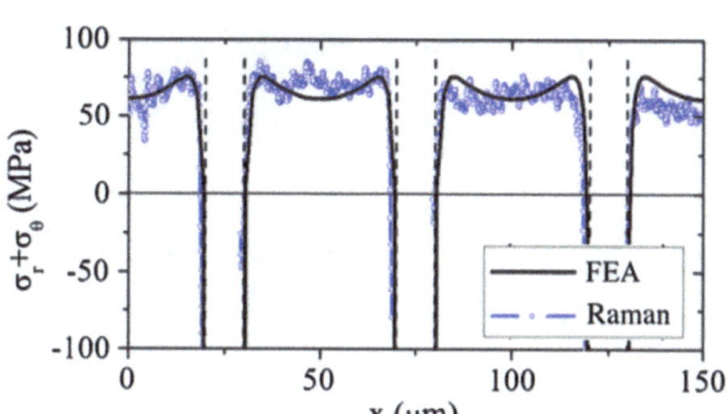

Fig. 4.7 Plot of the stress state in Si near the ends of TSVs. Three TSVs are shown, each as a vertical band with intervening Si [38]

assemblies. The radial stress at the Cu-filler/Si interface generally increases with the length of the TSV for a given diameter [17]. The TSV diameter also seems to affect the stress state in the TSV assembly, although the dependence of the hydrostatic stress on TSV diameter is not monotonic [64].

The stresses induced near the surface of Si next to the Cu-filled TSVs have an adverse impact on the electrical performance of devices in the immediate neighborhood of the TSVs due to induced piezoresistivity, which results in degradation of carrier mobility [1, 35]. This necessitates a keep-out zone (KOZ) in the Si, typically a few micrometers in width, in the immediate vicinity of each TSV, where active devices cannot be placed. The KOZ, which scales with the square of the TSV diameter, is also greater for high aspect-ratio TSVs, and places a significant overhead on the area available in the chip for active devices, particularly with the increasing density of TSVs in chips. Based on FEA of the stress state in Si next to TSVs, it has been inferred that Cu microstructures with smaller yield strengths reduce the KOZ size, which increases with increasing yield strength until the via yields plastically, and then remains stable [39].

FEA of TSVs in free-standing chips shows that after both heating and cooling, maximum von Mises stress, and hence plastic deformation of the copper occurs near the interface close to TSV-ends, which is a root cause of copper pumping [12]. It is further observed that the maximum principal stress (tensile) is in the radial direction at the interface near the middle of the TSV, and in the circumferential dielectric layer surrounding the TSV [49]. Evidence of dielectric cracking at the circumferential Cu-SiO_2-Si interface, as well as roughening (due to voiding) of the metal line at the top of the TSV has been noted after thermal cycling, as shown in Fig. 4.8 [63]. However, FE analysis of a stacked-die package shows that each die undergoes convex curvature when the package is cooled from the stress-free temperature (150 °C), which fundamentally alters the stresses and displacements relative to the free-standing die [49]. The convex curvature alters the radial interfacial stress to compressive, reducing the proclivity towards interfacial fracture as well as dielectric cracking, and shifting the critical failure location to the copper-pillar near the solder-microbump interface in 3D packages [49]. In fact, the failure probability near the silicon-Cu pillar-microbump

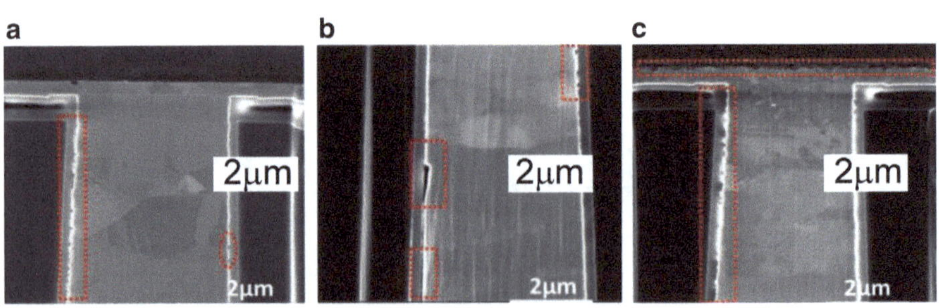

Fig. 4.8 TSV (**a**) prior to cycling, and after (**b**) 500 and (**c**) 2000 cycles between 30 and 150 °C, showing development of dielectric cracks and top-line roughening [63]

junction (as indicated by the maximum equivalent strain), as well as the required KOZ diameter, increases with increasing diameters of both the TSV and the copper-pillar in a 3D package [37].

In addition to producing defects (e.g., voids) in TSVs, including complications in electrical performance and potentially causing interfacial and or dielectric fracture, induced stresses also give rise to a plasticity-related phenomenon commonly referred to as copper pumping, which can have serious reliability implications. Because of the pervasiveness of this phenomenon, it is discussed separately in Sect. 4.3.1.3.

4.3.1.2 Microstructure and Stresses

Direct measurement of stresses in Cu-TSVs with X-ray micro-diffraction reveal a significant tensile hydrostatic stress ($\sim$234 MPa) at room temperature, which goes to compression ($-$ 196 MPa) during annealing at 200 °C, and becomes a smaller tensile stress when the sample is cooled to the ambient (167 MPa). The large initial tensile stress is caused by grain boundary elimination during self-annealing and device fabrication, and as noted previously, is undesirable from the reliability perspective, since this causes large stresses in Si. A subsequent annealing treatment lowers the tensile hydrostatic stress in Cu, even though the zone of initially larger grains expands during annealing, as shown in Fig. 4.9 [7], possibly because of relaxation associated with plasticity and creep at the high temperature.

EBSD studies have shown that during annealing, large grains remain stable when there is a preponderance of $\Sigma 3$ twin boundaries, but grains without twins and smaller grains grow rapidly [75]. It has also been noted that the Cu-TSV has random texture both before and after annealing. However, microvoids or small cracks have been noted to form during annealing, thereby reducing stress, as shown in Fig. 4.10. This is possibly because of vacancy diffusion to pre-existing defects

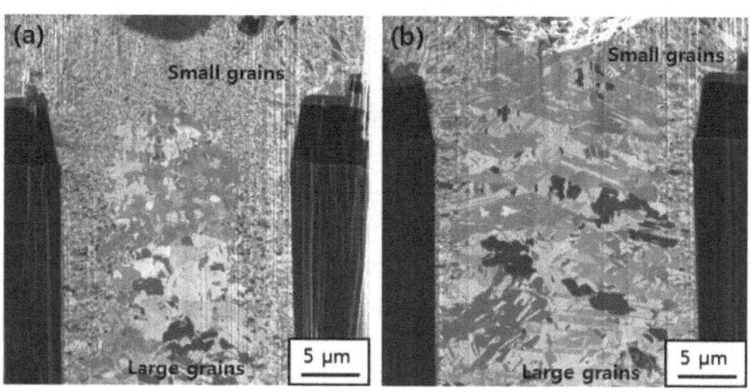

Fig. 4.9 Cross-sections of Cu-filled TSVs in: (**a**) the as-received state, showing a zone of large grains in the center surrounded by smaller grains; and (**b**) after annealing, showing an expanded central zone with large grains [7]

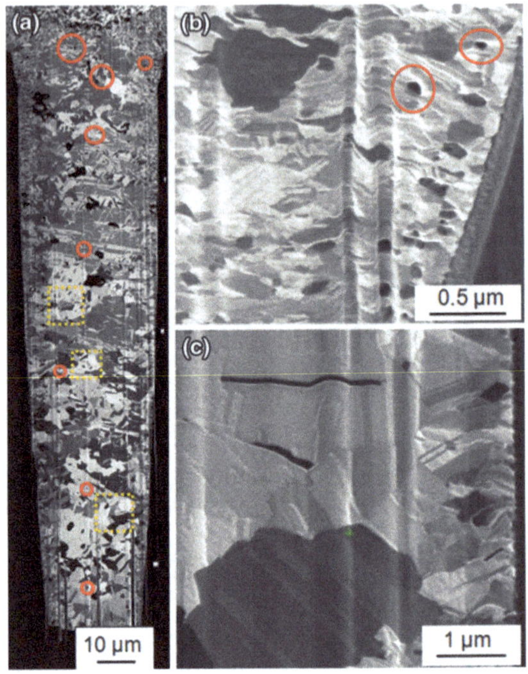

Fig. 4.10 (**a**) Voids and cracks form at various locations in a TSV after annealing (*red circles*). (**b**) Voids near the top of the TSV, and (**c**) grain boundary cracks near one of the side-walls [75]

under hydrostatic stress gradients within the TSV during annealing, as noted in the discussion associated with Fig. 4.5.

FEA-based modeling work has also reported the linkage between the microstructure of copper grains and the stress in Cu-TSVs. The results from a linear elastic mechanical model clearly demonstrate that the stress distribution is rather heterogeneous inside the TSV filler, as shown in Fig. 4.11, considering the anisotropy of the elastic compliance tensor of copper. Depending on the texture, morphology and distribution of the copper grains, stress concentrations may occur at the grain boundaries [89]. Elastoplastic models have also been conducted directly on copper grain structures with the aim of explaining the formation of copper pumping [40, 55]. However, the drawbacks of this kind of model are clear. First, although the mechanical analysis is conducted directly on copper microstructure, the microstructure is assumed to remain unchanged during the mechanical loading process. Second, the plastic deformation mechanisms at the atomic scale, e.g., the motion of dislocations, have not been considered. However, the dislocation involved processes such as recovery, recrystallization, and grain growth have been reported to occur in electrodeposited copper and can significantly influence its mechanical properties [54]. Recrystallization has also been suspected to account for the copper-pumping phenomenon [2, 61, 64]. Considering the limitation of the current modeling techniques, more robust simulation methodologies that take into account the polycrystalline nature of Cu TSVs and their complex deformation and stress relaxation mechanisms are needed [64]. Section 4.4 will introduce atomistically informed modeling techniques to address this limitation.

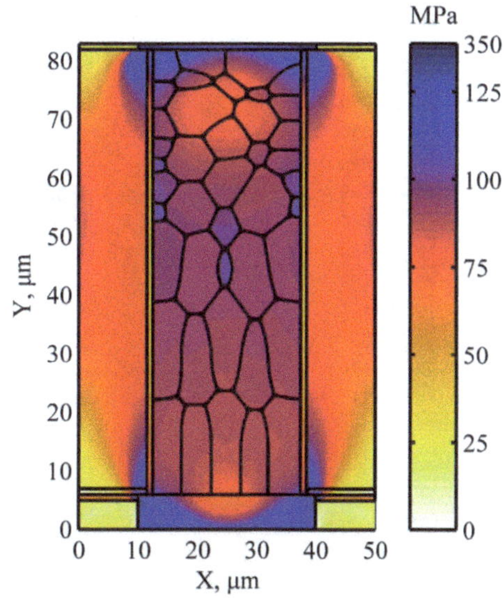

Fig. 4.11 Distribution of the von Mises stress in a Cu-TSV structure from a linear elastic mechanical model with Cu grain microstructure [89]

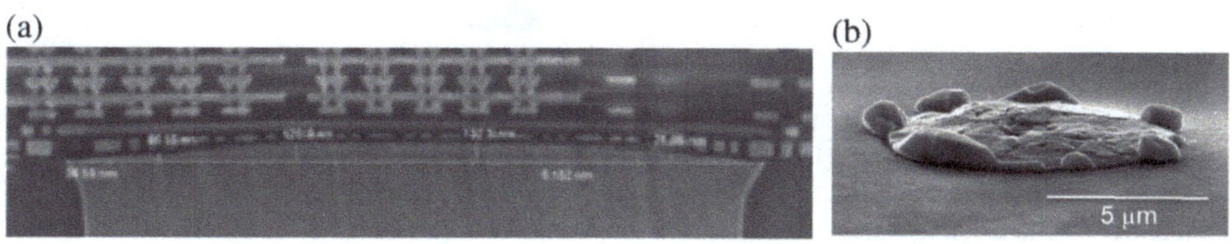

Fig. 4.12 Examples of Cu pumping. (**a**) Uniform pumping, showing a cross-sectional image displaying curvature of the *top* of a TSV, resulting in deformation of the RDL structure above [76]; (**b**) Non-uniform pumping showing a few Cu grains protruding out of the Si surface [53]. The sample in (**a**) was annealed at 400 °C for 60 min, and the sample in (**b**) was annealed at 425 °C for 90 min

4.3.1.3 Metal Pumping: Extrusion or Intrusion of TSVs

During service and consequent thermal cycling, the CTE mismatch between Cu and Si results in residual stresses in each material (σ_{Cu}, σ_{Si}), as well as significant interfacial shear stresses (τ_i) near the extremities of the via. The induced stresses, if high enough, may cause plastic deformation of the filler [39, 40, 60, 76] with the differential straining between Cu and Si sometimes accommodated by interfacial sliding [18, 47, 53]. This results in extrusion (and sometimes, intrusion) of Cu relative to the Si. Generally, the extrusion of the Cu occurs due to plastic deformation of the metal near the via ends (typically, towards the top or mouth of the via), and can be enhanced by creep and grain boundary sliding. Two examples of this phenomenon are illustrated in Fig. 4.12.

When heated up relatively rapidly and kept at a constant elevated temperature for a period, the protrusion observed is generally associated with plastic

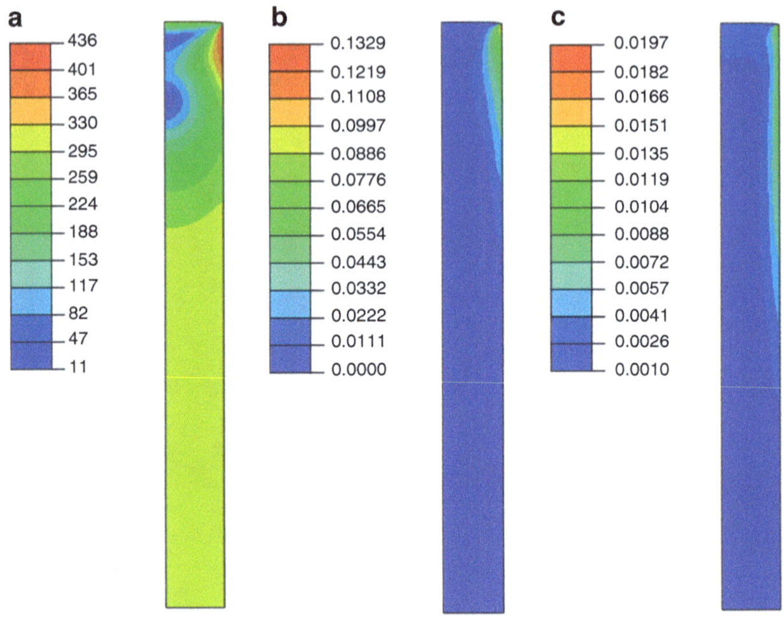

Fig. 4.13 Von Mises stress in MPa (**a**), and equivalent plastic (**b**) and creep strains (**c**) in a Cu-TSV following fabrication at 425 °C and cooling to −25 °C (prior to thermal cycling). The figures show one quadrant of the TSV, with the interface at the *right*, the axis of the TSV at the *left*, and TSV mid-plane at the *bottom*. Although the plastic strain plot highlights only the strain concentration near the interface, most of upper part of the TSV, where Von Mises stress $> \sim 250$ MPa, has yielded

deformation of the copper, with no relative displacement observed at the interface between Cu and Si, as observed in Fig. 4.12. Generally, the extrusion increases with increasing temperature, and occurs during any pre- or post-CMP annealing, as well as heating associated with dielectric-deposition for fabricating the back end of line (BEOL) or redistribution layer (RDL). Finite element modeling shows that following heating to a high annealing/fabrication temperature (425 °C) and cooling down, the Von Mises stress exceeds the yield strength over the entire upper part of the TSV, resulting in yielding of the top part of the via (Fig. 4.13a). As a result, inelastic (plastic and creep) strains accrue in the top part of the via, with strain concentrations adjacent to the interface near the TSV-end (Fig. 4.13b and c). This accrual of plastic and creep strain at the top of the TSV leads to copper pumping, or protrusion of the Cu from the top. This is particularly acute for blind vias, where any expansion of the Cu must necessarily occur at the open end of the via.

EBSD studies show that Cu grains in a TSV typically have random grain boundary misorientations, and the via has little crystallographic texture, and no correlation between orientation and extruded grains has been noted [75]. This is evident from Fig. 4.14, which clearly shows a random texture in the TSV. It has been suggested that the protrusion observed is uniform when the grain size near the

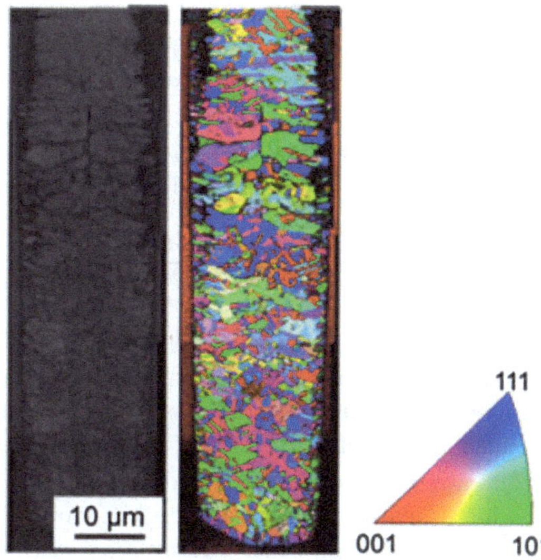

Fig. 4.14 EBSD orientation map of a TSV, showing a random grain texture [75]. This is typical of most TSV structures reported in the literature

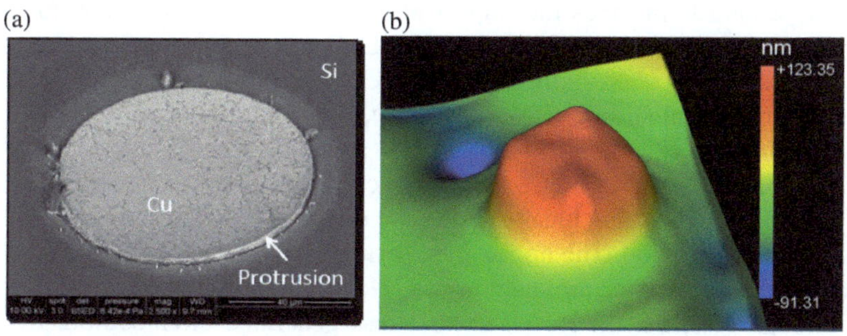

Fig. 4.15 Evidence of relative displacement between Cu and Si at the interface. (**a**) SEM image of Cu protrusion in an 80 μm diameter via, and (**b**) scanning white light interferometry image of Cu protrusion in a 10 μm diameter via. In both, the majority of protrusion occurs at the interface [53]

TSV ends is large, but non-uniform when there are smaller grains near the TSV ends [40]. In general, it is noted that the extent of protrusion is larger when there are high angle grain boundaries at the TSV ends that promote creep processes, but not when there are coherent twin boundaries [15].

It is useful to note that for adherent Cu–Si interfaces, when TSVs undergo uniform protrusion, the top of the Cu typically attains a convex curvature, as shown in Fig. 4.12a, with minimal or no relative displacement between the Cu and Si (or the barrier layer) at the interface. This is particularly true when the die is either cycled rapidly, or held at a constant temperature. On the other hand, when the die is cycled slowly, shear stresses are repeatedly generated and relieved at the interface near the ends of the TSV, allowing sufficient time to drive diffusionally accommodated interfacial sliding, which results in a significant, step-like interfacial displacement [18, 47, 53], as illustrated in Fig. 4.15.

Diffusionally accommodated interfacial sliding at hetero-interfaces (i.e., interfaces between dissimilar materials) is akin to grain boundary sliding, and may be driven by interfacial shear stresses (τ_i) that occur at the extremities of a TSV. In addition, it may be enhanced or mitigated by an electric current flowing through the TSV due to associated electromigration along the interface. The resultant displacement rate is given by [45, 46, 65]:

$$\dot{U} = \frac{8\Omega\delta_i D_i}{kTh^2}\tau_i + \frac{4\delta_i D_i}{kTh}Z^* eE \qquad (4.2)$$

where Ω is the atomic volume, h is the roughness of the topographically periodic interface, k and T are the Boltzaman constant and temperature, respectively, D_i is the interfacial diffusion coefficient, δ_i is the thickness of the interfacial region, and Z^* and e are the effective charge number of the diffusing ion and the charge of an electron, respectively. Thus, $\dot{U}$ depends linearly on both τ_i and the electric field E (which equals $j\,\rho$, where ρ = resistivity of the filler and j is the current density), the relative signs of which determining whether they augment or mitigate each other's contributions. In Eq. (4.2), the first term gives the effect of shear stress, while the second term gives the impact of superimposed electromigration (EM). Since Z^* is negative, the second term (due to EM) acts against the first term (due to stress-driven creep) when both driving forces (τ_i and E) are positive, and thus reduce the interfacial sliding rate. Conversely, if τ_i and E have opposite signs, the two terms in Eq. (4.2) augment each other and increase the interfacial sliding kinetics. Fig. 4.15 shows the effect of interfacial sliding due to τ_i only (i.e., when $E = 0$), and Fig. 4.16 schematically shows the origin of τ_i in through and blind vias after a thermal excursion. The impact of superimposed current on interfacial sliding is discussed separately, in conjunction with electromigration in Sect. 4.3.2.

In through-vias, the interfacial shear stress generated after cooling from the fabrication temperature is symmetric about the length of the via and is concentrated at both ends. As a result, the interfacial sliding, and hence copper extrusion, would also be symmetric at the two ends for straight (i.e., untapered) vias. For blind vias, the interfacial shear stress rises from the blind end towards the open end, and therefore, the resultant protrusion at the open end is greater. It should be noted that interfacial sliding, which occurs due to interfacial diffusion under the applied shear stress, is a mechanism that accommodates differential deformation of the Cu and Si at the interface and therefore a thermal excursion as well as a high enough temperature and sufficient time to make diffusional processes active is needed. This is why, under rapid cycling conditions, Cu pumping typically occurs uniformly due to inelastic deformation of Cu without significant contribution from interfacial sliding, whereas when cycled slowly over a larger temperature range, steps emerge due to sliding at the interface.

Since the thermal expansion coefficient of Cu is much larger than that of Si, when cooled from an elevated temperature, significant relative shrinkage of Cu can occur, which can also be accommodated at the interface by sliding. In this case, instead of protrusion, the ends of Cu vias may intrude into the Si with a sharp step at

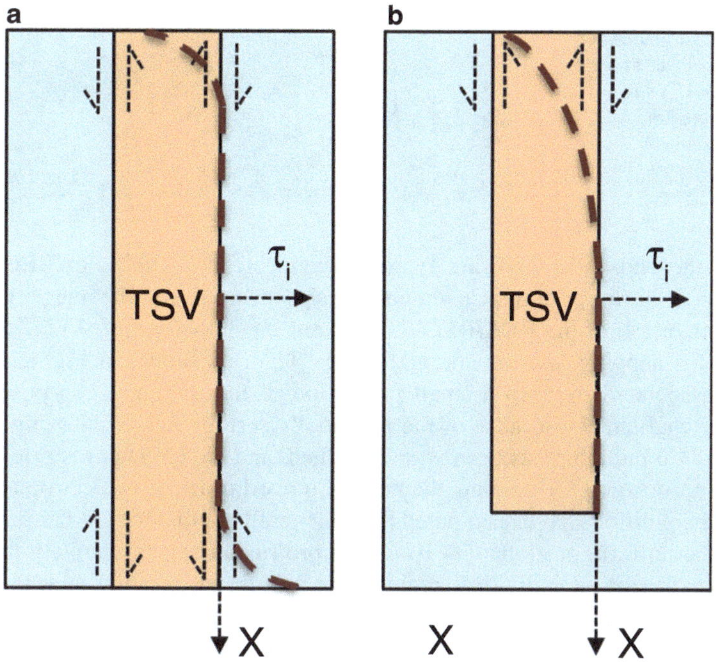

Fig. 4.16 Schematics showing shear stress distribution at the TSV-Si interface in (**a**) through-via, and (**b**) blind-via. The interfacial shearstress (*dashed line*) is symmetric about the length of the TSV for the through-via (**a**), and is non-zero only near the TSV-ends. τ_i is zero at the blind end of the via in (**b**), and rises monotonically towards the open end

Fig. 4.17 The end of a TSV intruding into the Si, following five thermal cycles from −25 to 150 °C at 0.1 °C/min. Prior to cycling, the TSV end protruded out by ∼ 30 nm from the Si surface

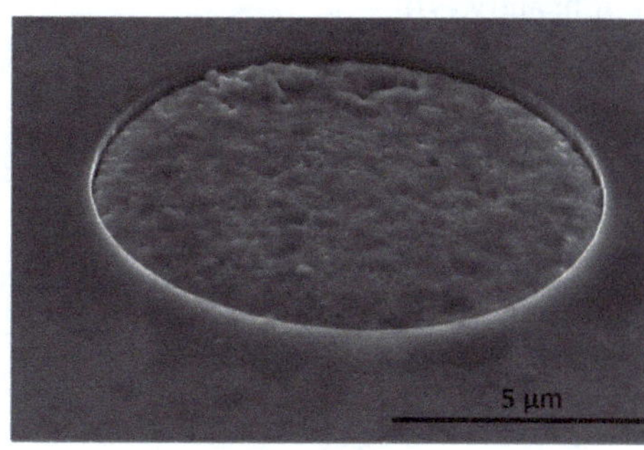

the interface, as illustrated in the SEM image in Fig. 4.17. Again, the shrinkage is due to deformation of the Cu, but the interfacial step is due to diffusionally accommodated interfacial sliding. It should be noted, however, that the rate of protrusion or intrusion of the via ends during thermal cycling decreases with increasing number of cycles, as the stresses generated in the metal filler reach saturation due to progressive strain hardening. Therefore, after several cycles, protrusion/intrusion due to Cu pumping levels off [18, 47, 53].

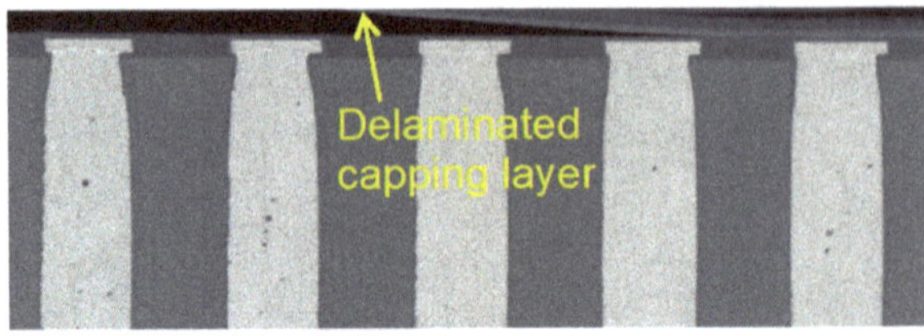

Fig. 4.18 SEM micrograph showing delamination of the capping layer at the *top* of a row of TSVs following thermal excursions

Since the ends of a TSV are typically connected to a RDL or BEOL layer, protrusion or intrusion of the Cu via poses a significant risk to the integrity of these layers. Distortion of the RDL/BEOL structures, as shown in Fig. 4.12a, or delamination of a capping layer at the end of the TSV, as shown in Fig. 4.18 due to stresses associated by even a small protrusion of the via ends, can pose serious reliability challenges. Because of the potentially serious reliability complications, the role of Cu pumping has been widely studied, and the effects of various process parameters such as TSV spacing, diameter, Cu overburden after electroplating, and annealing conditions have been noted [76]. Generally, a majority of Cu-pumping is noted to be uniform or global ($\sim$ 10–30 nm protrusion), with relatively few TSVs showing extrusion of individual grains after a high temperature anneal. Spacing appears to have little if any effect, but the pumping is greater on average for larger via diameters [14, 76]. However, although the average protrusion of an array of TSVs is larger for larger TSVs, the maximum copper pumping appears to be independent of the TSV diameter, which suggests that TSV diameter has little effect on BEOL reliability [14]. Copper overburden typically has no effect, but a higher pre-CMP annealing temperature ($\sim$ 430 °C) reduces pumping. Finally, an additional annealing step, following pre-CMP anneal, reduces pumping significantly [76].

4.3.2 Electromigration Related Effects

Although electromigration (EM) is a significant reliability issue in metallic interconnects in electronics, particularly in BEOL structures, TSVs are generally less susceptible to EM induced failures. This is primarily due to their relatively large cross-sections, which reduce the current density. However, the combination of electric current and complicated stress states above and below the TSV, where it joins the BEOL or RDL structures, can cause substantial diffusional effects, and give rise to EM-related void growth. Finite element modeling of the effects of stress-gradient, potential gradient, and temperature gradient on the atomic flux divergence (AFD), which correlates with diffusional flow and hence electromigration, has noted AFDs are typically high where the top and bottom metallizations meet the TSV [80]. A large proportion of the AFD is due to the stress gradient generated due to Joule heating, with relatively little due to the potential gradient. Still EM damage and void growth can occur. Generally, strategies to reduce stress gradients will reduce void growth.

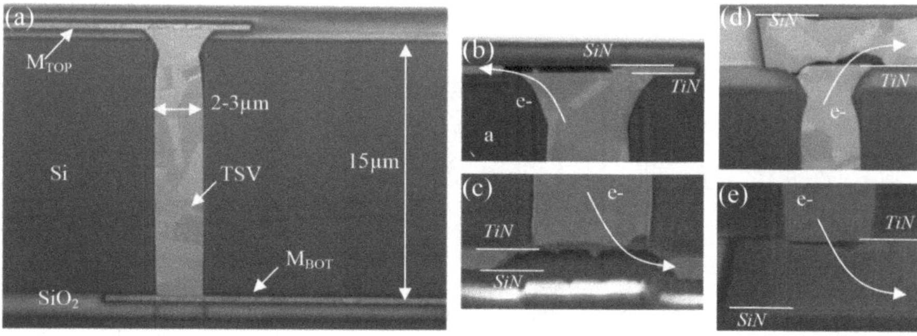

Fig. 4.19 (**a**) Chip showing a TSV with metal lines at *top* and *bottom*. Voids form due to EM at 300 °C at exits of electron flow at *top* and *bottom* for thin metal lines (**b** and **c**) and thick metal lines (**d** and **e**) [25]

EM experiments on devices with thin as well as thick metal layers at the top and bottom of TSVs generally show no void formation due to EM within the TSVs, irrespective of the current flow direction [25, 26]. However, voids form downstream of the direction of electron flow, at the intersection of TSV-end and the metal layer in the RDL. It has been hypothesized that these voids nucleate because migration of Cu atoms from the TSV to the void is prevented by the TiN barrier layer, but migration out of the void region can occur into the Cu lines at the end of the TSV. Thus, the layer of TiN, by being effective as a diffusion barrier, actually becomes the root cause behind EM void nucleation. In thin Cu lines, voids occupy the entire thickness of the line (between the TiN barrier and SiN capping layer). Current crowding at the location where the electron current exits the TSV is greater for the thicker lines, and therefore, thicker lines do not mitigate voiding. In thick lines, the voids are at the interface between TiN and the Cu-line. Experimental electromigration studies [13] have also shown voiding in backside Cu lines below the TSV downstream of the electron flow direction, but at the SiN–Cu interface, as opposed to at the TiN–Cu interface seen in Fig. 4.19. However, small voids present prior to EM experiments inside the TSV remained unaltered. Thus, even though the TSVs themselves are not susceptible to EM damage, the BEOL and RDL structures in the device can be significantly affected by EM.

A different type of complication is posed by interfacial sliding under electromigration conditions, as indicated in Sect. 4.3.1.3. As evident from Eq. (4.2), an applied electric current may enhance, or mitigate the kinetics of interfacial sliding due to the interfacial shear stress, depending on the direction of the applied field. Even when there is little or no shear stress at the interface, EM can drive interfacial sliding along the TSV-Si (actually, TSV-barrier layer) interface. Evidence of this is shown in Figs. 4.20 and 4.21, where the end of a TSV is seen to protrude in the direction of electron flow, and intrude opposite to electron flow for samples under EM with thermal cycling conditions, and at a constant temperature under a constant current. Under applied current, diffusional flow of Cu occurs along the interface in the direction of electron flow, and this can result in a time-dependent

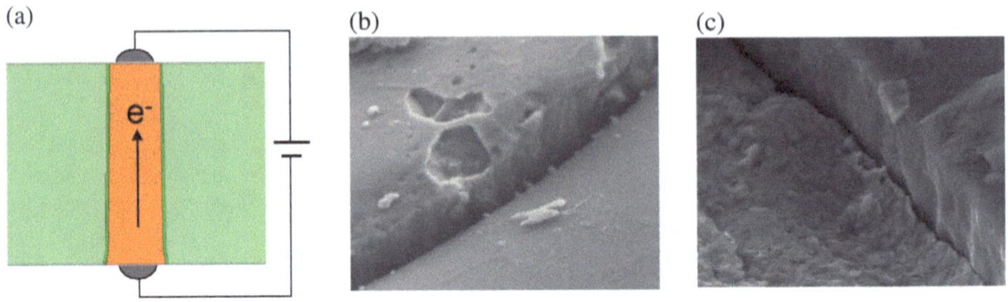

Fig. 4.20 (**a**) Schematic of experimental arrangement for electromigration cum thermal cycling (25–425 °C) experiments. (**b**) TSV edge showing protrusion due to three thermal cycles + EM due to a current density of 5.2×10^4 A/cm^2, with upward electron flow through Cu. (**c**) The same TSV following a reverse current density of -5.2×10^4 A/cm^2, showing that the Cu now intrudes into Si [18, 47]

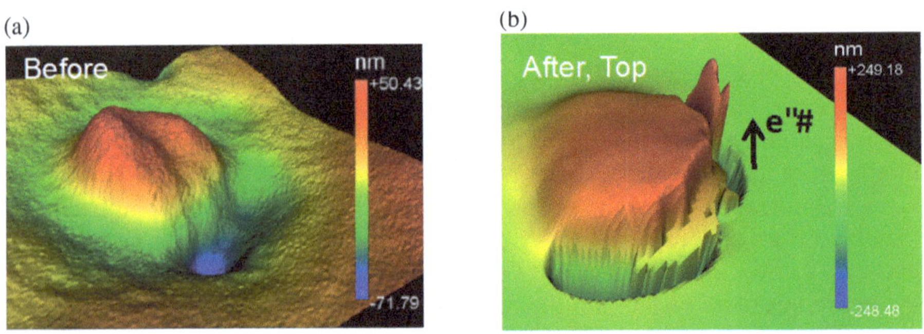

Fig. 4.21 Scanning white light interferometry (SWLI) images of the end of a TSV before and after EM experiments, showing protrusion in the direction of electron flow. A current was passed through Cu thin films deposited on the *top* and *bottom* of the chip through several TSVs simultaneously, for 62 h at 170 °C. The current density through each TSV was 5×10^5 A/cm^2 [53]

shift of the position of the Cu-filler relative to Si [18, 47, 53]. Since EM induced interfacial sliding is non-symmetric (i.e., it causes protrusion of the via downstream of the direction of electron flow and intrusion at the opposite end), and unlike under thermal cycling conditions, it accrues continuously [18, 47, 53], it may pose a potentially serious reliability challenge, particularly as the current density through the vias increases with decreasing TSV diameter. It is noted that while this via-migration under EM conditions is noted in experiments conducted on samples with through-vias with no RDL layer, the presence of an RDL or BEOL dielectric can constrain such migration and therefore mitigate it. However, because of the low elastic modulus of the RDL/BEOL dielectric, the effect of any imposed constraint may be limited, and therefore, the effect of this phenomenon on the stability of the RDL or BEOL layer needs further study.

4.4 Towards Atomistically informed Reliability Modeling of TSVs

As discussed in Sects. 4.2.2.4 and 4.3.1.2, more robust modeling techniques are needed to complement the FEA and advanced experimental characterization in order to provide a complete picture of the microstructure and stress inside the TSVs. This section introduces two methods, i.e., the crystal plasticity finite element (CPFE) method and the phase field crystal (PFC) method, and discusses their capabilities for atomistically informed reliability modeling of TSVs.

4.4.1 The CPFE Method

The challenge of managing mechanical stress is not new, and a number of simulators do exist and have a long track record of use in the electronics industry. Most of the proven simulators are based on FEA, or derivatives of that class of modeling technique. The established FEA stress simulators have typically been used for addressing the traditional chip-package interactions, and have therefore mostly modeled physical deformations, such as cracking, delaminating, or fracturing [69]. In this class of analyses Si dies and TSV fillers are typically modeled as monolithic bricks and isotropic material models based on empirical equations are used, even though it has been known since 1934 that crystalline materials deform plastically by the slip of dislocations on discrete slip systems [70, 81, 82]. However, a physically based CPFE method is relevant in order to address the polycrystalline nature of Cu TSVs and their complex deformation and stress relaxation mechanisms.

Early CPFE models used phenomenological constitutive equations and considered dislocation slip as the only deformation mechanism [70]. In a phenomenological constitutive model, a critical resolved shear stress, $\tau_c{}^{\alpha}$, is used as a state variable for each slip system α. The shear rate, $\dot{\gamma}_{\alpha}$, is formulated as a function of the resolved shear stress and the critical resolved shear stress:

$$\dot{\gamma}^{\alpha} = f(\tau^{\alpha}, \tau_c^{\alpha}) \tag{4.3}$$

The evolution of the material state is then formulated as a function of the total shear, γ, and the shear rate, $\dot{\gamma}^{\alpha}$:

$$\tau_c^{\alpha} = g(\gamma, \dot{\gamma}) \tag{4.4}$$

The first CPFE simulations were performed in 1982 using a simplified setup of two symmetric slip systems to study the tensile behavior of a single crystal [66]. The technique was later extended to a polycrystalline scenario using a 2D setup with two or three slip systems [32, 33]. Simulations on a face-centered cubic (FCC)

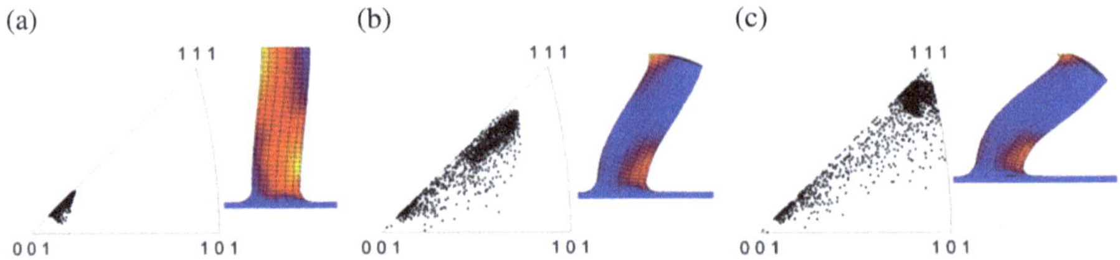

Fig. 4.22 Evolution of crystallographic orientation and geometry of a Cu single crystal pillar with a diameter-to-length ratio of 0.29 under compression with a strain (an engineering thickness reduction) of (**a**) 0.05, (**b**) 0.15, and (**c**) 0.25. *Color* coding represents accumulated plastic shear ranging from *blue* (*low*) to *yellow* (*large*). Starting single crystal orientation: [1 1 1 2] compression axis (unstable) [68]

crystal with 12 slip systems was reported in 1991 [3]. Applying the aforementioned phenomenological constitutive laws to small-scale deformation, interface mechanics, and twinning and/or deformation-induced phase transformations are often found inadequate [70]. To address size effects, strain gradient theories [57] can be introduced into the CPFE framework. As strain gradients can be associated with geometrically necessary dislocations (GNDs), new internal-variable constitutive formulations were developed to incorporate dislocation densities as physically based state variables replacing the strain variables. This latter class of constitutive models also allows the flexibility to incorporate additional metallurgical mechanisms such as grain boundary mechanics or damage initiation into the model [70]. To deal with additional deformation mechanisms such as occurring in TWIP (twinning-induced plasticity) or TRIP (transformation-induced plasticity) steels, the CPFE framework was further extended [41].

The result from a unidirectional compression on a copper single crystal using the CPFE method is shown in Fig. 4.22 [68]. The simulation was implemented in the Düsseldorf Advanced MAterial Simulation Kit (DAMASK) framework [71]. Although the mechanical boundary conditions and the material are both different from the real Cu-TSV structures, the result highlights the usefulness of the CPFE method in capturing atomistically informed deformation and deformation induced crystallographic orientation evolution. The body centered cubic (BCC) tungsten single crystal subjected to uniaxial loading has also been recently studied using the DAMASK framework [10].

4.4.2 The PFC Method

The phase field (PF) methodology is an atomically diffuse interface method for modeling of complex microstructures in solidification, precipitation, and strain-induced phase transformations [11, 67]. The PF may be seen as describing the degree of crystallinity or atomic order/disorder in a phase [67]. More recently, a

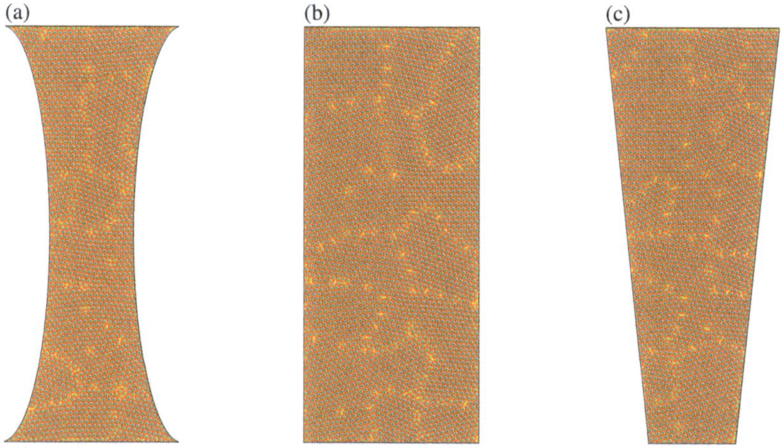

Fig. 4.23 Simulations on atomistic scale microstructural formation in (**a**) *hourglass*, (**b**) *rectangle*, and (**c**) *trapezoid* shaped TSVs using the phase field crystal model

new class of PF models has been developed, called the PFC models, which describes the thermodynamics and dynamics of phase transformations through an atomically varying order parameter field that is loosely connected to the atomic density field [20, 21, 74]. PFC models naturally capture most of the salient physics of nucleation, polycrystalline solidification, grain boundaries [20, 28], and solidification in multicomponent and multiphase systems [22, 58]. In addition, PFC models also capture, in the context of a single order parameter, elasticity and plasticity phenomena relevant to solid-state processes such as dislocation source creation, dislocation stability [4, 5], and creep [6]. The original PFC model was predominately used for the study of 2D triangular and 3D BCC crystal symmetries [20, 21]. Later models introduced multipeaked two-point correlation kernels in the nonlocal part of the free energy that allowed for a simple yet robust approach to simulate most of the common metallic crystal structures (2D square, BCC, FCC, HCP) in phase transformations [29, 88]. These so-called structural PFC (XPFC) models were later generalized to binary and multicomponent and multiphase alloys [30, 58].

Figure 4.23 illustrates a 2D PFC simulation of grain formation in TSVs with different geometries. To study the copper pumping phenomenon, mechanical loads have to be applied on such samples with corresponding grain structures. Since the PFC method does not model a solid-vacuum interface, traction boundary conditions in the PFC model using a penalty term are introduced. In deformation simulations, dislocation creation and annihilation are emergent characters of the PFC model. Therefore, applying the PFC model to TSV filler allows the investigation of the dislocation dynamics responsible for the metal extrusion or intrusion problem. The recorded dislocation dynamics can also be used to formulate dislocation-based constitutive laws for the CPFE method. In addition, progress has been made to couple the atomic-scale PFC density field to order parameters that describe ferromagnetic and ferroelectric ordering [73]. As such, the PFC models can be used to study the role of external magnetic or electric fields on the evolution of atomic scale defect structures on diffusion time scales. This provides an opportunity to study electromigration related reliability issues of TSVs using the PFC models.

4.5 Summary

Some of the key reliability issues related to TSVs are discussed, starting with the origin and nature of stresses in TSVs, their effects on microstructure development and evolution, and their role on device performance and BEOL or RDL reliability. The main reliability complications are (1) void growth to relieve internal stress gradients during pre or post-CMP annealing, (2) copper pumping during processing or service due to differential thermal expansion mismatch and associated inelastic deformation of the copper that may be accommodated by sliding at the interface, and associated distortion of the BEOL/RDL structure, (3) piezoresistive effects that lead to device performance loss and require a KOZ adjacent to TSVs, and (4) electromigration induced void growth downstream of the electron flow direction in the interconnect lines outside the TSV. These effects have been reviewed in the context of the associated physical mechanisms. Finally, two promising methods, i.e., the crystal plasticity finite element method and the phase field crystal method, to enable atomistically informed reliability modeling of TSVs are introduced.

Acknowledgements The editors would like to thank Prof. Tengfei Jiang from University of Central Florida for her critical review of this chapter. The authors (PK and ID) acknowledge financial support for some of the reported work by the National Science Foundation (DMR-0513874 and DMR-1309843), Cisco Research Council, and the Semiconductor Research Corporation. The contributions of, and collaborations with several colleagues (Dr. Lutz Meinshausen, formerly of Washington State University, and currently at Global Foundries, Dresden, Germany; Dr. Tae-Kyu Lee, formerly of Cisco Systems, and currently at Portland State University; Dr. Ravi Mahajan of Intel Corporation; Dr. Vijay Sarihan of Freescale Semiconductor, and Professor Muhannad Bakir of Georgia Tech) are gratefully acknowledged. The assistance of current and former colleagues (Dr. Hanry Yang of Washington State University, and Dr. Zhe Huang, formerly of Washington State University, and currently at Seagate Technologies) with the literature survey is also gratefully acknowledged. The author (ZH) acknowledges financial support for his research by the Pearl River Science and Technology Nova Program of Guangzhou under grant no. 2012J2200074, the National Natural Science Foundation of China (NSFC) under grant no. 51004118, and Guangdong Natural Science Foundation under grant no. 2015A030312011. The author (ZH) also acknowledges useful discussions with Dr. F. Roters of Max Planck Institute for Iron Research on the CPFE method and Prof. N. Provatas of McGill University on phase crystal models.

References

1. K. Athikulwongse, A. Chakraborty, J.-S. Yang, D.Z. Pan, S.K. Lim, Stress-driven 3D-IC placement with TSV keep-out zone and regularity study, in: *International Conference on Computer-Aided Design (ICCAD)*, San Jose, CA, Nov 2010 (IEEE/ACM, 2010), p. 669
2. P. Bayat, D. Vogel, R.D. Rodriguez, E. Sheremet, D.R.T. Zahn, S. Rzepka, B. Michel, Thermo-mechanical characterization of copper through-silicon vias (Cu-TSVs) using micro-Raman spectroscopy and atomic force microscopy. Microelectron Eng. **137**, 101–104 (2015)
3. R. Becker, J.F. Butler, H. Hu, L.A. Lalli, Analysis of an aluminum single crystal with unstable initial orientation (0 0 1) [1 1 0] in channel die compression. Metall. Trans. A **22**, 45–48 (1991)
4. J. Berry, N. Provatas, J. Rottler, C.W. Sinclair, Defect stability in phase-field crystal models: stacking faults and partial dislocations. Phys. Rev. B **86**, 224112 (2012)
5. J. Berry, N. Provatas, J. Rottler, C.W. Sinclair, Phase field crystal modeling as a unified atomistic approach to defect dynamics. Phys. Rev. B **89**, 214117 (2014)
6. J. Berry, J. Rottler, C.W. Sinclair, N. Provatas, Atomistic study of diffusion-mediated plasticity and creep using phase field crystal methods. Phys. Rev. B **92**, 134103 (2015)
7. A.S. Budiman, H.-A.-S. Shin, B.-J. Kim, B.-J. Kim, S.-H. Hwang, H.-Y. Son, M.-S. Suh, Q.-H. Chung, K.-Y. Byun, N. Tamura, M. Kunz, Y.-C. Joo, Measurement of stresses in Cu and Si around through-silicon via by synchrotron X-ray microdiffraction for 3-dimensional integrated circuits. Microelectron Reliab. **52**, 530–533 (2012)
8. H.J. Bunge, R.A. Schwarzer, Orientation stereology—a new branch in texture research. Adv. Eng. Mater. **3**, 25–39 (2001)
9. H.J. Bunge, *Texture Analysis in Materials Science—Mathematical Methods* (Butterworth & Co, London, 1982)
10. D. Cereceda, M. Diehl, F. Roters, D. Raabe, J.M. Perlado, J. Marian, Unraveling the temperature dependence of the yield strength in single-crystal tungsten using atomistically-informed crystal plasticity calculations. Int. J. Plasticity **78**, 242–265 (2016)
11. L.Q. Chen, Phase-field models for microstructure evolution. Ann. Rev. Mater. Res. **32**, 113–140 (2002)
12. E.J. Cheng, Y.L. Shen, Thermal expansion behavior of through-silicon-via structures in three-dimensional microelectronic packaging. Microelectron Reliab. **52**, 534–540 (2012)
13. H.-J. Choi, S.-M. Choi, M.-S. Yeo, S.-D. Cho, D.-C. Baek, J. Park, An experimental study on the TSV reliability: electromigration (EM) and time dependant dielectric breakdown (TDDB). in: *IITC: IEEE International Interconnect Technology Conference*, San Jose, CA, June 2012 (2012)
14. J. De Messemaeker, O.V. Pereira, B. Vandevelde, H. Philipsen, I. De Wolf, E. Beyne, K. Croes, Impact of post-plating anneal and through-silicon via dimensions on Cu pumping, in: *Electronic Components & Technology Conference* (IEEE, 2013), p. 586
15. J. De Messemaeker, O.V. Pereira, H. Philipsen, E. Beyne, I. De Wolf, T. Van der Donck, K. Croes, Correlation between Cu microstructure and TSV Cu pumping, in: *Electronic Components & Technology Conference* (IEEE, 2014), Piscataway, New Jersey, US. p. 613
16. I. De Wolf, K. Croes, O. Varela Pedreira, R. Labie, A. Redolfi, M. Van De Peer, K. Vanstreels, C. Okoro, B. Vandevelde, E. Beyne, Cu pumping in TSVs: effect of pre-CMP thermal budget. Microelectron Reliab. **51**, 1856–1859 (2011)
17. I. De Wolf, V. Simons, V. Cherman, R. Labie, B. Vandevelde, E. Beyne, In-depth Raman spectroscopy analysis of various parameters affecting the mechanical stress near the surface and bulk of Cu-TSVs, in: *Electronic Components & Technology Conference* (IEEE, 2012), p. 331
18. I. Dutta, P. Kumar, M.S. Bakir, Interface-related reliability challenges in 3-D interconnect systems with through-silicon vias. JOM **63**:70–77 (2011)
19. K.E. Elder, M. Grant, Modeling elastic and plastic deformations in nonequilibrium processing using phase field crystals. Phys. Rev. E **70**, 051605 (2004)

20. K.R. Elder, M. Grant, Modeling elastic and plastic deformations in nonequilibrium processing using phase field crystals. Phys. Rev. E **70**, 051605 (2004)
21. K.R. Elder, M. Katakowski, M. Haataja, M. Grant, Modeling elasticity in crystal growth. Phys. Rev. Lett. **88**, 245701 (2002)
22. K.R. Elder, N. Provatas, J. Berry, P. Stefanovic, M. Grant, Phase-field crystal modeling and classical density functional theory of freezing. Phys. Rev. B **75**, 064107 (2007)
23. M. Faheem, R.R. Giridharan, Y. Liang, P. van Der Heide, Micro-XRD characterization of a single copper filled through-silicon via. Mater. Lett. **161**, 391–394 (2015)
24. X. Feng, T. Huang, M. Li, The influence of annealing on characteristics of copper in TSV. ECS J. Solid State Sci. Technol. **4**, P451–P455 (2015)
25. T. Frank, S. Moreau, C. Chappaz, L. Arnaud, P. Leduc, A. Thuaire, L. Anghel, Electromigration behavior of 3D-IC TSV interconnects, in: *Electronic Components & Technology Conference* (IEEE, 2012), p. 326
26. T. Frank, S. Moreau, C. Chappaz, P. Leduc, L. Arnaud, A. Thuaire, E. Chery, F. Lorut, L. Anghel, G. Poupon, Reliability of TSV interconnects: electromigration, thermal cycling, and impact on above metal level dielectric. Microelectron Reliab. **53**, 17–29 (2013)
27. K.J. Ganesh, A.D. Darbal, S. Rajasekhara, G.S. Rohrer, K. Barmak, P.J. Ferreira, Effect of downscaling nano-copper interconnects on the microstructure revealed by high resolution TEM-orientation-mapping. Nanotechnology **23**, 135702 (2012)
28. L. Granasy, F. Podmaniczky, G.I. Toth, G. Tegze, T. Pusztai, Heterogeneous nucleation of/on nanoparticles: a density functional study using the phase-field crystal model. Chem. Soc. Rev. **43**, 2159–2173 (2014)
29. M. Greenwood, N. Provatas, J. Rottler, Free energy functionals for efficient phase field crystal modeling of structural phase transformations. Phys. Rev. Lett. **105**, 045702 (2010)
30. M. Greenwood, N. Ofori-Opoku, J. Rottler, N. Provatas, Modeling structural transformations in binary alloys with phase field crystals. Phys. Rev. B **84**, 064104 (2011)
31. J.M.E. Harper, C. Cabral Jr., P.C. Andricacos, L. Gignac, I.C. Noyan, K.P. Rodbell, C.K. Hu, Mechanisms for microstructure evolution in electroplated copper thin films near room temperature. J. Appl. Phys. **86**, 2516–2525 (1999)
32. S.V. Harren, R.J. Asaro, Nonuniform deformations in polycrystals and aspects of the validity of the Taylor model. J. Mech. Phys. Solids **37**, 191–232 (1989)
33. S.V. Harren, H.E. Deve, R.J. Asaro, Shear band formation in plane strain compression. Acta Metall **36**, 2435–2480 (1988)
34. A. Heryanto, W.N. Putra, A. Trigg, S. Gao, W.S. Kwon, F.X. Che, X.F. Ang, J. Wei, R.I. Made, C.L. Gan, K.L. Pey, Effect of copper TSV annealing on via protrusion for TSV Wafer Fabrication. J. Electron Mater. **41**, 2533–2542 (2012)
35. P.S. Ho, S.K. Ryu, K.H. Lu, Q. Zhao, J. Im, R. Huang, Reliability challenges for 3D interconnects: A material and design perspective. Presentation at the 3D Sematech Workshop, Burlingame, 17 March 2011 (2011)
36. T. Hrncir, J. Dluhos, L. Hladik, E. Moyal, Advances in FIB-SEM analysis of TSV and solder bumps—approaching higher precision, throughput and comprehensiveness, in: *ISTFA 2014: Proceedings of the 40th International Symposium for Testing and Failure Analysis*, Houston, TX, November 2014 (2014)
37. S.-H. Hwang, B.-J. Kim, H.-Y. Lee, Y.-C. Joo, Electrical and mechanical properties of through-silicon vias and bonding layers in stacked wafers for 3D integrated circuits. J. Electron Mater. **41**, 232–240 (2012)
38. T. Jiang, S.-K. Ryu, Q. Zhao, J. Im, R. Huang, P.S. Ho, Measurement and analysis of thermal stresses in 3D integrated structures containing through-silicon-vias. Microelectron Reliab. **53**, 53–62 (2013)
39. T. Jiang, S.K. Ryu, J. Im, H.-Y. Son, N.-S. Kim, R. Huang, P.S. Ho, Impact of material and microstructure on thermal stresses and reliability of through-silicon via (TSV) structures, in: *IITC: IEEE International Interconnect Technology Conference*, Kyoto, June 2013 (2013)

40. T. Jiang, C. Wu, J. Im, R. Huang, P.S. Ho, Effect of microstructure on via extrusion profile and reliability implication for copper through-silicon vias (TSVs) structures, in: *IITC/AMC: International Interconnect Technology Conference/Advanced Metallization Conference*, San Jose, CA, May 2014 (IEEE, 2014), p. 377
41. S.R. Kalidindi, Incorporation of deformation twinning in crystal plasticity models. J. Mech. Phys. Solids **46**, 267–290 (1998)
42. S.R. Kalidindi, M. De Graef, Materials data science: current status and future outlook. Annu. Rev. Mater. Res. **45**, 171–193 (2015)
43. L.W. Kong, J.R. LIoyd, K.B. Yeap, E. Zschech, A. Rudack, M. Liehr, A. Diebold, Applying X-ray microscopy and finite element modeling to identify the mechanism of stress-assisted void growth in through-silicon vias. J. Appl. Phys. **110**, 053502 (2011)
44. Krause, M., et al., Characterization and failure analysis of TSV interconnects: from non-destructive defect localization to material analysis with nanometer resolution, in: *Electronic Components & Technology Conference* (IEEE, 2011), p. 1452
45. P. Kumar, I. Dutta, Influence of electric current on diffusionally accommodated sliding at hetero-interfaces. Acta Mater **59**, 2096–2108 (2011)
46. P. Kumar, I. Dutta, Effect of substrate surface on electromigration-induced sliding at hetero-interfaces. J. Phys. D **46**, 155303 (2013)
47. P. Kumar, I. Dutta, M.S. Bakir, Interfacial effects during thermal cycling of Cu-filled through-silicon vias (TSV). J. Electron Mater. **41**, 322–335 (2012)
48. L.E. Levine, C. Okoro, R. Xu Full elastic strain and stress tensor measurements from individual dislocation cells in copper through-Si vias. IUCrJ **2**, 635–642 (2015)
49. X. Liu, Q. Chen, V. Sundaram, M. Simmons-Matthews, K.P. Wachtler, R.R. Tummla, S.K. Sitaraman, Thermo-mechanical behavior of through silicon vias in a 3D integrated package with inter-chip microbumps, in: *Electronic Components & Technology Conference* (IEEE, 2011), p. 1190
50. K.H. Lu, S.-K. Ryu, Q. Zhao, K. Hummler, J. Im, R. Huang, P.S. Ho, Temperature-dependent thermal stress determination for through-silicon-vias (TSVs) by combining bending beam technique with finite element analysis, in: *Electronic Components & Technology Conference* (IEEE, 2011), p. 1475
51. L.B. Mauer, J. Taddei, R. Yousself, Wet silicon etch process for TSV reveal, in: *Electronic Components & Technology Conference* (IEEE, 2014), Piscataway, New Jersey, US. p. 878
52. C. McDonough, B. Backes, W. Wang, R. Caramto, R.E. Gree, Thermal and spatial dependence of TSV-induced stress in Si, in: *Interconnect Technology Conference and 2011 Materials for Advanced Metallization (IITC/MAM)*, Dresden, May 2011 (2011)
53. L. Meinshausen, M. Liu, T.K. Lee, I. Dutta, L. Li, Reliability implications of thermo-mechanically and electrically induced interfacial sliding of through-silicon vias in 3D packages, in: *ASME 2015 International Technical Conference and Exhibition on Packaging and Integration of Electronic and Photonic Microsystems*, San Francisco, CA, July 2015 (2015)
54. H.D. Merchant, Thermal response of electrodeposited copper. J. Electron Mater. **24**, 919–925 (1995)
55. N. Nabiollahi, N. Moelans, M. Gonzalez, J. De Messemaeker, C.J. Wilson, K. Croes, E. Beyne, I. De Wolf (2015) Microstructure simulation of grain growth in Cu through silicon vias using phase-field modeling. Microelectron Reliab. **55**, 765–770
56. National Science and Technology Council, Materials genome initiative for global competitiveness. National Science and Technology Council, Washington, DC (2011), http://www.mgi.gov. Accessed 21 June 2016
57. W.D.Nix, J.R. Greer, G. Feng, E.T. Lilleodden, Deformation at the nanometer and micrometer length scales: effects of strain gradients and dislocation starvation. Thin Solid Films **515**, 3152–3157 (2007)
58. N. Ofori-Opoku, V. Fallah, M. Greenwood, S. Esmaeili, N. Provatas, Multicomponent phase-field crystal model for structural transformations in metal alloys. Phys. Rev. B **87**, 134105 (2013)

59. C. Okoro, K. Vanstreels, R. Labie, O. Luhn, B. Vandevelde, B. Verlinden, D. Vandepitte, Influence of annealing condition on the mechanical and microstructural behavior of electroplated Cu-TSV. J. Micromech. Microeng. **20**, 045032 (2010)
60. C. Okoro, C. Huyghebaert, J. Van Olmen, R. Labie, K. Lambrinou, B. Vandevelde, E. Beyne, D. Vandepitte, Elimination of the axial deformation problem of CuTSV in 3D integration. AIP Conf. Proc. **1300**, 214 (2010)
61. C. Okoro, R. Labie, K. Vanstreels, A. Franquet, M. Gonzalez, B. Vandevelde, E. Beyne, D. Vandepitte, B. Verlinden, Impact of the electrodeposition chemistry used for TSV filling on the microstructural and thermo-mechanical response of Cu. J. Mater. Sci. **46**, 3868–3882 (2011)
62. C. Okoro, L.E. Levine, R. Xu, K. Hummler, Y. Obeng, X-ray micro-beam diffraction measurement of the effect of thermal cycling on stress in Cu TSV: a comparative study, in: *Electronic Components & Technology Conference* (IEEE, 2014), p. 1648
63. C. Okoro, J.W. La, F. Golshany, K. Hummler, Y.S. Obeng, A detailed failure analysis examination of the effect of thermal cycling on Cu TSV reliability. IEEE Trans. Electron Dev. **61**, 15–22 (2014)
64. C. Okoro, L.E. Levine, R. Xu, Y.S. Obeng, Experimentally, how does Cu TSV diameter influence its stress state? in: *Electronic Components & Technology Conference* (IEEE, 2015), p. 54
65. K.A. Peterson, I. Dutta, M.W. Chen, Diffusionally accommodated interfacial sliding in metal-silicon systems. Acta Mater. **51**, 2831–2846 (2003)
66. D. Peirce, R.J. Asaro, A. Needleman, An analysis of nonuniform and localized deformation in ductile single crystals. Acta Metall. **30**, 1087–1119 (1982)
67. N. Provatas, K. Elder, *Phase-Field Methods in Material Science and Engineering* (Wiley-VCH, Weinheim, 2010)
68. D. Raabe, D. Ma, F. Roters, Effects of initial orientation, sample geometry and friction on anisotropy and crystallographic orientation changes in single crystal microcompression deformation: a crystal plasticity finite element study. Acta Mater. **55**, 4567–4583 (2007)
69. R. Radojcic, M. Nowak, M. Nakamoto, TechTuning: stress management for 3D through-silicon-via stacking technologies. AIP Conf. Proc. **1378**, 5–20 (2011)
70. F. Roters, Advanced material models for the crystal plasticity finite element method: development of a general CPFEM framework. Habilitation Thesis, RWTH Aachen University (2011)
71. F. Roters, P. Eisenlohr, C. Kords, D.D. Tjahjanto, M. Diehl, D. Raabe, DAMASK: the Düsseldorf Advanced MAterial Simulation Kit for studying crystal plasticity using an FE based or a spectral numerical solver. Procedia IUTAM **3**, 3–10 (2012)
72. S.K. Ryu, T. Jiang, K.H. Lu, J. Im, H.-Y. Son, K.-Y. Byun, R. Huang, P.S. Ho, Characterization of thermal stresses in through-silicon vias for three-dimensional interconnects by bending beam technique. Appl. Phys. Lett. **100**, 041901 (2012)
73. M. Seymour, F. Sanches, K. Elder, N. Provatas, Phase-field crystal approach for modeling the role of microstructure in multiferroic composite materials. Phys. Rev. B **92**, 184109 (2015)
74. M. Seymour, N. Provatas, Structural phase field crystal approach for modeling graphene and other two-dimensional structures. Phys. Rev. B **93**, 035447 (2016)
75. H.-A-S. Shin, B.-J. Kim, J.-H. Kim, S.-H. Hwang, A.S. Budiman, H.-Y. Son, K.-Y. Byun, N Tamura, M Kunz, D.-I. Kim, Y.-C. Joo, Microstructure evolution and defect formation in Cu through-silicon vias (TSVs) during thermal annealing. J. Electron Mater. **41**, 712–719 (2012)
76. D. Smith, S. Singh, Y. Ramnath, M. Rabie, D. Zhang, L. England, TSV residual Cu step height analysis by white light interferometry for 3D integration, in: *Electronic Components & Technology Conference* (IEEE, 2015), Piscataway, New Jersey, US. p. 578
77. M. Song, K.R. Mundboth, J.A. Szpunar, L. Chen, R. Feng, Characterization of local strain/stress in copper through-silicon via structures using synchrotron X-ray microdiffraction, electron backscattered diffraction and nonlinear thermomechanical model. J. Micromech. Microeng. **25**, 085002 (2015)

78. P. Stefanovic, M. Haataja, N. Provatas, Phase field crystal study of deformation and plasticity in nanocrystalline materials. Phys. Rev. E **80**, 046107 (2009)
79. G.G. Stoney, The tension of metallic films deposited by electrolysis. Proc. R. Soc. Lond. A **82**, 172–175 (1909)
80. Y.C. Tan, C.M. Tan, X.W. Zhang, T.C. Chai, D.Q. Yu, Electromigration performance of through silicon via (TSV) – a modeling approach. Microelectron Reliab. **50**, 1336–1340 (2010)
81. G.I. Taylor, The mechanism of plastic deformation of crystals. Part I. Theoretical. Proc. R. Soc. Lond. A **145**, 362–387 (1934)
82. G.I. Taylor, The mechanism of plastic deformation of crystals. Part II. Comparison with observations. Proc. R. Soc. Lond. A **145**, 388–404 (1934)
83. T. Tian, R. Morusupalli, H. Shin, H.-Y. Son, K.-Y. Byun, Y.-C. Joo, R. Caramto, L. Smith, Y.-L. Shen, M. Kunz, N. Tamura, A.S. Budiman, On the mechanical stresses of Cu through-silicon via (TSV) samples fabricated by SK Hynix vs. SEMATECH Enabling robust and reliable 3-D interconnect/integrated circuit (IC) technology. Procedia Eng *139*, 101–111 (2016)
84. A.D. Trigg, L.H. Yu, C.K. Cheng, R. Kumar, D.L. Kwong, T. Ueda, T. Ishigaki, K. Kang, W.S. Yoo, Three dimensional stress mapping of silicon surrounded by copper filled through silicon vias using polychromator-based multi-wavelength micro Raman spectroscopy. Appl. Phys. Exp. **3**, 086601 (2010)
85. V.H. Vartanian, R.A. Allen, L. Smith, K. Hummler, S. Olson, B.C. Sapp, Metrology needs for through-silicon via fabrication. J. Micro/Nanolith. MEMS MOEMS **13**, 011206 (2014)
86. H. Wang, P. Cheng, H. Wang, R. Liu, L. Sun, Q. Rao, Z. Wang, T. Gu, G. Ding, Effect of current density on microstructure and mechanical property of Cu micro-cylinders electrodeposited in through silicon vias. Mater. Charact. **109**, 164–172 (2015)
87. B. Wu, A. Kumar, S. Pamarthy, High aspect ratio silicon etch: a review. J. Appl. Phys. **108**, 051101 (2010)
88. K.A. Wu, M. Plapp, P.W. Voorhees, Controlling crystal symmetries in phase-field crystal models. J. Phys.: Condens. Matter **22**, 364102 (2010)
89. Z. Wu, Z. Huang, Y. Ma, H. Xiong, P.P. Conway, Effects of the microstructure of copper through-silicon vias on their thermally induced linear elastic mechanical behavior. Electron Mater. Lett. **10**, 281–292 (2014)
90. H. Xiong, Z. Huang, P. Conway, A method for quantification of the effects of size and geometry on the microstructure of miniature interconnects. J. Electron Mater. **43**, 618–629 (2014)
91. Q. Zhao, J. Im, R. Huang, P.S. Ho, Extension of micro-Raman spectroscopy for full-component stress characterization of TSV structures, in: *Electronic Components & Technology Conference* (IEEE, 2013), Piscataway, New Jersey, US. p. 397

Chapter 5
Fundamentals and Failures in Die Preparation for 3D Packaging

Hualiang Shi and Erasenthiran Poonjolai

5.1 Introduction

Over the past several decades, semiconductor industry has been following Moore's law to improve chip performance. Currently, major semiconductor company such as Intel Corp. is developing the 10 nm technology node for mass production. However, with reaching the physical limit, the further scaling of Silicon Front-End-of-Line (FEOL) becomes more and more challenging. Alternatively, some novel packaging approaches such as Package-on-Package (POP) and 3D packaging with Through-Silicon-Via (TSV) have been proposed to further reduce the package form factor, decrease power consumption, and improve interconnect speed. This has generated extensive research interests in the field of new material, process, and equipment.

This chapter is focused on the fundamentals and failures in die preparation for 3D package with TSV. Because several review articles have already summarized the TSV wafer fabrication processes [1–5], including temporary adhesive wafer bonding, high aspect ratio silicon etch, wafer singulation, etc., this chapter starts with a brief overview of general TSV wafer fabrication and singulation processes. Then, several key process issues which have not been discussed in previous review articles are investigated. First, the device wafer buckling or wrinkling post wafer bonding and thinning is discussed in details. This issue might induce yield loss at the downstream processes such as lithography patterning. The fundamental mechanism behind this issue and several potential solutions are investigated. Second, wafer debonding process is evaluated. Based on the simple viscosity definition and wafer geometry, a closed-form analytical solution is proposed for the thermal sliding wafer debonding process, which can be used for process control and

H. Shi (✉) • E. Poonjolai
Intel Corporation, 5000 W. Chandler Blvd, Chandler, AZ 85226, USA
e-mail: shi@gmail.com

Y. Li, D. Goyal (eds.), *3D Microelectronic Packaging*, Springer Series in Advanced Microelectronics 57, DOI 10.1007/978-3-319-44586-1_5

throughput optimization. The next two processes discussed are laser scribe and saw dicing, which impact die chipping and delamination. A closed-form solution of chipping induced by saw dicing is also investigated. The last process discussed in this chapter is the challenges and solution options for die pick and place.

5.2 Brief Overview of TSV Wafer Fabrication Processes

Different TSV wafer fabrication approaches have been reported [6, 7], including Via First, Via Middle, and Via Last. In Via First, TSV is fabricated before Front-End-of-Line process. In Via Middle, TSV is fabricated after Front-End-of-Line process and before Back-End-of-Line process. In Via Last, TSV is fabricated after Back-End-of-Line process. Please refer to Chap. 3 for details. The advantages and disadvantages of different TSV fabrication approaches have been reviewed by several groups and are beyond the scope of this chapter. In the following, Via Last approach is used as case study to highlight some key process issues and failure modes.

As indicated in Fig. 5.1, post-C4 bumping process, device wafer is attached to carrier wafer by using temporary bonding adhesive [1, 2, 8]. Carrier wafer can be silicon wafer or optical transparent wafer such as sapphire and quartz. Temporary adhesive can be thermal setting polymer, thermal plastic polymer, or UV curing polymer. The material properties of bonding adhesive need to be optimized to improve process yield and bonding quality [9–12]. The bonding adhesive should have good thermal resistance and chemical resistance to withstand acid, alkalis, and solvents used at elevated temperature during different downstream processes, have good adhesion with both device wafer and carrier wafer to survive downstream processes, and be easy to clean post debonding process. In the beginning of wafer bonding, both carrier wafer and device wafer need to be cleaned and dried to remove particles, contaminants, and moisture, which might induce wafer crack, adhesive bubbling or voiding and delamination later. Liquid or semiliquid adhesive is applied by spin coating approach to cover the wafer surface topography such as

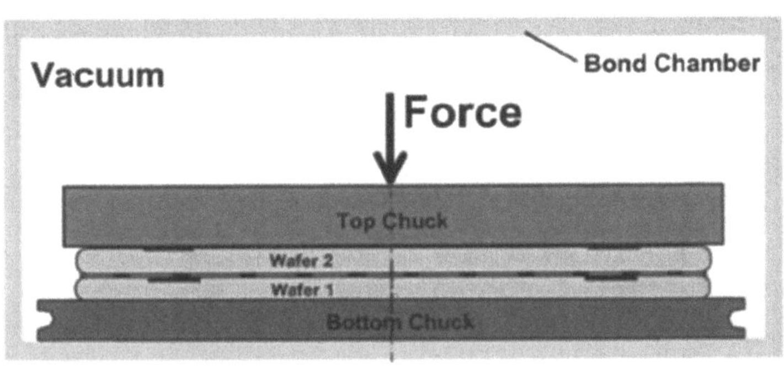

Fig. 5.1 Schematic drawing of wafer bonding

C4 bumps. If UV curing polymer is used, a laser-sensitive releasing coating layer is applied first. Soft baking is employed to remove solvents and volatile substances, which might induce adhesive bubbling or voiding later. Inside vacuum chamber, device wafer and carrier wafer are bonded together under controlled bonding pressure and elevated temperature. Bonding pressure deforms adhesive and brings device wafer and carrier wafer in contact. Bonding temperature and ramping profile as well as chamber pressure need to be tuned to reduce the adhesive bubbling or voiding and delamination. The total thickness variation (TTV) of adhesive also needs to be minimized, impacting the uniformity of wafer thinning.

Post temporary adhesive bonding process, device wafer gets edge trimming first to avoid yield loss due to defects such as wafer edge crack, chipping, and delamination, especially in the region with adhesive coverage issue [13]. Then, coarse grinding, fine grinding, and stress releasing steps are applied sequentially to thin down device wafer from bare silicon side [14–19]. Grinding process is based on grinding wheel containing diamond particles and might generate various defects, such as scratch, crack, chipping, and amorphous or polycrystalline surface damage layer. These defects reduce the silicon fracture toughness and increase wafer bow. Coarse grinding removes bulk silicon while fine grinding reduces the extent of damage induced by coarse grinding. By removing the grinding-induced damage layer, different types of stress-releasing approaches have been applied to increase the silicon strength, including Chemical Mechanical Polishing (CMP), dry polishing, wet etch, and dry etch. Both CMP and dry polishing use polyurethane-based polish pad. CMP also needs slurry which is a mixture of abrasive particle, water, and base. Dry etch uses fluorine- and oxygen-based plasmas (i.e., CF_4, SF_6, O_2, etc.) to etch silicon. Wet etch uses acid (such as hydrofluoric, nitric acid, etc.) to etch silicon. If the thickness and mechanical modulus of temporary bonding adhesive are not optimized and the residual compressive stress inside FEOL and BEOL is too high, buckling or wrinkling of device wafer might occur, causing challenges in the downstream processes. The mechanism and solutions for this issue will be discussed in details in the following section.

Once target silicon thickness is achieved, photolithography process is used for TSV patterning. Because silicon is transparent in the range of infrared (IR), IR camera is used to detect the metal fiducials at Back-End-of-Line through bulk silicon. Different types of alignment keys have been proposed to improve the overlay placement resolution [20, 21]. If device wafer has buckling or wrinkling issue, this TSV patterning process might be challenging. Post-TSV patterning, TSV silicon etch or removal is done by either plasma dry etch or laser drilling [3–5]. Among various high aspect ratio silicon etching processes, Bosch process is a common method adopted for TSV silicon etch, which alternates etching (SF_6) and passivation (C_4F_8). By considering the kinetics of plasma radical diffusion, reaction and recombination inside pore, via, and trench, Shi et al. proposed a Plasma Altered Layer model to investigate the chemical effect of plasma interaction with porous low-k dielectrics in the low plasma energy region [22, 23]. Based on this Plasma Altered Layer model, there are more reactions at the outer edge or top surface of pore, via, and trench. Shi et al. also introduced Sputtering Yield model to evaluate the physical effect of plasma interaction with porous low-k dielectrics in the high plasma energy region [24]. Based on this Sputtering Yield model, the sputtering yield is proportional to the square root of DC bias voltage and also depends on the material density and ion incident angle. Similar concepts can be extended to study TSV silicon etching. One issue related to Bosh process is the TSV sidewall scallop, which causes challenges in the downstream processes. By using Depth-Resolved Photoemission Microscopy, IR-Optical-Beam-Induced Resistance Change, Lock-In Thermography (LIT), and Electron Beam Absorbed Current (EBAC), the diode-like TSV leakage or short issue has been investigated by several groups [25–29]. Transmission Electron Microscopy (TEM) revealed local contact between TSV metal filling and bulk silicon. By optimizing the TSV etch process, the sidewall roughness was reduced and TSV leakage failure decreased. Another issue related to Bosch process is the silicon etching uniformity across wafer. Inductively Coupled Plasma (ICP) is usually used to generate plasma. However, the intrinsic nonuniformity issue of plasma density distribution might induce TSV open and short yield loss. If the TSV etching process is targeting at perfect etching in the center of wafer, there might be remaining silicon at the edge of wafer, causing TSV open failure there; if the TSV etching process is targeting at perfect etching at the edge of wafer, there might be overetching in the center of wafer, causing TSV leakage or short failure there.

Post-TSV silicon etch, a dielectric liner layer needs to be deposited to serve as insulation layer between bulk silicon and TSV metal filling. Similar as Si FEOL and BEOL, the RC delay of TSV interconnect might impact device performance, which has generated extensive research interests on the investigation of TSV capacitance [30]. As indicated in Fig. 5.2, high-frequency TSV capacitance consists of insulator capacitance and depletion capacitance. The TSV insulator capacitance increases with TSV diameter, TSV length, and liner dielectric constant and decreases with the increase of liner thickness. TSV depletion capacitance increases with TSV diameter, TSV length, liner thickness, and temperature and decreases with the increase of Si resistivity. Due to the limitation of temporary bonding adhesive, the deposition temperature of dielectric liner layer should be low. Different types of dielectric liner have been reported, including silicon oxide, silicon nitride, and polymers. Several

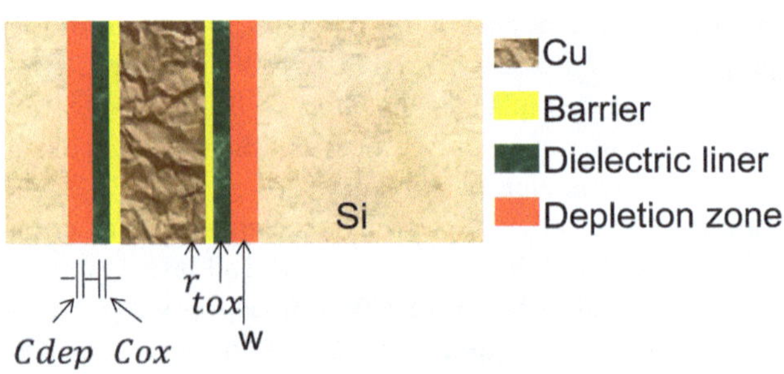

$$C_{ox}^{TSV} = \frac{2\pi\varepsilon_{ox}L}{\ln\frac{r+tox}{r}}$$

$$C_{dep}^{TSV} = \frac{2\pi\varepsilon_{s}L}{\ln\frac{r+tox+w}{r+tox}}$$

$$w_m = 2\sqrt{\frac{\varepsilon_s kT\ln\frac{N_a}{n_i}}{q^2N_a}}$$

Fig. 5.2 Schematic drawing of TSV capacitance

types of deposition method have been utilized, including Chemical Vapor Deposition (CVD) and Plasma-Enhanced Chemical Vapor Deposition (PECVD). The process parameters such as precursor chemistry and ratio as well as temperature impact liner permittivity. Some groups also reported that the microcrack inside the dielectric liner might contribute to the TSV leakage failure too [27, 28]. Finite Element Analysis (FEA) revealed high stress at groove point in liner and barrier at TSV corner due to the CTE mismatch of TSV metal, dielectric liner, metal barrier, and bulk Si.

Post dielectric liner deposition, a plasma breakthrough etch process such as CHF_3/SF_6 is applied to remove the dielectric material at the bottom of TSV cavity. If this dielectric breakthrough etch is not enough, there is remaining dielectric at the bottom of TSV, causing TSV open yield loss. On the other hand, if this dielectric breakthrough etch is too aggressive, the local interconnect inside BEOL might be destroyed, causing TSV leakage or short yield loss.

After cleaning process, a metal barrier layer such as Ta/TaN or Ti is deposited by Physical Vapor Deposition (PVD) to prevent the TSV metal diffusion into bulk silicon. Usually, copper is used as TSV metal filling material. Similar as traditional Back-End-of-Line (BEOL) process, copper deposition process includes copper seed layer deposition by PVD and copper electroplating. The plating chemistry and process need to be optimized to reduce the risk of TSV copper void and extrusion [31]. Compared with Via Last, the TSV copper in Via Middle can go through the high temperature BEOL copper annealing process, which reduces the risk of TSV copper extrusion. Some groups also developed electroless barrier layer deposition process to save cost [32]. Once TSV copper deposition process is done, Chemical Mechanical Polish (CMP) process is employed to remove the TSV copper overburden. Similar as indentation effect, low-k dielectric delamination or crack underneath TSV might occur during the CMP process. After CMP process, similar as traditional BEOL process and C4 bumping process, redistribution layer and microbump can be formed. Then, the thinned device wafer can be debonded from carrier wafer by thermal sliding, UV release with peeling, thermal decomposition with peeling, or solvent dissolution [9–12]. In thermal sliding, heat and mechanical force are used to slide off device wafer relative to carrier wafer. In UV release with peeling, light penetrates the optical transparent wafer and burns off the laser-sensitive release coating layer which has carbon black. In thermal decomposition with peeling, adhesive decomposes at elevated temperature. In solvent dissolution, solvent diffuses through the perforated wafer to attack the adhesive. The mechanism behind thermal sliding debonding is discussed in detail in the following.

Wafer singulation is one of the key assembly steps. In order to meet throughput, yield, and reliability targets, efficient process is required. Generation after generation, the requirement becomes more challenging due to the enhancement in package design, such as package form factor reduction and complex low-k ILD materials. In the case of 3D interconnect, the material properties, ILD stack up, and metal density in the wafer street region makes the singulation process extremely challenging. Typical process flows for wafer singulation are shown in Fig. 5.3. As indicated in

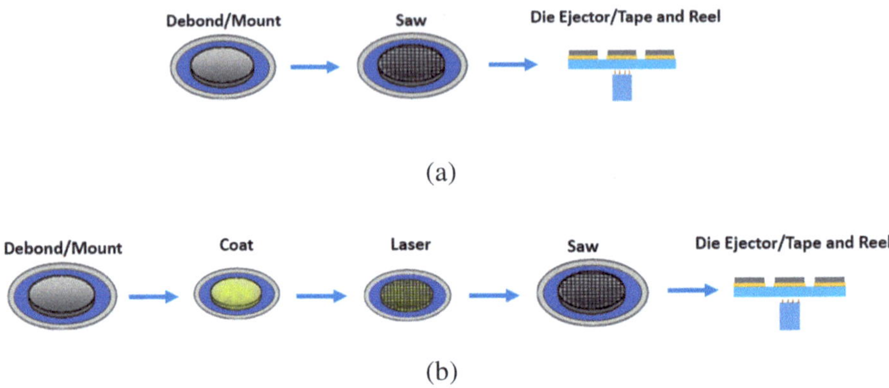

Fig. 5.3 Typical wafer singulation process flow: (**a**) without laser scribing or grooving and (**b**) with laser scribing or grooving

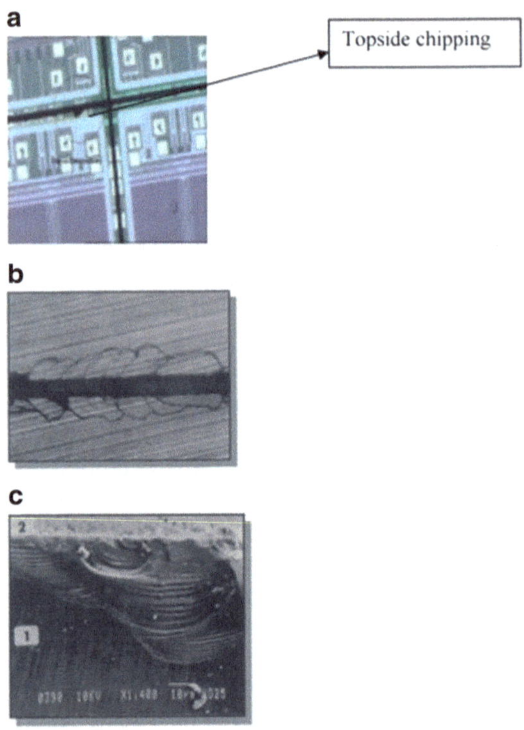

Fig. 5.4 Chipping and delamination of low-k ILD layers caused by diamond dicing saw blade: (**a**) topside chipping, (**b**) backside chipping, (**c**) SEM photo of die edge after removal from tape [33]

Fig. 5.3a, post wafer debond process, the thin device wafer is mounted at the top of dicing tape. The wafer is then singulated using a diamond saw blade. The blade cuts through the wafer street center region to separate active dies.

Figure 5.4 shows typical defects induced by mechanical saw dicing process [33]. The optimization of saw blade type and process condition is crucial in reducing these defects. Prior to die attach, die ejector is employed to pick the chips from dicing tape and place them into tape and the reel or direct chip attach. The ejector pin or blade design and configuration, and process conditions are optimized to reduce the die chipping and crack risks. The sensitivity of these factors is critical especially for thin die pick and place as it can lead to severe die crack. The flow in Fig. 5.3b has wafer coat and laser scribe processes prior diamond saw cutting process to meet technology scaling and material challenges. Further scaling of technology node and new low-k ILD materials introduced, singulation with saw blade becomes challenging. Due to the weak mechanical property, these low-k ILD materials poses risk of crack and delamination during the mechanical saw dicing process. In this scenario, laser scribe is applied to remove the low-k ILD and ETEST structures in the street area such that the saw blade interacts with ILD layer near active die during mechanical saw dicing. During laser scribe process, Si and ILD debris might be generated and redeposited on the wafer surface, which can induce nonwet or noncontact open yield loss during chip attach process. This also may lead to epoxy delamination and solder joint crack reliability failures. Wafer coat is applied before laser scribe to protect the wafer surface and will be cleaned during saw dicing. The laser scribe process and the die ejection process requirement depend on the complexity of silicon stack up (ILD layers, scribe structures, and crack stop or guard ring design), yield requirement, reliability issues, and cost consideration.

Figure 5.5 shows a comparison of cutting speed between saw dicing and laser thru cut (singulated by laser) which shows that die thickness below 150 μm can achieve faster processing speed than saw [34]. Laser thru cut is also being used depending on die thickness and stack complexity. The key advantage of laser thru cut is elimination of front side and back side chipping. The laser scribe process, saw dicing process, and die ejection process will be discussed in details in section 5.5–5.7.

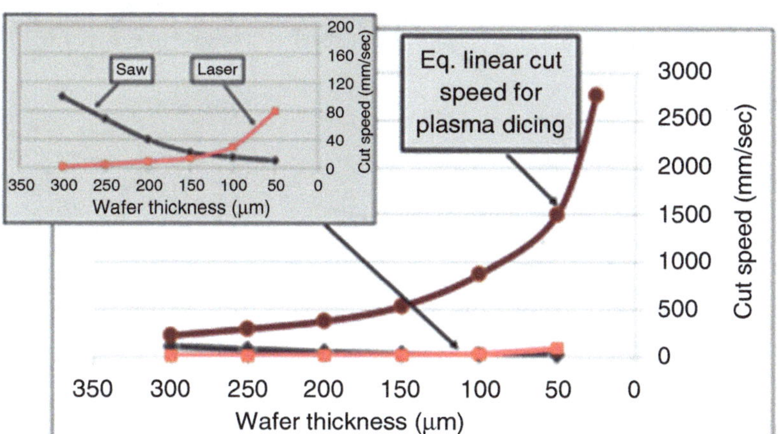

Fig. 5.5 Effective die singulation speeds for mechanical saws, lasers, and plasma singulation method versus wafer thickness (Used with permission of Plasma-Therm LLC) [34]

5.3 Wafer Buckling and Wrinkling

Mei and Huang have investigated the wrinkling and buckle delamination of elastic thin films on elastic substrate, compliant substrate, and viscoelastic substrate [35–37]. As indicated in Fig. 5.6, the critical compressive stress for wrinkling increases with stiffness ratio between substrate and film, increases with film thickness, and decreases with substrate thickness; the wrinkle amplitude increases with film strain and decreases with substrate modulus; the wrinkle wavelength increases with stiffness ratio. As indicated in Fig. 5.7, the critical compressive stress for buckle delamination increases with stiffness ratio and decreases with initial crack length; the transition of buckling modes depends on stiffness ratio and initial crack length.

Post-TSV device wafer thinning, the stack up of device wafer, temporary adhesive, and carrier wafer can be simplified as elastic film (device wafer) and viscoelastic substrate (temporary adhesive). During FEOL and BEOL and wafer thinning process, a residual stress exists on device wafer. Depending on the dielectric deposition condition, metal plating condition, and wafer thinning condition, the residual stress can be compression or tension. Different metrologies including synchrotron X-ray radiation, micron Raman, and bending beam have been employed to check the wafer residual stress. Based on Figs. 5.6 and 5.7, in order to reduce the risk of wafer buckling or wrinkling, different approaches can be applied, including (1) reducing compressive stress on device wafer by optimizing Si BEOL design and process and wafer thinning process, (2) increasing the modulus of temporary adhesive, (3) reducing the thickness of temporary adhesive, and (4) increasing device wafer thickness.

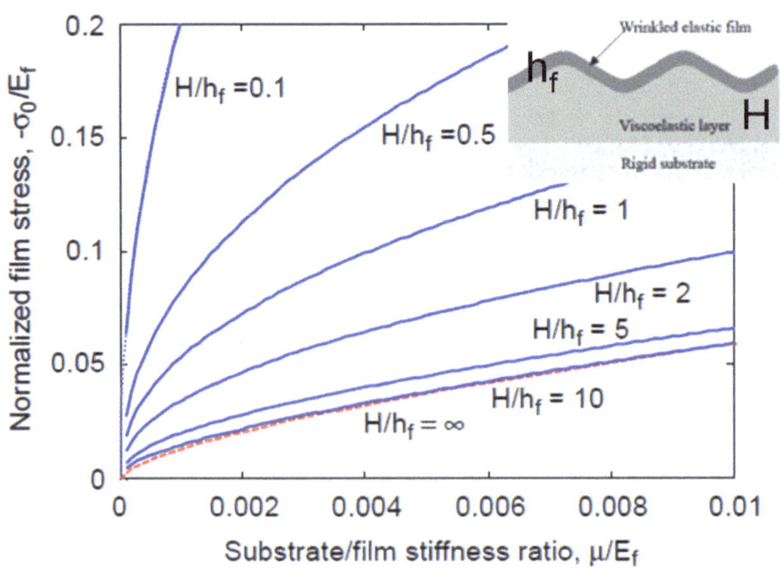

Fig. 5.6 The critical compressive stress for wrinkling

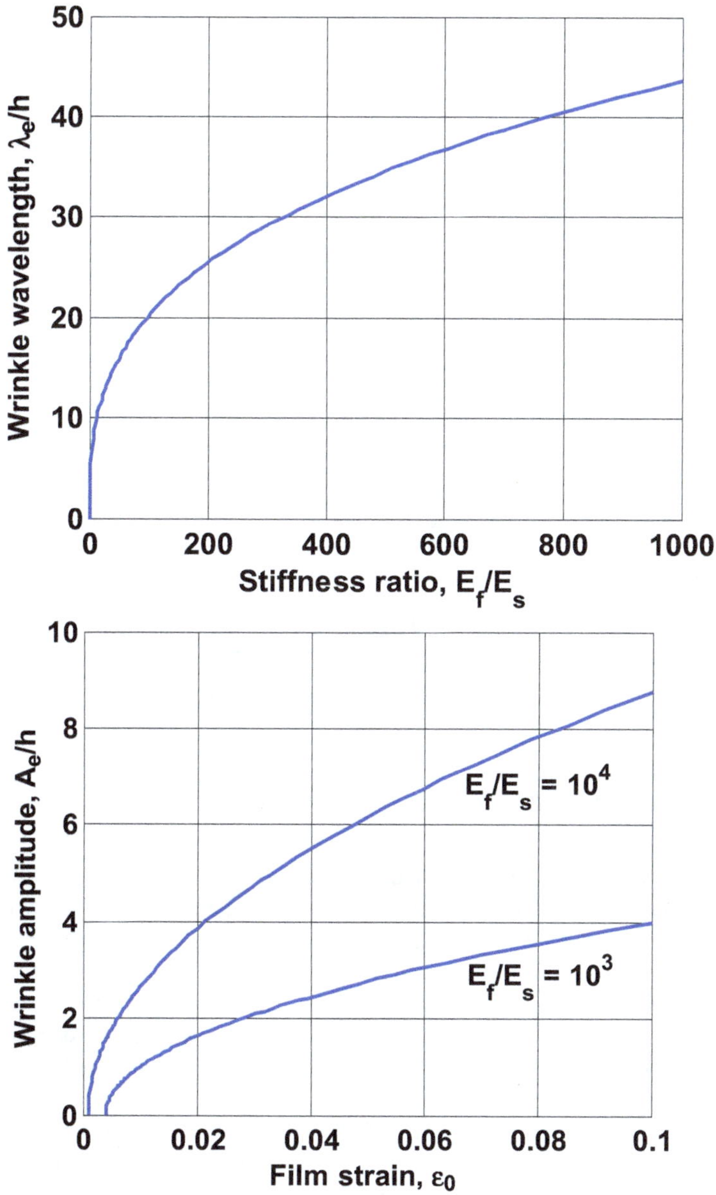

Fig. 5.6 (continued)

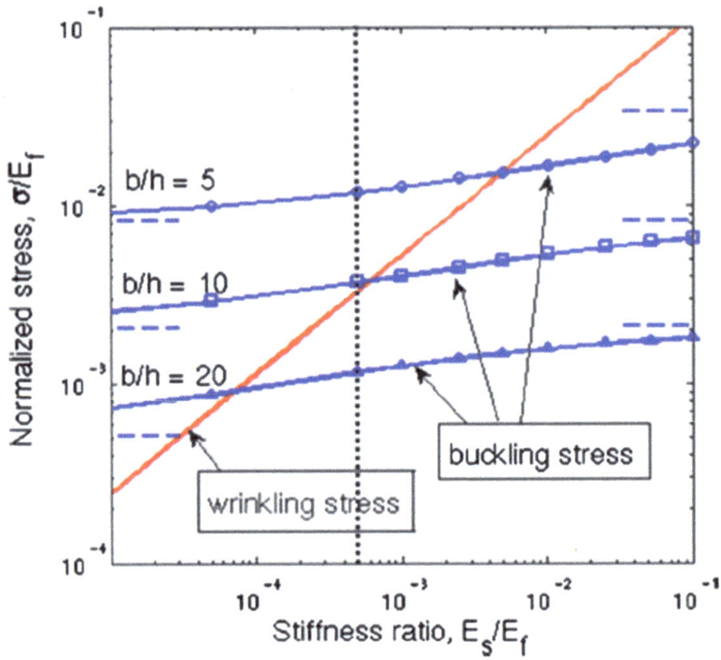

Fig. 5.7 The critical compressive stress for buckle delamination

5.4 Thermal Sliding Wafer Debonding

Inside the debonding chamber, device wafer and carrier wafer are held by two compliant vacuum chucks. The parallelism and cleanness of these two vacuum chucks are very important for thermal sliding debonding process. If particle exists at the surface of vacuum chuck, wafer crack, wafer scratch, and contamination might occur. Once device wafer and carrier wafer are heated up uniformly, the viscosity of thermal plastic adhesive decreases. At debonding temperature, adhesive becomes liquid and flowable. When sliding shear force is applied, device wafer is slided off relative to carrier wafer.

Based on viscosity definition and wafer geometry, Privett et al. investigated the kinetics of thermal sliding debonding [39]. In the following, their approach is further extended to derive a kinetic model for thermal sliding debonding process and material control. The side view and top view of thermal sliding process was shown in Fig. 5.8. Based on viscosity definition, we have

$$\frac{F}{A} = \eta = \frac{\upsilon}{H} \tag{5.1}$$

Based on geometry relationship, we have

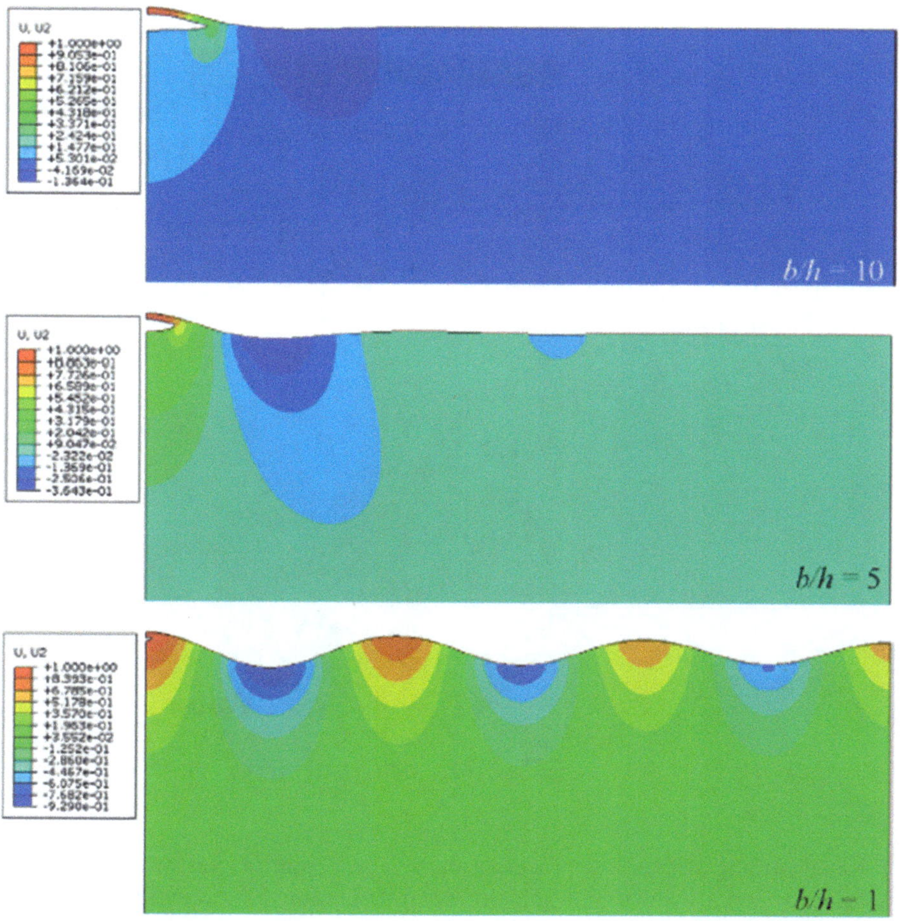

Fig. 5.7 (continued)

$$A = r^2(\theta - \sin\theta) \tag{5.2}$$

$$X = 2r\cos\left(\frac{\theta}{2}\right) \tag{5.3}$$

Based on velocity definition, we have

$$v = \frac{dX}{dt} = -r\sin\left(\frac{\theta}{2}\right)\frac{d\theta}{dt} \tag{5.4}$$

By combining equations (5.1)–(5.4), we can get

$$\frac{FH}{\eta r^3} = \sin\left(\frac{\theta}{2}\right)(\sin\theta - \theta)\frac{d\theta}{dt} \tag{5.5}$$

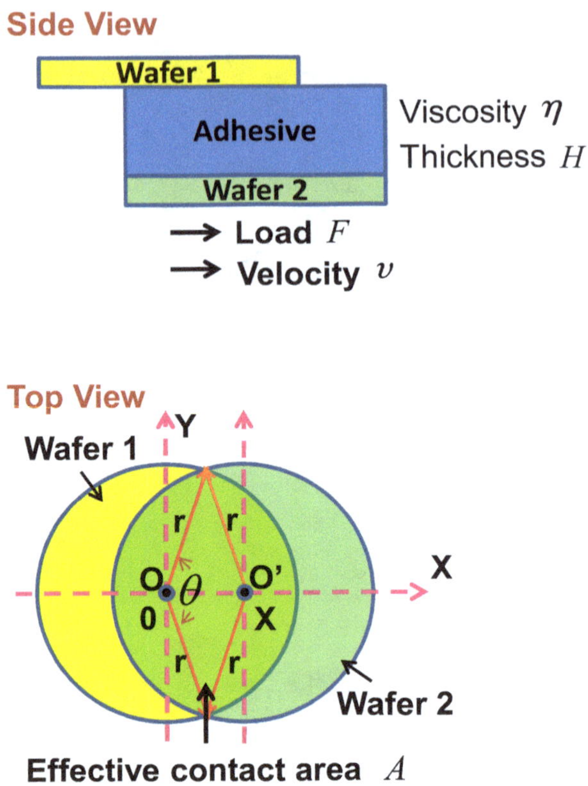

Fig. 5.8 Schematic drawing of thermal sliding debonding process

Thermal sliding debonding process can be force control, velocity control, or mixture of force control and velocity control. During force control debonding process, a constant force is applied. During velocity control debonding process, a constant velocity is applied. If debonding process starts as force control at time zero, by solving Eq. (5.5) and use initial condition

$$\theta(t = 0) = \pi \tag{5.6}$$

we can have a closed-form solution

$$t = \frac{\eta r^3}{FH}\left[2\theta \cos\left(\frac{\theta}{2}\right) - 4 \sin\left(\frac{\theta}{2}\right) + \frac{4}{3} \sin^3\left(\frac{\theta}{2}\right)\right] + C_0 \tag{5.7}$$

By combining Eqs. (5.3) and (5.7), we can get the equation governing the relationship between thermal sliding debonding time (t) and debonding position (x)

$$t=\frac{\eta r^3}{FH}\left[\frac{2x}{r}\arccos\left(\frac{x}{2r}\right)-4\sin\left(\arccos\left(\frac{x}{2r}\right)\right)+\frac{4}{3}\sin^3\left(\arccos\left(\frac{x}{2r}\right)\right)+\frac{8}{3}\right] \tag{5.8}$$

When debonding position (x) is equal to wafer diameter, t is the total debonding time. As indicated by Eq. (5.8), with the decrease of adhesive viscosity or with the increase of debond force or adhesive thickness, debonding time decreases. If debonding process starts with velocity control at time zero, by using Eqs. (5.1) and (5.4), we can have

$$F=\frac{\eta v r^2}{H}(\theta-\sin\theta) \tag{5.9}$$

By combining Eqs. (5.3) and (5.9), we can have

$$F=\frac{\eta v r^2}{H}\left(2\arccos\left(\frac{x}{2r}\right)-\sin\left(2\arccos\left(\frac{x}{2r}\right)\right)\right) \tag{5.10}$$

The closed-form solution, Eqs. (5.8) and (5.10) can be used to control thermal sliding process and material.

Post thermal sliding debonding, both device wafer and carrier need to be cleaned. Solvent is dispensed at the top of device wafer and carrier wafer. Post soaking, wafers are spun such that solvent dissolves the polymer and cleans the wafers.

One of the issues reported by several groups is the wafer bow post debond. Compared with traditional wafer, TSV wafer has interconnect structures on both sides. If the residual stresses on both sides are unbalanced, wafer bow might occur. This problem can be simulated by using Stoney equation. As indicated in Fig. 5.9, based on geometry, we can have

$$R^2=(R-H)^2+\mathrm{DNP}^2 \tag{5.11}$$

Stoney equation is

$$\sigma_f=\frac{B_S h_S^2}{6h_f R} \tag{5.12}$$

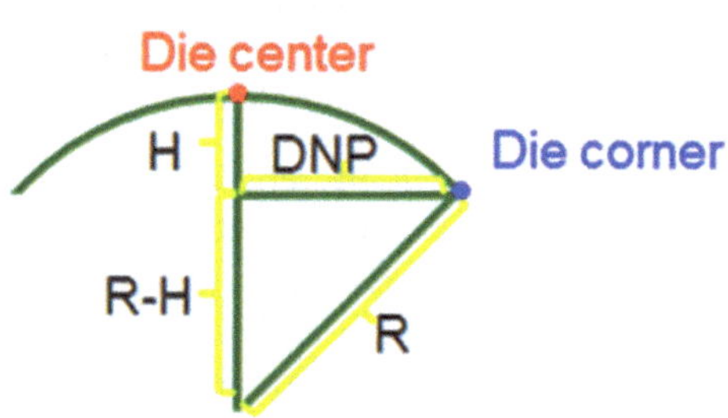

Fig. 5.9 Schematic drawing of wafer warpage

By combining Eqs. (5.11) and (5.12), wafer bow is

$$H \sim \frac{3h_f \sigma_f \mathrm{DNP}^2}{B_S h_S^2} \tag{5.13}$$

In order to reduce the wafer bow (H), we need to reduce the residual stress from the two sides of wafers or increase wafer thickness.

5.5 Wafer Laser Scribe

With the scaling of Si technology, the incorporation of advanced materials, and the growth of small handheld electronics business, assembly technologies are being developed to enable small form factor packages with advanced capabilities [40]. This poses significant challenges in current manufacturing technology and requires novel ideas. Lasers have made major contributions to modern manufacturing technology in heavy industries, medical applications, and space and military systems since the early 1970s. Laser dicing and scribing have been in focus for silicon technology, especially for MEMS, MOEMS, and other microscale devices [33]. However, the use of lasers is still in its early stage for most semiconductor applications, especially in high-volume packaging manufacturing. This is mainly due to the fact that advanced laser process is relatively slow above certain chip thickness and still not fully mature to meet the cost target. Moreover, optics technology is still growing with continuous advancement ongoing in the optoelectronics field to meet high volume manufacturing reliability and cost demand. The immediate potential applications of laser include laser dicing for thin wafer, and laser scribe or grooving for wafer with complex or challenging ILD stack before mechanical saw dicing.

The key fundamental advantages of laser process include [41]:

1. Laser cuts smaller street width due to narrow beam diameter and hence can enable narrow street width which can increase number of chips per wafer.
2. Laser singulation can enable non-Cartesian street. This can enable multichip size vehicle and harvest various chip sizes from same wafer.
3. Laser thru cut for thin wafer reduces or eliminates die edge defects and cracks.
4. Laser dicing or scribe is a "dry" process in contrast with saw process and hence potentially eliminates any corrosion and moisture-related issue. This advantage may be limited due to wafer coat requirement and post laser singulation wash. However, the time used is significantly less compared to that during saw process.

Multiple publications reported that although saw technology continues to progress in semiconductor dicing, there are challenges and difficulties in meeting yield and reliability requirement when the die thickness are reduced below 100 μm [40–43]. In order to reduce backside chipping (see Fig. 5.10), saw speed needs to be reduced significantly lower or use alternate blade type. As wafer thickness decreases, the

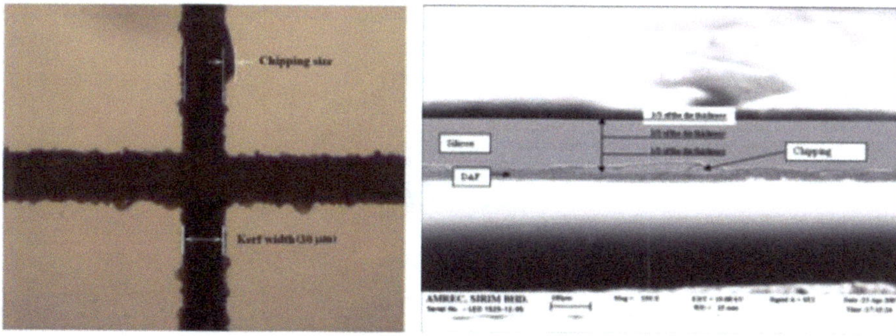

(a) (b)

Fig. 5.10 Backside chipping of silicon from mechanical sawing process: (**a**) typical BSC chip [72], (**b**) side wall image of BSC on thin die [58] (Color figure online)

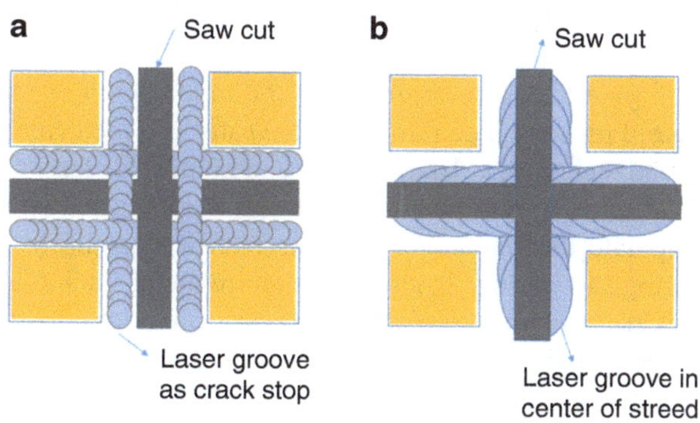

Fig. 5.11 Hybrid laser-saw process for wafers with advanced low-k dielectric and complex street structures: (**a**) two grooves scribed by laser at edges of street prior to saw dicing, (**b**) grooves scribed by laser in the center of street and saw cuts through the ablated region [41]

through cutting speed achievable with lasers increases [34] as shown in Fig. 5.5. In order to truly achieve higher run rate using laser thru cut, higher laser power and potentially an ultrafast laser system are required to further improve the laser dicing quality.

Wafers fabricated with low-k ILD material and high number of street metal structures may require hybrid laser-saw singulation process or alternate emerging singulation technology such as Stealth Dicing [46, 47]. This will ensure no metal stack interaction with saw blade and hence reduces stress induced by blade at die edge. This minimizes the die edge defect during saw singulation. Typically, ILD stack with SiO_2 is more appropriate with saw process compared to ILD stack with low-k materials [34]. Mechanical saw cut through metal layers and low-k dielectric material causes high chipping due to blade loading and hence higher blade bending which can lead to die edge defects.

A key solution for this issue is using laser scribe process to isolate the active die area prior to mechanical saw dicing process. As indicated in Fig. 5.11a, two tight grooves are made by laser at the edges of street; as shown in Fig. 5.11b, a wide

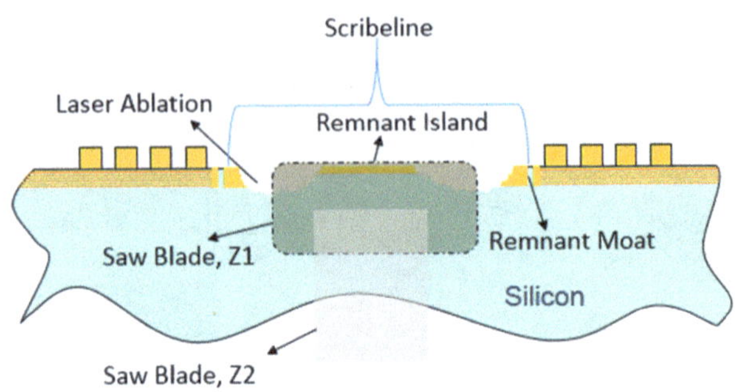

Fig. 5.12 Laser ablation and saw mechanism

groove is made by laser in the center of street [41]. In order to meet the yield and reliability requirement, key laser attributes are wavelength, laser power or fluence, and scribing technique. Multiple wavelengths can be used for scribing which has its own advantages and disadvantages. A typical wavelength generally being used is 355 nm scribe in nanosecond region. Also a traditional laser scribe processing moves a pulsed laser beam across the substrate [41].

Figure 5.12 further interprets the fundamental mechanism behind laser and saw hybrid process. The remaining ILD stack in the center of street is dictated by laser scribe technology, beam width, and number of laser passes. The idea of ILD stack ablation is to remove most of metal stacks in the street to allow the saw blade to singulate the remaining Si. This technique will ensure blade loading is reduced during saw dicing. Typically the entire metal stack is ablated during the laser scribe process which is then followed by saw singulation which can be a two- or one-step process [34]. The two-step process uses two blade widths to cut through remaining Si while the one-step process uses only one blade width to cut through the Si. The distance between laser scribe edge and saw cut edge is dictated by street width, scribe width, and saw blade width. This distance is one of the key process parameters because microcracks that are generated during laser and saw process can propagate during reliability stress and causes reliability issue.

There are multiple publications that discuss laser–solid interaction [48–50]. One of the key parameters to focus on in understanding laser–material interactions is the characteristic time for electrons and lattice to reach thermal equilibrium [51]. In nanosecond scale, the energy transport can be described by the Fourier conduction model [52, 53]. In this case, the deposition of laser energy is instantaneous compared to picosecond or femtosecond laser regime. In the latter case, the electrons and material lattice do not attain thermal equilibrium [51].

A typical heat conduction equation is valid for any spatial and temporal scales as in Eq. (5.14):

$$C\frac{\partial T}{\partial t} = -\nabla \cdot q'' + S \tag{5.14}$$

where C is the heat capacity (J/m^3 K), q'' is the heat flux vector, and S is the internal heat source (W/m^3) that can be caused by volumetric heating or coupling between electron. The heat capacity is $C = \rho c_p$ and the heat flux is related to the temperature gradient by the Fourier's law in Eq. (5.15):

$$q'' = -k\nabla T \tag{5.15}$$

Substituting Eqs. (5.15) into (5.14) results in the following heat conduction equation:

$$\rho c_p \frac{\partial T}{\partial t} = \nabla \cdot (k\nabla T) + S \tag{5.16}$$

Further information on ultrafast laser such as picosecond and femtosecond laser is well explained by a few researchers particularly by Y. Zhang [54].

When a laser irradiance exceeds 10^9 W/cm^2 for a nanosecond laser pulse, the temperature near the surface of a solid target can exceed the boiling point, thus a superheated liquid layer is formed [55]. It was suggested that explosive boiling takes place where homogeneous vapor bubble nucleation occurs when the target material reaches ~$0.9T_{tc}$ (T_{tc} is the thermodynamic critical temperature) [55]. As a consequence, the target material makes an abrupt transformation from superheated liquid into a mixture of liquid droplets and vapor which are then ejected from the target [56–60].

It is important to consider the ILD stack layers, material properties, and thickness prior selecting the right laser systems (wavelength, pulse duration, and frequency) to ensure scribe and singulation quality is acceptable for reliability and yield. Although historically, UV nanosecond laser is used for scribing and singulation, it can become a challenge to maintain similar yield and reliability standard of processing future advance low-k and stack complexity. This may require shorter pulse duration or ultrafast laser system or any other novel technology such as stealth dicing, plasma singulation, or laser water jet. The interaction of ultrafast laser pulses with a target material happens at a small timescale (picosecond or less) and a small spatial scale along the laser beam irradiation direction (tens of nanometers), accompanied with strong nonlinear, nonequilibrium, optically, thermally, and mechanically coupled processes [61]. Because the ablation mechanism is nonlinear, the energy coupling during ablation of a multistack low-k layer is highly efficient compared to nanosecond laser.

5.6 Wafer Saw Process

Wafer saw or dicing technology is a well-established process and being used generation after generation for dicing semiconductor materials. As indicated in Fig. 5.13, typical saw dicing process involves two blades in step-cut process and single blade in single thru cut process. The choice of saw dicing process depends on

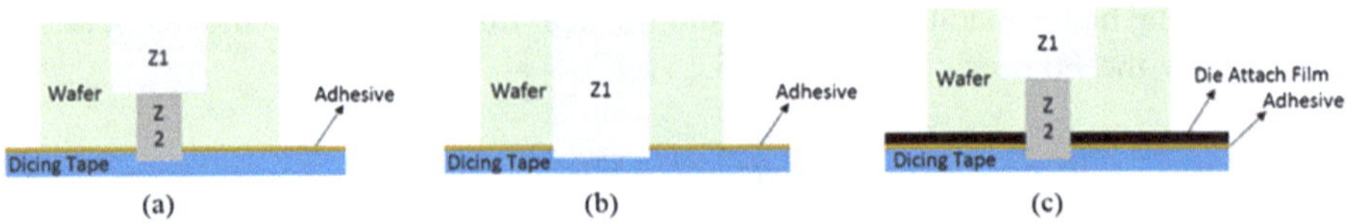

Fig. 5.13 Typical saw dicing process: (**a**) step-cut, (**b**) single thru cut, and (**c**) step-cut with DAF material

multiple parameters, including chip thickness, structures in the street, laser ablation, dicing tape, and the die attach film (DAF) [33, 45, 62, 63]. In general, step-cut with two blades is preferred in most cases because it can control front side chipping (FSC) with first blade (Z1) and back side chipping (BSC) with second blade (Z2).

The saw dicing is an abrasive process using diamond particles embedded in a solid binder and cutting through the material [33]. In this case, the abrasive material used is diamond embedded in solid binder. A typical failure mode observed during saw dicing is front side chipping (FSC) and back side chipping (BSC). The nature of chipping can be a single large chip or daisy chain of small or large chips. This is mainly due to the interaction of saw blade to the ILD stack material which forms FSC and the blade interaction with the dicing tape or DAF which can form BSC and DAF stringers or film burr [64]. The FSC and BSC also heavily dependent on street structures in a given wafer. High metal content streets can increase the blade loading and increase the number of chips and size formed in the street. It may also lead to microcracks, originating from Si sidewall. The chipping size, occurrence, die thickness and residual stress of the die can dictate the reliability risk. A typical failure in reliability is die crack due to assembly process, material and test interaction. In some cases, it may lead to the crack of molding compound or epoxy fillet. One of the challenges for BSC detection is the DAF material in certain products as it prevents BSC to be monitored using commonly used optical measurement tools. In general, the BSC is acceptable in most cases as long as the chip depth and width is minimal and not extending toward the active area.

Multiple papers discussed the factors impacting FSC or BSC [33, 45, 62, 63, 65]. In summary, the key factors to be considered in eliminating or reducing the chipping are as follows:

1. Blade type:
   (a) Grit size of a blade can impact the chipping size and frequency. If the grit size is increased, the cutting power increases [65] and hence reduces blade loading. However, the study also shows that the backside chipping increases if the grit is larger on wafer thickness less than 200 μm [65].
   (b) Higher diamond concentration results in less chipping due to a smaller load on each grit.
   (c) A softer bond material allows the blade for self-sharpening and hence reduces the chipping. However, the softer bond material can lead to faster blade wear and hence increases the frequency of blade change.

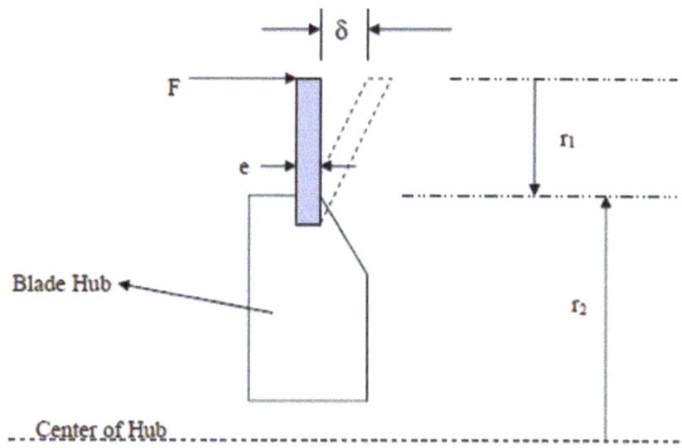

Fig. 5.14 Saw blade deflection during saw process [33]

2. Blade thickness: It is generally dictated by scribe width. However, this factor needs to be assessed carefully to understand the right size. Wider blade may cause the force applied to the dies is large and causes higher backside chipping. On the other hand, the smaller blade can tend to wobble during dicing and can cause higher chipping. The wobbling can be associated to die thickness and also metal structure density in street.
3. Process conditions such as precut dressing conditions [62], coolant and surfactant flow [66] and direction respective to saw cut location and blade setup are key factors to reduce FSC or BSC. It is important to monitor these parameters, and setup condition to ensure on the blade is maintained during dicing. Blade torque can be monitored online to ensure process stability is maintained and meets current challenges of the dicing process [66].

As indicated in Fig. 5.14, in order to calculate the stress level on a chip, the blade deflection distance δ can be calculated [33] using Eq. (5.17):

$$\delta = \frac{6Fr_1^3}{Er_2e^3} \tag{5.17}$$

where δ: displacement at the blade tip,
F: total side thrust force in N which is equal to ql,
q: loading per unit length due to side thrust,
l: length of a die,
r_1: blade exposure length,
r_2: radius of blade hub,
E: elastic modulus of blade,
e: blade thickness.

As indicated in Fig. 5.15, the maximum bending stress can be defined as [33]:

$$\sigma_x = \frac{M_x}{I(^h/_2)} = \frac{3ql^2}{bh^2} \tag{5.18}$$

where M_x is max bending moment at "x,"
I is second moment of inertia,

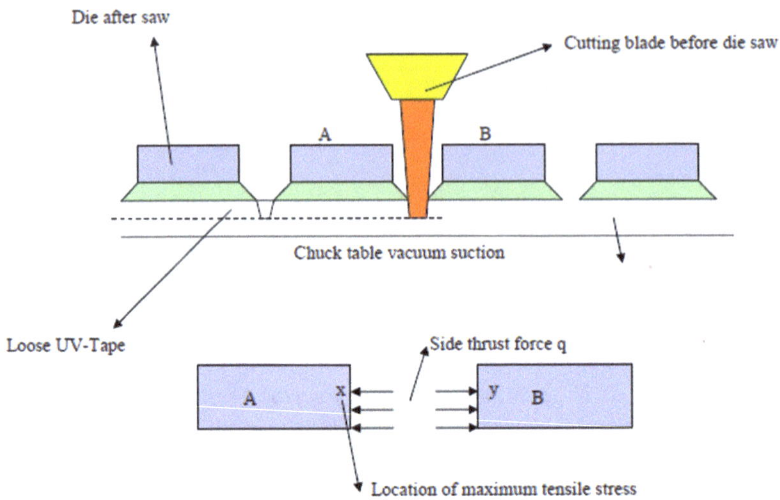

Fig. 5.15 Blade force on dies during saw process [33]

Fig. 5.16 Blade force vs. feed rate [33]

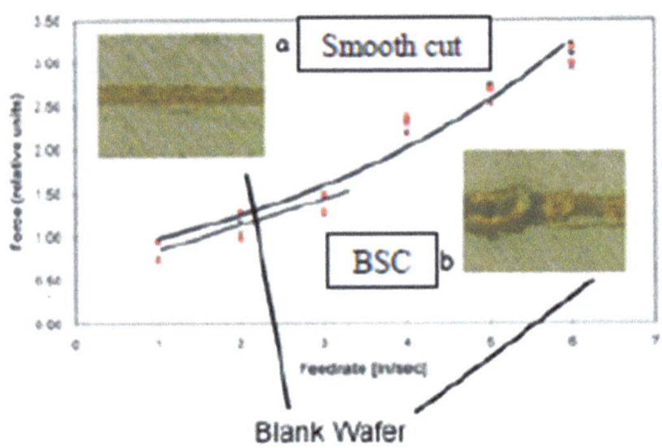

h is width of die A (see Fig. 5.15),
l is length of die A (see Fig. 5.15),
q is loading per unit length due to side thrust force,
b is thickness of die A (see Fig. 5.15).

As indicated in Fig. 5.16, the correlation between dicing force and feed rate was revealed [33]. A chip size can be calculated by the following equation [67–69]:

$$h_{\max} = \left[\frac{3}{C\tan\theta}\left(\frac{v_{\mathrm{w}}}{v_{\mathrm{s}}}\right)\left(\frac{a}{d_{\mathrm{s}}}\right)^{1/2}\right]^{1/2} \tag{5.19}$$

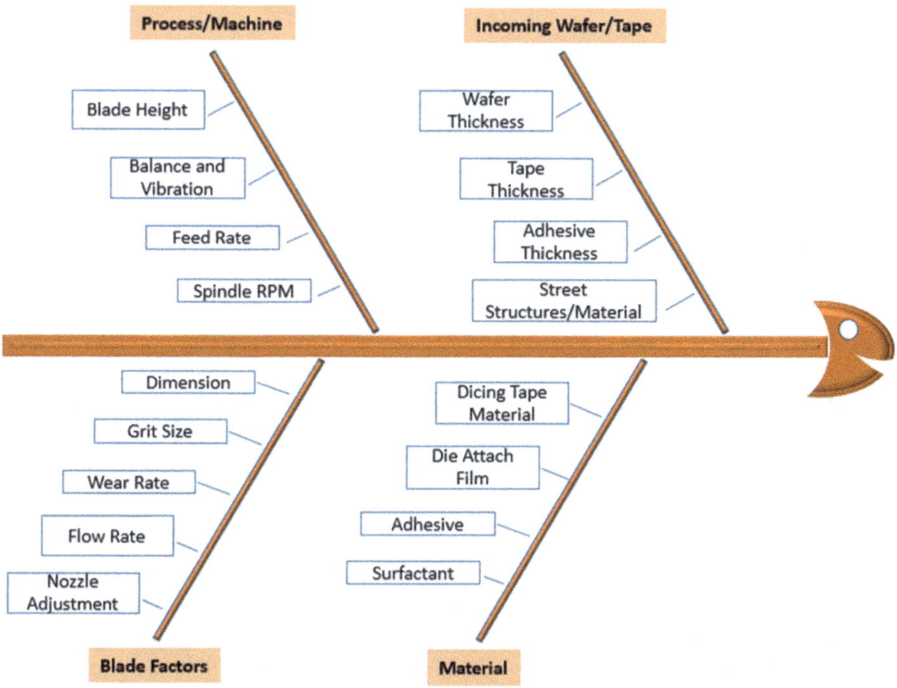

Fig. 5.17 Key process parameters for wafer dicing process (color figure online)

where C is the active grain number per unit area on the blade surface, θ is the cross-sectional semiincluded angle of an undeformed chip, v_w is the velocity of work-piece, v_s is the velocity of the blade, a is the feedrate of workpiece, and d_s is the diameter of the blade. C can be calculated as [70]

$$C = \frac{4f}{d_g^2 \left(\frac{4\pi}{3v}\right)^{2/3}} \tag{5.20}$$

where d_g is the equivalent diameter of the diamond grits within the blade and f is the active fraction of the diamond grits in the blade [70].

d_g is expressed as [69]:

$$d_g = 28 \times M^{-1.1} \tag{5.21}$$

where M is the mesh size of grains.

In summary, the FSC and BSC can be optimized by blade type, blade concentration and grit size, feed rate, blade RPM, and cut depth which require characterization in order to reduce and eliminate chipping and hence die crack. Figure 5.17 shows fishbone diagram of key process parameters that affect the saw performance [33, 70].

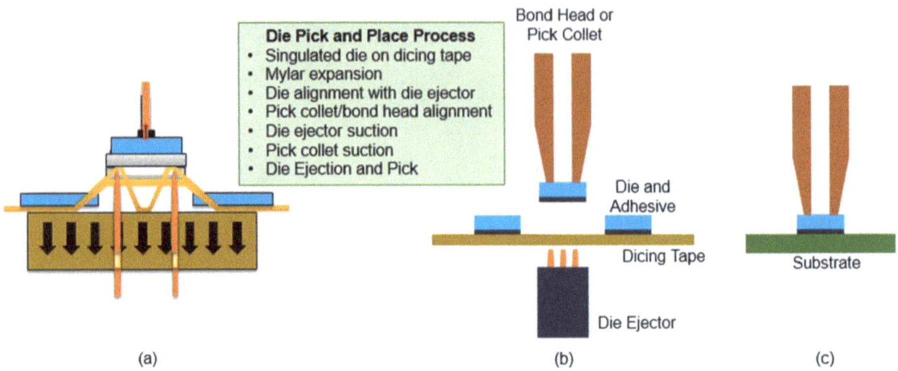

Fig. 5.18 Die pick-and-place process: (**a**) Die pick mechanism and dicing tape peeling process, (**b**) Die pick with bond head, (**c**) Die place on a substrate (Color figure online)

5.7 Wafer Die Ejector

After saw dicing process, the singulated dies are adhered to the dicing tape. At the next operation, the dies will be picked and placed onto a substrate or another die for the die attach process. Most of the semiconductor industry uses the pick-and-place process at the die attach assembly operation. In some cases, dies are picked and placed in tape and reel and shipped to die attach assembly line for assembly packaging. The mechanism and theory for each case is very much similar. In this section, examples are used from die pick and place in the die attach assembly line, typically called as wafer die ejector. As indicated in Fig. 5.18, the die pick-and-place process uses the vacuum pick collect and die ejector mechanism [70]. The die ejector typically has ejector pins or blades under the tape. The vacuum collects, picks the die, and places on a substrate for the die attach process.

The key process factors include chip thickness and size, tape adhesive modulus and stickiness, die ejector needle type and configuration, and die ejector parameters.

These factors are highly critical for thin die, especially below 100 μm. An unoptimized setting and material selection can lead to die crack. Multiple publications discuss the fundamental and modeling on die pick and place to identify critical parameters and die stress sequence during die pick process [71–74]. The tape stress will act on the die edge first and move to the top of ejection needles last [71].

As indicated in Fig. 5.19, the impact of single-needle and multineedle die ejector assembly process on die crack was studied [72]. In single-needle technology, the peeling energy release rate decreases as chips become thinner and larger, causing the die harder to peel off from the dicing tape [72]; in multineedle technology, pick ability was improved and the stress on die was reduced.

Cheng et al. [73] discuss die failure mechanism by experimental and finite element model. Figs. 5.20 and 5.21 show the contours of stress and displacement, respectively [73]. The key finding from the modeling data is the stress at the corner of the die is lower than the eject point due to piercing force; however, the

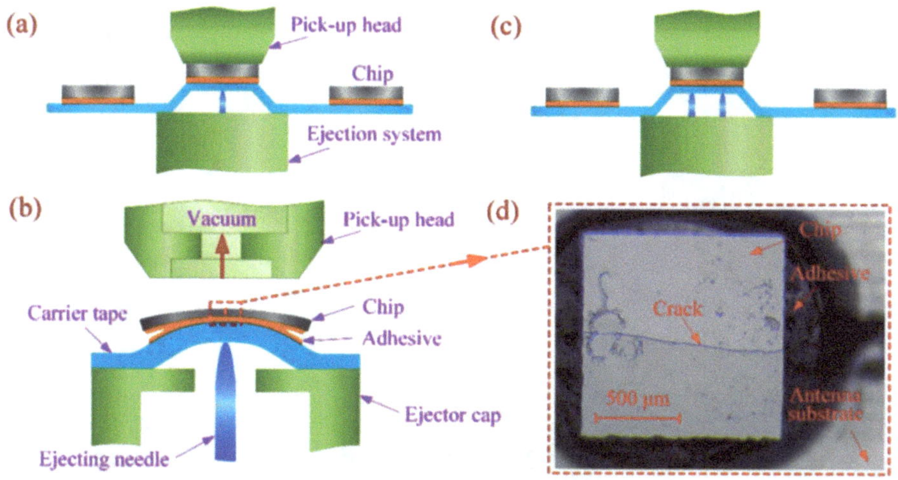

Fig. 5.19 Schematic drawing of die pick: (a) single needle, (b) potential die bending, (c) multineedle assembly with more even ejection, and (d) shows the die crack due to die bending [72] (Color figure online)

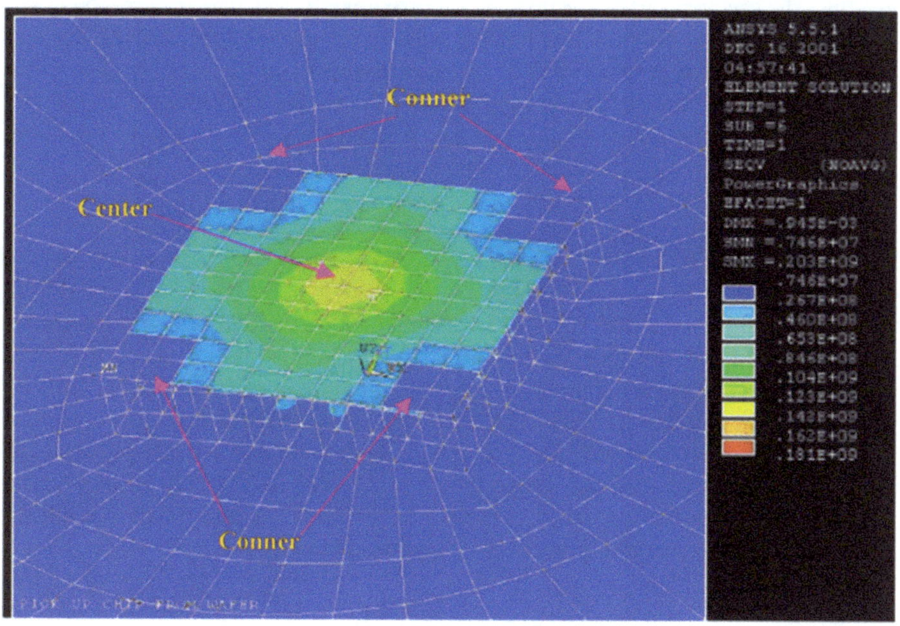

Fig. 5.20 Von Mises stress contour for the IC chip [73] (Color figure online)

displacement is higher at corner compared to the center of the die [73]. The die will start peeling off from the dicing tape from corner toward the center of the die.

In summary, selecting the right die ejector (pin or needless) and pin configuration and process settings are crucial in minimizing stress on die. Typically, the

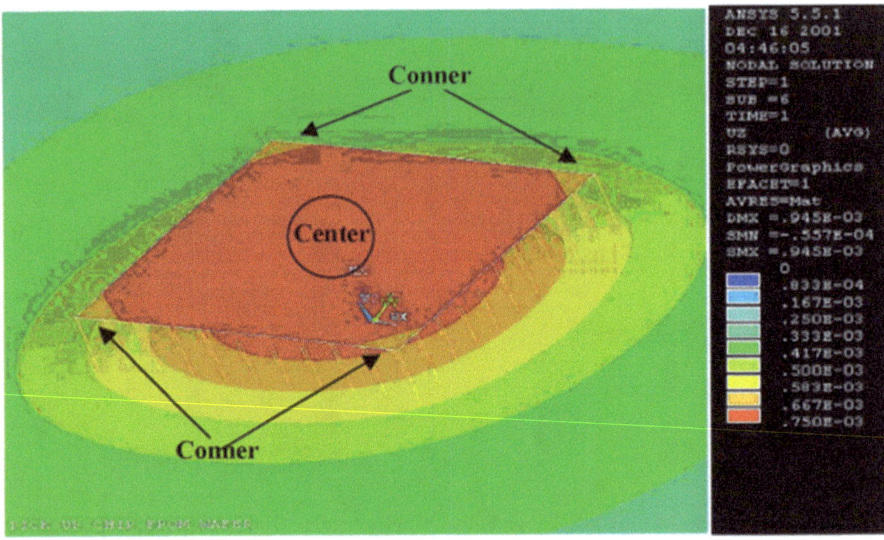

Fig. 5.21 Displacement contour for the IC chip [73] (Color figure online)

needless die ejector is used for thin die (approximately less than 75 μm). New die ejector designs and die pick mechanism need to be explored for stacked die and TSV pick-and-place process such as piston and slider ejector [71–73]. The other crucial parameters that need to be considered while setting process conditions is the electrostatic discharge (ESD) charge as this will severely damage the TSV silicon.

5.8 Conclusions

In summary, this chapter provides an overview of TSV wafer fabrication processes mainly focusing on the kinetics of wafer buckling or wrinkling, thermal sliding debonding, and die preparation processes including laser ablation, wafer saw dicing, and die pick mechanism. Understanding the debonding and die preparation mechanism and fundamental is key to deliver high yield and reliable material for TSV packaging. It must also be noted that the ESD charges not only during the die pick process is another important parameter to be observed and monitored during TSV package fabrication to prevent the circuitry damage.

Acknowledgments The authors would like to acknowledge Dr. Sairam Agraharam of Intel Corp. for valuable discussions. The editors would like to thank Guotao Wang of Intel Corp. for his critical review of this Chapter.

References

1. F. Niklaus, G. Stemme, J.-Q. Lu, R.J. Gutmann, Adhesive wafer bonding. J. Appl. Phys. **99**, 031101 (2006)
2. B.G. Yacobi, S. Martin, K. Davis, A. Hudson, M. Hubert, Adhesive bonding in microelectronics and photonics. J. Appl. Phys. **91**, 6227 (2002)
3. B. Wu, A. Kumar, S. Pamarthy, High aspect ratio silicon etch: a review. J. Appl. Phys. **108**, 051101 (2010)
4. V. Jansen, M.J. de Boer, S. Unnikrishnan, M.C. Louwerse, M.C. Elwenspoek, Black silicon method X: a review on high speed and selective plasma etching of silicon with profile control: an in-depth comparison between Bosch and cryostat DRIE processes as a roadmap to next generation equipment. J. Micromech. Microeng. **19**(033001) (2009)
5. J.P. Gambino, S.A. Adderly, J.U. Knickerbocker, An overview of through-silicon-via technology and manufacturing challenges. Microelectron. Eng. **135**, 73–106 (2015)
6. D. Henry, F. Jacquet, M. Neyret, X. Bailin, T. Enot, V. Lapras, C. Brunet-Manquat, J. Charbonnier, B. Aventurier, N. Sillon, Through silicon vias technology for CMOS image sensor packaging. *Electronic Components and Technology Conference*, 556–562, 2008
7. M. Puech, J.M. Thevenoud, J.M. Gruffat, N. Launay, N. Arnal, P. Godinat, Fabrication of 3D packaging TSV using DRIE. *Design, Test, Integration and Packaging of MEMS/MOEMS* (2008)
8. J. Lu, J. Mcmahon, R. Gutmann, 3D integration using adhesive, metal, and metal/adhesive as wafer bonding interfaces. *MRS Fall Meeting Symposium E*, 2008
9. R. Puligadda, S. Pillalamarri, W. Hong, C. Brubaker, M. Wimplinger, S. Pargfrieder, High-performance temporary adhesive for wafer bonding applications. Mater. Res. Soc. Proc. **970** (2007). 0970-Y04-09
10. M.H. Shungwu, D.L.W. Pang, S. Nathapong, P. Marimuthu, Temporary bonding of wafer to carrier for 3D-wafer level packaging. *Electronics Packaging Technologies Conference*, 405–411, 2008
11. J. Charbonnier, S. Cheramy, D. Henry, A. Astier, J. Brun, N. Sillon, Integration of a temporary carrier in a TSV process flow. *Electronic Components and Technology Conference*, 865–871, 2009
12. S. Pargfrieder, J. Burggraf, D. Burgstaller, M. Privett, A. Jouve, D. Henry, N. Sillon, 3D integration with TSV: temporary bonding and debonding. *Solid State Technology* (March 2009), 38–43
13. D. Bai, X. Zhong, R. Puligadda, J. Burggraf, D. Burgstaller, C. Lypka, J. Verzosa, Edge protection of temporary bonded wafers during backgrinding. ECS Trans. **18**(1), 757–762 (2009)
14. J.P. Gambino, Thin silicon wafer processing and strength characterization. *20th IEEE International Symposium on the Physical and Failure Analysis of Integrated Circuits*, 199–207, 2013
15. M.K. Grief, J.A. Steele Jr., Warpage and mechanical strength studies of ultra thin 150 mm wafers. *IEEE/CPMT Int'l Electronics Manufacturing Technology Symposium*, 190–194, 1996
16. W. Kroninger, F. Mariani, Thinning and singulation of silicon: root causes of the damage in thin chips. *Electronic Components and Technology Conference*, 1317–1322, 2006
17. S. Chen, I.G. Shih, Y.N. Chen, C.Z. Tsai, J.W. Lin, E. Wu, How to improve chip strength to avoid die cracking in a package. *IEEE Inter Society Conference on Thermal Phenomena*, 268–273, 2004
18. V.L.W. Sheng, N. Khan, D. Kripesh, Y.S. UK, Ultra thinning of wafer for embedded module. *IEEE Electronics Packaging Technology Conference*, 837–842, 2006
19. L. Wetz, J. White, B. Keser, Improvement in WL-CSP reliability by wafer thinning. *IEEE Electronic Components and Technology Conference*, 853–856, 2003
20. S. Farrens, Wafer and die bonding technologies for 3D integration. *MRS Fall 2008 Proceedings E*

21. L. Marinier, W.V. Noort, R. Pellens, B. Sutedja, R. Dekker, H.V. Zeijl, Front- to back-side overlay optimization after wafer bonding for 3D integration. Elsevier Science
22. H. Shi, H. Huang, J. Bao, J. Im, P.S. Ho, Y. Zhou, J.T. Pender, M. Armacost, D. Kyser, Plasma altered layer model for plasma damage characterization of porous OSG films. *IEEE Internationa Interconnect Technology Conference*, 78–80, 2009
23. H. Shi, H. Huang, J. Im, P.S. Ho, Y. Zhou, J.T. Pender, M. Armacost, D. Kyser, Minimization of plasma ashing damage to OSG low-k dielectrics. *IEEE International Interconnect Technology Conference*, 1–3, 2010
24. H. Shi, Mechanistic study of plasma damage to porous low-k: process development and dielectric recover, PhD Dissertation, The University of Texas at Austin, 2010
25. C. Cassidy, F. Renz, J. Kraft, F. Schrank, Depth-resolved photoemission microscopy for localization of leakage currents in through Silicon Vias (TSVs). *16th IEEE International Symposium on the Physical and Failure Analysis of Integrated Circuits*, 35–740, 2009
26. Y. Mizushima, H. Kitada, K. Koshikawa, S. Suzuki, T. Nakamura, T. Ohba, Novel through silicon vias leakage current evaluation using infrared-optical beam irradiation. Jpn. J. Appl. Phys. **51**(5S) (2012)
27. N. Ranganathan, D.Y. Lee, L. Youhe, G. Lo, K. Prasad, K.L. Pey, N. Ranganathan, D.Y. Lee, L. Youhe, G. Lo, K. Prasad, K.L. Pey, Influence of Bosch etch process on electrical isolation of TSV structures. IEEE Trans. Compon. Packag. Manuf. Technol. **1**(10), 1497–1507 (2011)
28. T. Nakamura, H. Kitada, Y. Mizushima, N. Maeda, K. Fujimoto, T. Ohba, Comparative study of side-wall roughness effects on leakage currents in through-silicon via interconnects. *3D Systems Integration Conference (3DIC), 2011 I.E. International*, 1–4, 2012
29. J.C. Lin, W.C. Chiou, K.F Yang, H.B. Chang, Y.C. Lin, E.B. Liao, J.P. Hung, Y.L. Lin, P.H. Tsai, Y.C. Shih, T.J. Wu, W.J. Wu, F.W. Tsai, Y.H. Huang, T.Y. Wang, C.L. Yu, C.H. Chang, M.F. Chen, S.Y. Hou, C.H. Tung, S.P. Jeng, D.C.H. Yu, High density 3D integration using CMOS foundry technologies for 28 nm node and beyond. *IEEE International Electron Devices Meeting (IEDM)*, 2.1.1–2.1.4, 2010
30. T. Bandyopadhyay, R. Chatterjee, D. Chung, M. Swaminathan, R. Tummala, Electrical modeling of through silicon and package vias. *IEEE International Conference on 3D System Integration*, 1–8, 2009
31. J. An, K. Moon, S. Lee, D. Lee, K. Yun, B. Park, H. Lee, J. Sue, Y. Park, G. Choi, H. Kang, C. Chung, Annealing process and structural considerations in controlling extrusion-type defects Cu TSV. *IEEE International Interconnect Technology Conference*, 1–3, 2012
32. F. Inoue, T. Shimizu, R. Arima, H. Miyake, S. Shingubara, Electroless deposition of barrier and seed layers for via last Cu-TSV metalization. *IEEE International Meeting for Future of Electron Devices*, Kansai (IMFEDK), 1–3, 2012
33. M. Vagues, Analysing backside chipping issues of the die at wafer saw, in *Partial Fulfillment of MatE 234*, 10 May 2003, 23 p
34. D. Lishan, T. Laserand, K. Mackenzie, D. Pays-Volard, L. Martinez et al., Wafer dicing using dry etching on standard tapes and frames. *International Symposium on Microelectronics*, vol. 2014, no. 1. International Microelectronics Assembly and Packaging Society, 2014
35. H. Mei, R. Huang, H. Mei, R. Huang et al., Buckling modes of elastic thin films on elastic substrates. Appl. Phys. Lett. **90**, 151902 (2007)
36. R. Huang, Kinetic wrinkling of an elastic film on a viscoelastic substrate. J. Mech. Phys. Solids **53**, 63–89 (2005)
37. H. Mei, R. Huang, Concomitant wrinkling and buckle-delamination of elastic thin films on compliant substrates. Mech. Mater. **43**, 627–642 (2011)
38. W.S. Lei, A. Kumar, R. Yalamanchi, Die singulation technologies for advanced packaging: A critical review. J. Vac. Sci. Technol. B **30**(4), 040801-1-27 (2012)
39. M. Privett, F. Murauer, J. Burggraf, S. Pargfrieder, TSV thinned wafer debonding proces optimization. *IWLPC (Wafer-Level Packaging) Conference Proceedings*, 144–148, 2008

40. J. Li, H. Hwang, E. Ahn et al., Laser dicing and subsequent die strength enhancement technologies for ultra-thin wafer. *Electronics Components and Technology Conference*, 761–766, 2007
41. C.M. Dunsky, Laser material processing in microelectronics manufacturing: status and near term-term opportunities. *Proc SPIE 5713, Photon Processing in Mircroelectronics and Photonics IV*, 200–214, 2005
42. S.Y. Luo, Z.W. Wang, Studies of chipping mechanisms for dicing silicon wafers. Int. J. Adv Manuf. **35**, 1206–1218 (2008)
43. K.W. Shi, K.Y. Yow, The characteristics and factors of a wafer dicing blade and its optimized interactions required for singulation high metal stack low-k wafers. *IEEE 15th Electronics Packaging Technology Conference (EPTC 2013)*, 208–212
44. J.W. Lin, M.H. Cheng, Investigation of chipping and wear of silicon wafer dicing. J. Manuf. Process **16**, 373–378 (2014)
45. S. Abdullah, S.M. Yusof, A. Jalar, M.F. Abdullah et al., Step cut for dicing laminated wafer in a QFN package. Solid State Sci. Technol. **16**(2), 198–206 (2008)
46. M. Kumagai, T. Sakamoto, E. Ohmura, Laser processing of dobed silicon wafer by the Stealth Dicing, IEEE 1-4244-1142-4/07 (2007)
47. W.H. Teh, D.S. Boning, R.E. Welsh, Multi-strata stealth dicing before grinding for singulation-defects elimination and die strength enhancement: Experiment and simulation. IEEE Trans. Semicond. Manuf. **28**(3), 408–423 (2015)
48. E. Fogarassy, S. Lazare, *Laser Ablation of Electronic Materials* (Elsevier, North Holland, 1992)
49. J.C. Miller, *Laser Ablation*, Springer Series, 1994
50. R.E. Russo, X.L. Mao, O.V. Borisvo, Laser ablation sampling. Trends Anal. Chem. **17**(8–9), 461–469 (1988)
51. S.S. Mao, *Experimental and Theoretical Studies of Picosecond Laser Interactions with Electronic Materials-Laser Ablation* (University of California, Berkeley, CA, 2000)
52. J.R. Ho, C.P. Grigoropoulos, J.A.C. Humphrey, Computational study of heat transfer and gas dynamics in the pulsed laser evaporation of metals. J. Appl. Phys. **78**, 4606–4709 (1995)
53. G. Callies, P. Berger, H. Hugel, Time-resolved observation of gas-dynamic discontinuities arising during excimer laser ablation and their interpretation. J. Phys. D **794–806** (1995)
54. Y. Zhang, D.Y. Tzou, J.K. Chen, Micro- and nanoscale heat transfer in femtosecond laser processing of metals. Comput. Phys. **1–45** (2015)
55. A. Okano, K. Takayanagi, Laser-induced fluorescene from collisionaly excited Si atoms in laser ablation plume. J. Appl. Phys. **86**, 3964–3972 (1999)
56. X. Zeng, X. Mao, R. Greif, R.E. Russo, Ultraviolet femtosecond and nanosecond laser ablation of silicon: ablation efficiency and laser-induced plasma expansion. *Proc SPIE 2004*, 5448, 1150
57. C. Pasquini, J. Cortez, L.M.C. Silva, F.B. Gonzaga, Laser induced breakdown spectroscopy. J. Braz. Chem. Soc. **18**(3), 463–512 (2007)
58. L.J. Radziemski, D.A. Cremers, *Handbook of Laser Induced Breakdown Spectroscopy* (Wiley, New York, NY, 2006)
59. G.M. Weyl, *Laser-Induced Plasmas and Applications* (Marcel Dekker, New York, NY, 1989)
60. C.A. Sacchi, Laser-induced electric breakdown in water. J. Opt. Soc. Am. B **8**(2), 337–345 (1991)
61. M.S. Amer, M.A. El-Ashry et al., Femtosecond versus nanosecond laser machining: comparison of induced stresses and structural changes in silicon wafers. Appl. Surf. Sci. **242**, 162–167 (2005)
62. A.T. Cheung, Dicing advanced materials for microelectronics. *International Symposium on Advanced Packaging Materials: Processes, Properties, and Interfaces*, 149–152, 2005
63. K.W. Shi, Y.B. Kar et al., Optimization of wafer singulation process on copper/low-k materials for semiconductor device assembly. Aust. J. Basic Appl. Sci. **8**(22), 6–11 (2014)

64. T.J. Su, C.L. Chiu, Y.F. Chen et al., Improvement of wafer saw film burr issues. Int. J. Model. Optim. **5**(5), 345–348 (2015)
65. Disco Corporation, *The Cutting Edge: Technical Newsletter*, no 5, 2002
66. I. Weisshaus, D. Shi, U. Efrat, Wafer dicing. *Solid State Technology: Insight for Electronics Manufacturing*
67. Z.Y. Zhang, F.W. Huo et al., Grinding of silicon wafers using an ultrafine diamond wheel of a hybrid bond material. Int. J. Mach. Tools Manuf. **51**(1), 18–24 (2011)
68. Z.Y. Zhang, Y.Q. Wu et al., Phase transformation of single crystal silicon induced by grinding with ultrafine diamond grits. Scripta Mater **64**(2), 177–180 (2011)
69. S. Malkin, C.S. Guo, *Grinding Technology: Theory and Applications of Machining with Abrasives*, 2nd edn. (Industrial Press, New York, 2008), pp. 54–79
70. H. Zhou, S. Qiu, Y. Huo, N. Zhang, High-speed dicing of silicon wafers conducted using ultrathin blades. Int. J. Adv. Manuf. Technol. **66**, 947–953 (2013)
71. P.J. Kim, Y.D. Ha, H.H. Park, J.H. Park, Development of die-bonder with multi and matrix picker and placer to increase production capacity. *Proceedings of the World Congress on Engineering and Computer Science*, vol. 1 (2012), 978–988
72. Z. Liu, Reliable peeling of ultrathin die with multineedle ejector. IEEE Trans. Compon. Packag. Manuf. Technol. **4**(9), 2156–3950 (2014)
73. T.H. Cheng, C.C. Du, C.H. Tseng, Study in IC chip failure during pick-up process by using experiemntal and finite element methods. J. Mater. Process. Technol. **172**, 407–416 (2006)
74. N. Saiki, K. Inaba, K. Kishimoto, H. Seno, K. Ebe, Study on peeling behavior in pick-up process of IC chip with adhesive tape. J. Solid Mech. Mater. Eng. **4**(7), 1051–1060 (2010)

Chapter 6
Direct Cu to Cu Bonding and Other Alternative Bonding Techniques in 3D Packaging

Tadatomo Suga, Ran He, George Vakanas, and Antonio La Manna

6.1 Introduction

Reliable, high-throughput, low-cost, and low-temperature bonding processes have always been in high demand in industrial, electronics, and semiconductor environments. Multiple techniques and processes have been demonstrated in different domain applications, with distinct advantages and disadvantages that we will explore in this chapter. With the trend towards further device scaling and applications-driven roadmaps, three-dimensional (3D) integration has evolved with bandwidth, power, performance, and form factor advantages over planar, side-by-side, electronic packaging. To enable 3D technology, high-density interconnects are required (sub-40 μm pitch) and stacking/bonding processes that are reliable and with high-throughput are in highly demand. Cu–Cu bonding is a promising candidate for 3D interconnects, since Cu is a well-understood metallurgy for back-end-of-line (BEOL) layers and for through-silicon vias (TSVs) and offers mechanical, electrical, and thermal advantages. Comparing solder-based pillar/bump processes with solder reflow or thermo-compression bonding in temperature range of 250–275 °C, the Cu–Cu bonding process should target lower temperature range (from room temperature to 250 °C).[1]

[1] Process time targets depend on stacking process, e.g., die-to-die (D2D) = chip-to-chip (C2C) and die-to-wafer (D2W) = chip-to-wafer (C2W) or wafer-to-wafer (W2W) processes, equipment/process throughput and manufacturing models. For Cu-Cu process to be viable in manufacturing, it needs to offer new capabilities and/or lower cost versus established solder-based processes.

T. Suga (✉) • R. He
The University of Tokyo, Tokyo, Japan
e-mail: suga@pe.t.u-tokyo.ac.jp

G. Vakanas
Intel Corporation, Chandler, AZ, USA

A. La Manna
IMEC Consortium, Leuven, Belgium

Y. Li, D. Goyal (eds.), *3D Microelectronic Packaging*, Springer Series in Advanced Microelectronics 57, DOI 10.1007/978-3-319-44586-1_6

The study of Cu–Cu bonding started at universities around 2000 [1, 2], soon after the announcement of Cu interconnects for integrated circuits by IBM in the late 1990s. For high-density interconnects, Cu–Cu bonding has been demonstrated since 2006 by Morrow et al. of Intel (with smallest Cu pads of 5 × 5 μm^2) [3] and Suga's group at the University of Tokyo (with Cu pads of 3 μm in diameter) [4], respectively. For 3D stacking in case of extremely thinned dies with 10 μm pitch TSVs, this has been demonstrated in 2006 by Swinnen et al. of IMEC [5]. Cu–Cu bonding for 3D prototypes such as memories, sensors, processors, and memory/processor stacks has been demonstrated by Tezzaron [6] and Intel [3]. Cu/SiO_2 hybrid bonding of Ziptronix (now owned by Tessera) has also been demonstrated on 3D stacking by Fermilab [7, 8], Sony [9], and Tezzaron/Novati in recent years.

In this chapter, we introduce the pros and cons of Cu–Cu bonding and stacking/bonding schemes for different applications. We review various methods of Cu–Cu bonding: (a) thermo-compression bonding (diffusion bonding), (b) Cu–Cu bonding with passivation capping layers, (c) surface activated bonding (SAB), and (d) alternative bonding methods (Cu/dielectric hybrid and Cu–Cu insertion bonding). We also review the effects of surface activation, surface microstructures and characteristics, and surface passivation for Cu–Cu bonding to understand how the bonding behavior depends on Cu surface cleanness, diffusion, temperature, compression pressure, and bonding atmosphere. Lastly, we summarize state of the art and recommendations for future directions.

6.2 Solder-Based vs. Solder-Less Bonding: Pros and Cons

Solder-based bonding is prevalent in 3D interconnects usually employing one-sided scheme of solder microbump (e.g., Cu/x/solder, where x would be a diffusion barrier like Ni and solder would be SnAg, SnCu, or other plated binary solder) bonded on metal pad (e.g., Cu with a passivation layer or other capping layer to prevent oxidation). The key advantages of solder-based schemes are process robustness given that solder is malleable and more forgiving in terms of bump height uniformity and co-planarity. However, a solder-based process is time-consuming and has higher cost due to the additional material stack up (usually involving photoresist, lithographic exposure and development, plating, etch, and clean/strip processes). Moreover, a solder-based process is more prone to handling damage of solder bumps. Also, the process temperatures for solder-based processes are driven by the melting point of the solder (for instance, 232 °C for 100 % Sn and ~221 °C for eutectic SnAg). The solder-based bonding forms intermetallic compounds (IMCs) and results in an IMCs/solder/IMCs structure, as shown in Fig. 6.1a. By controlling the thickness of the solder, bonding can be achieved through solid–liquid (solid–liquid interdiffusion [SLID] bonding) or solid-state (solid-state interdiffusion [SSID] bonding) reactions between Cu and solder, consuming all the solder and forming IMCs at the bonding interface, as shown in Fig. 6.1b. Comparing to the conventional solder reflow and SLID bonding, the SSID bonding

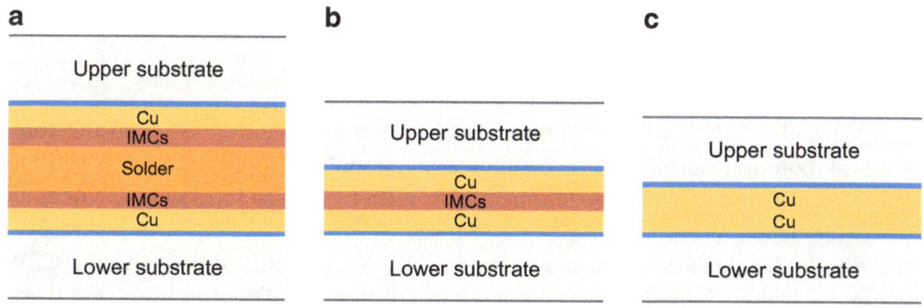

Fig. 6.1 Schematic diagram of the bonded structures using (**a**) soldering, (**b**) SLID/SSID bonding, (**c**) Cu–Cu bonding methods

is performed at lower temperatures (below the melting point of solder) but under higher bonding pressures (~50–150 MPa).

Solder-based bonding limits the electrical conductivity, reliability (since cracks are prone to occur at solder-IMCs interface or inside IMCs), and the minimum pitch of connects. Direct Cu–Cu bonding without use of solder and formation of IMCs at the bonding interface, as shown in Fig. 6.1c, has been developed to solve these concerns. Comparing to other direct metal bonding such as Al–Al bonding, Cu–Cu bonding interconnects promise better electrical conductivity, power consumption, resistive-capacitive (RC) delay, and electromigration resistance. In addition, Cu interconnects also provide excellent heat dissipation and thermal reliability, which meets the requirements of some applications such as power electronics with an operation temperature as high as 250 °C [10].

Thus, solder-less, Cu–Cu bonding technology has the prospects to simplify processing, lower costs, and result in higher reliability and performance than solder-based bonding. A main challenge of Cu–Cu bonding is that Cu surface is readily oxidized by O_2 and H_2O during exposure to air, and the resulted thick (>10 nm) Cu oxides (CuO and Cu_2O) prevent bonding formation at below 300 °C. Depending on the bonding environment (vacuum, inert, reducing, or ambient atmosphere), surface preparation and passivation, and handling of warpage and flatness, Cu–Cu bonding processes have been demonstrated at temperatures lower than the usual lead-free solder melting temperatures, down to 100–150 °C range with some demonstrations also down to room temperature. Variants of the Cu–Cu bonding processes include: thermo-compression bonding (or called diffusion bonding), surface activated bonding (SAB), Cu–Cu bonding with assist processes (surface cleaning, treatment, and passivation with noble and non-noble capping layers), and hybrid bonding and insertion bonding. We will revisit such process conditions in more details in subsequent sections of this chapter.

6.3 Stacking and Bonding Schemes, Technologies, and Applications

Stacking and bonding schemes and technologies are classified based on the form factor of bonding surfaces, the geometry and topology of the active electronic devices and packages and target applications. With respect to form factor, processes are distinguished as chip-to-chip (C2C) (or die-to-die, D2D), chip-to-wafer (C2W) (or die-to-wafer, D2W), and wafer-to-wafer (W2W). Benefits of the C2C and C2W bonding schemes include use of known-good-dies (KGDs) for high yield and integration of chips of different sizes with high flexibility. However, the disadvantages are low throughput (especially C2C) and low alignment accuracy. W2W enables high-throughput and high alignment accuracy; however, it suffers from yield loss due to lack of KGDs and inflexibility with different sizes of stacked layers. To solve the drawbacks, multichip-to-wafer (mC2W) bonding has been developed for bonding of multiple chips temporary assembled on a carrier wafer (e.g., through liquid assisted self-assembly [11]) onto a wafer.

Bonding schemes can also be distinguished based on how the active surfaces (having devices) are brought into contact, e.g., face-to-face (F2F), and back-to-face (B2F) (or face-to-back, F2B), and back-to-back (B2B). The face side refers to the front side of wafers, where active devices are fabricated; backside refers to the opposite side of the face side and is usually obtained after thinning and insulation/metallization. In case of passive interposer that has no active devices, the surface firstly processed is commonly termed as face side and the opposite backside. F2F bonding uses Cu pads on the top metal layer above the active devices, and wafers are brought into contact between the face sides of both wafers, as illustrated in Fig. 6.2a. F2F bonding is widely used for two-layer 3D stacking. Morrow and coworkers at Intel employed F2F bonding for 3D stacking of wafers having active devices such as 65-nm MOSFETs and 4-MB SRAMs using Cu bonding pads with size ranging between 5 μm × 5 μm and 6 μm × 40 μm [3]. Subsequently, one of the F2F bonded wafers can be thinned and processed for insulation/metallization from the backside for additional bonding through either B2F or B2B bonding schemes. The B2F bonding brings wafers into contact between the backside and face side of the two wafers, as illustrated in Fig. 6.2b; the B2B bonding brings wafers into contact between backsides. A wafer can also be thinned and backside processed with help of temporary bonded carrier before the B2F bonding, as illustrated in Fig. 6.2c, or B2B bonding. The B2F and B2B bonding is usually used for 3D stacking of more than two layers together with the F2F bonding.

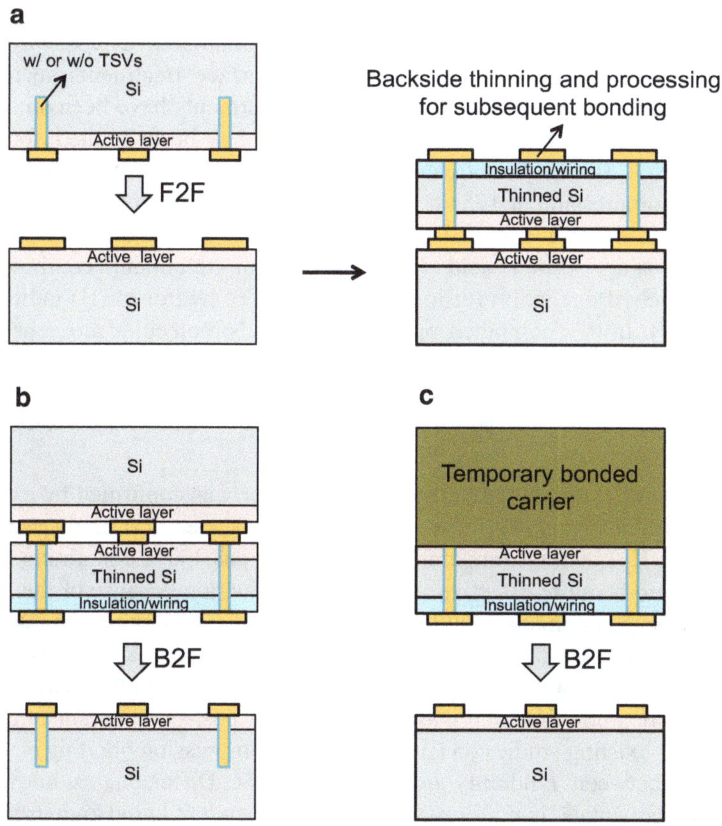

Fig. 6.2 Stacking schemes: (**a**) F2F before wafer backside processing, (**b**) B2F for bonding of third layer, (**c**) B2F with temporary bonding

6.4 Thermo-Compression Bonding (Diffusion Bonding): Material Fundamentals and Microstructure Effects

The basis of Cu–Cu thermo-compression bonding is interdiffusion and self-diffusion at elevated temperatures and under an external compression, depending on the cleanliness of the mating Cu surfaces and potentially additional passivation or capping layers. The compression pressure, which is mainly dependent on the surface topography and roughness, can be in order of 100–150 MPa for as-plated Cu films/pillars, and <2.5 MPa for thin smooth Cu films or CMP (Chemical-Mechanical Polishing) smoothened Cu films. Typically, Cu–Cu thermo-compression bonding is performed at 300–400 °C in vacuum or protecting/reducing gas environment, followed by a post-bonding annealing at 300–400 °C to improve the bond strength [12]. However, such high process temperatures may cause large thermal expansion and stress, which degrade or even damage thermal- and stress-sensitive materials and devices. Moreover, formation of interfacial voids were observed when bonding temperature is higher than 300 °C [13].

In order to lower the bonding temperature, surface treatments such as wet chemical cleaning and gas/vapor-phase thermal treatments have been studied. For instance, forming gas (H_2 + Ar or N_2) treatment has been studied for Cu–Cu bonding at below 200 °C. After treatment at 175 °C for 30 min, wafers were bonded with thermo-compression at 175 °C for 30 min followed by 200 °C 1 h annealing [14, 15]. W. Yang and coworkers at the University of Tokyo studied Cu–Cu bonding by using formic acid (HCOOH) vapor treatment combined with Pt-catalyzed dehydrogenation for in situ generation of hydrogen (H) radicals/molecules [16, 17]. In this combined process, H radicals/molecules are generated by heating Pt folis exposed to HCOOH vapor through the following reaction.

$$HCOOH \xrightarrow{Pt} CO_2 + 2H \tag{6.1}$$

The generation of H contained in the HCOOH vapor was confirmed by ex situ gas chromatography analysis [16]. The detected H_2 peak area was almost same with standard 0.1 % H_2 spectra. Compared to the forming gas, the H-containing HCOOH vapor appears to be more effective for Cu oxide reduction because of the presence of highly reactive H radicals, resulting in strong Cu–Cu bonding (shear strength of above 10 MPa) at 200 °C. As summarized in Table 6.1, comparing to the forming gas treatment, the HCOOH vapor treatment enables Cu–Cu bonding with very short treatment and bonding time and without the need of post-bonding annealing.

Most of the existing studies on Cu–Cu thermo-compression bonding were based on diffusion between randomly oriented Cu films. Diffusion, as known from materials fundamentals, is not only temperature dependent but also microstructure dependent. Cu–Cu bonding by using Cu layers that have special microstructures, such as Cu nanorod array [18], Cu particles [19], and highly (111)-oriented nanotwin Cu films [20], has also been investigated in order to lower the bonding temperatures. Table 6.2 summarizes data of Cu surface diffusion coefficients on

Table 6.1 Thermal budget of Cu–Cu bonding by forming gas and formic acid vapor surface treatments

	Cu treatment	Bonding and annealing
Forming gas (H_2 + Ar or N_2) [14, 15]	175 °C 30 min	175 °C 30 min + 200 °C 1 h anneal
Formic acid vapor [16, 17]	200 °C 1–10 min	200 °C 5 min without annealing

Table 6.2 Cu surface diffusion coefficients on three crystal planes as function of temperature (Reproduced from [20] under the Creative Commons Attribution 4.0 International License)

	Cu surface diffusivity, $D_{surface}$ (m^2/s)		
Temperature (°C)	(111)	(100)	(110)
150	6.85×10^{-10}	2.15×10^{-14}	6.61×10^{-16}
200	9.42×10^{-10}	1.19×10^{-13}	5.98×10^{-15}
250	1.22×10^{-9}	4.74×10^{-13}	3.56×10^{-14}
300	1.51×10^{-9}	1.48×10^{-12}	1.55×10^{-13}

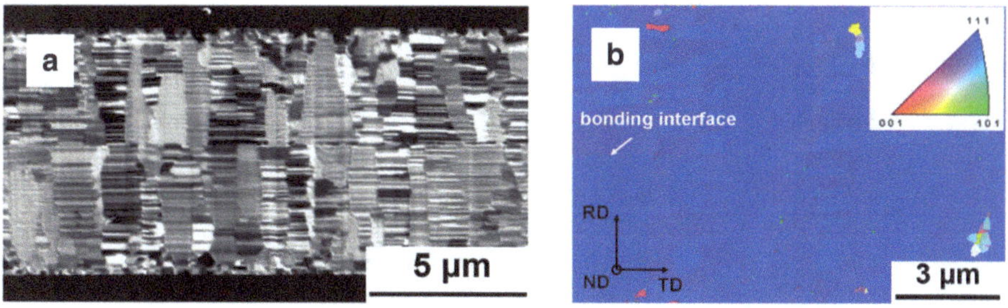

Fig. 6.3 Bonding between two electroplated (111)-oriented Cu films at 200 °C for 30 min: (**a**) TEM cross-sectional image and (**b**) electron backscattered diffraction (EBSD) orientation image (Reproduced from [20] under the Creative Commons Attribution 4.0 International License.)

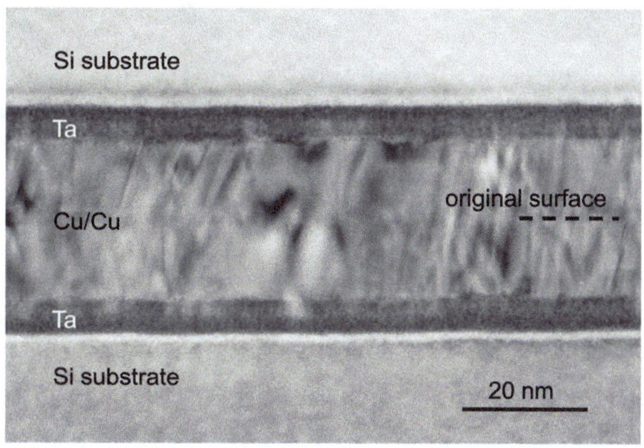

Fig. 6.4 TEM cross-sectional image of nanocrystalline Cu films bonded at room temperature in UHV immediately after sputter deposition. The original interface is invisible owing to Cu self-diffusion (Reproduced with permission from [21]; Copyright © 2010, AIP Publishing LLC.)

three crystal planes as function of temperature in the range of 150–300 °C [20]. Because of the larger surface diffusivity on (111) plane than (100) and (110) planes, the Cu–Cu bonding temperatures can be reduced by using (111)-oriented Cu surfaces. C.-M. Liu and coworkers studied the Cu–Cu bonding of highly (111)-oriented nano-twin Cu films and good bonding was obtained under thermo-compression at 150–250 °C for 10–60 min [20]. Figure 6.3a presents TEM image of the cross-sectional Cu–Cu interface at 200 °C for 30 min, showing a void-free bonding interface. Figure 6.3b presents an electron backscattered diffraction (EBSD) orientation image of the bonding interface, showing all the Cu grains near the bonding interface were oriented in the <111> direction. The surface diffusion also occurs even at room temperature if the surfaces are free of oxides. T. Shimatsu et al. demonstrated Cu–Cu bonding between clean nanocrystalline Cu films in ultra-high-vacuum (UHV) immediately after sputtering deposition [21]. TEM result showed that the bonding interface become invisible, as shown in Fig. 6.4, which is attributed to the rapid Cu self-diffusion at room temperature. The Cu nanocrystalline thin films were also successfully bonded in air at temperature as low as 100 °C [22].

Table 6.3 Various capping layers used for Cu surface passivation and the bonding temperatures

Capping layer	Bonding temperature (°C)	Features
SAM	250−300	Pre-bonding annealing is needed for SAMs desorption [24–28]
Ti	160–180	Ti diffuses away from the interface [29]
		Thickness of Ti is important for passivation, surface roughness, and TiO_x content (optimal 3 nm) [30]
Pd	150	Pd diffuses away from the interface
		Lower contact resistance than Ti passivation [31]
Au	250	Poor bond strength and formation of IMCs [25]

6.5 Passivation with Capping Layers (SAMs and Metals)

Since Cu is readily oxidized by O_2 and H_2O if exposed to air, Cu surface passivation with capping layers has been studied to protect Cu surfaces from oxidation and to improve the Cu–Cu bonding quality. The typical surface finishes such as electroless nickel/immersion gold (ENIG), immersion silver (ImAg), immersion tin (ImSn), and organic solderability preservatives (OSP) were mainly developed for solder-based bonding. Researchers have also developed the ENIG capping for solder-less chip-to-substrate assembly by using thermo-compression at <200 °C and under ~300 MPa for 2.5D packaging [23]. This section will focus on some emerging capping layers including organic self-assembled monolayers (SAMs) and metals such as sputtered Ti or Pd and electroless Ni- or Co-based alloys, as summarized in Table 6.3.

The SAMs are used as temporary capping layers for Cu film surface passivation. C.S. Tan and coworkers studied the SAM of alkane-thiol for Cu–Cu bonding at 250–300 °C [24–27]. The wafers were immersed into the solution of 1–hexanethiol [$CH_3-(CH_2)_4-CH_2-SH$, C-chain length of C6] after Cu film deposition. The thiol ($-SH$) head groups bind to Cu surface and form a densely packed SAM cap; the methyl ($-CH_3$) tail groups make the Cu surface hydrophobic [27]. After 3–5 days of storage, the SAM was desorbed with annealing at 250 °C for 10 min in vacuum or N_2 ambient to expose the Cu surfaces for bonding. The exposed Cu surface remained hydrophobic and clean for strong bonding, with shear strength of ~60 MPa comparing to ~10 MPa without use of the SAM.

Researchers at IMEC Consortium Belgium applied thiol-based SAMs as passivation of electroplated Cu pads/pillar/bumps in 3D interconnects. In reference [28], SAM types with different C-chain (length of C3, C10, and C18) were studied in flat and 3D patterned/microbumped samples at 40 μm pitch. Cyclic voltammetry was used to compare oxidation resistance qualitatively on standalone samples. After die-to-die stacking electrical probing was used to measure full daisy chain and sub-chain continuity and electrical resistance. It was found that the C18-SAM deposited in liquid phase (up to 24 h of immersion) on electro-deposited Cu demonstrated better layer stability and lower oxidation compared to the

C10-SAM, which was in turn better than the C3-SAM. C18-SAM passivation degrades fast in ambient at room temperature conditions regardless of immersion time (based on comparison of 1 day vs. post-1 week cyclic voltammetry). Thus, "sit times" of materials in the manufacturing process queue should be of the order of days; otherwise, materials need to be stored in vacuum/inert atmosphere and "time critical loop" needs to be set up and controlled. Based on electrical probing of two-die stacks, C18-SAM passivation resulted in lower electrical resistance ~500 Ω compared to three times higher resistances for C10 SAM stacks with order-of-magnitude wider variance. C18-SAM samples also yielded electrically ~20 % higher in terms of connected daisy chains compared to C10 SAM. It was also found that microwave plasma cleaning prior to SAM deposition is more effective than citric acid cleaning, based on voltammetry on flat samples and electrical resistance and daisy chain continuity of two-die stacks.

Unlike SAMs that are removed before bonding, metal capping layers are present and involved in interfacial reaction during the bonding. Y.-P. Huang and coworkers studied Cu–Cu bonding by using sputtered Ti and Pd capping layers [29, 31]. Due to lower activation energy at the surface, Cu has a tendency to diffuse toward the bonding interface. In contrast, $Ti(TiO_x)$ diffuse toward Si substrate.[29] This diffusion behavior results in a $Ti(TiO_x)$/Cu–Cu/$Ti(TiO_x)$ bonded structure. Similar behavior was also found by using the Pd capping layer. Results of TEM images and EDX composition profiles of the bonded structure show that the interface mainly contains Cu and the oxygen content using Pd capping is smaller than that using Ti capping [29, 31]. Electrical measurements also exhibit much lower contact resistance with Pd than Ti [31]. A. K. Panigrahi et al. also investigated the influence of the thickness of the Ti capping layers on passivation and bonding results. They demonstrated that 3 nm Ti capping layer is effective for passivation of Cu surface with small surface roughness and low TiO_x content [30]. The Au capping layer is not suitable for Cu–Cu bonding because of the poor bond strength, which may be owing to formation of IMCs in the absence of barriers between Cu and Au layers [25].

Work at IMEC has demonstrated an approach to use non-noble capping layers (e.g., electroless NiB and CoB) to passivate Cu surfaces and enable better bonding [32]. The B in the non-noble capping layers fits into the spaces in the Ni or Co lattice and acts as an interstitial element thereby preventing oxidation of the Ni or Co present in the capping layer. As such, NiB or CoB alloy with an atomic concentration percentage of B from 10 to 50 % behaves as a noble metal for surface passivation but at a lower cost.

6.6 Surface Activated Bonding (SAB) Processes

The SAB method is a room-temperature bonding method by using pre-bonding surface activation in UHV. The origin of the SAB dates back to experiments on adhesion in UHV conducted by NASA in the 1970s. Professor Suga's group in the

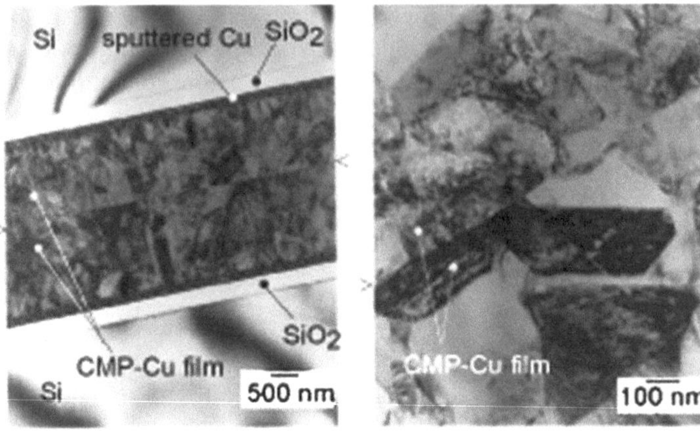

Fig. 6.5 TEM images of the interface between Cu films bonded at room temperature (Reproduced with permission from [34]; Copyright © 2005, Springer Science + Business Media, Inc.)

University of Tokyo made considerable progress in UHV bonding in the 1980s, and the technique was extended to apply to homo-/heterogeneous bonding between metals, Si/III–V semiconductors, glasses, and polymers.

In addition to the diffusion and reactions at elevated temperatures, there are also always chemical interactions between atoms on the mated clean surfaces. The origin of these interactions is the cohesive and adhesive energy of solids, which enables solid-state bonding even if there is no high-temperature reaction. The SAB method removes surface oxides and contaminants by utilizing a pre-bonding surface activation by Ar atom beam bombardment in UHV, which prevents rapid reoxidation and recontamination of the surfaces prior to bonding. Therefore, the SAB in principle can enable bonding of various materials (metal-to-metal, metals-to-ceramics, metals-to-semiconductors, and semiconductors-to-semiconductors) at room temperature, i.e., without either heating or post-bonding annealing [2, 33, 34]. Figure 6.5 shows the TEM images of Cu–Cu bonding interface prepared by using SAB at room temperature. The visible interface implies that no significant diffusion and Cu grain growth occurred across the bonding interface. It is also indicated that the SAB is less dependent on the Cu diffusion and microstructures of the Cu grains. The SAB avoids formation of oxides at the bonding interface and thermal-related issues such as thermal stress, thermal expansion, and expansion-induced bonding misalignment.

Based on the SAB method, T. Suga proposed the concept of bump-less interconnect in 2000 [35, 36], which is defined as two layer structures bonded directly with metallic interconnections and insulating layer in a coplaner plane, as illustrated in Fig. 6.6. The layer structures represent either combinations of LSI chip and substrate, two different devices (RF, digital, analog, logic, memory, etc.) or wiring layers and device layers. Especially, this structure is expected to be applied to (1) bonding of wiring layer for global interconnections and device layer on chip, (2) improvement of yield by dividing the wiring layer, (3) improvement of signal

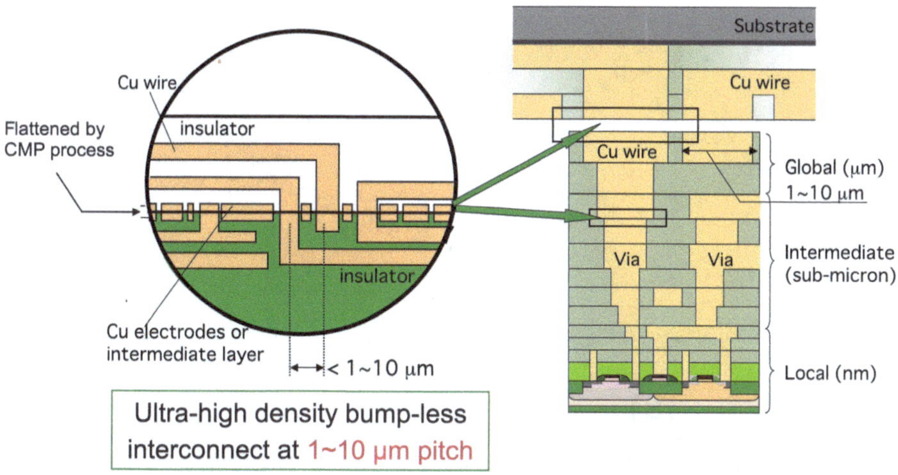

Fig. 6.6 Schematic representation of bump-less direct bonding

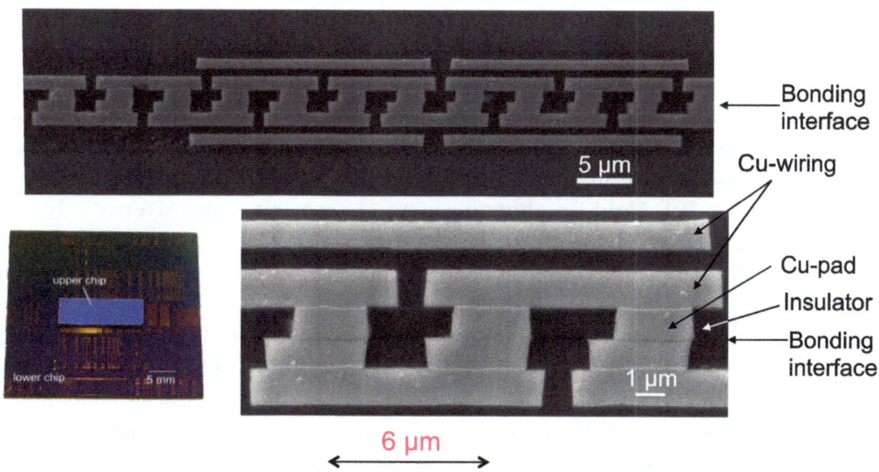

Fig. 6.7 SEM cross-sectional image of bump-less Cu–Cu bonded structure with 6 μm pitch

transmission rate by adopting adequate transmission line structure and shortening distance between devices, (4) interconnecting two different device layers, such as separation of analog and digital devices, and (5) bonding of optical device to Si substrate, and bonding for hetero-junction of semiconductors [35]. It is a generalized concept for what is often called today as (Cu/dielectric) hybrid bonding.

The SAB method demonstrated chip-scale bump-less Cu–Cu bonding interconnects of 1,000,000 electrodes at 3 μm critical dimension and 6 μm pitch, as shown in Fig. 6.7 [4, 37]. To date, the SAB method has also enabled high-volume industrial applications for metal laminates and MEMS packaging.

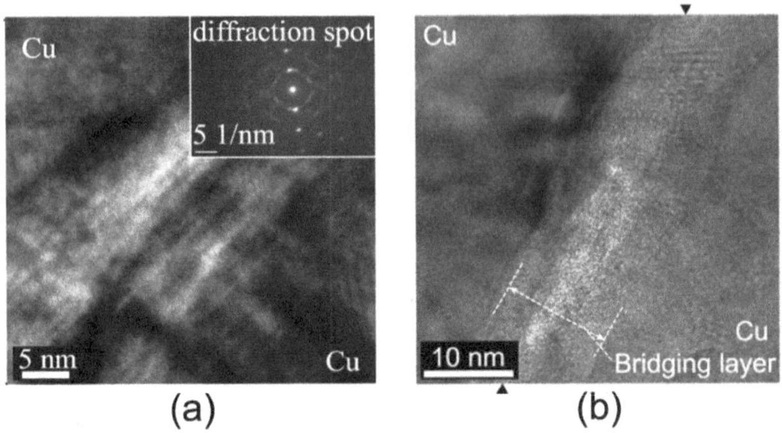

Fig. 6.8 TEM images of the Cu–Cu bonding interface prepared by the modified SAB processes at 150 °C: (**a**) modified diffusion bonding using dry O_2 (Reproduced with permission from [38]; Copyright © 2009, The Japan Society of Applied Physics) and (**b**) vapor-assisted SAB using humid N_2 gas (Reproduced with permission from [39]; Copyright © 2012, TMS)

SAB was also modified (namely modified diffusion bonding [38] and vapor-assisted SAB [39]) for Cu–Cu bonding in ambient air at 150 °C [38-41]. Figure 6.8 shows the TEM images of Cu–Cu bonding interfaces obtained with the two processes, by using Ar beam irradiation followed by dry O_2 or humid N_2 exposure, respectively [38, 39]. Void-free bonding with O-containing interlayer of ~15 nm thickness was achieved by both methods. The bonded interconnects exhibit low resistances that are only slightly higher than that obtained by using conventional SAB in UHV, as compared in Fig. 6.9 [39].

SAB also comes with disadvantages, mainly its difficulty of bonding some ionic materials to each other, like glass and silicon dioxide (SiO_2). The reason for this is still not clear, but it is assumed that the surface of ionic materials is spontaneously polarized at different levels by ion beam bombardment, which is performed prior to bonding. To overcome the challenge of traditional SAB, a modified method was developed. In this approach, the surfaces to be bonded are sputter cleaned by Ar beam and simultaneously deposited with Fe and subsequently Si layer. The metallic/silicon thin layer may shield the surface polarity of the ionic materials and enable room temperature bonding of SiO_2, glass, and various single crystalline wafers and polymer films at room temperature with high bond strength. Industrial application that were enabled by this modified SAB method are sealing of glass and polymer devices such as organic electro-luminescent display (OELD) or lightening devices since there is no other suitable method for good sealing agent against permeation of water and oxygen from the atmosphere into those devices.

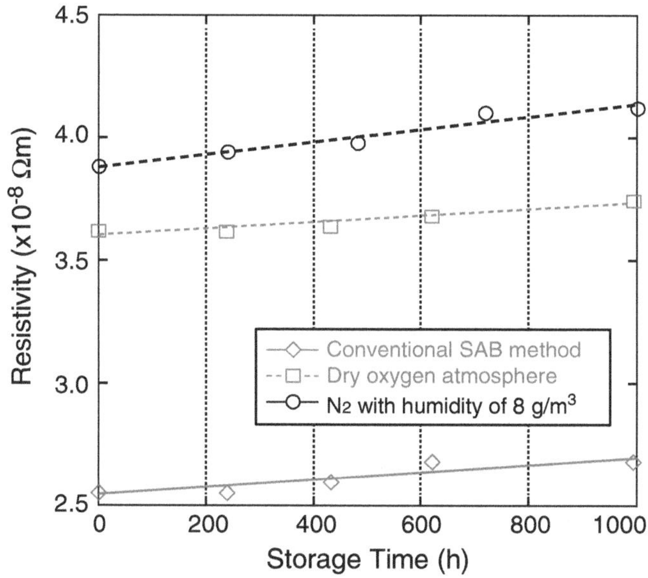

Fig. 6.9 Electrical resistivity of Cu–Cu bonding interconnects by using conventional SAB and modified SAB with dry O atmosphere (modified diffusion bonding [38]) and humid N_2 atmosphere (vapor-assisted SAB [39]) after high-temperature storage testing at 150 °C (Reproduced with permission from [39]; Copyright © 2012, TMS.)

6.6.1 Cu/Dielectric Hybrid Bonding

Bump-less interconnect promises high-density direct vertical electrical interconnects with very short length between 3D stacked chips or wafers [35, 36]. At the same time, the dielectric passivation area (e.g., oxide/nitride, polymer adhesives) should be bonded so to enhance the bond strength, heat dissipation, and Cu corrosion protection with a seamless interface. Although such hybrid bonding of metals and dielectric materials has been investigated by for instance, Au/adhesive, Au/SiO_2, and Cu-Sn/adhesive combinations, Cu/SiO_2 and Cu/adhesive hybrid bonding are the most promising options for high electrical performance.

The methods used for Cu/SiO_2 and Cu/adhesive hybrid bonding are different because of the different mechanisms of SiO_2–SiO_2 bonding and polymer adhesive bonding. The SiO_2–SiO_2 bonding is typically based on the hydrophilic bonding mechanism, which requires hydrophilic surface modification prior to bonding; the adhesive bonding usually uses thermosetting polymer adhesives, for instance, benzocyclobutene (BCB), polyimide (PI), and polybenzoxazole (PBO), and the bonding is based on thermo-compression bonding.

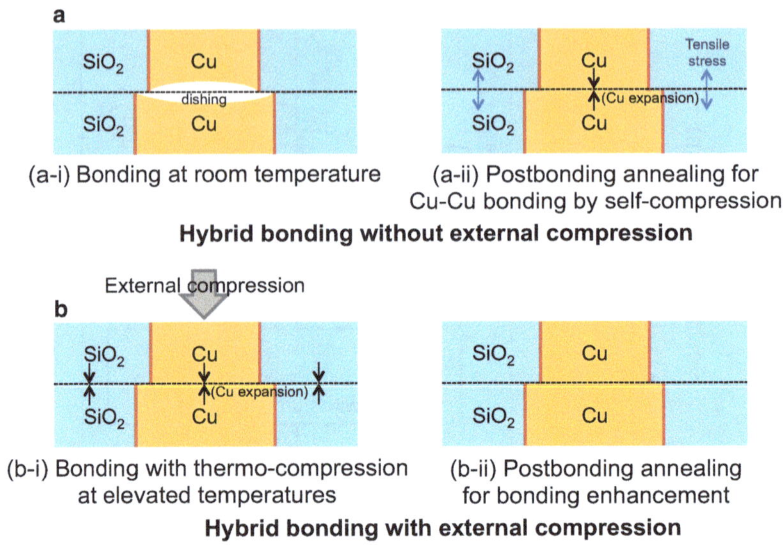

Fig. 6.10 Cu/SiO_2 hybrid bonding (**a**) without and (**b**) with external compression

6.6.2 Cu/SiO_2 Hybrid Bonding

Cu/SiO_2 hybrid bonding can be done through hydrophilic surface modification of the chips or the wafers followed by bonding and then post-bonding annealing. Because of possible CMP dishing of the Cu surface, chips or wafers can be initially bonded only through the SiO_2–SiO_2 bonding at room temperature, as shown in Fig. 6.10a-i. Post-bonding annealing (typically at 200–400 °C) is required to enhance the SiO_2–SiO_2 bonding and to induce thermal expansion of Cu for Cu–Cu bonding, as shown in Fig. 6.10a-ii. The as-bonded SiO_2–SiO_2 strength must be sufficiently high to sustain the stress induced by Cu thermal expansion. Cu/SiO_2 hybrid bonding can be also conducted with thermo-compression, in which the external compression is applied on both the Cu–Cu and SiO_2–SiO_2 interfaces during bonding (Fig. 6.10b-i), before post-bonding annealing to further enhance the bonding (Fig. 6.10b-ii).

Table 6.4 compares some methods that have been investigated for Cu/SiO_2 hybrid bonding. The efficiency of plasma activation for SiO_2–SiO_2 and Cu–Cu bonding is still questionable although it has been shown to be very effective for hydrophilic Si–SiO_2 bonding. By using the plasma activation, the bond strength of the SiO_2–SiO_2 pairs is significantly lower than the Si–SiO_2 pairs [42, 43]. For metal bonding, the plasma activation has been developed for low-temperature bonding of solder [44], Au films [45], and Au particles [46]. However, the benefit still remains unclear for Cu–Cu bonding at below 300 °C. The residual H_2O and O_2 in the plasma chamber, whose pressure is typically in the range of 0.1–100 Pa, could oxidize the Cu surfaces even though the Ar or N_2 plasma can be used. M. Park et al. reported

Table 6.4 Comparison of bonding methods for Cu/SiO_2 hybrid bonding

Methods	SiO_2–SiO_2	Cu–Cu
Plasma activation bonding	1. Significantly lower bond strength than Si–SiO_2 and Si–Si bonding at <300 °C 2. Difficult to remove interfacial H_2O	1. Cu oxide [47] and nitride [14, 48] formation 2. Surface roughening [47] 3. Sheet resistance increase [47] 4. Post-activation storage decreases the bond strength [48]
Direct Bond Interconnect (DBI®) [49, 50]	1. High bond strength at 200 °C 2. Small amount of interfacial H_2O [51, 52]	Low resistances are obtained at low temperatures, but technical details are rarely disclosed
Special CMP treatment of Leti [53–55]	1. Low bond strength at <300 °C 2. Difficult to remove interfacial H_2O 3. Bond strength at low temperature depends on the film deposition process [56]	1. Ultra-smooth surface is essential 2. Bond strength at room temperature depends on the Cu film deposition method [57] 3. Post-activation storage (e.g., more than 2 h) decreases the Cu–Cu bond strength [58]
Vapor-assisted SAB [39, 59]	Details not reported	High bond strength and low resistance with bonding at 150 °C
Combined SAB [60, 61]	1. Bonding in vacuum to reduce interfacial H_2O 2. High bond strength at 200 °C	1. High bond strength at 200 °C 2. Ultra-thin interfacial CuO_x for low resistance

formation of Cu_2O and increase in the electrical sheet resistance after Cu surface treatment by Ar plasma [47]. Their bonding results also showed poor bonding quality with obvious large voids with bonding at 300 °C after the Ar plasma activation [47]. Furthermore, some other influences of the plasma treatment on Cu surface, such as pimples and delamination induced by N_2 plasma treatment [14], needs further investigation.

The Direct Bond Interconnect (DBI®) is a famous Cu/SiO_2 (or SiN_x) hybrid bonding technique developed by researchers at Ziptronix, Inc. (acquired by Tessera Technologies, Inc. in 2015) [49, 50]. In this technique, after surface plasma activation and chemical treatments for bonding species (e.g., Si–OH and Si–NH_2 groups) termination, wafers are bonded in ambient at room temperature without external compression (resulting in considerable high SiO_2–SiO_2 strength through interfacial Si–O–Si and Si–N–N–Si bonds) [51, 52, 62], followed by post-bonding annealing at elevated temperatures (125–400 °C [63]) for Cu–Cu bonding facilitated by internal compression induced by Cu thermal expansion [49]. By using of fluorinated oxide, the oxide–oxide strength can be further improved because of the improved absorption of interfacial H_2O by the fluorinated oxide. This process and its variants have been used and licensed at IMEC, Fermilab, Tezzaron/Novati, and Sony. Fermilab and Sony, respectively, has applied it in 3D stacked image sensors in recent years [7–9]. A eight-layer wafer stack, containing 8 layers of transistors

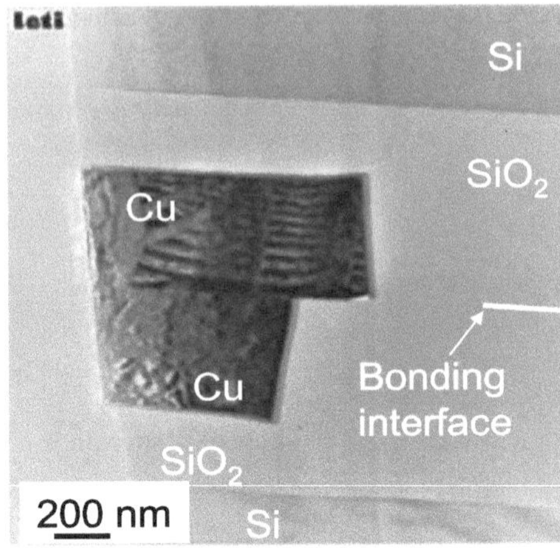

Fig. 6.11 TEM image of the Cu–Cu bonded interface obtained by CMP treatment (Reproduced with permission from [53]; Copyright © 2012, Wiley-VCH Verlag GmbH & Co. KGaA.)

and 80 layers of interconnect, bonded by the DBI® technique was demonstrated by Tezzaron Semiconductor and Novati Technologies in the 2015 IEEE 3DIC conference.

Researchers at CEA Leti developed a direct bonding method for Cu/SiO_2 hybrid bonding by making Cu and SiO_2 surfaces ultra-smooth and hydrophilic using optimized CMP [53–55]. The bonding is conducted in air at room temperature and without external compression. Figure 6.11 shows the interface of the Cu/SiO_2 hybrid bonded structure reported by researchers of CEA Leti [53]. The bonding behaviors were studied in detail by Cu–Cu and SiO_2–SiO_2 bonding experiments using blanket films. For Cu–Cu bonding, bond-strengthening behavior at low temperatures was found depending on the Cu film deposition method [57]. Using Cu films electrodeposited on Si wafers, the Cu–Cu bonding energy is around 0.8 J/m^2 as-bonded and increases to around 2.8 J/m^2 after 60 days of storage; using physical vapor deposited Cu films, the Cu–Cu bonding energy is around 0.5 J/m^2 as-bonded and slightly increases to 0.7 J/m^2 after 120 days of storage [57, 64]. The SiO_2–SiO_2 bonding energy is around 0.2 J/m^2 at room temperature. Typically, post-bonding annealing at 200–400 °C is employed to improve the bonding energy and to close the gaps between CMP-dished Cu surfaces [54], which is the same as the DBI's concept. After 200 °C annealing, the SiO_2–SiO_2 bonding energy is comparable to that by using plasma activation bonding but still lower than the Si bulk fracture energy [65]. Literatures suggest the SiO_2–SiO_2 bonding quality can be limited by the presence of excess interfacial H_2O molecules. It has been reported that voids are generated owing to the excess H_2O at the SiO_2–SiO_2 bonding interface [56, 66]. Furthermore, F. Fournel and coworkers reported that the SiO_2–SiO_2 bond strength can be decreased by the water stress corrosion effect induced by the interfacial H_2O, which is difficult to remove at temperatures below 400 °C [67]. In case of annealing at 400 °C, the strength of the Cu/SiO_2 hybrid bonded wafer is significantly improved [54], but voids are generated at the Cu–Cu bonding interface and in the Cu films [68, 69]. This technique has been demonstrated in a 3D stacked image sensor on a logic die [70].

It is also feasible to realize hybrid bonding by using the vapor-assisted SAB and combined SAB methods. Although the conventional SAB methods is effective for Cu–Cu bonding at room temperature, it was shown to be ineffective for the SiO_2–SiO_2 bonding [71]. The vapor-assisted SAB method was developed not only for low-temperature Cu–Cu bonding in ambient atmosphere, but also for hybrid bonding with material combinations of Cu, SiO_2, and polyimide [38, 39, 59]. However, this method also faces the concern of low SiO_2–SiO_2 bonding quality due to the water stress corrosion effect and generation of voids owing to trapping of excess H_2O. The combined SAB method was recently proposed to improve the SiO_2–SiO_2 bonding quality for Cu/SiO_2 hybrid bonding, based on bonding in vacuum for pre-bonding removal of excess H_2O molecules adsorbed on the wafers and for prevention of gas trapping [60, 61]. The combined SAB involves a combination of surface irradiation using a Si-containing Ar beam and pre-bonding attach–detach process prior to bonding in vacuum. The Si atoms added in the Ar beam are expected to increase the number of reactive Si sites on SiO_2 surface, while the pre-bonding attach–detach process is used to enhance the OH adsorption and to remove excess H_2O prior to bonding in vacuum. As a result of the combined procedure, high Cu–Cu, SiO_2–SiO_2, and SiO_2–SiN_x bond strength has been realized by bonding in vacuum of 10^{-2} Pa under an external compression of 2.5 MPa at 200 °C for 30 min followed by 200 °C annealing in ambient for 2 h [61]. The external compression is applied to ensure the wafers are tightly contacted even in the presence of large surface roughness, Cu dishing, and wafer warp and bow. Figure 6.12a shows the bond strength of various blanket bonded pairs, which is close to the Si bulk fracture strength of 2.5 J/m^2. Figure 6.12b shows the

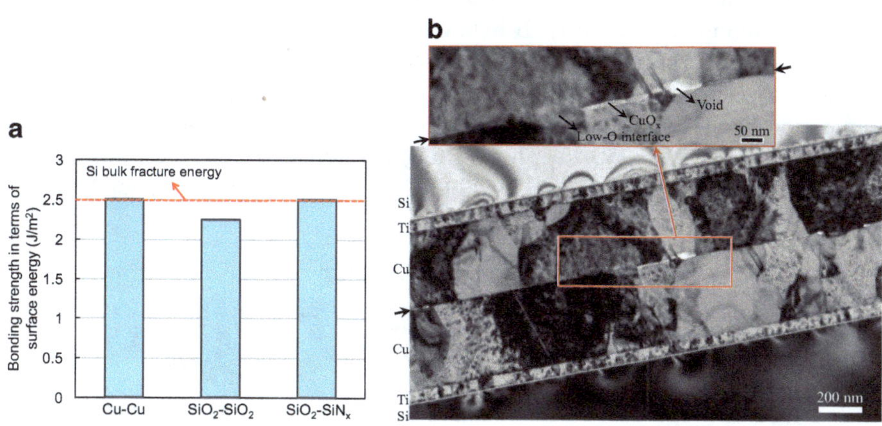

Fig. 6.12 Results of the combined SAB for (**a**) bond strength of various blanket films at 200 °C and (**b**) TEM image of the Cu–Cu bonded interface (Reproduced with permission from [60]; Copyright © 2012, The Electrochemical Society)

microstructure of the Cu–Cu bonding interface, containing low-O interface, ultrathin CuO_x interlayer, and some small voids.

Since it is also demonstrated that high SiO_2–SiO_2 bond strength can be obtained with bonding at room temperature followed by 200 °C annealing without thermo-compression [61], it can be interesting to optimize the combined SAB for Cu/SiO_2 hybrid bonding without compression. According to the authors' (T. Suga and R. He) experience, strong Cu–Cu bonding can also be achieved at 200 °C by combining Ar plasma activation (in low vacuum of 60 Pa) with the pre-bonding attach–detach process. Since the plasma activation has been widely studied for SiO_2–SiO_2 bonding, this combined approach also holds promise for the development of Cu/SiO_2 hybrid bonding without the use of high vacuum.

In short summary, Cu/SiO_2 hybrid bonding is much more complex than Cu–Cu bonding in the aspects of simultaneous surface activation of Cu and SiO_2, bonding conditions, and the presence of Cu dishing. Further researches are to be done to increase the bond strength obtained at <250 °C, to understand the influences of Cu dishing and wafer warp and bow, and to ensure high yield of low-resistance bonded interconnects.

6.6.3 *Cu/Adhesive Hybrid Bonding*

Besides Cu/SiO_2 hybrid bonding, Cu/adhesive hybrid bonding using polymer adhesives instead of SiO_2 has also been investigated for 3D integration. Researchers at IBM developed Cu/adhesive (polyimide) hybrid bonding by using lock-and-key bonding structures [72]. Researchers at RPI (USA) [73] and ASET (Japan) [74] developed Cu/adhesive hybrid bonding with benzocyclobutene (BCB) and polybenzoxazole (PBO) adhesives prepared by CMP, respectively. Figure 6.13 shows a micrograph of the Cu/BCB hybrid bonded structure [73].

Cu/adhesive hybrid bonding is typically conducted through an "adhesive-first" bonding approach, as shown in Fig. 6.14a, in which the adhesive is thermo-compression bonded and highly cured at a lower temperature (depending on the adhesive materials, e.g., 250 °C 1 h for BCB) before Cu–Cu thermo-compression bonding at a higher temperature (350–400 °C) [73, 75, 76]. This two-step bonding

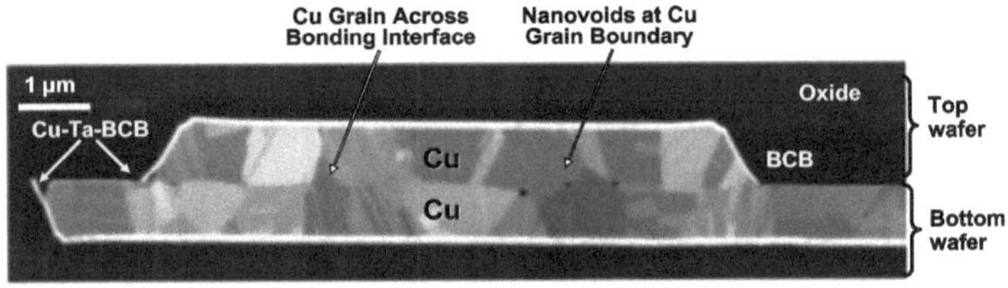

Fig. 6.13 SEM cross-sectional image of Cu/BCB hybrid bonded structure (Reproduced with permission from [73]; Copyright © 2012, Wiley-VCH Verlag GmbH & Co. KGaA)

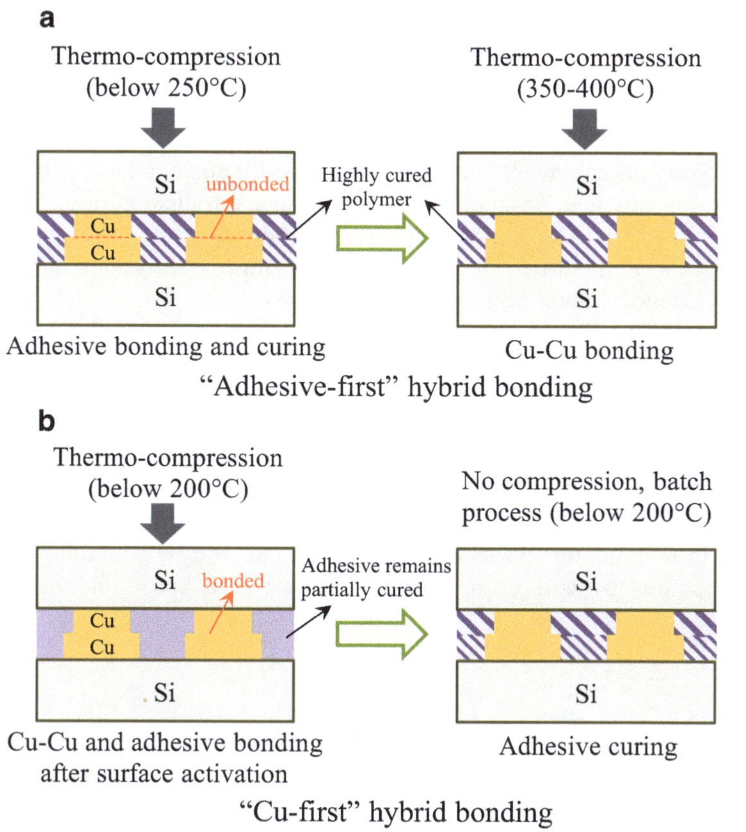

Fig. 6.14 Cu/adhesive hybrid bonding: (**a**) "adhesive-first" hybrid bonding process and (**b**) "Cu-first" hybrid bonding process in demand

sequence is employed because that the high Cu–Cu bonding temperature may damage the adhesive if it is not fully cured beforehand. The "adhesive-first" hybrid bonding approach has issues such as limitation of the choice of adhesive materials (having high thermal stability during high-temperature Cu–Cu bonding), low throughput due to long-time thermo-compression, and high thermal stress due to high Cu–Cu bonding temperature. Also, the mutual slip between the upper/lower substrates during the adhesive bonding/curing step may cause great misalignment of the final bonded structure [77]. To solve these issues, we suggest that it is highly desired to develop a "Cu-first" hybrid bonding approach, in which the Cu–Cu bonding is done at a low temperature and within a short time (e.g., $\leq$10 min) prior to the long-time adhesive curing, as shown in Fig. 6.14b. Therefore, it is essential to obtain the Cu–Cu bonding at reduced temperature that is comparable to or even lower than the adhesive bonding/curing temperature, which is typically below 250 °C or even 200 °C.

Effective surface activation methods for low-temperature (<250 °C) Cu/adhesive hybrid bonding are still rarely studied although various physical and chemical surface activation methods have been studied for Cu–Cu bonding, as described above in this chapter. The presence of adhesive in the Cu/adhesive hybrid bonding limits the surface activation needs to be adhesive compatible, i.e., introducing acceptable chemical/thermal damages and Cu contaminants to the adhesive materials. For instance, Ar atom beam and Ar plasma irradiations are considered to be unfriendly for the Cu/adhesive surface activation, which is mainly because they may induce Cu impurities on the adhesive owing to adsorption of physically sputtered Cu atoms onto the adhesive surface [78].

It seems promising to reduce the thermal budget for the Cu/adhesive hybrid bonding and to avoid the sputtering induced Cu impurities on adhesives by using the H-containing HCOOH vapor treatment. The H-containing HCOOH vapor treatment enables strong Cu–Cu bonding (shear strength of above 10 MPa) at 200 °C with considerably short pre-bonding treatment time (≤10 min) and thermo-compression time (e.g., 5 min) [16]. Since some adhesives can sustain 200 °C heating for a certain time, it should be possible to realize the "Cu-first" hybrid bonding through optimization of the H-containing HCOOH vapor treatment by controlling temperature and time for Cu/adhesive surface treatment.

6.7 Alternative Bonding Techniques: Insertion Bonding

Cu–Cu insertion is a bonding process approach for low-temperature bonding and has been applied to Cu-TSVs bonding schemes. The method relies on applying high shear stresses to yield large local plastic deformation of the Cu–Cu bonding surfaces [79]. The high shear stress is achieved by design of a sloped sidewall landing pad (instead of the usual flat pad) and by applying force/pressure on TSV nails inserted into the sloped pads (refer to Fig. 6.15).

To enable the insertion process, the landing pads have sloped sidewalls plated with Cu. This can be realized either by modifying the back-end-of-line (BEOL) passivation process, or by additional passivation steps. The latter approach can

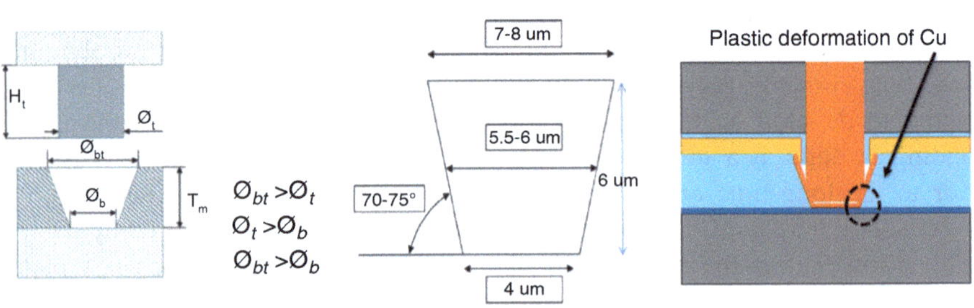

Fig. 6.15 Cu–Cu Insertion bonding (Cu–TSV scheme)

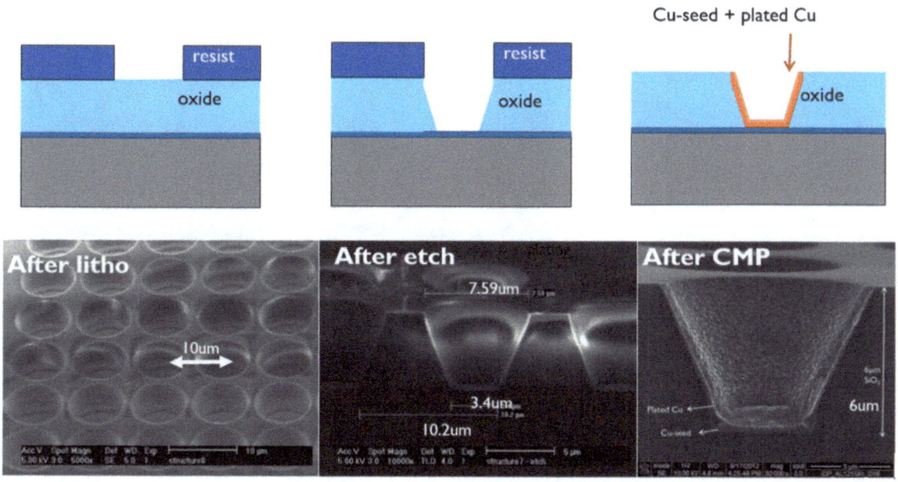

Fig. 6.16 Cu–Cu Insertion bonding: landing pad processing steps

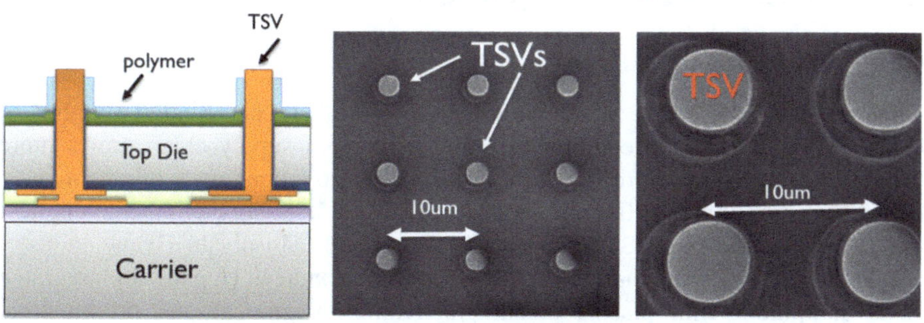

Fig. 6.17 Cu–Cu Insertion bonding: Top die prior stacking step

offer more flexibility at the price of additional lithography and etching steps. Figure 6.16 indicates images of the landing pads after key processing steps like lithography, etch, and final Cu CMP.

The processing of the top die is based on a temporary bonding system, with the wafer bonded on a temporary carrier while TSVs are exposed from backside. The wafer backside is standardly passivated with a SiN layer, this is necessary to avoid any possible Cu diffusion from exposed TSV. A polymer layer is than applied and developed in order to expose TSVs and act as filling layer between top and bottom dies during the stacking step (refer to Fig. 6.17).

A process with bonding temperature of 100 °C was demonstrated at IMEC/ Belgium with a seamless bond interface [79], while at room temperature the bonding interface was visible. Figure 6.18 shows some stacking images based on Cu–Cu insertion bonding where both dies have TSVs. Potential process

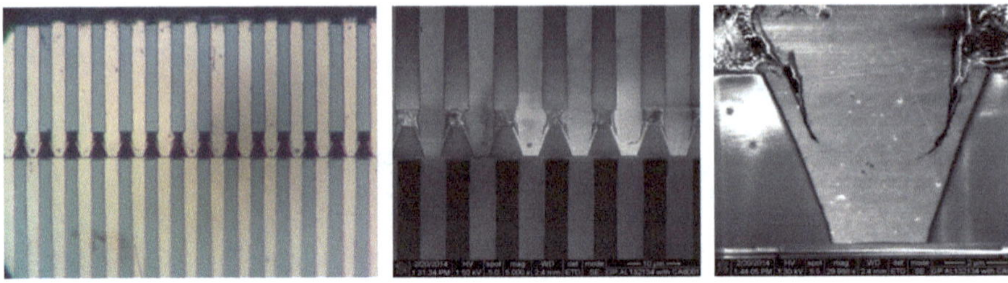

Fig. 6.18 Cu–Cu Insertion bonding: X-section after stacking of two dies with TSVs

improvements include Cu–Cu insertion bonding variant processes in the presence of cleaning agent, optimization of sidewall angle, and microstructure considerations. This approach can be used for different stacking schemes (die-to-wafer, die-to-die) and can also be extended to multi-die stacking.

6.8 Cu–Cu Bonding: Equipment Landscape and State of the Art

In terms of high-volume manufacturing (HVM), there are limited number of 300 mm W2W bonding equipment, while equipment supply chain grows more with C2W and further for C2C thermo-compression bonding equipment. Key equipment suppliers for 300 mm wafer-level Cu–Cu bonding equipment suppliers are: EVG (Austria) and Suss Microtech (Germany). Mitsubishi Heavy Industry (Japan) developed a standard SAB equipment and Bondtech (Japan), an integrated SAB equipment with plasma cleaning/activation, alignment and pre-bonding, bonding and heating chambers, both for 300 mm wafer bonding.

In terms of applications and commercialization, surface activated bonding (SAB) has been applied in volume production in metal laminates, MEMS packaging, and OELD devices to ensure hermetic sealing against permeation of water and oxygen from the atmosphere into these devices. Cu/SiO_2 hybrid bonding (DBI®) has been applied by Sony for 3D stacked back-illuminated image sensors (IMX260 used in Samsung Galaxy S7 Edge) [9], while new applications have been reported for hybrid bonding for 3D stacked hybrid pixel detectors for X-rays at Fermilab [80–82].

6.9 Chapter Summary and Future Recommendations

In this chapter, we reviewed various low-temperature Cu–Cu bonding methods and the fundamental aspects of the bonding mechanisms. The effects of surface activation, Cu diffusion, Cu film microstructures, and surface passivation by SAMs and metal capping layers was explained. The surface activation is of great importance to obtain good Cu–Cu bonding at below 250 °C or even at room temperature. Cu/dielectric hybrid bonding using the DBI®, CMP treatment and combined SAB methods was discussed. We also introduced the insertion bonding method for Cu–TSVs bonding. Equipment for Cu–Cu bonding for HVM has been introduced briefly. Scaling to large surface area chips, wafers, or panels with higher warpage and high-density interconnects remains a challenge especially in meeting thermal requirements, lower compression pressure, and lower process time. For researches to be done is to develop reliable, high-throughput, low-cost, low-temperature bonding processes for void-less bonding with high strength and reliability.

Acknowledgement The editors would like to thank Hualiang Shi from Intel Corporation for his critical review of this Chapter.

References

1. A. Fan, A. Rahman, R. Reif, Copper wafer bonding. Electrochem. Solid State Lett. **2**, 534–536 (1999). doi:10.1149/1.1390894
2. A. Shigetou, N. Hosoda, T. Itoh, T. Suga, Room-temperature direct bonding of CMP-Cu film for bumpless interconnection. In *51st Electronic Components and Technology Conference, 2001*, Orlando, pp. 755–760
3. P.R. Morrow, C.-M. Park, S. Ramanathan, M.J. Kobrinsky, M. Harmes, Three-dimensional wafer stacking via Cu-Cu bonding integrated with 65-nm strained-Si/low-k CMOS technology. IEEE Electron Dev. Lett. **27**, 335–337 (2006). doi:10.1109/LED.2006.873424
4. A. Shigetou, T. Itoh, M. Matsuo, N. Hayasaka, K. Okumura, T. Suga, Bumpless interconnect through ultrafine Cu electrodes by means of surface-activated bonding (SAB) method. IEEE Trans. Adv. Packag. **29**, 218–226 (2006). doi:10.1109/TADVP.2006.873138
5. B. Swinnen, W. Ruythooren, P.D. Moor, L. Bogaerts, L. Carbonell, K.D. Munck, B. Eyckens, S. Stoukatch, T. Tezcan, Z. Tokei, J. Vaes, J.V. Aelst, E. Beyne, 3D integration by Cu-Cu thermo-compression bonding of extremely thinned bulk-Si die containing 10μm pitch through-Si vias. In *2006 International Electron Devices Meeting (IEDM)* 2006, pp. 1–4
6. Tezzaron Company History. http://www.tezzaron.com/about-us/company-history. Accessed 18 June 2016
7. G.W. Deptuch, M. Demarteau, J.R. Hoff, R. Lipton, A. Shenai, M. Trimpl, R. Yarema, T. Zimmerman, Vertically integrated circuits at Fermilab. IEEE Trans. Nucl. Sci. **57**, 2178–2186 (2010). doi:10.1109/TNS.2010.2049659
8. R. Yarema, G. Deptuch, J. Hoff, F. Khalid, R. Lipton, A. Shenai, M. Trimpl, T. Zimmerman, Vertically integrated circuit development at Fermilab for detectors. J. Instrum. **8**, C01052 (2013). doi:10.1088/1748-0221/8/01/C01052
9. Chipworks (2016) Samsung Galaxy S7 Edge Teardown Report
10. M. Higashiwaki, K. Sasaki, T. Kamimura, M.H. Wong, D. Krishnamurthy, A. Kuramata, T. Masui, S. Yamakoshi, Depletion-mode Ga_2O_3 metal-oxide-semiconductor field-effect transistors on β-Ga_2O_3 (010) substrates and temperature dependence of their device characteristics. Appl. Phys. Lett. **103**, 123511 (2013). doi:10.1063/1.4821858
11. T. Fukushima, Y. Yamada, H. Kikuchi, M. Koyanagi, New three-dimensional integration technology using self-assembly technique. In *IEEE International Devices Meeting, 2005 IEDM Technical Digest*, Washington, 2005, pp. 348–351

12. Y.-S. Tang, Y.-J. Chang, K.-N. Chen, Wafer-level Cu–Cu bonding technology. Microelectron. Reliab. **52**, 312–320 (2012). doi:10.1016/j.microrel.2011.04.016
13. C.S. Tan, R. Reif, N.D. Theodore, S. Pozder, Observation of interfacial void formation in bonded copper layers. Appl. Phys. Lett. **87**, 201909 (2005). doi:10.1063/1.2130534
14. B. Rebhan, T. Plach, S. Tollabimazraehno, V. Dragoi, M. Kawano, Cu-Cu wafer bonding: an enabling technology for three-dimensional integration. In *2014 International Conference on Electronics Packaging (ICEP)*, 2014, pp. 475–479
15. B. Rebhan, S. Tollabimazraehno, G. Hesser, V. Dragoi, Analytical methods used for low temperature Cu–Cu wafer bonding process evaluation. Microsyst. Technol. **21**, 1003–1013 (2015). doi:10.1007/s00542-015-2446-2
16. W. Yang, M. Akaike, M. Fujino, T. Suga, A combined process of formic acid pretreatment for low-temperature bonding of copper electrodes. ECS J. Solid State Sci. Technol. **2**, P271–P274 (2013). doi:10.1149/2.010306jss
17. W. Yang, M. Akaike, T. Suga, Effect of formic acid vapor in situ treatment process on Cu low-temperature bonding. IEEE Trans. Compon. Packag. Manuf. Technol. **4**, 951–956 (2014). doi:10.1109/TCPMT.2014.2315761
18. P.-I. Wang, S.H. Lee, T.C. Parker, M.D. Frey, T. Karabacak, J.-Q. Lu, T.-M. Lu, Low temperature wafer bonding by copper nanorod array. Electrochem. Solid State Lett. **12**, H138–H141 (2009). doi:10.1149/1.3075900
19. T. Ishizaki, R. Watanabe, A new one-pot method for the synthesis of Cu nanoparticles for low temperature bonding. J. Mater. Chem. **22**, 25198–25206 (2012). doi:10.1039/C2JM34954J
20. C.-M. Liu, H.-W. Lin, Y.-S. Huang, Y.-C. Chu, C. Chen, D.-R. Lyu, K.-N. Chen, K.-N. Tu, Low-temperature direct copper-to-copper bonding enabled by creep on (111) surfaces of nanotwinned Cu. Sci. Rep. **5**, 9734 (2015). doi:10.1038/srep09734
21. T. Shimatsu, M. Uomoto, Atomic diffusion bonding of wafers with thin nanocrystalline metal films. J. Vac. Sci. Technol. B **28**, 706–714 (2010). doi:10.1116/1.3437515
22. T. Shimatsu, M. Uomoto, Room temperature bonding of wafers with thin nanocrystalline metal films. ECS Trans. **33**, 61–72 (2010). doi:10.1149/1.3483494
23. V. Smet, M. Kobayashi, T. Wang, P.M. Raj, R. Tummala, A new era in manufacturable, low-temperature and ultra-fine pitch Cu interconnections and assembly without solders. In *2014 64th Electronic Components Technology Conference (ECTC)*, pp. 484–489
24. C.S. Tan, D.F. Lim, S.G. Singh, S.K. Goulet, M. Bergkvist, Cu–Cu diffusion bonding enhancement at low temperature by surface passivation using self-assembled monolayer of alkane-thiol. Appl. Phys. Lett. **95**, 192108 (2009). doi:10.1063/1.3263154
25. D.F. Lim, J. Wei, K.C. Leong, C.S. Tan, Surface passivation of Cu for low temperature 3D wafer bonding. ECS Solid State Lett. **1**, P11–P14 (2012)
26. D.F. Lim, J. Wei, K.C. Leong, C.S. Tan, Cu passivation for enhanced low temperature (□300°C) bonding in 3D integration. Microelectron. Eng. **106**, 144–148 (2013). doi:10.1016/j.mee.2013.01.032
27. L. Peng, L. Zhang, J. Fan, H.Y. Li, D.F. Lim, C.S. Tan, Ultrafine pitch (6 μm) of recessed and bonded Cu-Cu interconnects by three-dimensional wafer stacking. IEEE Electron Dev. Lett. **33**, 1747–1749 (2012). doi:10.1109/LED.2012.2218273
28. E. Beyne E, The Minerals, Metals & Materials Society, Thiol-based Self-Assembled Monolayers (SAMs) as an alternative surface finish for 3D Cu microbumps. In *TMS 2015 Supplemental Proceedings*. Wiley, Orlando, 2015, pp. 1353–1360
29. Y.-P. Huang, Y.-S. Chien, R.-N. Tzeng, M.-S. Shy, T.-H. Lin, K.-H. Chen, C.-T. Chiu, J.-C. Chiou, C.-T. Chuang, W. Hwang, H.-M. Tong, K.-N. Chen, Novel Cu-to-Cu bonding with Ti passivation at 180 °C in 3-D integration. IEEE Electron Dev.Lett. **34**, 1551–1553 (2013). doi:10.1109/LED.2013.2285702
30. A.K. Panigrahi, S. Bonam, T. Ghosh, S.G. Singh, S.R.K. Vanjari, Ultra-thin Ti passivation mediated breakthrough in high quality Cu-Cu bonding at low temperature and pressure. Mater. Lett. **169**, 269–272 (2016). doi:10.1016/j.matlet.2016.01.126

31. Y.-P. Huang, Y.-S. Chien, R.-N. Tzeng, K.-N. Chen, Demonstration and electrical performance of Cu-Cu bonding at 150 °C with Pd passivation. IEEE Trans. Electron. Dev. **62**, 2587–2592 (2015). doi:10.1109/TED.2015.2446507
32. E. Beyne, V.J. De, J. Derakhshandeh, L. England, G. Vakanas, *Thin Nib or Cob Capping Layer for Non-Noble Metallic Bonding Landing Pads* (Springer, New York, 2015)
33. T.H. Kim, M.M.R. Howlader, T. Itoh, T. Suga, Room temperature Cu–Cu direct bonding using surface activated bonding method. J. Vac. Sci. Technol. A **21**, 449–453 (2003). doi:10.1116/1.1537716
34. A. Shigetou, T. Itoh, T. Suga, Direct bonding of CMP-Cu films by surface activated bonding (SAB) method. J. Mater. Sci. **40**, 3149–3154 (2005). doi:10.1007/s10853-005-2677-1
35. T. Suga, Feasibility of surface activated bonding for ultra-fine pitch interconnection—a new concept of bump-less direct bonding for system level packaging. In *Electronic Components Technology Conference. 2000 Proceedings 50th IEEE*, Las Vegas, pp. 702–705
36. T. Suga, K. Otsuka, Bump-less interconnect for next generation system packaging. In *Electronic Components and Technology Conference, 2001 Proceedings, 51st IEEE*, pp. 1003–1008
37. A. Shigetou, T. Itoh, K. Sawada, T. Suga, Bumpless interconnect of 6-μm-pitch Cu electrodes at room temperature. IEEE Trans. Adv. Packag. **31**, 473–478 (2008). doi:10.1109/TADVP.2008.920644
38. A. Shigetou, T. Suga, Modified diffusion bonding of chemical mechanical polishing Cu at 150 °C at ambient pressure. Appl. Phys. Express **2**, 056501 (2009). doi:10.1143/APEX.2.056501
39. A. Shigetou, T. Suga, Vapor-assisted surface activation method for homo- and heterogeneous bonding of Cu, SiO_2, and polyimide at 150°C and atmospheric pressure. J. Electron. Mater. **41**, 2274–2280 (2012). doi:10.1007/s11664-012-2091-9
40. A. Shigetou, T. Suga, Modified diffusion bond process for chemical mechanical polishing (CMP)-Cu at 150 °C in ambient air. In *59th Electronic Components and Technology Conference*, San Diego, 2009, pp. 365–369
41. A. Shigetou , T. Suga, Homo/heterogeneous bonding of Cu, SiO_2, and polyimide by low temperature vapor-assisted surface activation method. In *IEEE 61st Electronic Components and Technology Conference (ECTC)*, Lake Buena Vista, 2011, pp. 32–36
42. T. Plach, K. Hingerl, S. Tollabimazraehno, G. Hesser, V. Dragoi, M. Wimplinger, Mechanisms for room temperature direct wafer bonding. J. Appl. Phys. **113**, 094905 (2013). doi:10.1063/1.4794319
43. T. Suni, K. Henttinen, I. Suni, J. Mäkinen, Effects of plasma activation on hydrophilic bonding of Si and SiO_2. J. Electrochem. Soc. **149**, G348–G351 (2002). doi:10.1149/1.1477209
44. Y.-H. Wang, K. Nishida, M. Hutter, T. Kimura, T. Suga, Low-temperature process of fine-pitch Au–Sn bump bonding in ambient air. Jpn. J. Appl. Phys. **46**, 1961 (2007). doi:10.1143/JJAP.46.1961
45. K. Okumura, E. Higurashi, T. Suga, K. Hagiwara, Influence of air exposure time on bonding strength in Au-Au surface activated wafer bonding. In *2015 International Conference on Electronics Packaging and IMAPS All Asia Conference, ICEP-IACC*, 2015, pp. 448–451
46. H. Ishida, T. Ogashiwa, Y. Kanehira, S. Ito, T. Yazaki , J. Mizuno, Low-temperature, surface-compliant wafer bonding using sub-micron gold particles for wafer-level MEMS packaging. In *IEEE 62nd Electronic Components and Technology Conference*, 2012, pp. 1140–1145
47. M. Park, S. Baek, S. Kim, S.E. Kim, Argon plasma treatment on Cu surface for Cu bonding in 3D integration and their characteristics. Appl. Surf. Sci. **324**, 168–173 (2015). doi:10.1016/j.apsusc.2014.10.098
48. S.L. Chua, G.Y. Chong , Y.H. Lee, C.S. Tan, Direct copper-copper wafer bonding with Ar/N2 plasma activation. In *IEEE International Conference on Electron Devices and Solid-State Circuits (EDSSC)*, 2015, pp. 134–137
49. P. Enquist, G. Fountain, C. Petteway, A. Hollingsworth, H. Grady, Low cost of ownership scalable copper direct bond interconnect 3D IC technology for three dimensional integrated

circuit applications. In *IEEE International Conference On 3D System Integration,* 2009, pp. 1–6
50. P. Enquist, Metal/silicon oxide hybrid bonding, in *Handbook of Wafer Bonding*, ed. by P. Ramm, J.J.-Q. Lu, M.M.V. Taklo (Wiley, Weinheim, 2012), pp. 261–278
51. Y.-L. Chao, Q.-Y. Tong, T.-H. Lee, M. Reiche, R. Scholz, J.C.S. Woo, U. Gösele, Ammonium hydroxide effect on low-temperature wafer bonding energy enhancement. Electrochem. Solid State Lett. **8**, G74–G77 (2005). doi:10.1149/1.1857671
52. Q.-Y. Tong, G. Fountain, P. Enquist, Room temperature SiO_2/SiO_2 covalent bonding. Appl. Phys. Lett. **89**, 042110 (2006). doi:10.1063/1.2240232
53. L. Di Cioccio, S. Moreau, L. Sanchez, F. Baudin, P. Gueguen, S. Mermoz, Y. Beilliard, R. Taibi, In *Cu/SiO$_2$ Hybrid bonding*, ed. by P. Garrou, M. Koyanagi, P. Ramm. Handbook of 3D Integration, Wiley-VCH, Dresden, 2014. pp. 295–312
54. L.D. Cioccio, P. Gueguen, R. Taibi, D. Landru, G. Gaudin, C. Chappaz, F. Rieutord, F. de Crecy, I. Radu, L.L. Chapelon, L. Clavelier, An overview of patterned metal/dielectric surface bonding: mechanism, alignment and characterization. J. Electrochem. Soc. **158**, P81–P86 (2011). doi:10.1149/1.3577596
55. I. Radu, D. Landru, G. Gaudin, G. Riou, C. Tempesta, F. Letertre, L. Di Cioccio, P. Gueguen, T. Signamarcheix, C. Euvrard, others. In *IEEE International 3D Systems Integration Confer ence (3DIC), IEEE,* Munich, 2010, pp. 1–6
56. C. Sabbione, L.D. Cioccio, L. Vandroux, J.-P. Nieto, F. Rieutord, Low temperature direct bonding mechanisms of tetraethyl orthosilicate based silicon oxide films deposited by plasma enhanced chemical vapor deposition. J. Appl. Phys. **112**, 063501 (2012). doi:10.1063/1.4752258
57. P. Gondcharton, B. Imbert, L. Benaissa, V. Carron, M. Verdier, Kinetics of low temperature direct copper–copper bonding. Microsyst. Technol. **21**, 995–1001 (2015). doi:10.1007/s00542-015-2436-4
58. P. Gueguen, L. Di Cioccio, P. Gergaud, M. Rivoire, D. Scevola, M. Zussy, A.M. Charvet, L. Bally, D. Lafond, L. Clavelier, Copper direct-bonding characterization and its interests for 3D integration. J. Electrochem. Soc. **156**, H772 (2009). doi:10.1149/1.3187271
59. A. Shigetou , T. Suga, Modified diffusion bonding for both Cu and SiO_2 at 150 °C in ambient air. In *2010 Proceedings 60th Electronic Components and Technology Conference (ECTC)*, Las Vegas, 2010, pp. 872–877
60. R. He, M. Fujino, A. Yamauchi, Y. Wang, T. Suga, Combined surface activated bonding technique for low-temperature Cu/dielectric hybrid bonding. ECS J. Solid State Sci. Technol. **5**, P419–P424 (2016). doi:10.1149/2.0201607jss
61. R. He, M. Fujino, A. Yamauchi, T. Suga, Combined surface-activated bonding technique for low-temperature hydrophilic direct wafer bonding. Jpn. J. Appl. Phys. **55**, 04EC02 (2016). doi:10.7567/JJAP.55.04EC02
62. Q.-Y. Tong J.G.G. Fountain, P.M. Enquist, Method for low temperature bonding and bonded structure. US Patent 6,902,987
63. C. Sanders, Continued adoption of low temperature direct bond technology for high volume 3D commercial applications. In *3D Architectures for Semiconductor Integration and Packaging (3D ASIP)*, 2012
64. L.D. Cioccio, F. Baudin, P. Gergaud, V. Delaye, P.-H. Jouneau, F. Rieutord, T. Signamarcheix, Modeling and integration phenomena of metal-metal direct bonding technology. ECS Trans. **64**, 339–355 (2014). doi:10.1149/06405.0339ecst
65. C. Rauer, H. Moriceau, F. Fournel, A.M. Charvet, C. Morales, N. Rochat, L. Vandroux, F. Rieutord, T. McCormick, I. Radu, Treatments of deposited SiOx surfaces enabling low temperature direct bonding. ECS J. Solid State Sci. Technol. **2**, Q147–Q150 (2013). doi:10.1149/2.004309jss
66. C. Ventosa, C. Morales, L. Libralesso, F. Fournel, A.M. Papon, D. Lafond, H. Moriceau, J.D. Penot, F. Rieutord, Mechanism of thermal silicon oxide direct wafer bonding. Electrochem. Solid State Lett. **12**, H373–H375 (2009). doi:10.1149/1.3193533

67. F. Fournel, C. Martin-Cocher, D. Radisson, V. Larrey, E. Beche, C. Morales, P.A. Delean, F. Rieutord, H. Moriceau, Water stress corrosion in bonded structures. ECS J. Solid State Sci. Technol. **4**, P124–P130 (2015). doi:10.1149/2.0031505jss
68. P. Gondcharton, B. Imbert, L. Benaissa, M. Verdier, Voiding phenomena in copper-copper bonded structures: role of creep. ECS J. Solid State Sci. Technol. **4**, P77–P82 (2015). doi:10.1149/2.0081503jss
69. P. Gondcharton, B. Imbert, L. Benaissa, F. Fournel, M. Verdier, Effect of copper–copper direct bonding on voiding in metal thin films. J. Electron. Mater. **44**, 4128–4133 (2015). doi:10.1007/s11664-015-3992-1
70. S. Lhostis, A. Farcy, E. Deloffre, F. Lorut, S. Mermoz, Y. Henrion, L. Berthier, F. Bailly, D. Scevola, F. Guyader, F. Gigon, C. Besset, S. Pellissier, L. Gay, N. Hotellier, M. Arnoux, A.-L. Le Berrigo, S. Moreau, V. Balan, F. Fournel, A. Jouve, S. Chéramy , B. Rebhan, G.A. Maier, L. Chitu, Reliable 300 mm wafer level hybrid bonding for 3D stacked CMOS image sensors. In *IEEE 66th Electronic Components* and *Technology Conference*, Las Vegas, 2016, pp. 869–876
71. H. Takagi, J. Utsumi, M. Takahashi, R. Maeda, Room-temperature bonding of oxide wafers by Ar-beam surface activation. ECS Trans. **16**, 531–537 (2008). doi:10.1149/1.2982908
72. F. Liu, R.R. Yu , A.M. Young, J.P. Doyle, X. Wang, L. Shi, K.-N. Chen , X. Li, D.A. Dipaola , D. Brown, C.T. Ryan , J.A. Hagan, K.H. Wong, M. Lu, X. Gu, N.R. Klymko, E.D. Perfecto, A.G. Merryman , K.A. Kelly, S. Purushothaman, S.J. Koester, R. Wisnieff, W. Haensch, A 300-mm wafer-level three-dimensional integration scheme using tungsten through-silicon via and hybrid Cu-adhesive bonding. In *IEEE International Electron Devices Meeting (IEDM)* 2008, pp. 1–4
73. J.-Q. Lu, J.J. McMahon, R.J. Gutmann, Hybrid metal/polymer wafer bonding platform, in *Handbook of Wafer Bonding*, ed. by J.J.-Q. Lu, M.M.V. Taklo, P. Ramm (Wiley-VCH, Weinheim, 2012), pp. 215–236
74. K. Hozawa, M. Aoki, F. Furuta, K. Takeda, A. Yanagisawa , H. Kikuchi, T. Mitsuhashi, H. Kobayashi, 3D integration technology using hybrid wafer bonding and its electrical characteristics. In *13th International Symposium on Electronics* Packaging *(ICEP 2013)*, Osaka, pp. 118–122
75. J.J. McMahon, J.-Q, Lu, R.J. Gutmann, Wafer bonding of damascene-patterned metal/adhesive redistribution layers for via-first three-dimensional (3D) interconnect. In *Proceedings Electronic Components and Technology, 2005, ECTC '05*, vol. 1, 2005, pp. 331–336
76. Z.-C. Hsiao, C.-T. Ko, H.-H. Chang, H.-C. Fu, C.-W. Chiang, C.-K. Hsu, W.-W. Shen, W.-C. Lo, *Cu/BCB hybrid bonding with TSV for 3D integration by using fly cutting technology* (IEEE, Kyoto, 2015), pp. 834–837
77. T. Sakai, N. Imaizumi, S. Sakuyama, Hybrid bonding technology with Cu-Cu/adhesives for high density 2.5D/3D integration. In *IEEE,* Big Island, 2016, pp. 1–6
78. R. He, T. Suga, Effects of Ar plasma and Ar fast atom bombardment (FAB) treatments on Cu/polymer hybrid surface for wafer bonding. In *International Conference on Electronics Packaging (ICEP),* Toyama, pp. 78–81
79. C. Okoro, R. Agarwal, P. Limaye, B. Vandevelde, D. Vandepitte, E. Beyne, Insertion bonding: A novel Cu-Cu bonding approach for 3D integration. In *60th* IEEE International Conference on *Electronic Components* and *Technology Conference* (ECTC 2010) *Proceedings*, pp. 1370–1375
80. G.W. Deptuch, G. Carini, P. Grybos, P. Kmon, P. Maj, M. Trimpl, D.P. Siddons, R. Szczygiel, R. Yarema, Design and tests of the vertically integrated photon imaging chip. IEEE Trans. Nucl. Sci. **61**, 663–674 (2014). doi:10.1109/TNS.2013.2294673
81. G.W. Deptuch, G. Carini, T. Collier, P. Gryboś, P. Kmon, R. Lipton, P. Maj, D.P. Siddons, R. Szczygieł, R. Yarema, Results of tests of three-dimensionally integrated chips bonded to sensors. IEEE Trans. Nucl. Sci. **62**, 349–358 (2015). doi:10.1109/TNS.2014.2378784
82. G.W. Deptuch, G. Carini, P. Enquist, P. Gryboś, S. Holm, R. Lipton, P. Maj, R. Patti, D.P. Siddons, R. Szczygieł, R. Yarema, Fully 3-D integrated pixel detectors for X-rays. IEEE Trans. Electron. Dev. **63**, 205–214 (2016). doi:10.1109/TED.2015.2448671

Chapter 7
Fundamentals of Thermal Compression Bonding Technology and Process Materials for 2.5/3D Packages

Sangil Lee

7.1 Introduction

Three-dimensional (3D) packaging has moved to the forefront in the electronic packaging industry as the demands toward higher performance, smaller form factor, and heterogeneous system integration continue to grow. As a result, recent consumer electronic devices have shown adoption of 3D packaging structures such as system in package (SIP), package on package (PoP), and multi-chip module.

Wire bonding and Through Silicon Vias (TSVs) are interconnection technologies, the most widely used for the aforementioned 3D packages. The wire bonding technology imposes limitation on the manufacturing throughput and number of I/O counts due to point-to-point bonding process. In contrast, the TSV interconnects theoretically can provide high throughput and fine pitch interconnections because thin TSV microbumped die will be stacked to each other in order to create 3D packages. However, TSV technology still requires a lot of research to exhibit similar levels of technical maturity of flip chip technology in view of assembly throughput and reliability performance even after SK Hynix produced a high-performance 3D stacked DRAM, Bandwidth Memory (HBM), implemented on AMD Fiji GPU. The device manufacturers demonstrated the innovative TSV technology by adopting newly developed bonding technique, Thermal Compression Bonding (TCB), which controls force and heat applied to 3D packages. The in situ bonding technology can complete a bonding cycle even in several seconds. The thermal process finishing a cycle in few seconds is an extremely short process compared to traditional mass reflow process and the instantly changing thermal condition has pushed key process materials such as flux and underfill beyond its limits. To accommodate the fast changing process, an extensive change in material

S. Lee, Ph.D. (✉)
Invensas Corporation, 3025 Orchard Parkway, San Jose, CA 95134, USA
e-mail: salee@invensas.com

Y. Li, D. Goyal (eds.), *3D Microelectronic Packaging*, Springer Series in Advanced Microelectronics 57, DOI 10.1007/978-3-319-44586-1_7

formulation is required for flux and underfill directly affecting quality of assembly yield and reliability performance.

In detail, bond head integrated with a heater will compress a die to material dispensed a bonding site or material pre-applied a die to a bonding site while the heater is increasing the temperature of the head above solder melting point. After reaching the point, the head will cool down the heater as fast as possible so that the bonded package quickly reaches just below solder solidification point. Note that the ramp rate of TCB is 100 °C/s and mass reflow may have at most 2.0 °C/s. When reaching the setting temperature, the bond head disengages from a package and a bonding cycle will complete. Even with the reflection on TCB ramp rate only, it would be sufficient to contemplate how much effort packaging industry has been making to demonstrate 3D packaging technology including assembly processes, materials, and equipment. Due to the aforementioned technical challenges with placement accuracy requirement under a very rapidly changing thermal process, TCB will be the most difficult process step out of 3D package assembly processes. Even though the TCB is a fast assembly process, the process will be a bottleneck restricting overall throughput because a unit will be processed individually. The resultant of low throughput would pose a cost disadvantage in 3D packages.

Therefore, this chapter will review the TCB technique including process tolerance, major process materials, and feasible building methods, where limited number of literatures exists. In addition, the chapter would provide readers with insight of how to design process building blocks relevant to products, select materials compatible with the chosen building block in terms of processability, reliability, and throughput. In other words, this chapter will explore engineering sciences and fundamentals required for an enabler of 3D electronic packaging [1–10].

7.2 Background

In the current semiconductor industry, 3D packaging implies interconnection technology using Through Silicon Vias (TSVs) technology. The technology can shorten vertical interconnects, replacing the long electrical paths in 2D packaging, to improve electrical performance or to reduce timing delay. The TSV idea was firstly introduced in 1962 and products adopting the TSV technology appeared to electronic industry in 2014. Even now, packaging industry still faces technical challenges such as substantially low throughput resulting from premature building blocks associated with process materials and electrical performance drop due to heat dissipation issues. Complicated supply chain in TSV industry would amplify the technical challenges.

All building blocks currently under investigation for 2.5/3D packages selected TCB technology to meet placement accuracy requirement of fine pitch flip chip stacking of thin TSV microbumped die. The thin die (<100 μm) may exhibit high warpage even at room temperature and the amount of warpage will exceed the limit of traditional flip chip process via mass reflow process. Due to the warpage challenge, Intel Corporation introduced TCB technology in high volume manufacturing process even for 130 μm pitch and 22 nm nodes of flip chip architecture in 2014 in order to be ready for all foreseen product configurations [11].

This session offers overview of 3D packaging configuration by reviewing current 2.5/3D commercial products. Next, the session compares TCB technology with traditional flip chip bonding process using mass reflow. The comparison will be placed on basic concepts of assembly process with related engineering sciences as well as process limitations associated with package design and equipment. Then fundamentals of process materials will be discussed in detail to understand key parameters of material formulation, and to investigate pros and cons of each material. Thus, background session will provide essential engineering sciences and fundamentals required to achieve primary objectives of this chapter; to explore 2.5/3D packaging bonding technology, to comprehend existing building blocks, to understand how building blocks associated with package configurations are developed, and to select process materials relevant to package structure.

7.2.1 Overview of 3D Package Configuration

Figure 7.1 shows a typical schematic of 3D package, Apple A7, manufactured using PoP stacking technology, which stacks LPDDR3 memory on top of microprocessor. This is the most traditional 3D electronic package with wirebond and flip chip used for interconnection methods. Recently 3D packaging technology implies interconnects using Through Silicon Vias (TSVs). Figure 7.2 shows AMD Fiji GPU with SK Hynix High Bandwidth Memory (HBM). SK Hynix, the Korean memory chip manufacturer, announced its high bandwidth memory (HBM) product in early 2014, claiming it to be the world's first 8 GB module adopting 2 GB, 20 nm node, DDR4 SDRAM. Introduction of the HBM modules to the semiconductor market took a year after the module was claimed in 2014. Look at Fig. 7.3

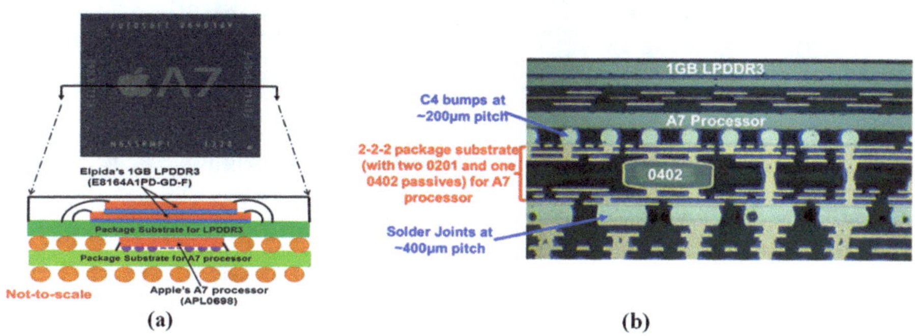

Fig. 7.1 Apple A7 PoP processor: (**a**) the top view of the A7 processor with a schematic of side view of the A7 PoP and (**b**) a micrograph of sectional view of the A7 PoP [12]

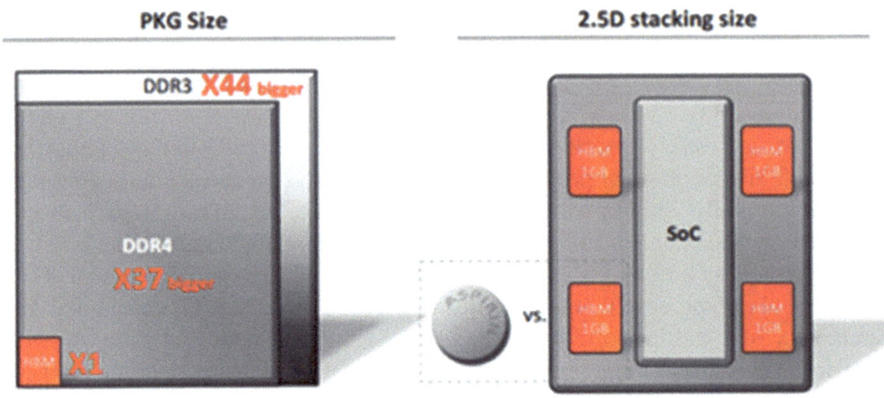

Fig. 7.2 Schematics of HBM size comparison (Source: SK Hynix)

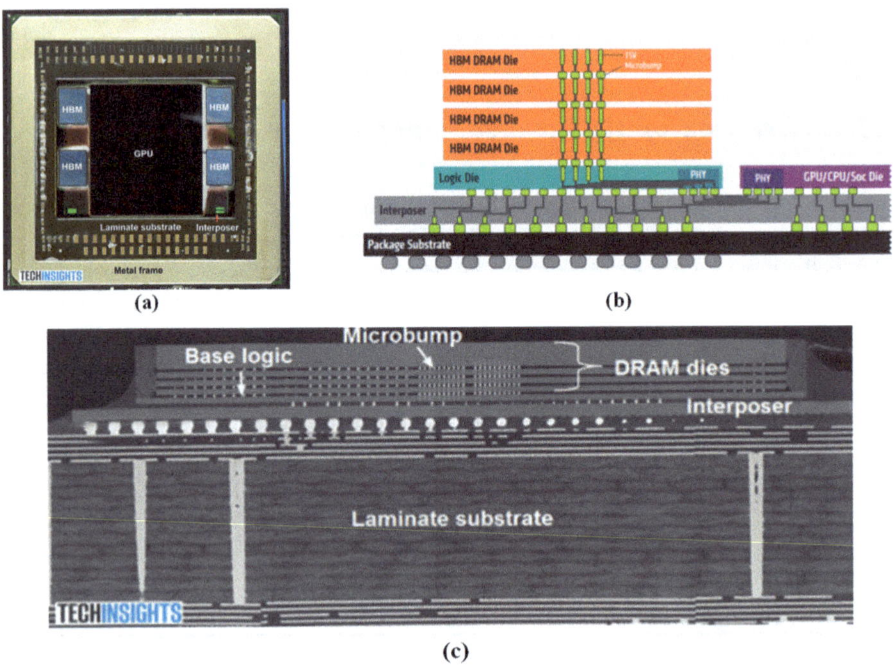

Fig. 7.3 AMD Fiji GPU with four SK Hynix HBM: (**a**) top view, (**b**) a schematic of side view, and (**c**) a SEM micrograph of AMD GPU [13]

comparing size of HBM with that of DDR3 and DDR4. 1 GB HBM is 44 times smaller than DDR3 and even smaller than one tablet of aspirin. The HBM uses TSVs to connect the DRAM dies and base logic die together and this is a new technology for DRAM stacking. It employs a base logic die as an interface between the four-DRAM die stack and an interposer that supports both the HBM modules and the AMD GPU. The HBM can be a 3D package, while the laterally spaced apart layouts for the GPU and HBM modules on the interposer construct a 2.5D package. Xilinx also released 2.5D FPGA using TSV interconnections, termed as Stacked Silicon Interconnect (SSI) technology. The SSI is a new generation of FPGA for Xilinx to deliver the largest FPGA with the highest bandwidth from their customers and the SSI FPGA has been in volume production. Through the TSV technology, Xilinx SSI combines die stack-up with four FPGA SLRs, silicon interposer, and package substrate as shown in Fig. 7.4a. The interposer provides tens of thousands of die-to-die connections to enable ultrahigh pin counts of interconnect. Scanned

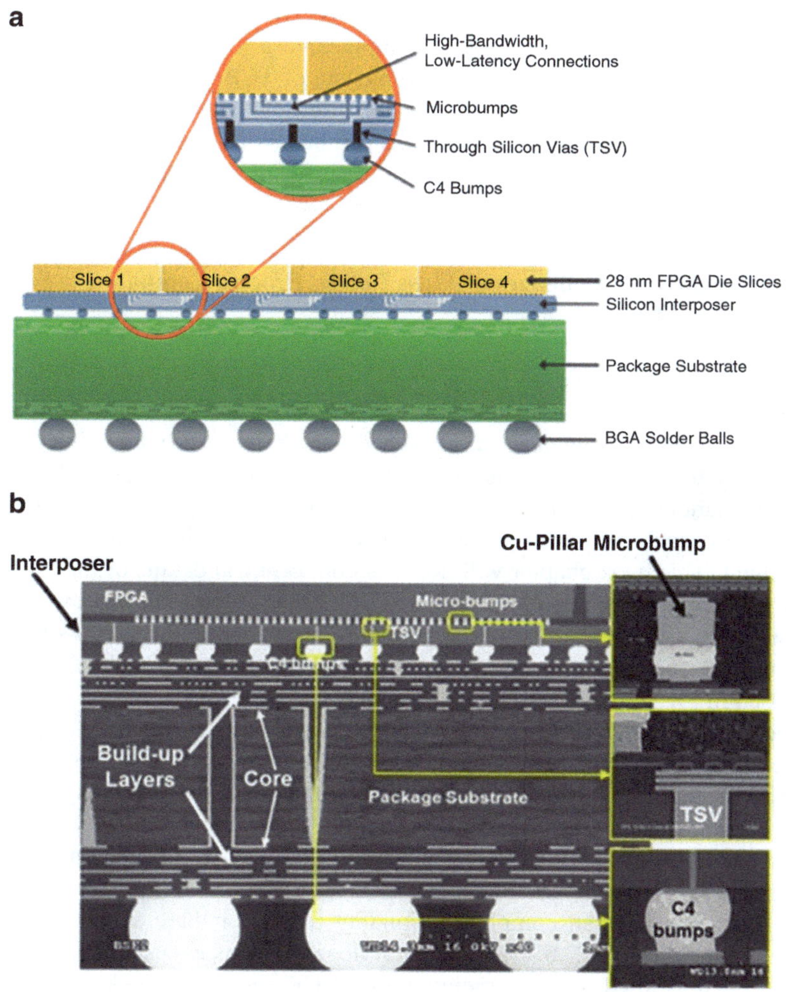

Fig. 7.4 2.5D Xilinx FPGA: (**a**) a schematic and (**b**) SEM micrographs of side view of FPGA [14]

Electron Microscope (SEM) Micrograph of 2.5D Xilinx FPGA (see Fig. 7.4b) depicts more details about SSI technology. TCB will attach microbumped FPGAs to a TSV interposer and the four FPGAs mounted on interposer will be reflowed to construct interconnections between the interposer and the organic substrate.

7.2.2 *Fundamentals of Thermal Compression Bonding Technology*

The purpose of this session is to provide some insight into Thermal Compression Bonding (TCB) required for interconnection for TSV die stacking; TCB bonding technology will be compared with traditional mass reflow in terms of process limitations and technical challenges; mechanism of TCB process will be explored with emphasis on functionality and configuration of TCB bonder. The technical insight shows how TCB assembly process can be characterized and developed by controlling key process parameters such as displacement, force, and temperature.

7.2.2.1 Technical Challenges of Mass Reflow Process Compared with TCB

Figure 7.5a shows traditional flip assembly process steps: dipping a die into flux reservoir or dispensing flux on substrate depending on products, attaching the die to the substrate, reflowing the die-attached substrate, removing flux residue, underfilling the gap between die and substrate, and curing underfill. In general, memory manufacturers prefer to use flux dipping but logic manufacturers use dispensing because of mostly assembly cost associated with equipment price and throughput. This book chapter will not focus on technical details of the traditional flux application implemented assembly process that at least six process steps are necessary to produce a single package of flip chip. Among the process steps, the mass reflow process establishes metallurgical interconnects through reflow profile described in Fig. 7.5b.

The thermal reflow process would take longer than several minutes, and the temperate ranges and duration times at each zone can be optimized through number of times of iteration in order to design high, stable yield process compatible with solder alloy composition and process materials of targeting products. More details can be found on the publications about traditional flip chip packaging assembly process [2, 3, 6].

In contrast, TCB can complete a bonding in two or three steps of assembly process as shown in Fig.7.6. Figure 7.6 describes TCB process using epoxy flux that is also called as filled Non-conductive Paste (NCP) or no-flow underfill. The TCB process can accomplish metallurgical interconnects as well as underfill curing at the same time. TCB offers apparently simpler process steps while complexity of the

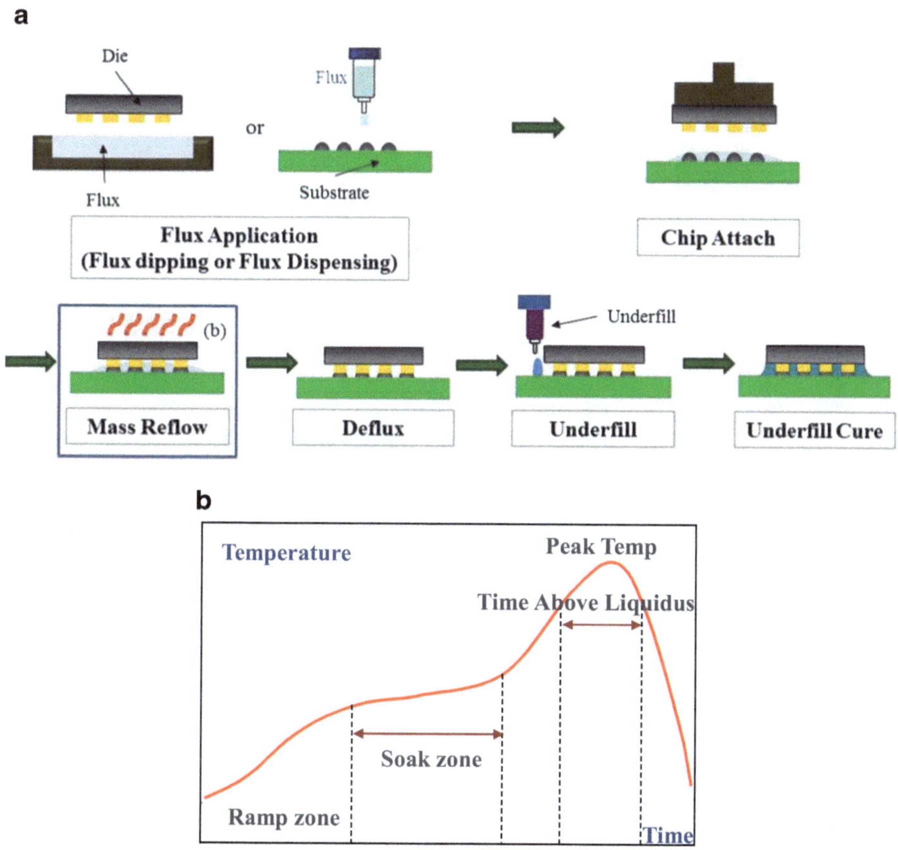

Fig. 7.5 Traditional flip chip assembly process: (**a**) a schematic of assembly process flow and (**b**) reflow profile plot

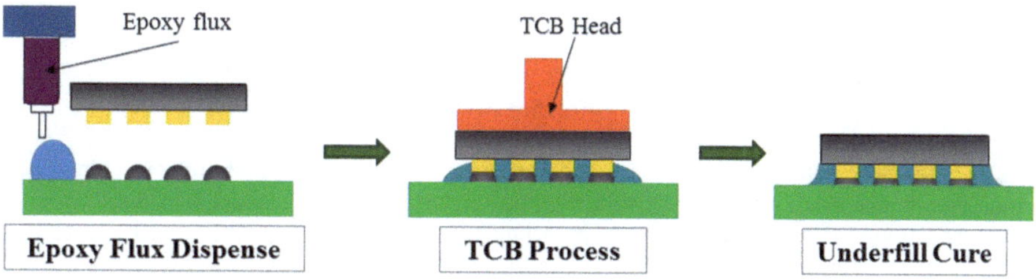

Fig. 7.6 Schematics of TCB using epoxy flux

TCB process development including process, material formulation, and equipment is much higher than that of conventional flip chip assembly process. The complexity results from simplifying or combining process steps that may be a compelling solution known to overcome technical challenges related to warpage. The warpage

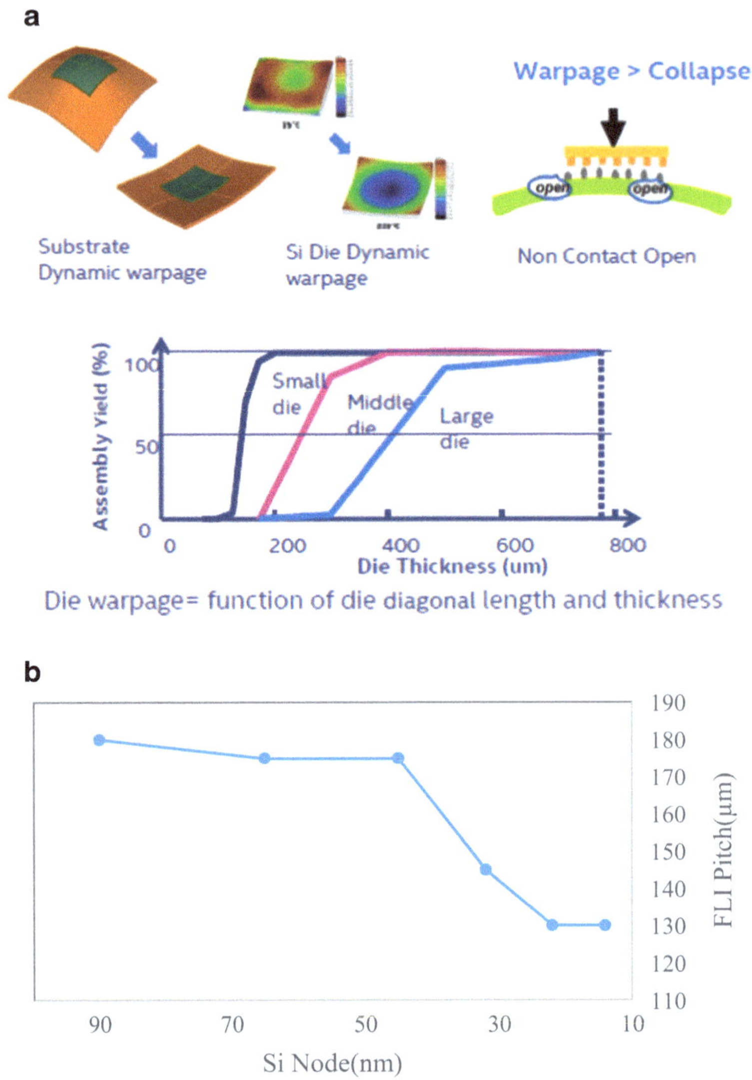

Fig. 7.7 (**a**) Schematics of dynamic warpage behavior of traditional solder collapse chip attach process with fine pitch and smaller solder bumps and (**b**) Intel pitch road map for first level interconnection with respect to silicon node technology [15]

issue addressed in Fig. 7.7 will explain why Intel starts to introduce TCB in mass production of flip chip products in 2014 even before the others consider expanding the TCB technology beyond 3D package interconnections to 2D boundary [15].

Dynamic warpage of silicon die and substrate during reflow process would not be a trivial issue any more as electronic packaging industry moves toward thinner die while reducing the pitch of first level interconnection in order to meet *Z*-height

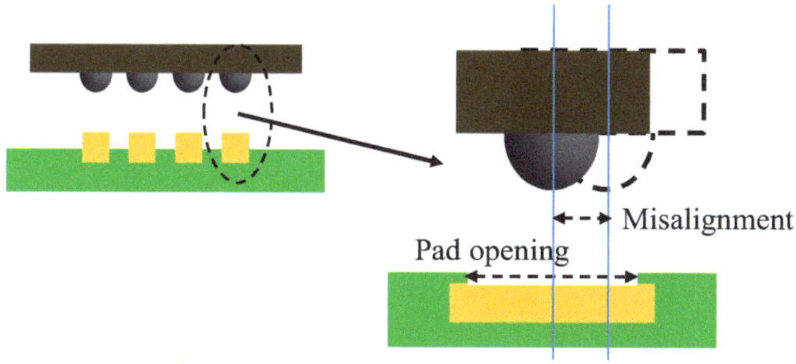

Fig. 7.8 A schematic of placement accuracy regarding design and self-alignment

requirements led by trend of consumer electronics. Suppose that die warpage is function of die diagonal length and thickness, a large die amplifies die thickness impact on assembly yield. A typical failure signature of yield loss due to warpage would be non-contact open that occurs when the amount of dynamic warpage is greater than that of solder collapse during solder reflow. To prohibit the open failures driven by warpage, TCB technology will accomplish metallurgical interconnections while controlled force and temperature are applied to both die and substrate under vacuum constrained head and stage respectively, possibly minimizing warpage of a silicon die and a substrate.

The TCB technology will require a finer placement accuracy and the accuracy of placement typically aims below ±2 μm in high volume manufacturing. The accuracy requirement can be roughly determined by taking pad opening into account (see Fig. 7.8), similar to flip chip process. Assuming that solder self-alignment can recover misalignment from chip attach process as long as bumps are placed on the half of pad opening prior to mass reflow, maximum misalignment allowable for flip chip attach with 60 μm of pads will be 30 μm. Converting into a range with respect to middle of half of pad opening, the flip chip attachment needs less than ±15 μm of placement accuracy. With the similar logic, TCB designed for 3D packages may require less than ±5 μm of placement accuracy reflecting approximate 20 μm pads of current TSV products. Most of equipment suppliers promotes that their TCB tool can guarantee even less than ±2 μm accuracy with 3σ.

Due to the finer placement accuracy requirement combined with in-situ heating and cooling mechanism of TCB, assembly throughput will significantly decrease. The fastest flip chip bonder may complete a cycle of flip chip bonding in 0.5 s, converting to 7200UPH while TCB process even using ideal equipment and materials may take much longer than flip chip. Figure 7.9 plots assembly throughput regarding cycle time and the fastest TCB process ever reported by Intel and ASM was 5 s, converting into 720UPH [16]. TCB technology remains undiscovered areas to both academia and industry in order to compete with traditional flip process in terms of cost and process stability.

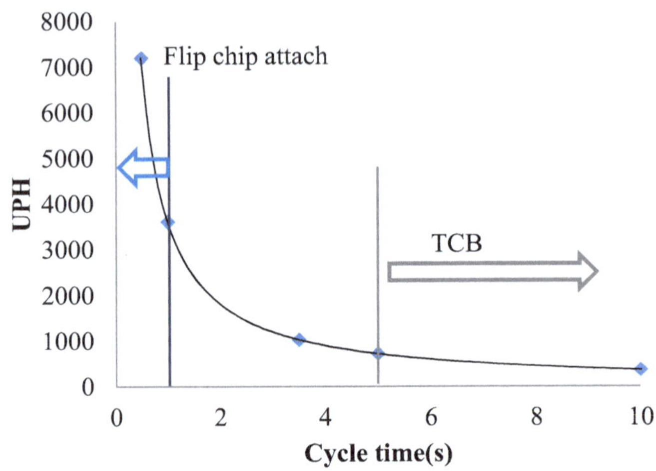

Fig. 7.9 Throughput plot: units per hour vs. cycle time

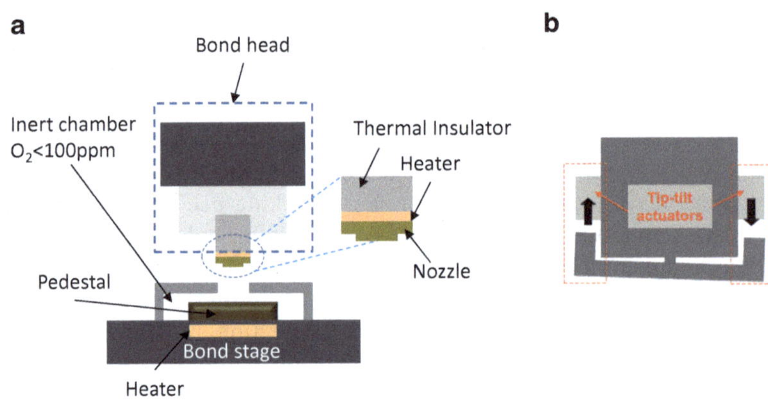

Fig. 7.10 Schematics of (**a**) TCB tool and (**b**) coplanarity adjustment

7.2.2.2 Thermal Compression Bonding Tool

The TCB technique originally was developed for metal-to-metal diffusion bonding and two metals are brought into atomic contact by applying force and heat simultaneously. The bonding technology was recently reconfigured for application of TSV die stacking and next generation flip chip packages. Figure 7.10 shows schematics of the TCB tool recently introduced to packaging industry. The bond head typically travels only in *z*-axis to guarantee the coplanarity of bond head to bond stage, an important factor to assembly yield. Intel and ASM TCB tool has been used for high volume manufacturing and the tool can ensure the coplanarity for each bonding location using tip-tilt actuators (see Fig. 7.10b) even at high operational temperature range. The head mainly comprises nozzle, heater, and insulation block. All three components should be manufactured with a high dimensional tolerance to confirm a consistent thermal expansion of the head over temperature cyclic process. Most assembly building blocks will prefer to accommodate inert environment maintained at least below 100 ppm O_2 level, and Intel and ASM TCB tool achieved a constant below 30 ppm O_2 level [16]. Heaters are implemented under pedestal where substrate, wafer, or carrier will be located during TCB process.

The hardware configuration will vary depending on machine suppliers. However, all machine manufacturers may focus on five major areas: position accuracy, coplanarity measurement and adjustment, die and substrate vacuum control, head rapid heating and cooling, and thermal management to demonstrate high, stable yield process at competitive manufacturing cost. The placement accuracy will be determined by layout of the stationary bond head and moving bond stage and vibration isolation from environment and fast-moving subcomponents is necessary for bond head to ensure robustness of placement accuracy. All moving axes with high resolution encoders and high performance of bearing can reduce motion errors. In addition, high resolution camera integrated with temperature controller will enhance alignment accuracy. *Z*-axis position accuracy of bond head at wide temperature range will be another critical aspect affecting tool performance. Figure 7.11 is examples of *Z* position accuracy of bond head within ±1 μm at 350 °C. This function is so significant to compensate variation of bump height and thermal expansion of the bond head in order to control bump height meeting a desired height of interconnections during local reflow process.

Advanced thermal management is as important as placement accuracy. Figure 7.12 shows temperature gradient of bond head heater set at 400 °C and surface temperature of bond stage maintained at 200 °C. Figure 7.12 reveals outstanding performance of insulators preventing heat transfer from heaters on head and stage.

Heating and cooling rates are additional heat-related parameters deciding TCB bonding time. Intel and ASM promote a rapid heating and cooling heater designed for a large die up to 33 mm × 22 mm and the heater demonstrates uniform temperature across the heater surface within ±5 °C at Cpk 1.33. Maximum heating without load is reported as greater than 125 °C/s in the air, expected to offer 100 °C/s heating rate with a full load comprising of nozzle and die in actual process. Cooling rate without load is reported as less than −50 °C/s in free air, expected to perform −50 °C/s with a full load. Constant vacuum pressure and real-time

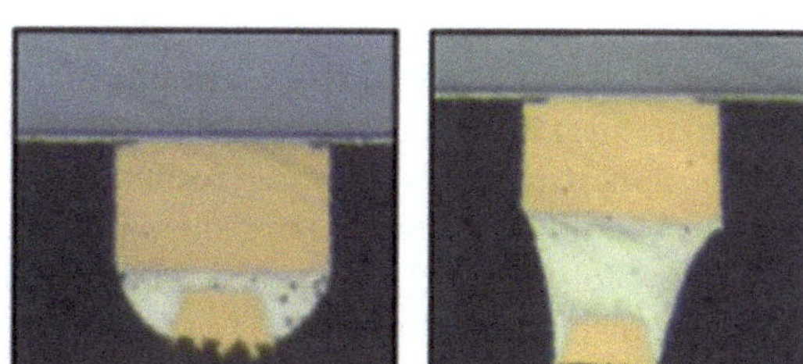

Fig. 7.11 Micrographs of microbump interconnection showing high accuracy of bond head in *z*-axis [16]

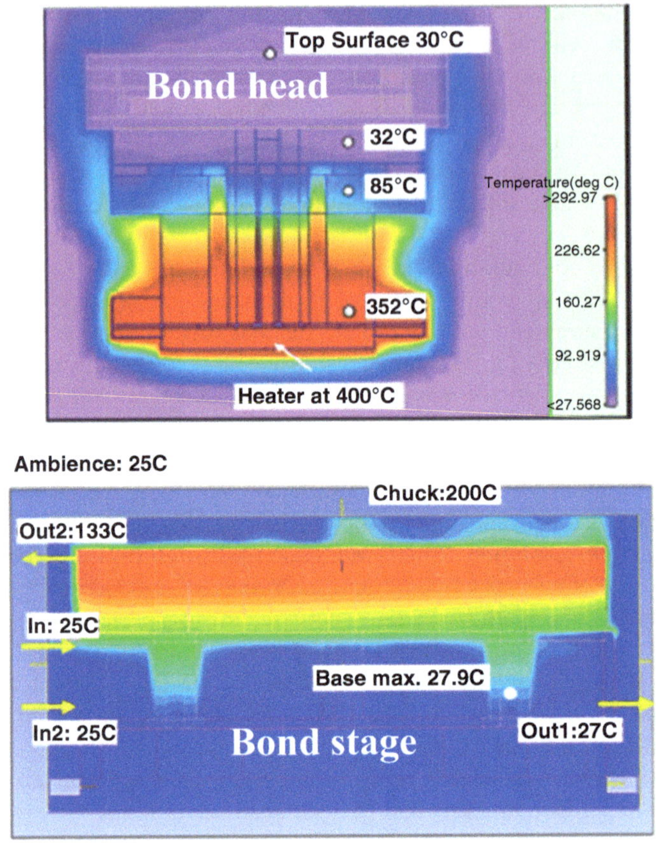

Fig. 7.12 Temperature gradient of bond head and bond stage [16]

monitoring system should be employed into TCB tool to ensure that die and substrate are held flat firmly during bonding. The monitoring system can be considered as a process gauge because vacuum leakage can indicate presence of foreign materials on bonding stage or abnormalities during bonding process.

7.2.2.3 Thermal Compression Bonding Process

Figure 7.13 shows a typical process profile of TCB head in a bonding cycle in terms of temperature, force, and displacement. These three factors are most significant process input parameters affecting assembly yields. Conversely, force and displacement also could be response parameters in assembly process.

On the plot of Fig. 7.13, the red, blue, and black curves indicate temperature, force, and displacement of a bond head, respectively. Table 7.1 explains the details about process parameters. The stage and head contact temperatures are identical

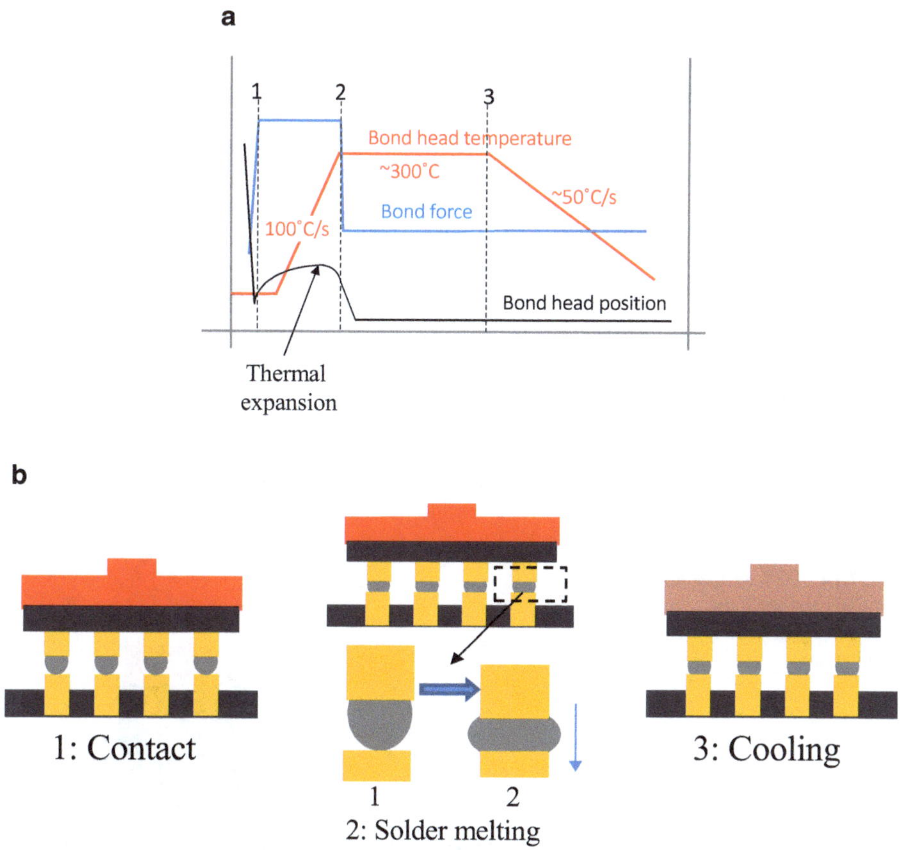

Fig. 7.13 (**a**) A typical TCB head profile including temperature, force, and displacement: (1). contact, (2) solder melting, and (3) cooling, and (**b**) schematics of significant events triggering to ramp head temperature, to activate position control mode (melting detection), and to release bond head after a cycle of bonding

and the temperature will be limited by building blocks, process materials. Ideally, the higher temperature is selected for stage and bond head contact, the shorter bonding process time is expected.

As soon as a bond head detects the bond head on contact (see status 1 of Fig. 7.13) when the head measures the contact force, the head temperature will rise up to the peak temperature that is much higher than solder melting point. A typical peak temperature for SAC 305 (96.5 % Sn, 3 % Ag, and 0.5 % Cu) is recommended as 300 °C because of a huge temperature delta across the package as displayed in Fig. 7.14. The temperature delta results from the extreme temperature difference between head and stage, maintained at 300 and 160 °C, respectively. Even though temperature of die surface adjacent to heater will be close to 300 °C, solder bumps on the outermost may hardly reach solder melting point in

Table 7.1 Process parameters

Process parameter	Recipe parameter	Comments
Force (input and response)	Contact force	Contact indicator between bump and pad
	Bond force	Maximum force
Temperature (input)	Bond head contact temperature	Process material dependent
	Bond head peak temperature	Maximum temperature
	Bond head release temperature	Sold alloy composition dependent
	Bond stage temperature	Process material dependent
	Bond head ramp rate	100 °C/s>
	Bond head cooling rate	−50 °C/s<
Time (optional input)	Dwell time at peak temperature of bond head	
Displacement (input and response)	Melting detection setting	~5 μm
	Z-height chase (optional)	Product dependent

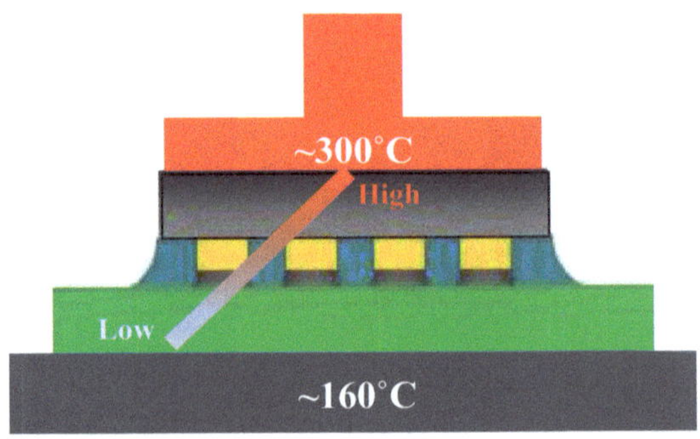

Fig. 7.14 TCB temperature delta across package between bond head and bond stage

case of large die packages. Namely, the heater temperature should increase substantially higher than solder melting point in order to ensure solder bumps on outermost to heat above solder melting point.

Suppose that the bond head contacts at about 200 °C, increases its temperature with a ramp rate of 100 °C/s up to 300 °C, holds the peak temperature in less than one second, and cools down to 200 °C with a cooling rate of −50 °C/s, theoretically total bonding time excluding material handling time can complete less than 4 s as Intel and ASM reported [11, 16].

The blue curve as shown in Fig. 7.13, bond head will maintain a constant force until solder becomes molten and resultant force of the head suddenly drops when entire solders reach melting point. Once the bond head typically has measured 5 μm (see Table 7.1) displacement after the sudden drop occurred on the head, indicating solder melting, TCB tool activates position control mode converting from force control mode and holds the head position because reaction force will not be measured anymore with molten solders. Note that bump height variation may be about 5 μm in flip chip bumping process, influencing selection of melting detection.

In addition, the bond head may be required to travel further down to avoid noncontact open failure depending on form factor. Then the bond head immediately decreases temperature below solder solidification point and disengages from the package. Remind that O_2 level is recommended to maintain at least below 100 ppm in order to create solder wetting friendly environment and produce less stingy flux residue [17]. The head temperature highlighted by red will rise up the peak temperature with controlled ramp rate as faster as heater can heat up after the bond head measures a certain reaction force, termed as a contact force determined by package configuration and process materials. As the blue curve shows, the reaction force will suddenly drop when solder reaches melting point and starts to collapse. As temperature of bond head increases, the bond head expands according to coefficient of thermal expansion (CTE) of each layer of package during the head maintains a constant force that means that system is under force control mode.

With the aforementioned fundamentals about TCB tool and process, let review details of process signatures observed on parts assembled using TCB process. SEM micrographs (see Fig. 7.15) compare TCB process with mass reflow process in terms of intermetallic (IMC) quality of solder interconnection. Continuous, consistent IMC was formed on the unit assembled using TCB even though thermal process is much shorter than mass reflow process. Remind that TCB may have several seconds of Time Above Liquidus (TAL) and in contrast, mass reflow may have at least 60s TAL.

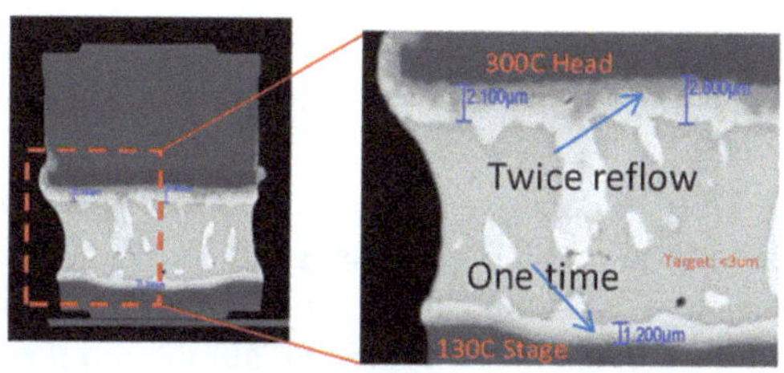

(a)

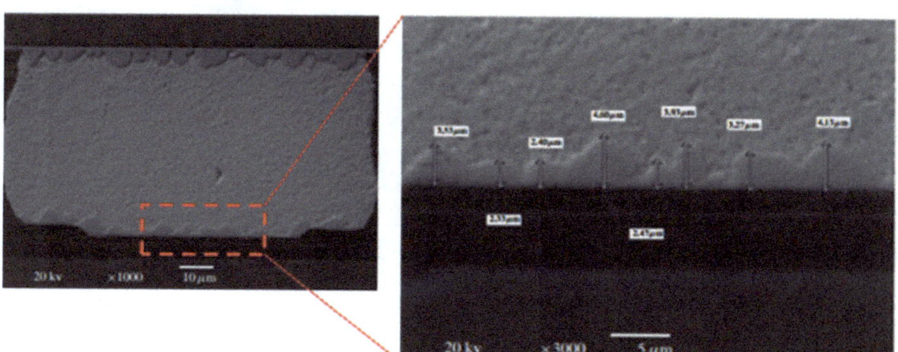

(b)

Fig. 7.15 SEM micrographs of intermetallic thickness measurement: (**a**) microbump assembled using TCB [18] and (**b**) C4 bump assembled using mass reflow process [17]

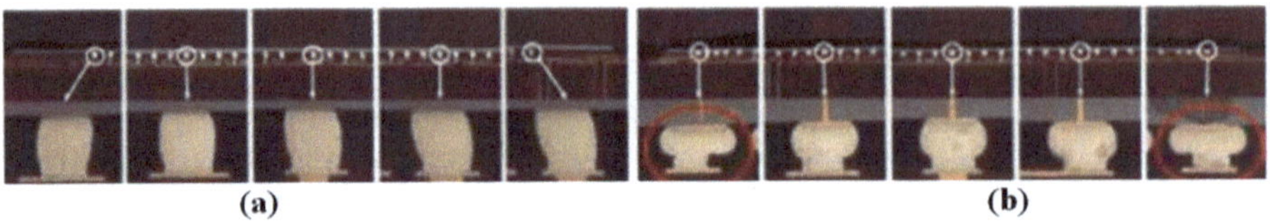

Fig. 7.16 Visual micrographs of microbump interconnection comparison for interposer attachment to organic substrate: (**a**) TCB and (**b**) mass reflow process [19]

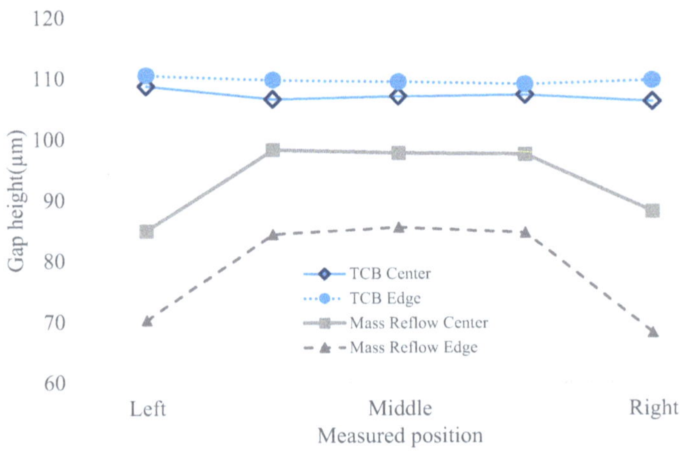

Fig. 7.17 Plot of measured gap height between interposer and BGA substrate [19]

Furthermore, Fig. 7.16 compares TCB with reflow process in gap height to emphasize the characteristic of TCB process. Both TCB and mass reflow attached an identical interposer to a BGA substrate. The gap height of interposer attached using TCB shows taller than that of interposer manufactured through reflow process because TCB can control displacement of bond head with high accuracy.

The gap height measured and plotted in Fig. 7.17 confirms that TCB offers a consistent gap height that was achieved by that TCB head controls the gap between die and substrate under vacuum support. The consistent height reflects warpage advantages of TCB-assembled package, related to in situ thermal process under high vacuum conditions. Remind that head temperature will rise up to 300 °C and stage temperature remains constant during TCB reflow. The stage temperature typically could be set at 160 °C, determined by process materials. The temperature difference would limit to design process parameters; on the other hand, it could compensate for Coefficient of Thermal Expansion (CTE) mismatch between silicon interposer (~3 ppm) and substrate (~20 ppm) because the temperature of bond stage, where higher CTE of substrate is placed, will be maintained lower than bond head [11]. In contrast, the mass reflowed unit exhibits about 10 μm difference of gap height between center and edge (see Fig. 7.17), which is generally observed on the unit assembled using mass reflow due to CTE mismatch in packages.

7.2.3 Fundamentals of Process Materials

7.2.3.1 Introduction

TCB technology is an enabler recently redeveloped for stacking microbumped 3D TSV dies in order to overcome technical challenges such as fine placement accuracy requirement and higher warpage associated with demands on thickness reduction. The warpage is expected on even 2D flip chip assembly process according to package roadmap and the warpage challenge has been pushing reflow process beyond its limit and thus Intel firstly announced TCB high volume production for 22 nm silicon nodes with 130 μm pitch flip chip in 2014 [11]. Because the in situ bonding technology needs a completely different type of material formulation, this session will provide fundamentals and engineering sciences about process materials with emphasis on flux and underfill. In general, the process material, assembly building block, can choose one from among water-soluble flux, no-clean flux, Non-Conductive Film (NCF), Non-Conductive Paste (NCP), and flux included underfill termed as epoxy flux. Fluxing action will be the most important functionality in all building blocks designed for TCB. How much residue remains after TCB is as important as fluxing capability. Regarding epoxy-based materials, void aspect also needs comprehensive investigations that cannot be performed only using analytical chemical metrology.

This session reviews analytical chemical metrology widely used to examine material properties impact on thermal behavior in order to characterize material formulation. The session also presents systematic fundamental studies using a simple structure of test vehicles to evaluate the materials in terms of voiding and wetting performance under thermal conditions close to actual assembly process while isolating unexpected side effects of equipment and process conditions [1–10]. The technical methods to evaluate material candidates, which will be introduced in this session, intend to help readers how to evaluate process materials and thus how to design preliminary process conditions effectively.

7.2.3.2 Basic Properties Measurement

Traditional thermal analysis is necessary to characterize material formulation and the analysis can be performed using Thermogravimetric Analysis (TGA), Differential Scanning Calorimetry (DSC), Dynamic Mechanical Analysis (DMA), and Thermomechanical analysis (TMA), which reveal quantitative results responding to material formulation, which are standard methods to identify properties of material (Table 7.2) [20–25].

TGA is a method of thermal analysis to measure the amount and rate of change in the weight of a material as a function of increasing temperature with a constant heating rate, or as a function of time with a constant temperature. The measurement technique has been used primarily to determine the composition of materials and to

Table 7.2 Thermal analyzer used for standard methods

Problems	Properties	Analysis	Standard method
Delamination	CTE	TMA	IPC TM-650 2.4.24.1
	Decomposition temperature	TGA	ASTM D3850
	Glass transition temperature	DSC	IPC TM-650 2.4.25C
		TMA	IPC TM-650 2.4.24C
		DMA	IPC TM-650 2.4.24.2
Through hole reliability	CTE (*Z*-axis)	TMA	IPC TM-650 2.4.24.1
Bad thermal stability	Glass transition temperature	DSC/DMA	IPC TM-650 2.4.25C
	Moisture content	TGA	IPC TM-650 2.4.24C
	Decomposition temperature	TGA	IPC TM-650 2.4.24.2
	Modulus	DMA	IPC TM-650 2.4.24.4
Size stability	CTE (*XY*-axis)	TMA	IPC TM-650 2.4.24C

Source: PerkinElmer

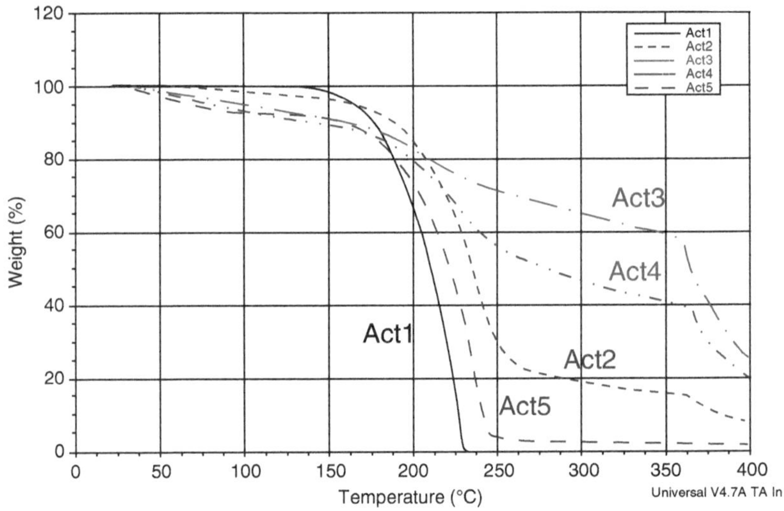

Fig. 7.18 Thermogravimetric analysis [26]

predict their thermal stability. The technique can characterize materials that exhibit weight loss or gain due to decomposition, oxidation, or dehydration. Namely, TGA can provide information about physical and chemical phenomena by measuring rate of change in the weight. Figure 7.18 plots mass loss of fluxes with five different types of activators in a common solvent as a function of increasing temperature and the activators are chemicals that react with oxides. Typical activators are carboxylic acids. Act1 is a dry compound, as only 1.7 % weight loss was observed at 150 °C. On the other hand, the slopes of the initial TGA curves of Act 3~5 are comparable. Act3 and Act4 pose different patterns of weight loss, where much slower decompositions were achieved in the 150~350 °C range. TGA curves of Act 1, 2, and 5 will be an ideal trend for no-clean flux, and curves of Act 3 and 4 close to traditional water-soluble flux.

The rate of mass loss will be influenced by base acid, acidity, solvent, etc., which also can affect wetting performance of flux. Unlike reflow process can directly leverage measurement data from TGA to decide preliminary process parameters; TCB cannot use the rate of mass loss from TGA to design process conditions because actual solder reflow time in TCB will be less than few seconds. No metrology exists to investigate behavior of material during thermal process occurring in few seconds. Therefore, wetting or voiding studies are necessary to follow in order to understand materials' capability under thermal condition similar to TCB process.

Figure 7.19 plots heat flow measured at different heating rates and different constant temperatures, and DSC measures the difference for heat required to increase sample temperature up to reference point as a function of temperature or time. The basic principle of this measurement technique allows identifying whether the process is exothermic or endothermic when the sample undergoes a physical transformation such as phase transitions (see Fig. 7.19c). Underfill curing generates heat, exothermic reaction, and solder wetting extracts heat from surrounding, endothermic reaction. This analysis could investigate homogeneity of flux and the measurement will be useful to understand thermal behavior of epoxy-based materials such as epoxy flux, nonconductive paste or film, and no-flow underfill. DSC will be conducted with different heating rate and constant temperatures to characterize material formulations.

Thermomechanical Analysis (TMA) is a technique that measures changes in sample size as a function of temperature, time, and load using testing geometries similar to those used in standard mechanical testing: (a) expansion which is used for both compression and CTE studies, (b) flexure or three-point bending, (c) extension or tensile, (d) penetration, and (e) dilatometry or bulk [28]. This measurement technique can be used to identify glass transition temperature T_g, where anamorphous material shows a phase change from rigid to flexible or fluid. The T_g temperature is taken as the intersection of the tangents from the CTE lines above and below the transition (Fig. 7.20). According to the sample geometry, a general trend of T_g is observed in Fig. 7.20.

Dynamic Mechanical Analysis (DMA) is widely used to characterize properties of material as functions of temperature, time, frequency, stress, atmosphere, or a combination of these parameters. DMA also typically determines T_g of polymer-based material from electronic package industry. Figure 7.21 shows how DMA detects T_g which is seen as a large drop (a decade or more) in the storage modulus and a concurrent peak in the tan delta reported as the T_g. The value reported as the T_g varies with industry with the onset of the E' drop, the peak of the tan delta, and the peak of the E' curve being the most commonly used.

The glass transition, T_g, is a range of behavior where a single temperature is accepted as the indicator per certain standards. Different industries have used different points from the same data set that can vary as much as 15 °C. DSC,

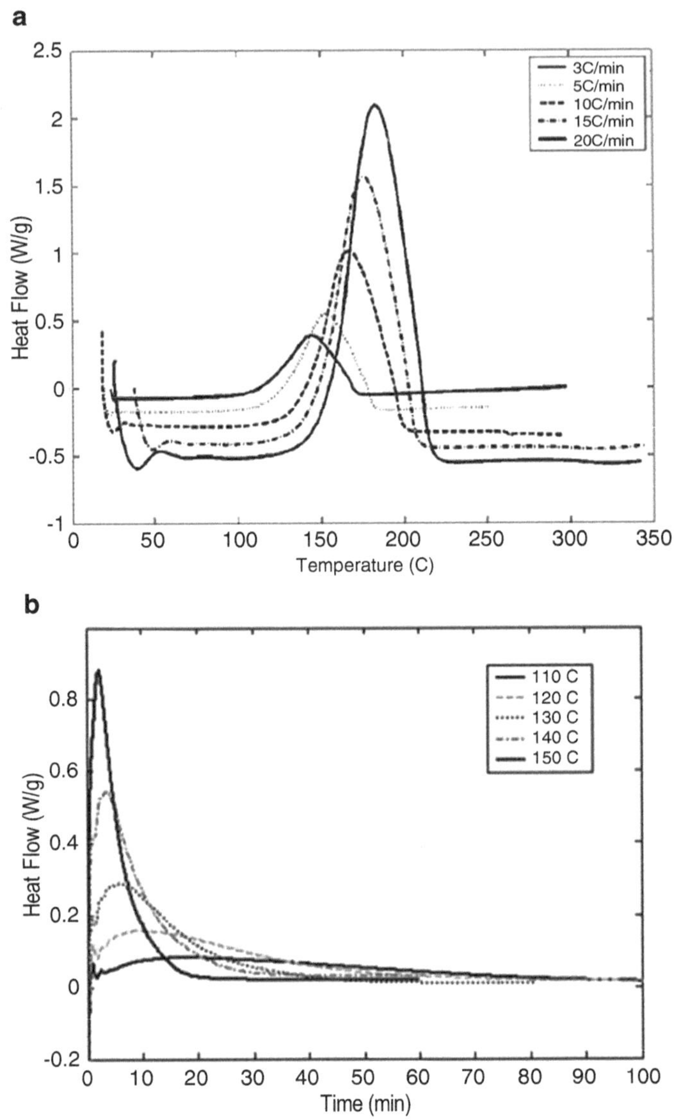

Fig. 7.19 DSC plots: (**a**) heat flow of epoxy flux measured at different heating rates, (**b**) different constant temperatures, and (**c**) solder wetting with no solder wetting in epoxy flux [1, 27]

TMA, and DMA measure different processes and therefore, the numbers vary a bit. The different instruments are sensitive to different properties and data from these experiments helps to predict how materials will respond when used in different temperature conditions. For example, DMA data is used to obtain modulus information while TMA gives coefficient of thermal expansion, or CTE. Both detect transitions, but DMA is much more sensitive.

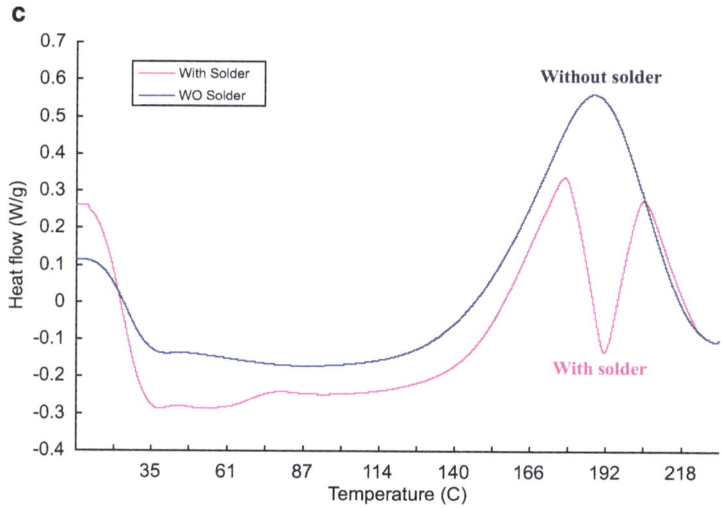

Fig. 7.19 (continued)

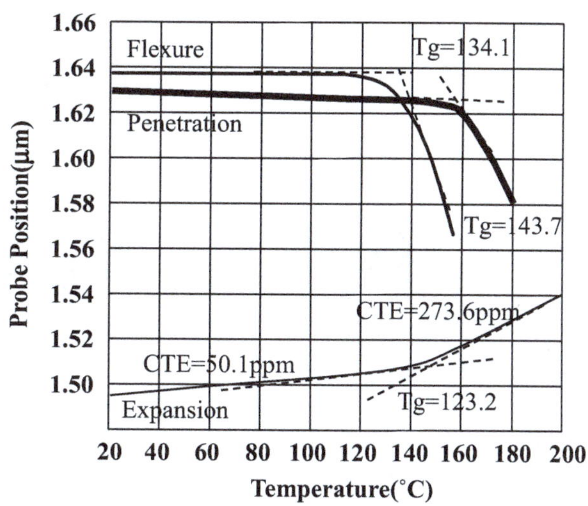

Fig. 7.20 Comparison of T_g by CTE, flexure, and penetration (*Source*: PerkinElmer)

7.2.3.3 Wetting Study

Wetting capability firstly should be investigated on the process materials regardless of assembly building blocks in TCB. This wetting study is designed to mimic rapidly heating TCB technology in order to investigate actual wetting performance of materials because existing metrology such as TGA, DSC, TMA, and DMA could not simulate thermal process less than few seconds. In addition, this experimentation could isolate the effect of equipment and test vehicle on wetting performance

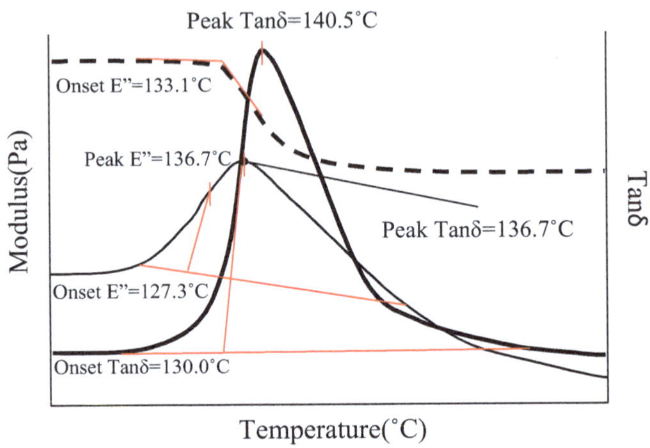

Fig. 7.21 DMA: The glass transition (T_g) temperature in storage modulus and tan delta (*Source*: PerkinElmer)

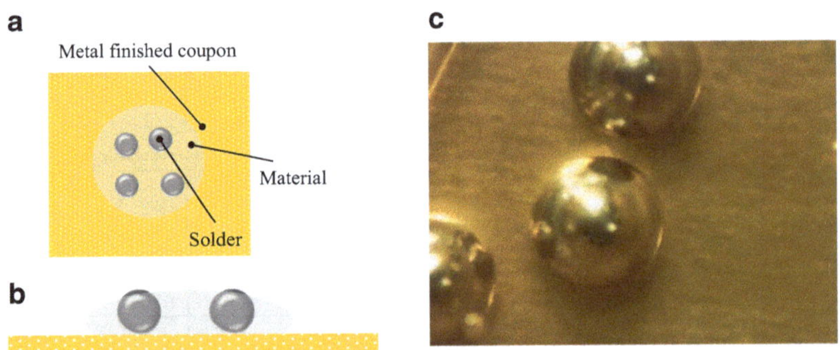

Fig. 7.22 A typical test vehicle in wetting study: schematics of (**a**) top view of test vehicle, (**b**) side view of test vehicle, and (**c**) visual micrograph of eutectic solders with epoxy flux on ENIG substrate

so that the wetting study only inspects material impact on solder wetting performance. To meet the objectives of this experiment, Fig. 7.22 proposes a test vehicle (TV) consisting of process material, solders, and metal finished coupon. TV will be prepared through two steps: dispense process materials such as flux or epoxy flux on metal finished coupon and place solders on the material-deposited TV. Electroless Nickel Immersion Gold (ENIG) finish is recommended because the gold finished surface will prevent oxidation [1–5].

TV shown in Fig. 7.22 will be placed on hotplate where temperature is set at above solder melting point under either air or nitrogen ambient. The instant exposure of TV to a high temperature mimics solder wetting process undergoing high ramp rate in TCB process. Remove TV from the hotplate in few seconds and then wetting metrics quantify wetting performance of material candidates.

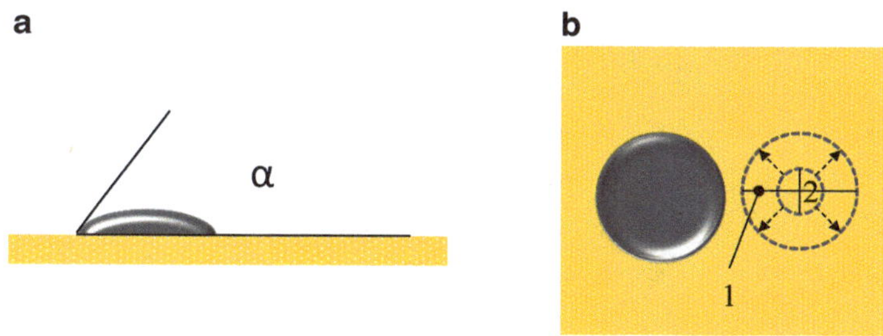

Fig. 7.23 Solder wetting metric: (**a**) contact angle and (**b**) wetting area

Figure 7.23 offers two metrics, which can be used for the wetting study and Fig. 7.23a is termed as contact angle representing wettability of solder associated with fluxing capability. Figure 7.23b describes measurement of solder wetting area ratio determined by dividing wetting area (1) by solder diameter (2) at time zero.

The results from wetting study provide preliminary process parameters and downselect underperformed materials. The well-performed candidates can be applied to mass production process to optimize process parameters compatible with assembly equipment and package configuration.

7.2.3.4 Void Formation Study

Two major memory companies, SK Hynix and Samsung, which ever released commercial TSV products, introduced HBM and 3D TSV DDR4 DRAM using TCB technology with Non-Conductive Film (NCF). The assembly building block is similar to other epoxy-based materials such as epoxy flux, no-flow underfill, and Non-Conductive Paste (NCP) in terms of basic material formulation. The material formulation mainly comprises epoxy, flux agent, curing agent, and fillers and therefore it is hard to differentiate among epoxy flux, NCP, and no-flow underfill. Voiding characteristics of the epoxy-based building blocks is a significant aspect to be examined for material selection. Voids can result from chemical reactions between solder wetting and epoxy curing or volatile chemical components exceeding its boiling point when exposing to heat in assembly process [1–10]. This void formation study suggests a material prescreening method using the test vehicle as shown in Fig. 7.24 to inspect voiding characteristics as a function of temperature and time [2–6]. TV prepared as shown in Fig. 7.24 is recommended to reflow on hotplate at a high temperature close to bond head peak temperature. In addition, the TV is necessary to be exposed to bonding stage temperature in order to identify its isothermal stability whether voids occur at even much lower than solder melting point. Investigation should be carried out with multiple levels of dwell time at each temperature.

Test vehicle can explore flux residue inducing underfill voids in assembly building block using water-soluble or no-clean fluxes. First, complete wetting

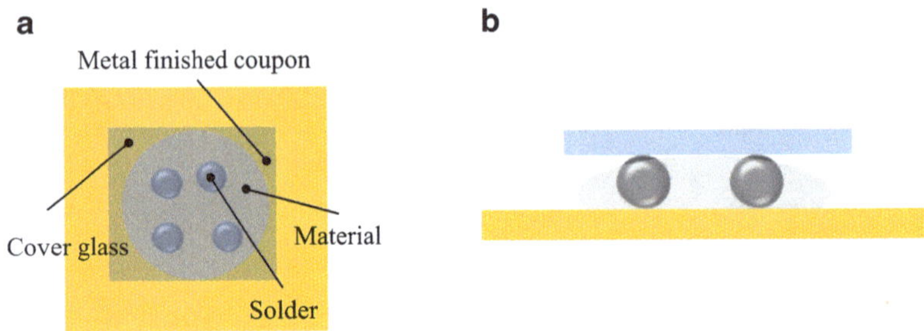

Fig. 7.24 A typical test vehicle in void formation study: (**a**) *Top* view of schematic illustration of test vehicle and (**b**) *side* view of schematic illustration of test vehicle

a

$$SnO + 2RCOOH \rightarrow Sn(RCOO)_2 + H_2O$$

$$PbO + 2RCOOH \rightarrow Sn(RCOO)_2 + H_2O$$

b

$$CuO + 2RCOOH \rightarrow Cu(RCOO)_2 + H_2O$$

$$Cu_2O + 2RCOOH \rightarrow Cu(RCOO)_2 + Cu + H_2O$$

Fig. 7.25 (**a**) Sn/Pb oxide and (**b**) Cu oxide reduction process

study to produce flux residue post solder reflow. Next dispense underfill on the same test vehicle reflowed through wetting study and a glass cover will be placed on the underfill-deposited TV. In case of water-soluble flux, deionized (DI) water should be applied to remove flux residue prior to dispensing underfill. Then cure the underfilled test vehicle and conduct a visual inspection on the test vehicle after curing to investigate voids nucleation.

7.3 Principles of Materials Formulation

Fluxing capability is most important aspect in all assembly building blocks, process materials, designed for 3D stacking process using TCB technology. The fluxing mechanism requires the right chemistry coupled with a proper heat cycle to remove the oxide and surface contamination. A synergetic combination between chemistry and processing can promote a clean and solder-wettable metal surface that is a prerequisite to achieve good metallurgical bonding. The outstanding fluxing performance in metallurgical bonding is equally important for traditional mass reflow process. Namely, the fluxing capability is common criteria on which material can be selected based for a building block with mass reflow as well as TCB. In the TCB process, materials perform oxide reduction even after exposure to a high temperature ranging 140–160 °C. In the oxide reduction, the flux agent removes oxide on the surface solder bumps and Cu pads (see Fig. 7.25). Figure 7.25 describes how oxide can be removed using carboxylic acid as a flux agent or activator to provide good wetting condition [1, 9].

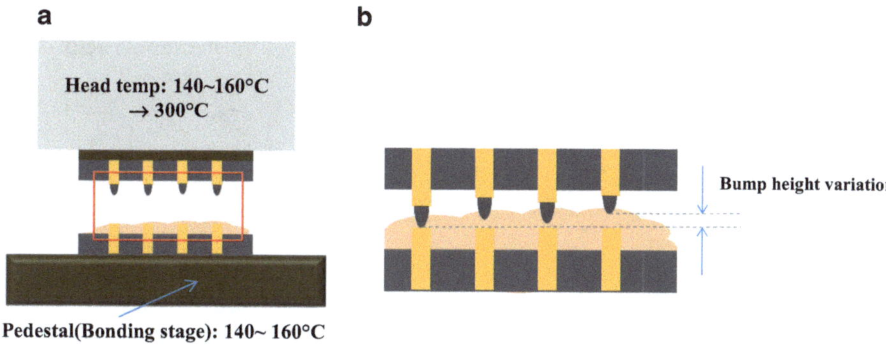

Fig. 7.26 (**a**) Schematics of TCB head and pedestal (bonding stage) and (**b**) bump height variation

In particular, TCB will not require the same level of tackiness in flux formula as expected in mass reflow process. The tackiness is an essential characteristic of flux designed for mass reflow to hold position of die attached to substrate within allowable limit for solder self-alignment. However, TCB will not let solder self-alignment engage assembly process by controlling position with a high accuracy in all three axis. Instead of the tackiness, materials are required to exhibit a longer isothermal stability in terms of fluxing performance and mechanical properties due to thermal boundary conditions as given in Fig. 7.26a. Figure 7.26b shows tolerance of bumps on die, which should be reflected on material formulation to guarantee material's functionality as a heat transfer media.

Epoxy-based building blocks with TCB need careful consideration on fillers in terms of size and concentration, affecting reliability performance and assembly yields. The materials should preserve lowest viscosity, expected to appear near T_g, when head and stage are at contact temperature prior to bonding a die to a bonding site during materials handling including die transfer and vision recognition. The viscosity, termed as isothermal viscosity, is a significant factor contributing to occurrence of filler entrapment that is a typical root cause of open failures in TCB with epoxy-based materials.

This session reviews major assembly process materials compatible with TCB technology with emphasis on formulation, technical challenges, and advantages in order to provide guidance for optimal process design. With the fundamental understanding, readers are expected to select a building block compatible with package configuration and equipment, and to understand how material formulation can be evaluated to optimize assembly process.

7.3.1 Water-Soluble Flux

Water-soluble flux is based on organic acid providing good cleaning performance. The flux will produce corrosive residue after solder reflow so pressurized water wash process is necessary to eliminate the flux residue. As the water-soluble flux

Table 7.3 Typical carboxylic acids as a flux agent

Acetic acid	Oxalic acid	Malonic acid	Pentanoic acid
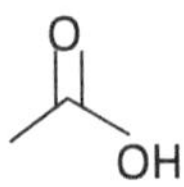	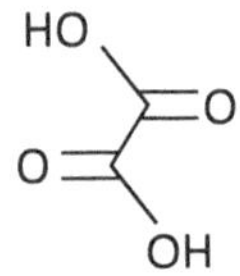	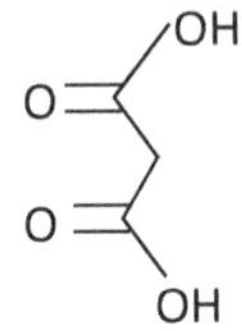	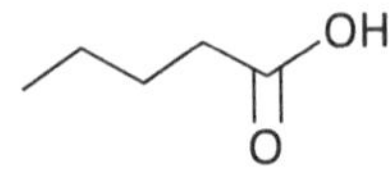

has been used for traditional flip chip assembly process with mass reflow process, the flux can be used to attach 2.5D interposer to substrate using mass reflow process. The flux also can be applied to TCB technology with no major changes on flux formulation [11]. The traditional water-soluble flux comprises of vehicles, activators, surfactants, thickening agents, and solvents. The vehicle uses high-temperature tolerant chemicals in the form of liquid or solid and the material will be generally liquid at targeted solder melting point. The vehicle materials dissolve by-products from reaction between activators and oxides, and carry the chemical components away from the surface of wetting. Flux activators remove oxides on metal surface to form a strong metallurgical bond. Typical activators are carboxylic acids and Table 7.3 below shows acids widely used for activators in flux (see Fig. 7.25).

In addition to chemical reaction, base material will perform as a heat transfer media. Surfactants help flux to spread well, namely decreasing surface tension of flux. In similar, solvents, which dissolve components, could control viscosity of flux to dispense flux.

7.3.2 No-Clean Flux

No-clean flux does not mean that no-flux residue remains after solder reflow. No-clean flux generally have the same aggressiveness as Rosin Mildly Activated (RMA) fluxes while achieving much lower residue by using lower solid content, compared to water-soluble flux. Solid content refers to the ratio of solvent thinner to solid component in the flux. A typical no-clean flux has less than 15 % solids, compared with 50 % in other types of flux. Unlike conventional solder reflow process, TCB technology for 3D package stacking can meet less than ±2.0 μm placement accuracy in process so that TCB does not require flux tackiness that is necessary to maintain position of die attached on substrate within allowable offset range for solder self-alignment during mass reflow process. On the other hand, no tackiness requirement allows simplicity of no-clean flux formulation and focuses on enabling no-clean flux to achieve low residue characteristics and consequently, deflux process can be skipped. To demonstrate no-deflux process, two conditions

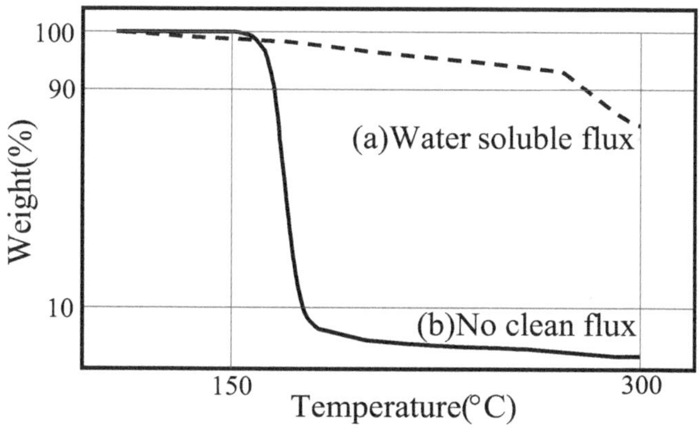

Fig. 7.27 Thermogravimetric analysis comparison: (**a**) water-soluble flux and (**b**) no-clean flux

should be met: dispensing a least amount of flux only barely to cover entire area of pads or bumps on substrate and controlling oxygen level of bonding ambient at least below or even further lower than 100 ppm. The wetting friendly condition will permit to dispense least amount of flux, minimizing source of flux residue, and remain less corrosive residue.

TGA (Fig. 7.27) shows two different types of mass loss plots as a function of temperature. (a) is a typical mass loss of water-soluble flux and (b) is an ideal trend of mass loss for no-clean flux designed for TCB technology. To demonstrate such significant mass loss while not losing wetting performance, no-clean flux in TCB is recommended to adopt high percentage of solvent content with least amount of acid-based activator. The majority of solvent will evaporate as soon as the flux is dispensed. After solvent evaporates, thin layer of acid will be created on the surface of solder bumps or pads, which would be sufficient to achieve metallurgical bonding under low concentration of O_2 less than 100 ppm. Another suggestion is to use noncorrosive solvent necessary to dispense activators. That is the way to establish no-clean flux TCB with no-deflux process by minimizing the inescapable corrosive flux residue.

7.3.3 Capillary Underfill

Capillary Underfill (CUF) is still an important process material for the building block with mass reflow process using water-soluble or no-clean fluxes in 2.5/3D process. The building block can be applied to attach a 2.5D interposer to a substrate. CUF could flow the extreme fine gap as shown in Fig. 7.28 as long as viscosity and filler size are well optimized. The figure even shows underfill encapsulation in a 3 μm Si and Si gap. Technically speaking, CUF has no barrier in filling a gap. However, CUF is limited to use below 120 μm pitch flip chip application due to flux

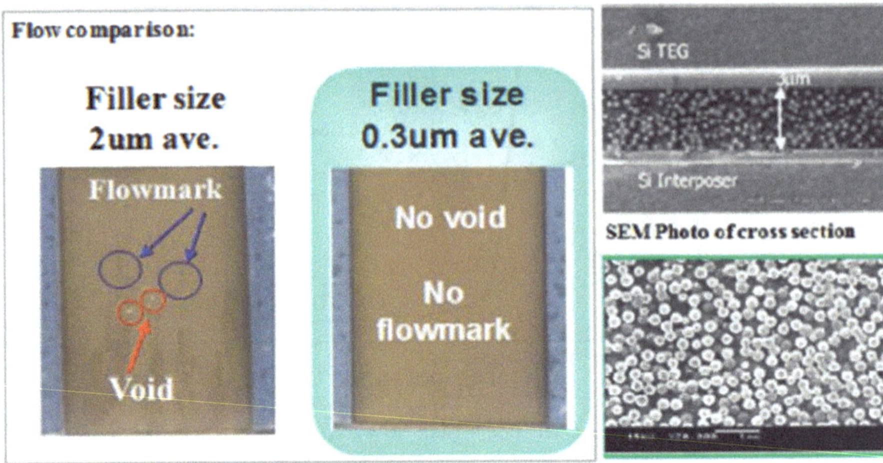

Fig. 7.28 Capillary underfill evaluation [29]

residue. Below 120 μm pitch, deflux process cannot guarantee flux residue free in high volume manufacturing process and the residue can impede flow characteristics, resulting in underfill voids and underfill delamination. Such defects will negatively affect reliability performance. Therefore, industry has been developing both advanced cleaning process and no-clean flux in order to overcome this issue [29].

Continuous demand on Keep Out Zone (KOZ) reduction combined with thin die thickness would be another technical challenge in CUF so that Molded Underfill (MUF) was introduced for alternative encapsulation technique in flip chip application. Due to above technical challenges regarding bump pitch, KOZ (or die-to-die space), and die thickness (~50 μm), 3D packages stacking cannot implement CUF process into microbumps interconnections. Instead, TCB technology assumes pre-applied epoxy materials or epoxy flux for 3D stacking to accomplish microbumps reflow and encapsulation through a single thermal process.

7.3.4 Epoxy Flux (No-Flow Underfill or Non-Conductive Paste)

Epoxy flux can be called as no-flow underfill or non-conductive paste. TCB technology with the process material will complete microbumped solder reflow and encapsulation together. The material performs function of both flux and epoxy-based underfill in process so that it is termed as epoxy flux and as a result, deflux will be eliminated. In general, TCB with epoxy flux can offer ideal building block thanks to process simplicity and assembly throughput. On the other hand, the

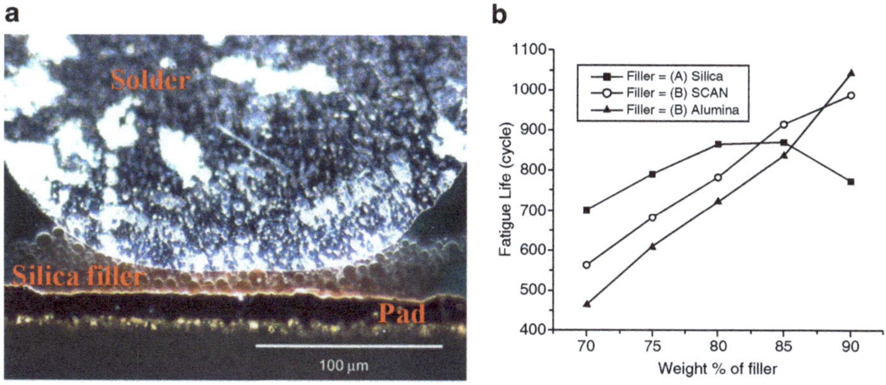

Fig. 7.29 (**a**) A visual micrograph of typical open failure due to filler entrapment and (**b**) plot of fatigue life with respect to weight percentage of filler [1]

material formulation limits its application because of concerns about reliability performance and assembly yield, associated with filler concentration and package configuration. An increase of filler concentration typically would improve solder joints reliability performance while the increasing filler concentration will increase the possibility of filler entrapment between bumps and pads (see Fig. 7.29), consequently resulting in electrical open failures.

In addition to solder joint reliability associated with filler concentration, underfill may need to protect fragile low-k dielectric layers. The demand on low-k protection can be affected by package configuration including die to die (silicon to silicon) or die to substrate or module to substrate. Die to package integration challenges are most often observed on the flip chip packages with a large die where the differences of CTE between the package and die become more pronounced [30]. In general, solder joint reliability requires stiff and rigid underfills while more compliant underfill properties are expected to protect low-k layers. Considerations for successful material candidates include CTE, modulus, Poisson's ratio, toughness, and T_g, which are covariant; a change to one property also changes the others. Underfill CTE and modulus are loosely correlated at temperatures less than T_g although the modulus is changed to a certain extent with fillers and additives (see Fig. 7.30). Higher T_g materials generally have higher modulus and lower CTE over the operating temperature. Mechanical properties also affect other characteristics such as processability and reliability. The processability of a flip chip underfill is impacted by viscosity and resin type while reliability is impacted by all mechanical properties.

Aw et al. demonstrated TCB assembly process using NCP for 30 μm pitch Cu pillar microbumps die on organic substrate with Cu bond pads [31]. They also address the technical challenge, filler entrapment-induced failure, in their research. Figure 7.31 shows a typical failure, induced by filler entrapment, on microbump interconnection.

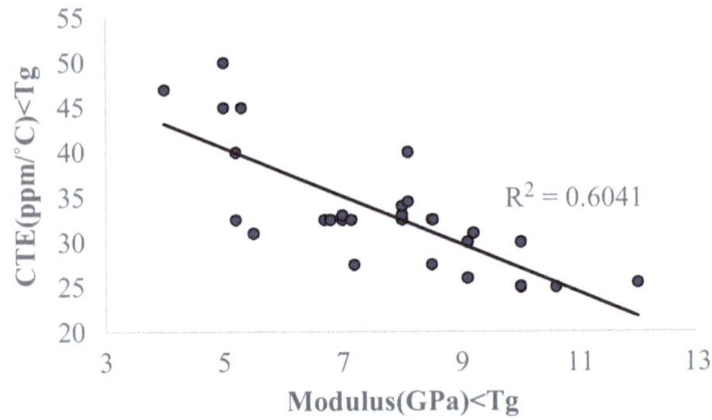

Fig. 7.30 Correlation between CTE ($<T_g$) and E ($<T_g$) [30]

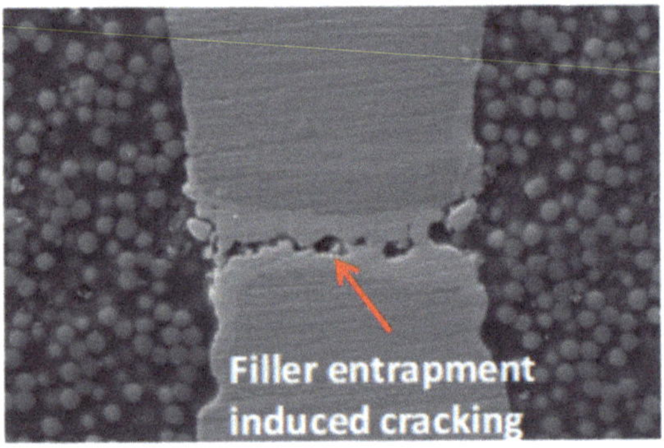

Fig. 7.31 A SEM micrograph of typical open failure due to filler entrapment [31]

Jin-Ye et al. investigated process conditions and TV design impacts on entrapment that is a well-known technical challenge in TCB assembly process using NCP (see Fig. 7.32) [32]. They attempted to discover the interaction bond force with bump design.

Apparently, epoxy flux process development looks simple. However, this process development will be most technically complicated building block to define process parameters in mass production. Solid understanding on polymer-based material and TCB technology are necessary to apply epoxy flux to actual assembly process. Even though a high yield process is achieved, poor reliability performance can prohibit its use for certain package configuration.

7.3.5 Pre-Applied Epoxy-Based Materials (Non-Conductive Film and B-Stage Material)

Non-conductive Film (NCF) and B-stage material also allow skipping flux cleaning process because encapsulation happens on wafer. In addition, this pre-applied materials include flux agent so that flux application is unnecessary like epoxy flux. Pre-applying materials on wafer benefits to assembly cost reduction from wafer-level assembly process.

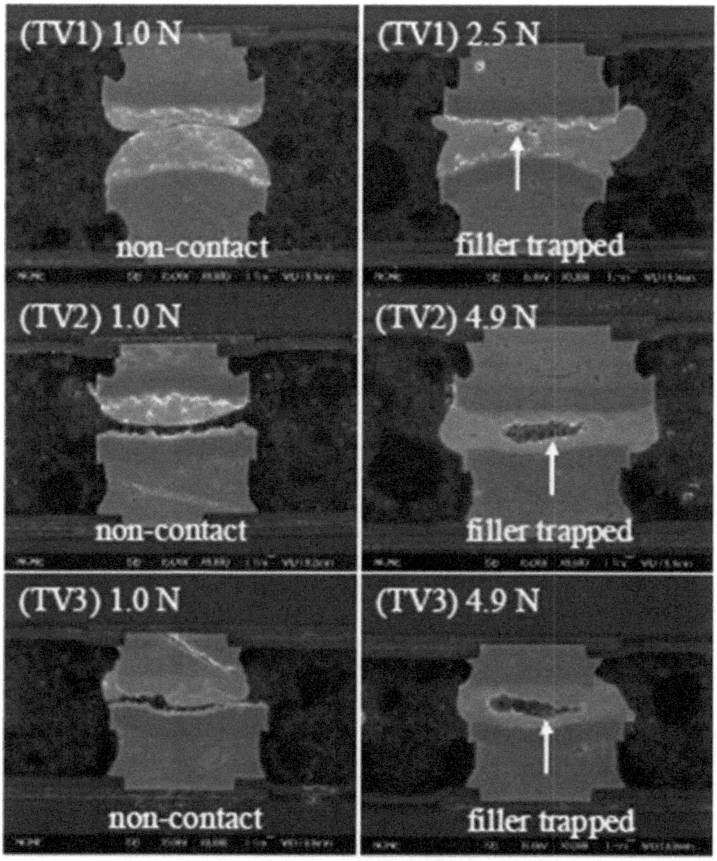

Fig. 7.32 SEM micrographs of open failure due to filler entrapment in TCB [32]

Figure 7.33 shows different configurations of pre-applied epoxy that has ever been introduced to packaging industry. Figure 7.33a–c were designed to avoid filler entrapment inducing open failures during TCB process. Both TSV HBM and RDIMM selected configuration (d), NCF, for building block with TCB technology. The NCF will be bonded to wafer through vacuum lamination process. The vacuum lamination process can produce voided NCF (see Fig. 7.34a) caused by improper process parameters or irrelevant material formulation. The pre-existing voiding can grow and form macro size voids during TCB process. Poor process parameters also can easily generate NCF voiding in bonding process as other epoxy-based materials do exhibit. Figure 7.34b, c illustrate typical failures that occur in TCB mass production. Indeed, Fig. 7.34b shows a unique NCF process signature, overflow contamination, or compressed fillet, that is resultant from NCF building block with TCB technology.

Assuming that epoxy flux is selected for the building block of memory or Application Processor (AP), assembly process requires the material to sustain at

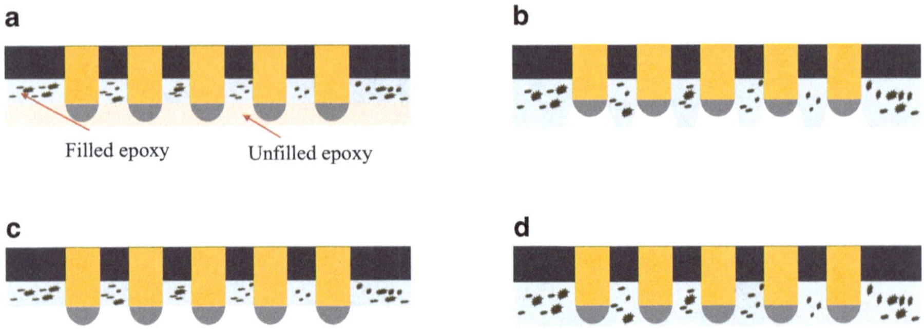

Fig. 7.33 Schematics of NCF and pre-applied epoxy-based materials: (**a**) filled and unfilled epoxy, (**b**) exposed bumps using cavity, (**c**) exposed bumps, and (**d**) typical NCF

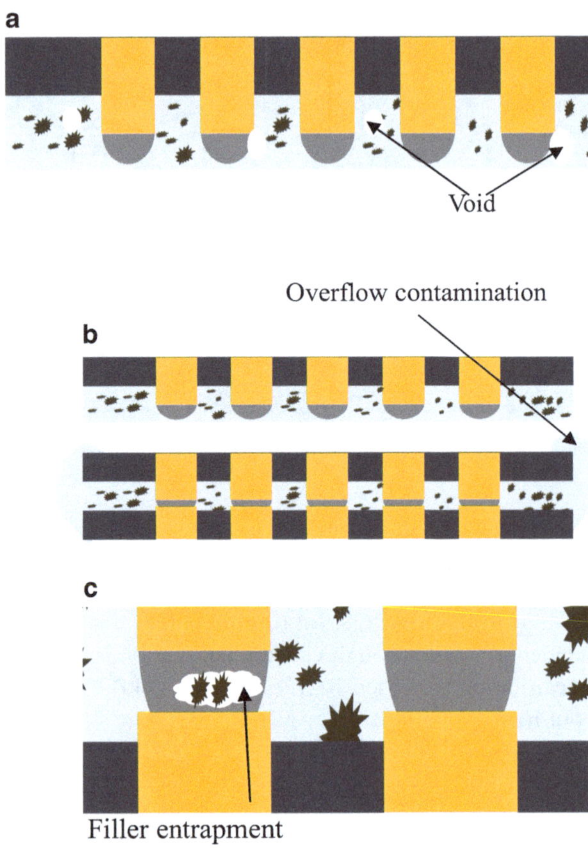

Fig. 7.34 Schematics of failures of NCF: (**a**) NCF voids in lamination process, (**b**) NCF overflow contamination in TCB, and (**c**) filler entrapment on interconnections

high temperature on bonding stage and to pose an extreme long sit-time because a unit will be bonded to multiple bonding sites on strip form substrate or wafer one by one. Dispensing epoxy flux on each site right before bonding can be considered to demonstrate the building block with TCB. Memory manufacturer has not reported introduction of epoxy flux in mass production yet.

7.4 Assembly Process Design

7.4.1 Introduction

Figure 7.35 shows a typical schematic of 2.5/3D flip chip package comprising vertically stacked High Bandwidth Memory (HBM) and an interposer with horizontally mounted HBM. The interposer is attached to GPU using traditional flip chip assembly process with mass reflow.

TCB technology adopts NCF to construct HBM by stacking TSV microbumped flip chip. The stacking process also can consider no-clean flux or epoxy flux for an alternative building block. Encapsulation material can be applied through wafer-level packaging process or by dispensing epoxy flux on substrate. Both are pre-applied methods possibly achieving encapsulation for even less than 100 μm pitch interconnection. Table 7.4 describes details on major encapsulation building blocks in terms of application, assembly process, and pitch limit.

This session explores major assembly building blocks compatible with TCB technology, and reviews assembly process and materials ever used for commercial 3D TSV products. With technical insight on bonding technology with process materials, this session will provide process development procedure in order to share how to design assembly building blocks relevant to package configuration and to optimize process parameters with the selected process materials.

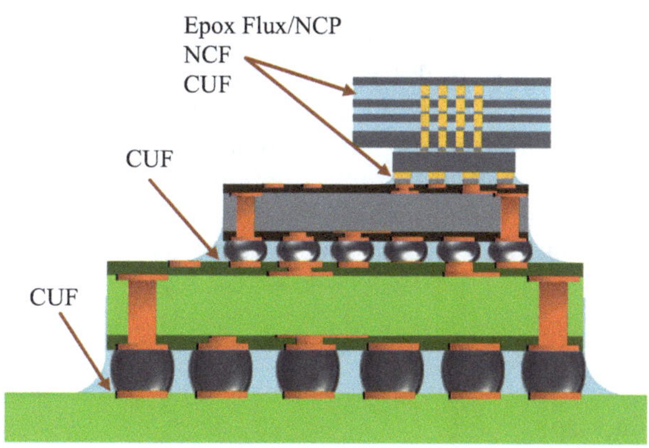

Fig. 7.35 A schematic of 2.5/3D flip chip package

Table 7.4 Major encapsulation building blocks and pitch limit

Process step	Pre applied		Post chip attach
Material	Non-conductive film (NCF)/B-stage Epoxy	Epoxy flux	Capillary underfill (CUF)
Application method	Lamination/spin coating on wafer	Dispensing on substrate	Dispensing on substrate
Assembly process	TCB	TCB	Reflow or TCB
Pitch limit	>10 μm	>10 μm	>100 μm

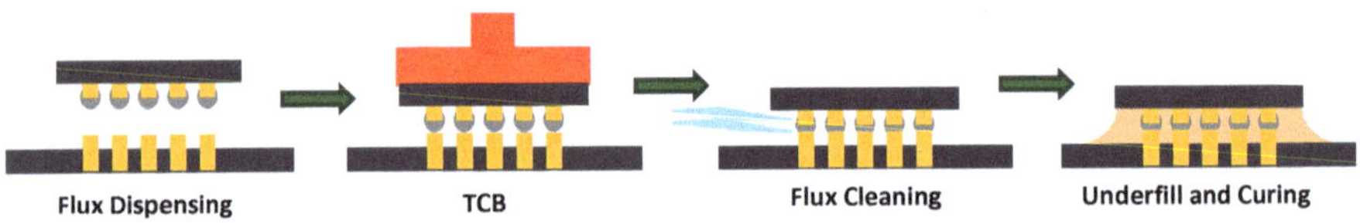

Fig. 7.36 A schematic of TCB building block: water-soluble flux

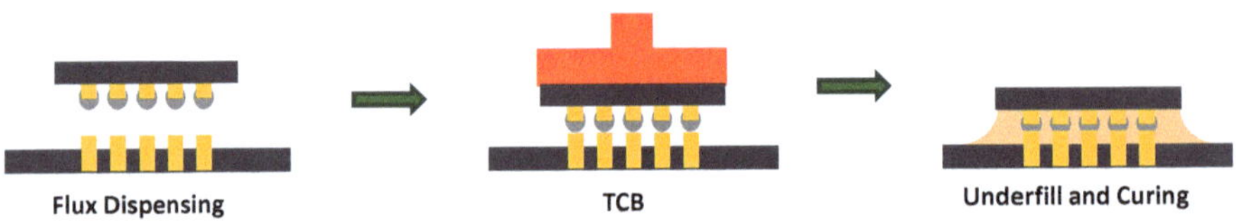

Fig. 7.37 A schematic of TCB building block: no-clean flux

7.4.2 TCB Assembly Building Block

Figure 7.36 shows the assembly building block using water-soluble flux that TCB technology can implement into mass production with no changes on its formulation. Even though the water-soluble flux is applicable to microbump interconnection for TSV stacking, this building block will not be adopted due to bump pitch limitation in deflux and underfilling processes. A bonding completes in four steps: flux dispensing, TCB, flux cleaning, and underfilling and curing.

Figure 7.37 shows the assembly building block using no-clean flux that requires intensive research to offer formulation compatible with TCB process. A key technical challenge results from enabling no-clean flux with no deflux process before underfill dispensing. To overcome the challenge, two items should be met: generating least amount of flux residue is first step to achieve and selecting underfill well matched with flux residue can dissolve the residue during cross-linking in order to avoid underfill delamination induced by flux residue. This building block is theoretically compatible with TSV interconnections but packaging industry has not reported technical data about the implementation of no-clean flux to 3D TSV stacking.

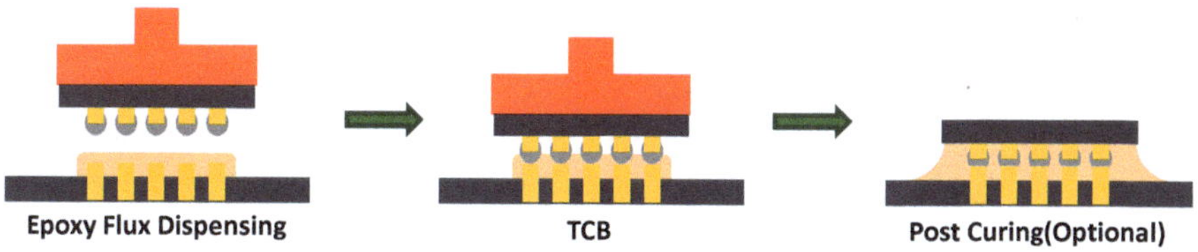

Fig. 7.38 A schematic of TCB building block: epoxy flux

Fig. 7.39 A schematic of epoxy flux deposited substrate on TCB bonding stage

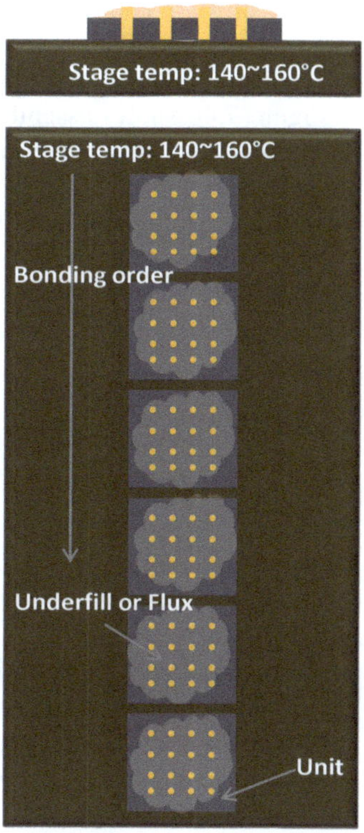

Figure 7.38 shows the assembly building block using epoxy flux. In general, epoxy-based building blocks are simpler than flux-based process in terms of process steps. The assembly process comprises two steps: epoxy flux dispensing and TCB. The two-step process completes a bonding cycle using underfill including flux agent. On the other hand, the epoxy flux is complicated to design high, stable yield process with a wide process window.

On exposing to heat, epoxy flux starts to cure (see Fig. 7.39). Even after the epoxy curing process is activated, the material should maintain viscosity and slow degree of cure prior to TCB bonding. Such time requirement is termed as sit-time. The longer sit-time is provided, the wider process window can be achieved. The sit-time with a bonding cycle time decides how many units can be placed on stage

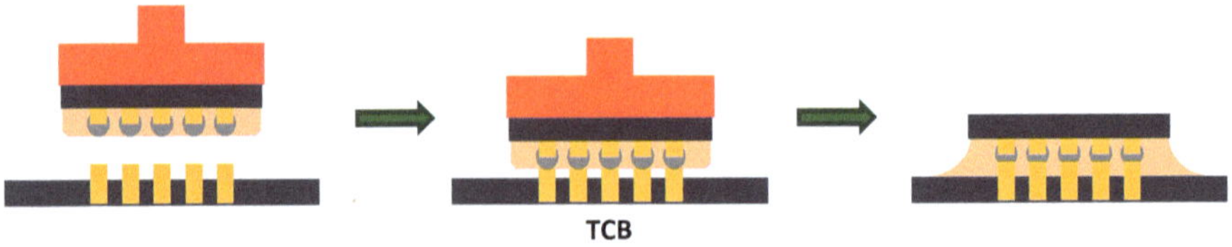

Fig. 7.40 A schematic of TCB building block: NCF

together or when epoxy flux can be dispensed. This building block is theoretically compatible with TSV interconnections but memory manufacturers have not reported its implementation to 3D TSV stacking.

Figure 7.40 shows assembly building block using NCF that is plan of record (POR) process that two memory manufacturers have released their products to the market. NCF will be attached to wafer through vacuum lamination process. TCB head picks up individual unit from the wafer and the NCF pre-applied die is bonded to die or substrate. Because of highly viscous NCF properties, the bonding cycle time has been known to very slower than other building blocks.

7.4.3 TCB Assembly Building Block Design and Development

Flow chart (see Fig. 7.41) recommends the procedure of process development for the 2.5/3D package, including material characterization and fundamental investigations.

First, simulation and historical data will define basic material properties such as T_g, viscosity, modulus, and filler concentrations. Note that the material properties are covariant and will be compromised to be suitable with package configuration. According to the preliminary results, material formulation is designed to meet initial reliability and process requirement. Next, metrology measures behavior of materials under thermal condition, predicting its performance in actual process. Then, fundamental investigations designed to examine wetting and voiding will characterize material formulation and provide baseline process parameters on TCB HVM. The preliminary process parameters are optimized to meet process requirements in terms of voids and electrical yield, and the optimal process parameters will be validated with large number of test vehicles, statistically sound number of units. Reliability tests follows to confirm its performance. Aforementioned systematic studies will be most effective method to demonstrate high, stable assembly process by reducing iteration between process and material developments.

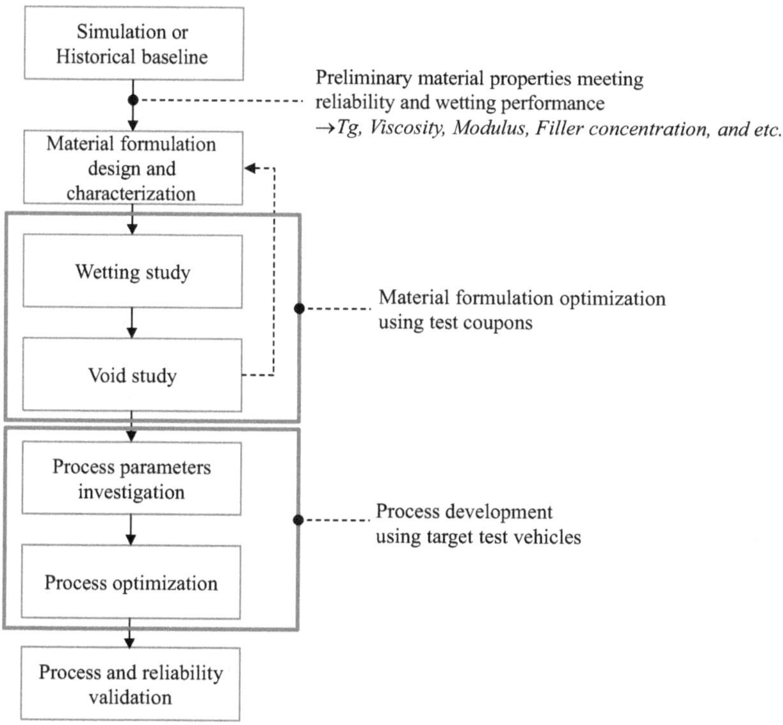

Fig. 7.41 Flowchart of material formulation characterization

7.4.3.1 TSV Memory Stacking

High Bandwidth Memory (HBM) will be first commercially available product and Table 7.5 describes its specification. NCF with TCB is current POR building block in mass production of top two DRAM manufacturers, Samsung and SK Hynix, who ever produced 3D TSV DRAM.

The fine pitch I/O and low gap height restrict to apply the traditional underfilling process for encapsulation of 3D TSV DRAM (see Fig. 7.42), resulting from the limit of deflux process, known as greater than 100 μm. Consequently, the underfilling process combined with the limit of flux cleaning process downselects two flux-based materials among four major materials designed for TCB assembly building blocks.

In addition, stacking die will expose interconnections to undergo multiple times of solder reflow. Namely, first bonded die can be more compressed in the following reflow process of second die if displacement is proportion to applied force when no encapsulation was made (see Fig. 7.43). As a result, stacking with no encapsulation can result in solder bridge during TCB process.

As Fig. 7.42 indicates, TCB can select epoxy flux or NCF for stacking process of fine pitch thin die with microbump and TSV. Enabling epoxy flux would create

Table 7.5 Samsung 3D TSV DRAM specification

Die size	<6.0 × 6.25 mm	Bump pitch	~60 μm
Die thickness	~50 μm	Gap height	~15 μm

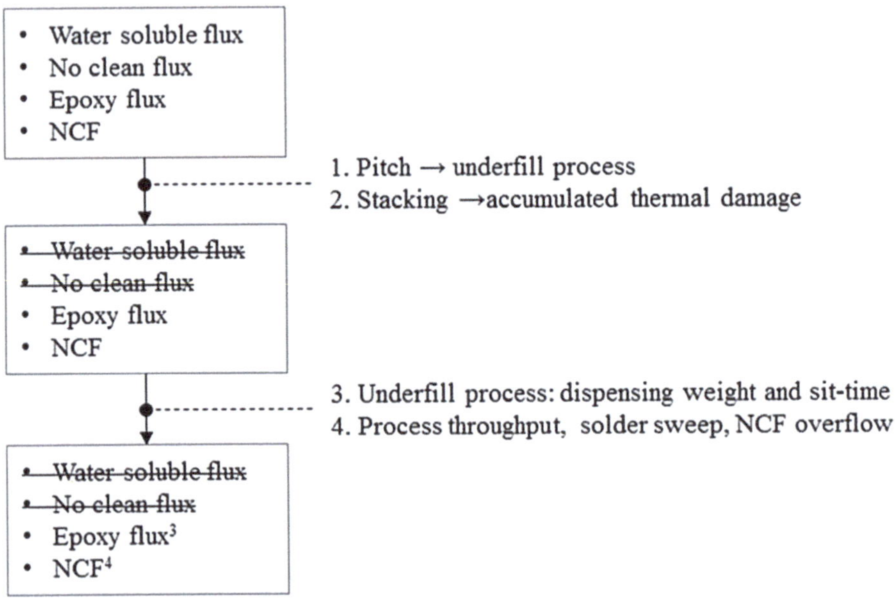

Fig. 7.42 Flowchart of building block selection

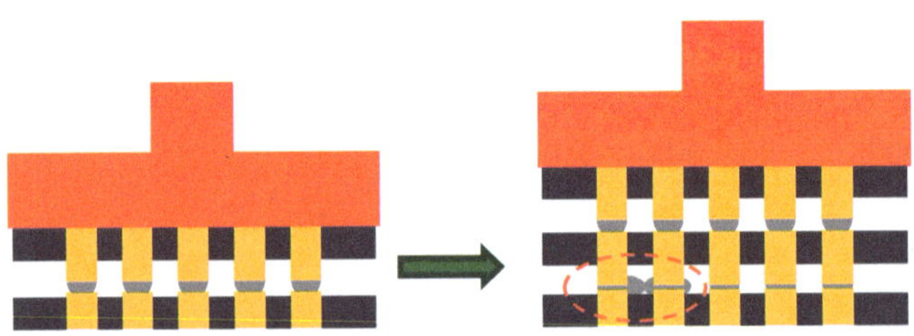

Fig. 7.43 A schematic of solder bridge during stacking due to multiple times of solder reflow

technical challenges mostly in underfill process: how to precisely dispense epoxy flux less than 2 mg enough to fill typical HBM package and how to design entire process sequence to compromise between sit-time of material and process throughput or to develop material to meet a long sit-time requirement in TCB process. Suppose that density of epoxy is 1.6 cm/g^3, calculate volume required for underfill dispensing process (Table 7.5) [10]. The amount required is roughly 1.7 mg to fill the gap height between die and bonding site. To dispense such small amount of epoxy flux with high repeatability and reproducibility not to cause underfill overflow on thin die would be an extremely difficult task compared with underfill

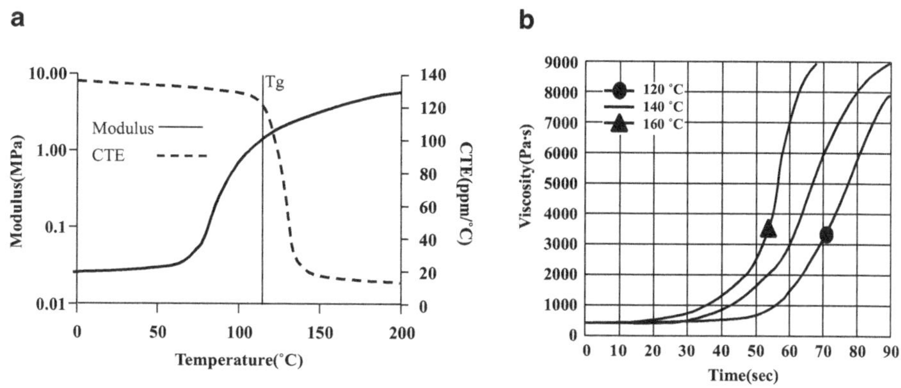

Fig. 7.44 Material property plot: (**a**) modulus and CTE as a function of temperature and (**b**) viscosity of epoxy flux verse time found by Rheometer [21]

process for typical flip chip packages. Dispensing excessive amount of epoxy will overflow on the top of die, possibly causing damage on TCB head, which is most important component in TCB equipment. While dispensing process is being optimized, fundamental investigations such as wetting and void studies are recommended to conduct on material candidates, meeting reliability requirements, to choose good materials. Use properties of a sample material to know how to define process parameters. First, review glass transition temperature of the materials (see Fig. 7.44a) to define baseline stage temperature range that is greater than and equal to 120 °C. With the initial thermal condition, sit-time impact on viscosity (Fig. 7.44b) can determine the stage temperature. Suppose that a bonding cycle takes 5 s and ten bonding sites exist on stage, total 50 s are expected for sit-time, possibly allowing bonding stage to heat up to 160 °C. The temperature should be reevaluated on TCB tool in order to optimize process parameters with interactive effects. NCF also can use the same method to determine temperature at bonding stage.

Table 7.6 proposes baseline process parameters of epoxy flux on the properties of material, described as Fig. 7.44, in TCB process. Unidentified recipe parameters can be optimized through series of Design of Experiment (DOE). The DOEs should be statically designed to decouple thermal impact on mechanical parameters such as bond force and *Z*-height chase. All mechanical events should occur sequentially well to isolate interactive effects between thermal and mechanical parameters. In detail, *Z*-height chase should not be made when bond head will expand according to CTE of the bond head while the head ramps up to the peak temperate. The *Z*-height chase is recommended to happen at constant temperatures. DOEs still count on typical process metrics: electrical continuity, voids, and solders joint integrity.

Aforementioned procedure could be applied to NCF process development as well. Figure 7.45 shows micrographs of Samsung RDIMM assembled using TCB with NCF, which two leading DRAM manufacturers ever developed TSV products selected in mass production process. Figure 7.45a shows flat underfill fillet which is

Table 7.6 Recommended baseline process parameters

Process parameter	Recipe parameter	Value
Force	Contact force	~3N recommended
	Bond force	DOE
Temperature	Bond head contact temperature	<160 °C
	Bond head peak temperature	300 °C
	Bond head release temperature	DOE/Above or equal to contact temperature
	Bond stage temperature	<160 °C
Time	Dwell time at peak temperature of bond head	DOE
Displacement	Melting detection setting	~5 μm recommended
	Z-height chase (optional)	DOE

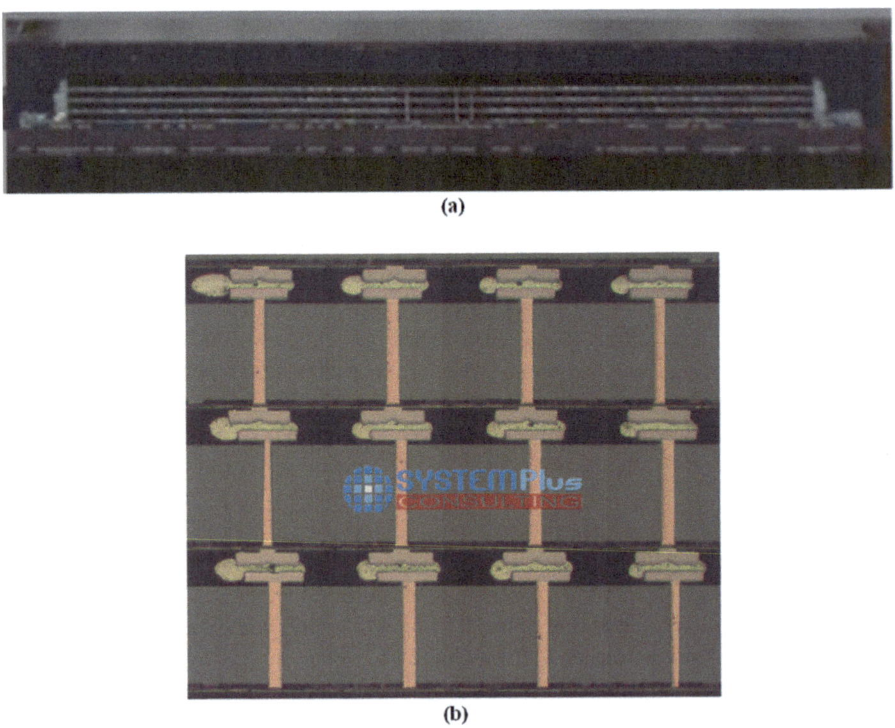

Fig. 7.45 Visual micrographs of side view of Samsung TSV RDIMM : cross-sectioned (**a**) TSV module and (**b**) microbump interconnection [11]

unordinary shape and has not seen in flip chip packages. The shape of fillet may be a unique signature of TCB process with NCF to avoid NCF contamination of the top surface of 50 μm thickness of die because highly viscous NCF will be squeezed out

Table 7.7 Configuration of test vehicles

Die	Dimensions (mm)	10 × 12 × 0.2
	Bump pitch (μm)	60/120
	No. of bumps	7744
	Bump height (μm)	35
Interposer	Dimensions (mm)	27.05 × 19.25 × 0.1
	Bump pitch (μm)	180
	TSV dimensions (μm)	10 × 100

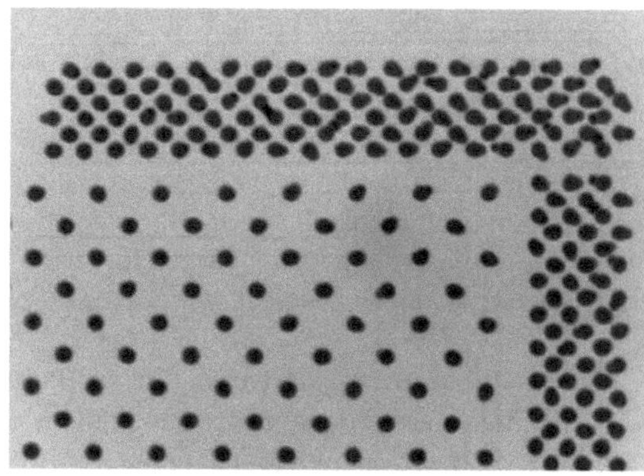

Fig. 7.46 X-ray image showing solder bridging of the microbump die after TCB to the Si interposer in the periphery array [18]

of packages while TCB head is applying a controlled heat to packages with a high force. The overflow could be prevented by compressing the NCF extrusion using an oversized collet mounted on TCB head. Solder sweeping observed on Fig. 7.45b also confirms that NCF flow is so viscous to push molten solders to the direction of NCF flow. Due to the high viscosity of NCF, a high bond force compared to other building blocks is expected to apply to package during TCB bonding process in order to ensure good contact between microbumps and pads by penetrating NCF and to avoid filler entrapment inducing open failures. As a result, process cycle time was known to take much longer than other building blocks.

7.4.3.2 Memory Module to Logic or Silicon Interposer Attachment

This session reviews another building block, TCB with no-clean flux, to attach microbump die to the Si interposer. Woychik and Lee et al. [18] demonstrate no-clean flux process with the 60 μm pitch periphery array of the microbump die to the interposer. See table for details about test vehicles used for their studies (Table 7.7).

Figure 7.46 shows an X-ray image of solder bridges in the 60 μm pitch periphery array of the microbump die after assembly to the Si interposer. In the 3D stacking processes development using no-clean flux, Woychik and Lee et al. encountered solder bridges that are typically induced by the failure at decoupling thermal and mechanical process parameters in TCB process. To achieve the final acceptable

Table 7.8 DOE1: thermal process parameters

Leg	Stage temp (°C)	Dwell time at peak temp 305 °C (s)	Head release temp (°C)	Yield
1	100	5	80	1 open in 1 die
2	130	5	80	1 open in 2 die
3a	130	1	80	2 opens on 1 die
3b	130	1	130	2 opens on 1 die

process window, they performed two different DOEs: to identify significant thermal and mechanical process parameters affecting solder bridging of the microbump die in order to design high yield assembly process and to validate stability of the assembly process with large number of builds.

In the first task, the three variables to evaluate were stage temperatures, time at peak temperature, and cool down temperature (see Table 7.8) and the other parameters selected baseline process parameters used for parts showing solder bridges in Fig. 7.46. For example, the peak temperature was 305 °C and melting detection was 5 μm.

The results of this task have shifted the failure mode to solder opens. Namely, this DOE designed to decouple thermal and mechanical parameters accomplished the primary objective and found upper limit of thermal process not resulting in solder bridges. With the findings, second DOE investigated a primary mechanical parameter, displacement of bond head in z-axis, to avoid open failures by travelling further deeper. For all four legs of the DOE there were solder opens as shown in Table 7.8, which implies that the thermal process window is stable enough not to result in short failures and mechanical process window is independent of thermal process parameters because no short failures were observed at even higher temperature and longer peak. From this build cross sectioning of the microbump solder interconnects was done and it was found that there was a slight bow of the die as shown in Fig. 7.47. Figure 7.47 shows the edge microbump die solder joints having a height of 40 μm and the center die solder joints having a height of 42 μm. It is believed that this bow is due to a temperature gradient across the microbump die that causes an uneven gap height.

DOE2 selected upper limits of thermal process window identified in DOE1, which do not induce short failures, to guarantee stable solder interconnection. The mechanical process condition, melting detection, to eliminate solder opens is described in Table 7.9. In order to prevent total or insufficient collapse of molten solders, there is a z-displacement setting on the TCB tool to control the amount of z-displacement of the head. Remind that this amount of controlled z-displacement after solder reflow terms the "Melting detection." This is a tool parameter sensitive to yield and the parameter mainly depends on equipment's characteristics. Here the major variable was to control the melting detection height ranging from 6 to 10 μm. From this work, the electrical testing confirmed no solder opens and no shorts when using a melt setting height of 10 μm

Only a melt setting of 10 μm yielded a nominal joint height of about 32 μm without any opens or bridges. Figure 7.48 shows a comparison of the nominal MBD

Fig. 7.47 Comparison of edge and center microbump die interconnect height using DOE1 bonding conditions [18]

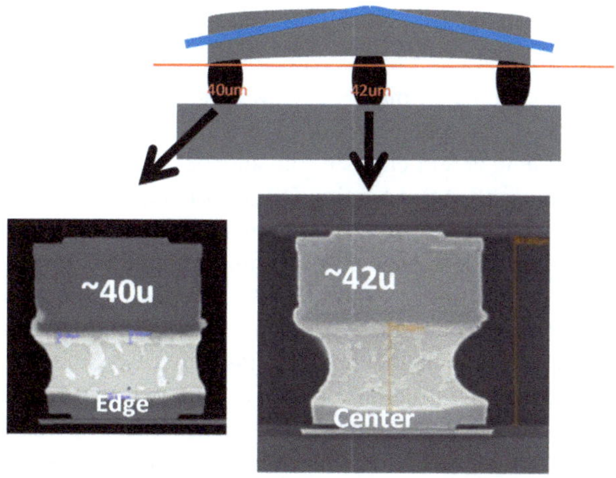

Table 7.9 DOE2: mechanical process parameters

Leg	Stage temp (°C)	Contact temp (°C)	Dwell time at peak temp 305 °C (s)	Head release temp (°C)	Bond force (*N*)	Melting detection (μm)	Yield (%)
1	130	130	5	130	2.8	6	≈90 (open)
2	130	130	5	130	2.8	8	≈90 (open)
3	130	130	5	130	2.8	10	100

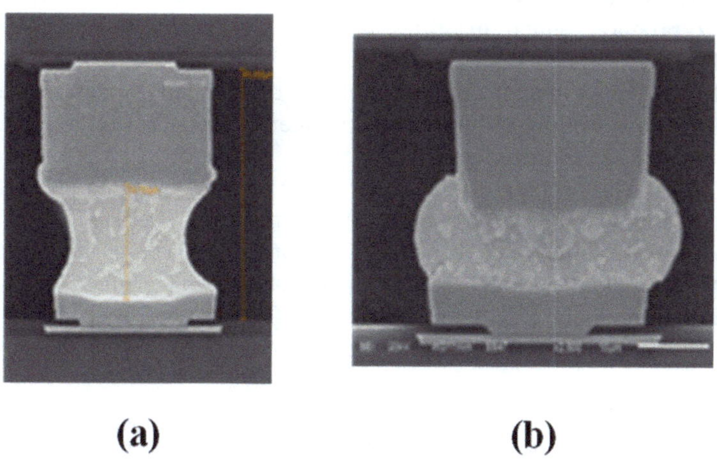

Fig. 7.48 Comparison of microbump die nominal joint heights: (**a**) melting detection of 6 μm and (**b**) 10 μm using DOE2 [18]

interconnect heights when a melt setting of 6 and 10 μm was used. For the 6 μm melt setting a nominal joint height of 42.6 μm was achieved, and for a 10 μm melt setting a nominal joint height of 31.8 μm was achieved (Fig. 7.48).

7.5 Summary and Discussion

This book chapter mainly reviews bonding technology, TCB, designed for 3D TSV stacking with perspectives of process, equipment, and materials to provide its fundamentals and engineering sciences that still remain undiscovered and insufficient even after two leading DRAM manufacturers such as Samsung and SK Hynix released the TSV products into market. Therefore, comparison study between TCB and the conventional process supplies unique process signatures, which can be found on units only assembled using the TCB process in order to emphasize process characteristics and technical challenges of 3D stacking process.

Aforementioned technical details are also limited because a few manufacturers only have infrastructure of TSV manufacturing process, which is indispensable for research and development. To minimize the technical gap between the leading manufacturers and the others, this chapter explores four major building blocks, water-soluble flux, no-clean flux, epoxy flux, and NCF, providing advantages and disadvantages of each process materials. The understanding on major assembly building blocks could help to improve assembly process quality in terms of throughput and cost, close to traditional flip chip process using mass reflow. Furthermore, the understanding with proposed methodology to develop assembly process will benefit how to design a building block that is most compatible with package configuration. Indeed, NCF is the building block that two leading DRAM manufacturers selected for TSV 3D stacking technology. A primary drawback from TCB with NCF is poor throughput because of a long bonding cycle time induced by NCF material properties, so that several multimillion dollar capital investment is required to produce wanted quantities, consequently increasing overall assembly cost. In detail, a high bond force is necessary to apply to package to penetrate highly viscous NCF by microbumps on die in order to contact full to pads, consequently guaranteeing good wetting and no filler entrapments. To improve throughput of 3D stacking process, TCB using epoxy flux can be considered as a good alternative building block. This chapter also reviews 2.5/3D research using no-clean flux with TCB to guide how to develop assembly process through systematic DOEs.

Figure 7.49 illustrates a concept of hybrid process attaching die using a high accuracy TCB tool in less than a second to guarantee high placement accuracy and reflowing the pre-aligned multiple die together followed by low-cost multi-heads flip chip bonders. The hybrid assembly process could overcome the throughput

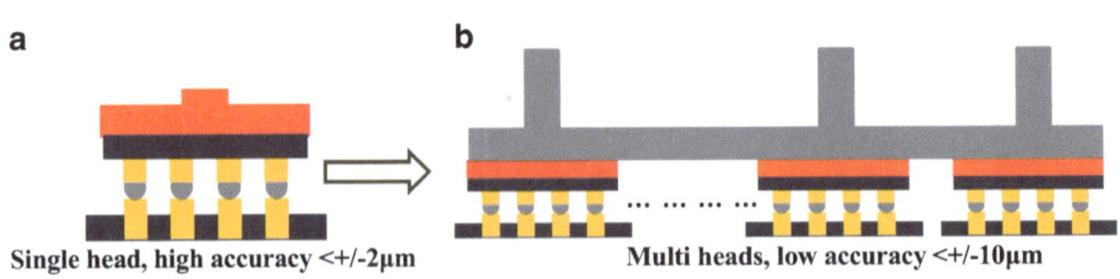

Fig. 7.49 Schematic of hybrid assembly process: (**a**) pre-alignment placement and (**b**) post TCB

challenge while advantages of NCF are still maintained [33]. Namely, the assembly process will help move the industry toward a process that can deliver advanced assembly design rules at a cost competitive position.

This chapter also highlights the mostly unexplored area in 3D bonding technology, interactive effects among process, material, and equipment. A few leading packaging companies which drive both material and equipment suppliers can conduct research and development (R/D) efforts on investigation of the interaction to improve 3D TSV bonding technology including suggested hybrid process research. Academia is recommended to provide fundamentals of all three fields to answer technical challenges resulting from intrinsic mechanism of TCB, fasting heating and cooling. For example, academia can advise how to eliminate the temperature gradient on package during TCB bonding, possibly lowering a peak temperature beneficial to reduction of a bonding cycle time. Comprehensive studies intend to examine orientation and size of solder grains under reflow condition with one directional heat flux from head to stage. This chapter may deliver fundamentals of bonding technology and process materials, which are contingent just leaving R/D phase, and lead packaging industry to enable high-throughput, low-cost 3D assembly process even when compared to conventional 2D technology.

Acknowledgement The editors would like to thank Paul Vianco from Sandia National Laboratory, Yonghao Xiu and Yuying Wei from Intel Corporation for their critical review of this chapter.

References

1. S. Lee, *Fundamental Study of Underfill Void Formation in Flip Chip Assembly* (Georgia Institute of Technology, Atlanta, 2009)
2. S. Lee, R. Master, D.F. Baldwin, Assembly yields characterization of high I/O density, fine pitch flip chip in package using no-flow underfill. In *Electronic Components and Technology Conference*, Reno, 2007, p. 35
3. S. Lee, R. Master, D.F. Baldwin, Assembly yields characterization and failure analysis of Flip chip in package using no-flow underfill. In *International Wafer Level Packaging Congress*, 2007, pp. 169–175
4. S. Lee, R. Master, D.F. Baldwin, Void formation study of high I/O density, fine pitch flip chip in package using no-flow underfill. In *Surface Mount Technology Association International*, 2007, pp. 525–530
5. S. Lee et al., Void formation study of flip chip in package using no-flow underfill. IEEE Trans. Electron. Packag. Manuf. **31**(4), 297–305 (2008)
6. S. Lee et al., Assembly yield characterization and void formation study on high I/O density and fine pitch flip chip in package using no-flow underfill. In *Surface Mount Technology Association International*, 2008, p. 673
7. S. Lee, M.J. Yim, D. Baldwin, Void formation mechanism of flip chip in package using no-flow underfill. J. Electron. Packag. **131**, 0310141–0310145 (2009)
8. S. Lee et al., Near void-free assembly development of flip chip using no-flow underfill. IEEE Trans. Electron. Packag. Manuf. **32**(2), 106–114 (2009)
9. S. Lee, D. Baldwin, Heterogeneous void nucleation study in flip chip assembly process using no-flow underfill. ASME J. Electron. Packag, **136**(1), 011005-011010 (2014)

10. S. Lee, H.-M. Zhou, D. Baldwin, A numerical study of void nucleation and growth in flip chip assembly process. Model. Simul. Mater. Sci. Eng. **18**(6), 065005–065025 (2010)
11. A. Eitan, K.Y. Hung, Thermo-compression bonding for fine-pitch copper-pillar flip-chip interconnect—tool features as enablers of unique technology. In *Electronic Components and Technology Conference (ECTC), 2015 I.E. 65th*, 2015
12. J.H. Lau, The future of interposer for semiconductor IC packaging. Chip Scale Rev. **18**(1), 32–36 (2014)
13. Package Analysis of the SK-Hynix High Bandwidth Memory (HBM). 2015
14. W.-S. Kwon et al. Enabling a manufacturable 3D technologies and ecosystem using 28 nm FPGA with stack silicon interconnect technology. In *International Symposium on Microelectronics, International Microelectronics Assembly and Packaging Society*, 2013
15. K. Ichikawa, *Key Technology Challenges in Computing Package and Assembly* (Assembly Technology Development Japan, Intel Corporation, Chandle, 2014)
16. S. Lau, Thermo-compression bonding for fine-pitch copper pillar flip chip interconnect. In *SEMICON Advanced Packaging Symposium*, 2014, ASMPT, Taiwan
17. Z. Li et al. Sensitivity analysis of Pb free reflow profile parameters toward flip chip on silicon assembly yield, reliability and intermetallic compound characteristics. In *Electronic Components and Technology Conference (ECTC), 2010 Proceedings 60th*, Las Vegas, 2010
18. C.G. Woychik et al., in *New approaches to develop a scalable 3D IC assembly method*, San Diego, 2015
19. D.S. Patterson, 2.5 D/3D Packaging enablement through copper pillar technology. Chip Scale Rev. **16**(3), 20–26 (2012)
20. C. Wong, S.H. Shi, G. Jefferson, High performance no-flow underfills for low-cost flip-chip applications: material characterization. IEEE Trans. Compon. Packag. Manuf. Technol. Part A **21**(3), 450–458 (1998)
21. C.P. Wong et al., Characterization of a no-flow underfill encapsulant during the solder reflow process. In *Electronic Components and; Technology Conference*, 48th IEEE, Seattle, 25–28 May 1998
22. C.P. Wong, S.H. Shi, G. Jefferson, High performance no-flow underfills for low-cost flip-chip applications: materials characterization. IEEE Trans. Compon. Hybrids Manuf. Technol. **21** (3), 450–458 (1998)
23. S. Shi, D. Lu, C.P. Wong, Study on the relationship between the surface composition of copper pads and no-flow underfill fluxing capability. IEEE Trans. Electron. Packag. Manuf. **22**(4), 268–273 (1999)
24. C.P. Wong, S.H. Shi, No-flow underfill of epoxy resin, anhydride, fluxing agent and surfactant US 6180696 B1, G.T.R. Corporation, 2001
25. H. Li et al., Syntheses and characterizations of thermally degradable epoxy resins. III. J. Polym. Sci. A Polym. Chem. **40**(11), 1796–1807 (2002)
26. Y. Shi, X. Wei, B. Tolla, Smart chemistry towards highly efficient soldering material formulation. In *Proceedings of SMTA International,* Rosemont, 2014, pp. 436–443
27. Z. Zhang, E. Beatty, C. Wong, Study on the curing process and the gelation of epoxy/ anhydride system for no-flow underfill for flip-chip applications. Macromol. Mater. Eng. **288** (4), 365–371 (2003)
28. H. O'Neal et al., Comparison of Tg values for a graphite epoxy composite by differential scanning calorimetry (DSC), thermomechanical analysis (TMA), and dynamic mechanical analysis (DMA). J. Adv. Mater. **26**(3), 49–54 (1995)
29. B. Schmaltz, Packaging materials for 2.5/3D technology. Int. Symp. Microelectron. **2013**(1), 000276–000284 (2013)
30. A. Lucero, G. Xu, D. Huitink, Low-κ-package integration challenges and options for reliability qualification. In *Reliability Physics Symposium (IRPS), 2012 I.E. International*, Anaheim, 2012

31. J.L. Aw et al., Thermal compression bonding with non-conductive adhesive of 30 μm pitch Cu pillar micro bumps on organic substrate with bare Cu bondpads. In *Electronics Packaging Technology Conference (EPTC), 2014 I.E. 16th*, Singapore, Dec 2014
32. J. Jing-Ye et al., The development of high through-put micro-bump-bonded process with non-conductive paste (NCP). In *Microsystems, Packaging, Assembly and Circuits Technology Conference (IMPACT), 2012 7th International*, Taipei, Oct 2012
33. D. Hiner et al., Multi-die chip on wafer thermo-compression bonding using non-conductive film. In *2015 I.E. 65th Electronic Components and Technology Conference (ECTC)*, San Diego, 2015

Chapter 8
Fundamentals of Solder Alloys in 3D Packaging

Kwang-Lung Lin

8.1 The Microbumping Process

The microbump material reactions may occur during assembly process and performance. The assembly process is typically the reflow operation which involves liquid solder/solid metallization interaction. A microbump experiences a couple of times of reflow throughout the production of the bump itself and the afterwards joining process. The performance, i.e., the product functioning, of the microbumps may encounter thermal cycling and longtime thermal ageing. Besides, the microbumps certainly experience electrical current stressing during product application. There is a chapter dealing with electromigration. Thus, the discussion in this chapter will not involve the materials reaction under electric current stressing. Readers are referred to Chap. 7 for electromigration-related subjects.

Flip chip bonding technology is one of the major interconnect technologies of chip level packaging. The other two are wire bonding and Tape Automated Bonding (TAB). Being an area array bonding design, the flip chip technology deserves the highest I/O (input/output) numbers possible among the three designs gives rise to larger pitch size than the peripheral bonding design of the wire bond and TAB. These three chip level packaging technologies are adopting different bonding materials. Au wire, coated Cu wire, and silver alloy wire of the dimension of around 20 μm in diameter are the joining materials being used for the wire bonding process. The discussion on the adoption of different materials is beyond the scope of this chapter. Au bump produced on the electrode terminal (I/O) of the chip is being used for the TAB bonding process. Both wire bonding and TAB bonding are fabricated through thermal compressing. However, the flip chip bonding is

K.-L. Lin (✉)
Department of Materials Science and Engineering, National Cheng Kung University, Tainan, Taiwan R. O. C.
e-mail: matkllin@mail.ncku.edu.tw

Y. Li, D. Goyal (eds.), *3D Microelectronic Packaging*, Springer Series in Advanced Microelectronics 57, DOI 10.1007/978-3-319-44586-1_8

fabricated via reflow process which involves the melting of solder bump to perform liquid/solid interaction. Accordingly, the production of the solder bump is the prerequisite for enabling the flip chip bonding process. In light of its highest I/O capability, the flip chip bonding technology became the only and must choice for the 3D package technology. The microbump is a modification based on the conventional flip chip bump structure.

The conventional flip chip solder bump (or C4 bump, following the original IBM-designed "control collapsed chip connection") faces the challenge of current crowding at the entry of electric current to the solder bump. The current crowding may result in the formation of micro-voids which aggregate to form cracks propagating at the interface between the solder bump and the UBM (under bump metal) layer. A thought in the line of the C4 solder bump design by reducing the bump height and the UBM area would be a simple approach to produce the microbump. Nevertheless, the miniaturization of the solder joint in the 3D microbump faces the challenge of increasing current density due to the shrinking contact area. It would largely reduce the life time as a result of the high current density and the reducing volume of the microbump. A very promising structure being adopted nowadays, however, significantly eliminates the influence of current crowding behavior by enlarging the thickness of the UBM. The Cu UBM of conventional C4 solder bump is now replaced by Cu pillar in 3D microbump. The thickness of the Cu pillar can be in the range of 10–30 μm instead of the thin film Cu UBM with shrinking diameter of around 25 μm. The diameters of the Cu pillar and the associated UBM, for example, Ti thin film, on the trace is much smaller than that of the conventional C4 solder bump. However, the high electrical conductivity of Cu and the high volume of the Cu pillar enable the fast dissipation of the electric current through the Cu pillar so as to avoid the current crowing at the UBM/solder interface.

Figure 8.1 presents an example of the bumping process [1] for producing microbump. The process begins with the producing of RDL (redistribution line) on the chip. The conductive RDL was produced with plating process on the adhesive and seed layers Ti/Cu. After incorporation of the necessary passive oxide layer, the Ti/Cu layer was sputtered again as the seed layer for the solder plating. The Cu pillar and solder, Sn herein this case, were then produced with plating. Appropriate UBM may also be applied between Cu pillar and solder to reduce Cu consumption. The solder will be reflowed after plating to form a solder cap for bonding height control. Figure 8.2 presents the in-line bump and the magnified image of the microbump of which the bump dimension, including Cu pillar, is less than 10 μm.

The microbumps on chip are bonded to the bonding pad or trace line on the substrate in the 3D packaging process. There are various structures available for the substrate metallization. Regardless of the design, the metallization layers on substrates are mostly following the structures of the conventional C4 substrate or even the BGA (ball grid array) substrate. These may include OSP (organic solder preservative) Cu, Cu with barrier layers such as Ni/Pd/Au (ENEPIG, electroless nickel electroplating palladium immersion gold), etc. The top layers of these structures, either Cu or Au, are not feasible for direct bonding with the Cu pillar.

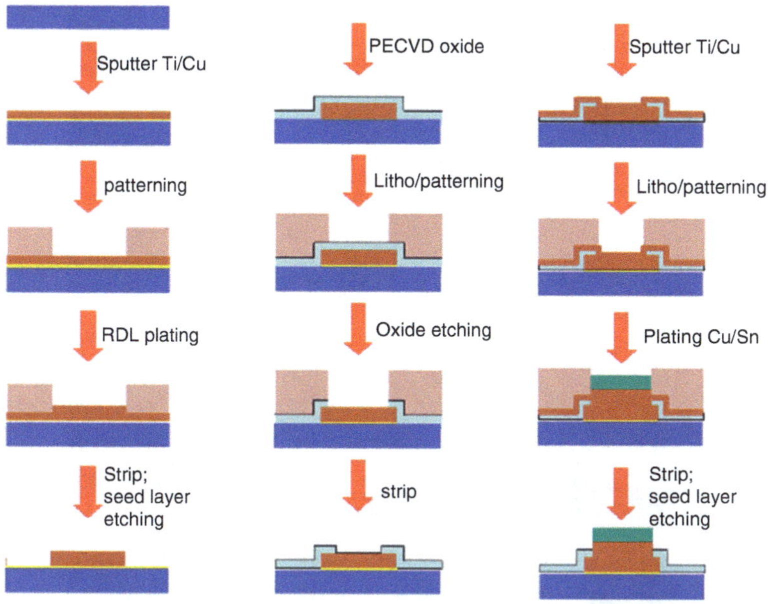

Fig 8.1 Example of the microbumping process for producing Sn microbump [1]

Research is aiming for Cu to Cu direct bonding. A few options have been proposed and are being moved towards the larger scale try run. Before the feasible industrial process of Cu to Cu direct bonding becomes true, the Cu pillar is still joined with the substrate metallization with solder. For the interest of miniaturization, the thickness of solder layer is being kept as small as possible to be within ~10–30 μm. For the convenience of incorporating with the Cu pillar manufacturing process and the concerns of productivity and technology feasibility, the solder layer is being produced with electroplating. The plated solder is reflowed to form sub-hemisphere dimension, mostly like arc, and is termed as solder cap. The entire microbump produced on the chip is thus consisted of the adhesive/conductive layers (in the thickness range of hundred nanometers) on top of Al trace, the Cu pillar, and the solder cap.

Microbump can be joined on TSV (through silicon via) Cu, substrate Cu pad, or Cu trace on substrate. The bump on trace process applied thermal compression bonding to join the microbump on Cu trace line. The Cu trace line being proposed may adopt OSP-Cu. The OSP can be removed with plasma treatment or with the embedded flux in the non-conductive paste (NCP). Figure 8.3a presents the Cu trace on substrate. The FIB-SEM image, Fig. 8.3b, shows the thickness of the OSP layer is around 300~450 nm [2].

Figure 8.4 presents the Cu pillar on the top and the Cu trace line on the bottom substrate [2]. Thermal compression operation melts the solder which forms

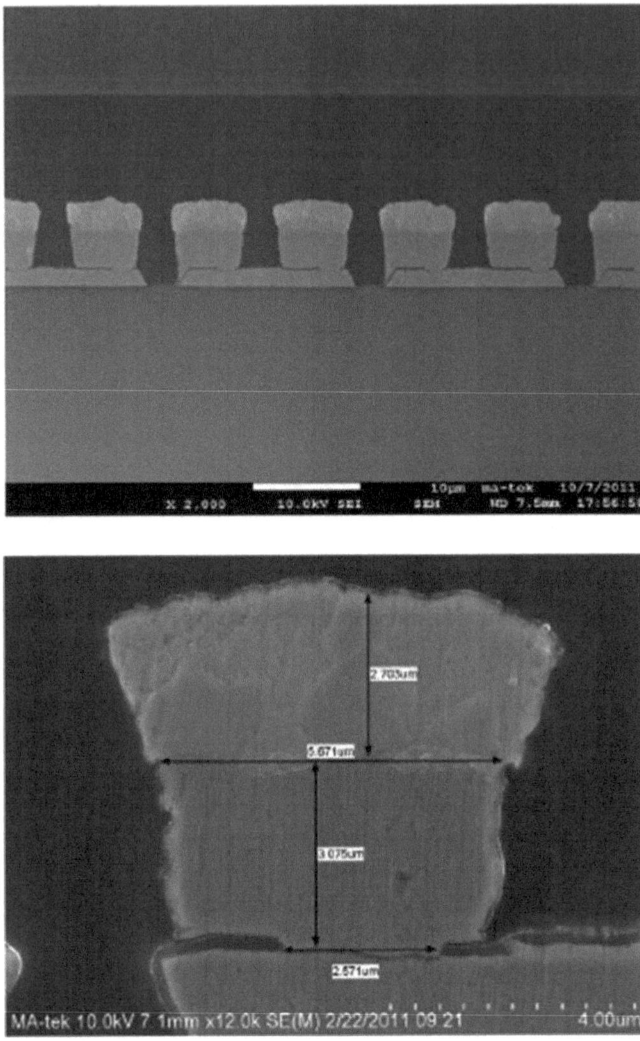

Fig. 8.2 The SEM image of the Cu/Sn microbump produced following the process of Fig. 8.1 [1]. (**a**) In-line microbumps, (**b**) magnified image of microbump

intermetallic compounds with the Cu pillar and Cu trace. The table below Fig. 8.4 presents the compositions of the compounds formed and indicates the formation of Cu_3Sn and Cu_6Sn_5 in the As-bonded microjoint. The thin Cu_3Sn was formed between Cu_6Sn_5 and Cu (pillar or trace). Filler was introduced by the paste of the TCNCP (thermal compressive non-conductive paste) process.

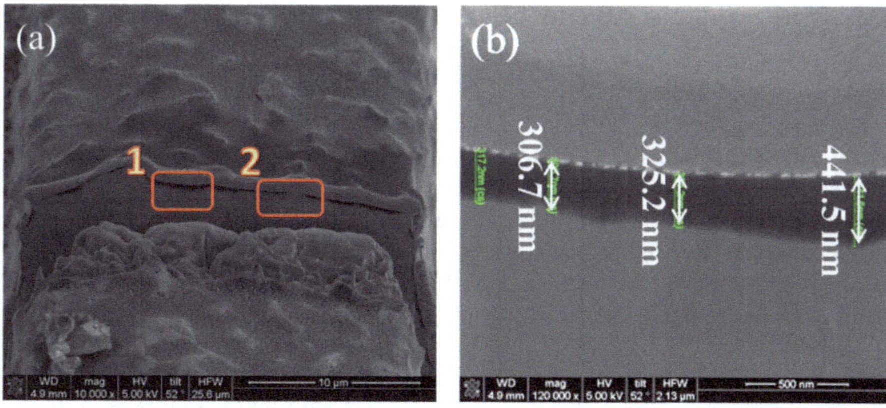

Fig. 8.3 (**a**) The FIB-SEM image shows the Cu trace on substrate with OSP coverage, the thin dark line region across the cut trace is the OSP. The top thick coating on the OSP is the Pt deposit for FIB purpose, (**b**) the enlarged image of the region 1 of (a) shows the thickness of the OSP layer [2]

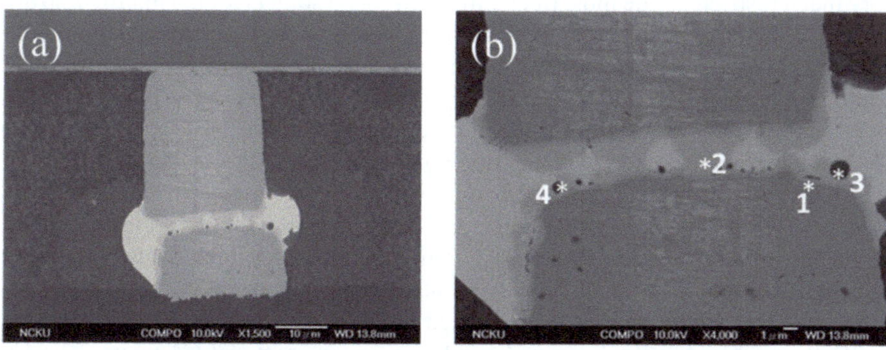

	EDX analysis result
Point 1	Cu_3Sn
Point 2	Cu_6Sn_5
Point 3 & 4	Filler particle

Fig. 8.4 The TCNCP process bonds Cu pillar/solder microbump on Cu trace on substrate [2]

8.2 The Solder Alloys in Microbump

A variety of solder alloys have been commercialized and in use for the conventional C4 solder bump and the BGA solder ball. The solder balls are discrete products of which the compositions are adjusted by molten alloying. The alloying technologies are feasible for producing whatever the desired compositions. Accordingly, a number of solder ball compositions have been commercialized, especially after the patent concern of the Sn-Ag-Cu solder no longer exists. The C4 solder bump is being produced via WLP (wafer level packaging). The WLP can adopt either stencil printing using solder paste or electroplating. The solder in the stencil printing process is in the form of solder powder of which the composition is generally alloyed by molten technology and thus can feature a variety of composition. However, the composition of the electroplated C4 solder bump and similarly the solder cap of 3D microbump produced by electroplating is restricted by the redox potential of the elements especially when alloying is proposed. Accordingly, there are limiting compositions of the solder cap reported. Table 8.1 lists the solder alloys reported for microbump. It is apparent that the composition of the microbump is very much restricted. The most frequently mentioned solders for the microbump are pure Sn or Sn-xAg. The plating process of Sn and Sn-xAg is commercially mature in terms of plating bath and plating conditions. Pure Sn bump is rarely used for C4 bump, while the Sn-xAg with relatively low Ag content has been familiar to the electronic industry especially for the ductility consideration.

Sn-Ag is eutectic system with intermetallic compound. The eutectic composition is Sn3.5Ag. In the high Ag content region, with Ag content greater than 3.5 %, the system will form Ag_3Sn intermetallic compound. A high fraction of the compound in the solder tends to downgrade the ductility of the solder joint. Thus, the Sn-Ag solders being adopted are in the low Ag content range, e.g., with Sn2.5Ag or less Ag contents. Some investigations even applied pure Sn. The Sn-Ag-Cu ternary solder in 3D IC practice is seen less mentioned than in conventional C4 bump. The Cu content may complicate the intermetallic compound formation. The Sn-Bi solder is applied only when very low bonding temperature, reflow or thermal compressing bonding is needed. The incorporation of Bi may raise the concern of embrittlement.

8.3 The Formation of Intermetallic Compounds in the As-Produced Microbump

In comparison with traditional solder joints, the solder compositions of the microbump are relatively simplified as shown in Table 8.1. However, the metallizations of the chip terminal and substrate may still similar to what have been adopted in the pass. There is not really much new metallization combinations beyond the reported C4 and BGA systems for the microbump application. On the one hand, the reaction products formed in the microbump joint are more or less

Table 8.1 The solder alloys and structures for microbump

Microbump structure	Bonding technology	Ref.
5.4 μm Cu/3.6 μm Sn/9.5 μm Cu	TCB 240, 260 °C	[3]
Cu/Sn/Cu	TCB	[4]
20 μm Cu/SnAg/3 μm Ni	Reflow	[5]
Cu/Ni/Sn2.5Ag/Ni/Cu	TCB	[6]
Cu/Ni/SnAg/Ni/Cu	Reflow	[7]
300 Å Cr/500 Å Cu/35 μm Cu /35 μm SAC405	Not available	[8]
Ti/Cu(seed)/Cu(wire)/3.0 μm Cu/3.5 μm SnAg	TCB at 280 °C	[9]
Cu/Ni/Sn	TCB at < 250 °C	[10]
Cu/Ni/SnAg	TCB	[11]
30 μm Cu/3 μm Ni/25 μm SnAg	Reflow	[12]
33 μm Cu/25 μm SnAg	Reflow	[12]
50 μm Cu/20 μm SAC305	TCB	[13]
42 μm Cu/3 μm Ni/35 μm Solder	TCB	[14]
28 μm Cu/2 μm Ni/15 μm SnAg	TCB	[15]
Cu/Ni/Sn 2.5Ag	TCB	[16]
5 μm Cu/2 μm Ni/8 μm Sn57Bi	Reflow at 180 °C	[17]
5 μm Cu/2 μm Ni/8 μm SAC305	Reflow	[17]
Cu/Ni/Sn	Not available	[18]
100 nm Ti/200 nm Cu/12 or 8 μm Cu/10 or 25 μm SAC305	Reflow	[19]
100 nm Ti/200 nm Cu/12 μm Cu/2 μm Ni/10 μm SAC305	Reflow	[19]
NiFe/Pb-free solder (undefined)	Reflow	[20]
NiFe/Cu/Pb-free solder (undefined)	Reflow	[20]
8 μm Cu/SnAg	Reflow	[21]
8 μm Cu/Ni/SnAg	Reflow	[21]
5 μm Cu/3 μm Ni/5 μm Sn 2.5Ag	TCB	[22]
2 μm Cu/5 μm Sn (on substrate)	TCB	[23]
Cu/Sn/Ni/Cu (TSV)	TCB	[24]
Cu/Ni/Sn/Ni/Cu (TSV)	TCB	[24]
Cu/Sn/Cu (TSV)	TCB	[25]
Cu/Sn/Au/Ni/Cu (TSV)	TCB	[25]
Cu/Ni/Sn/Au/Ni/Cu (TSV)	TCB	[25]
50 nm Ti/120 nm Cu/3 μm Cu/3 μm Sn	TCB 260 °C	[26]
10 μm Cu/10 μm SnAg	TCB 250 °C	[27]
Cu/Sn/Cu (TSV)	Reflow	[28]
Cu/Sn/Ni/Cu (TSV)	Reflow	[28]
Cu/Ni/Sn/Ni/Cu (TSV)	Reflow	[28]
10 μm Cu/10 μm SnAg	Reflow	[29]
5 μm Cu/3 μm Ni/5 μm Sn2.5Ag	TCB 300°C	[30]
16 μm Cu/10 μm SnAg/Cu	Reflow 240 °C	[31]
2 μm Ni/1 μm Cu/4 μm Sn	Reflow 245 °C	[32]
2 μm Cu/5 μm Sn	TCB 300°C, 350 °C	[33]
25 μm Cu/15 μm Sn58Bi	TCB 170 °C	[34]
20 μm Cu/12 μm Sn	TCB	[35]

TCB thermal compression bonding, *TSV* through silicon via; *SAC* SnAgCu predictable. On the other hand, the miniaturized joint volume may affect the relative proportion of compounds as well as the microstructure of the joint.

The joining of microbump is generally carried out through traditional reflow process or thermal compressing bonding (TCB) as stated in Table 8.1. Both processes involve solid (Cu pillar and metallization on substrate)/liquid (solder) interaction. Some reports referred this as solid liquid interdiffusion (SLID) [3]. The interactions in the microbump system generally belong to the same reactions as were observed for BGA and C4 bump. However, the solder volume in the microbump is much reduced. For instance, the volume of 80 μm C4 bump is 55 times as that of a microbump of 25 μm diameter and 10 μm height [3]. In light of the same reaction conditions, time and temperature, the volume ratio of the intermetallic compound in the microbump will be much more than that in the C4 bump and the BGA solder joint. The Sn cap of the 9.5 μm Cu/3.6 μm Sn microbump bonded on Cu substrate will form thin Cu_3Sn at the Cu surface, large volume of Cu_6Sn_5 between Cu_3Sn and Sn, and residual Sn at the center, Fig. 8.5a, when bonded for 10 s. The Sn almost completely converted to Cu_6Sn_5 after 1 min of bonding operation at 240 °C, Fig. 8.5b [3]. Tiny amount of Sn may exist within the intermetallic compound regions which form voids after longer reaction time, Fig. 8.4c, d.

In the case of Cu/Sn/Cu joint, the Cu_3Sn intermetallic compound will be formed attaching to the Cu at longer reaction time, meanwhile, the Cu_6Sn_5 became porous. The pore volume grows at higher bonding temperature. The joint starts forming

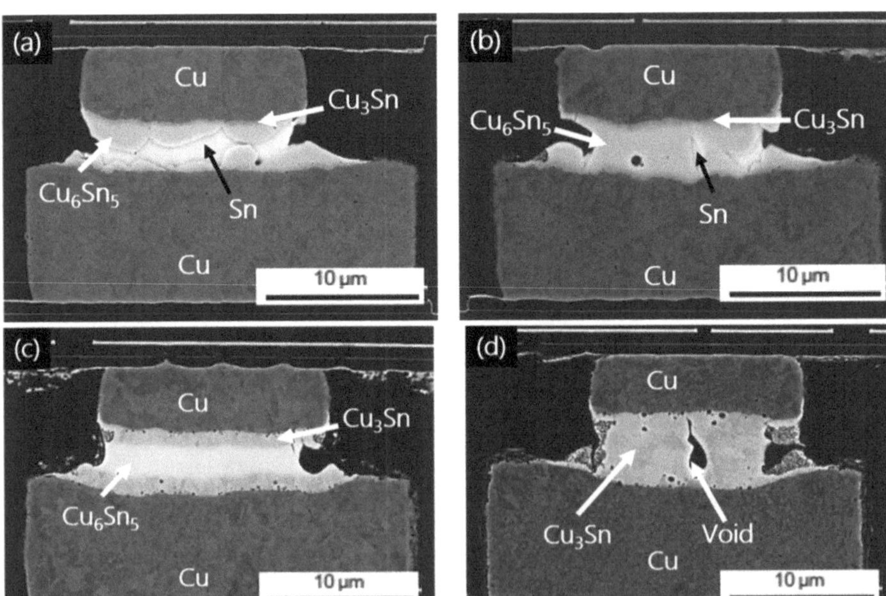

Fig. 8.5 The formation of IMC in the Cu/Sn/Cu microbump produced by thermal compression bonding at 240 °C for (**a**) 10 s; (**b**) 1 min; storage after bonding for (**c**)10 min; (**d**) 3 h [3]

Cu_6Sn_5 within the bulk Sn solder and Cu_3Sn at the solder/Cu boundaries. The Cu_6Sn_5 converts to Cu_3Sn during longer bonding time which left behind pores within the Cu_6Sn_5 region. The solder joint with both intermetallic compounds formed at shorter reaction time or lower reaction temperature will further exhibit the transformation from Cu_6Sn_5 to Cu_3Sn during high temperature storage, Fig. 8.5c. The conversion will result in the formation of void at the central region where the top and bottom conversion reactions meet, Fig. 8.5d. The pore volume formed corresponds to the volume shrinkage (theoretically 40.9 %) for the Cu_6Sn_5 to Cu_3Sn transformation [3].

Ni layer is a common diffusion barrier for reducing the interaction between Cu and solder during reflow and afterwards operations. The Ni deposit on both Cu ends of the SnAg microbump will form Ni_3Sn_4 [7] intermetallic compound after thermal compression bonding. The growth of the Ni_3Sn_4 through the thermal compression bonding process has an activation energy of 127.8 kJ/mol. The thickness of the Ni_3Sn_4 grows exponentially with respect to the bonding time in the temperature range of 250–300°°C. The mechanical properties of the Ni_3Sn_4 formed shows elastic anisotropy. The compound exhibits much less elastic anisotropy in the a–c plane than others. The Young's modulus and Linear Coefficient of Thermal Expansion of the compound are highly temperature dependent, while not the Poisson's ratio [7]

Ni exhibits face centered cubic structure, with similar atomic dimension and continuous solid solubility with Cu [36]. The solder joints with Ni or Cu metallization commonly formed $(Cu,Ni)_6Sn_5$ or $(Ni,Cu)_3Sn_4$ intermetallic compound depending on the combination. For instance, The SnAgCu solder in contact with Ni metallization may form $(Ni,Cu)_3Sn_4$ at the interface. In an asymmetric metallization structure of Ni and Cu on counter chips, respectively, the as-reflowed Sn-Ag solder joint formed $(Cu, Ni)_6Sn_5$ at the Ni/solder interface and Cu_6Sn_5 at the Cu/solder interface after reflow [5, 12], Fig. 8.6. The Cu inclusion in the $(Cu, Ni)_6Sn_5$ on the Ni metallization side is resulted from the diffusion of Cu from the trace or substrate, where no Ni barrier layer was provided. The short diffusion distance within the microbump allows the migration of Cu to reach the Ni layer

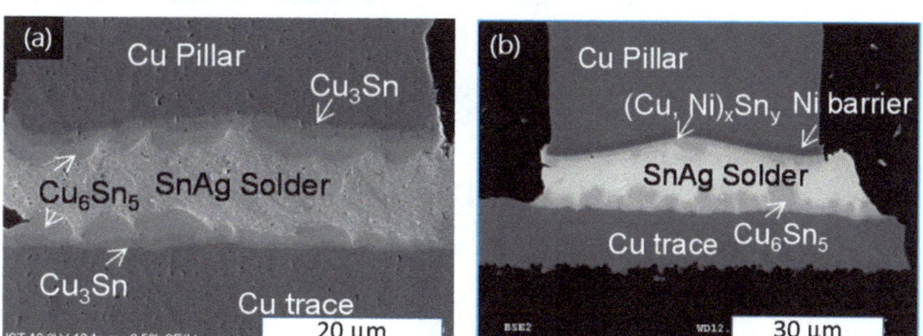

Fig. 8.6 The IMC formation in the bump on trace of microbump joint (**a**) without Ni barrier on the Cu pillar, (**b**) with Ni barrier on the Cu pillar; no barrier on the Cu trace [12]

surface during bonding that results in the formation of the ternary intermetallic compound. The Cu_3Sn was formed only when no Ni layer was incorporated [12], Fig. 8.6a. Ag_3Sn particles will also form during the reflow process for microbump [19]. The satisfactory Ni barrier layer in a microbump between Cu(5 μm) and Sn2.5Ag(3 μm) forms Ni_3Sn_4 at the interface after reflow. The microbump still consists of residual solder at the center of the joint [30].

While Ni has been widely recognized as good diffusion barrier, there were efforts to modify the under bump metal (UBM) or ball limiting metal (BLM) to further suppress the growth of the intermetallic compound. NiFe was introduced in a microbump study [20]. A 2 μm Ni/2 μm NiFe combination as BLM showed effectively suppress the formation of Cu-Ni-Sn intermetallic compound. A thin layer of $FeSn_2$, less than 0.2 μm, was formed at the interface between solder and BLM that provides the function of reaction barrier.

The low temperature eutectic Sn57Bi solder joints can be reflowed at 180 °C for a fine pitch assembly. The dimension of the bump was found to affect the microstructure of the as-reflowed solder joint. The microbump consists of Cu pillar/Ni on Si chip while Cu pad/Ni/Au on the Si substrate [17]. The 250 μm diameter BGA solder joint, Fig. 8.7a, and the 100 μm diameter Cu pillar solder joint, Fig. 8.7b,

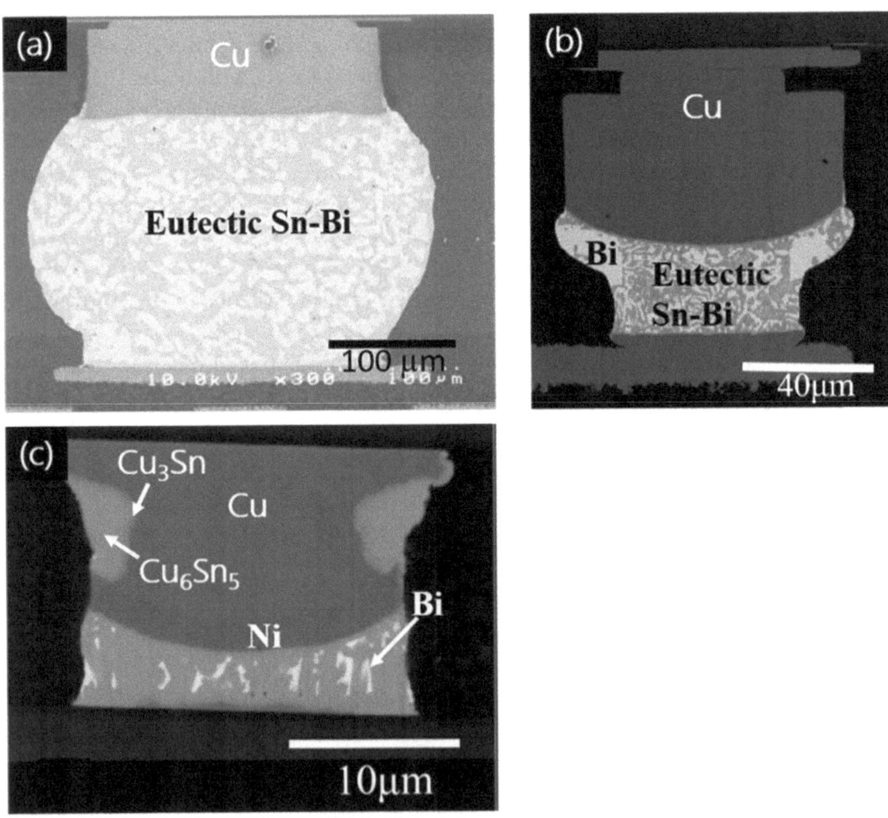

Fig. 8.7 The effect of dimension on the microstructure of the as-reflowed Sn57Bi solder joint (**a**) 250 μm BGA solder ball, (**b**) 100 μm diameter Cu pillar solder bump, (**c**) 25 μm diameter Cu pillar microbump [17]

exhibit eutectic microstructure after reflow. However, the fine pitch 25 μm diameter microbump form Cu_3Sn and Cu_6Sn_5 by consuming all the Sn atoms of the Sn57Bi solder, Fig. 8.7c [17]. Bi phases precipitate in the Cu_6Sn_5 region [17]. Bi and Cu exhibits nearly zero mutual solubility [36]. Bi and Sn form eutectic at the composition of Sn57Bi, yet exhibit very limiting mutual solubility [36]. The small solder volume of the microbump allows the rapid consumption of Sn to form intermetallic compounds. The Bi, as exhibiting nearly zero solubility in Cu and Sn, are expelled to precipitate in the large Cu_6Sn_5 intermetallic phase, Fig. 8.6c. An Sn58Bi microbump formed Cu_6Sn_5 on Cu pillar side and Ni_3Sn_4 on the Ni(5 μm)/Au (0.3 μm) metallization of the substrate. Meanwhile, the Bi-rich phases were dispersed in the joint [34].

The electroless Au deposit was applied as wetting layer in conventional C4 bump or BGA solder joint. The thin Au layer provides fast wetting interaction so as to assist the formation of the joint. In the microbump, however, the volume of the Au layer becomes an important factor in determining the microstructure of the solder joint [22]. A Cu(5 μm)/Ni(3 μm)/Sn2.5Ag(5 μm) top chip was bonded on Si substrate with TCB, where the metallization was either Cu(5 μm)/Ni(3 μm)/Au (0.5 μm) or Ni(2–3 μm)/Pd(0.05–0.1 μm)/Au(0.02 μm). The Cu/Ni/Au metallizations gave rise to $(Ni, Au)_xSn_y$ compound in the joint. The extent of Au in the compound decreases from the Cu/Ni/Au deposited substrate to the Cu/Ni deposited chip. The rapid reaction and the small volume of the microbump render the solder completely transformed into the ternary compound after bonding. The microjoint becomes intermetallic dominated joint after bonding [22]. The Ni/Pd/Au combination in the investigation has much thinner Au layer. Instead, it applied a slightly thick Pd layer. The Au and Pd layers were all consumed after bonding. The Pd layer retards the growth of Ni_3Sn_4 layer on the substrate side while forms large (Pd, Ni) Sn_4 IMC in the joint [22].

8.4 Microstructure Variation of Microbump Under Thermal Mechanical Conditions

The thermal mechanical conditions referred herein include the high temperature storage and thermal cycle treatment of the joint. The phase transformation induced by thermal ageing is influential on the final microstructure of the microbump due to the small solder volume. The Cu/Sn/Cu microbump formed $Cu/Cu_3Sn/Cu_6Sn_5/Sn/Cu_6Sn_5/Cu_3Sn/Cu$ structure after reflow [3]. The residual Sn is of limiting volume which may not exist for some joints. Thermal ageing of the joint at 240 °C gradually converted the Cu_6Sn_5 into Cu_3Sn. The long time storage, 3 h, made the microbump a Cu_3Sn only joint, Fig. 8.5b [3]. A Cu pillar bump with Sn4Ag0.5Cu solder cap forms scallop Cu_6Sn_5 IMC in the as fabricated microbump. The Cu_6Sn_5 IMC transform from scallop structure to planar structure after ageing of the microbump at 180 °C. The discontinuous Cu_3Sn IMC formed between the Cu_6Sn_5 and Cu pillar.

Voids were formed between Cu_3Sn and Cu pillar, where the fracture occurs after the shearing test [8].

The high volume fraction of the intermetallic compound in the microbump indicates that the intermetallic compound may be determinant for the performance of the joint. The thermal fatigue life, assessed by thermal cycle testing, of a microbump joint was found to be affected by the microstructure and types of the intermetallic compound. A theoretical simulation indicates that the thermal fatigue life is disproportional to the Young's modulus of the IMC [7]. However, a combinatory effect of the modulus and the coefficient of thermal expansion (CTE) seems to actually determine the fatigue life. The microbump of high volume of IMC may show fatigue life in the order of Ni_3Sn_4 (136 GPa, 13.7 ppm) > Cu_6Sn_5 (124 GPa, 19.0 ppm) > Cu_3Sn (143 GPa, 18.2 ppm), the values in the parenthesis are (Young's Modulus, CTE) [7]. The volume fraction of the IMC affects the fatigue life of the microjoint. An increasing IMC thickness or fraction in the microjoint will increase the thermal fatigue life of the joint. The thermal fatigue life reaches the minimum value at 42.8 % volume fraction of the Ni_3Sn_4 IMC [7].

The growth of intermetallic compound in the microbump is controlled by the intrinsic diffusion coefficients of the elements. The diffusion coefficient of element in dilute solution can be measured with tracer method. However, it is difficult to measure the diffusion coefficient in concentrate volume such as intermetallic compound with similar approach. A simulated annealing numerical method was developed for estimating the diffusion coefficients in intermetallic compounds. The intrinsic diffusion coefficients of the constituent elements in the commonly encountered intermetallic compounds and the pure metals are listed in Table 8.2. These values indicate that diffusion of Cu is slightly faster than Sn in Cu_3Sn, while of similar speed in Cu_6Sn_5. The diffusion of Ni in Ni_3Sn_4 is faster than Sn. During high temperature annealing of a Cu/Sn/Ni/Cu microbump, the Cu_6Sn_5 formed at the first place after reflow will convert to Cu_3Sn at the expense of Cu. The diminishing of

Table 8.2 Intrinsic diffusion coefficients of elements in different phases

Phase	Element	$D(m^2/s)$, 170 °C	Ref.
Cu	Cu	1.07×10^{-29}, 3.46×10^{-29}	[38]
	Sn	3.98×10^{-26}	[39]
Sn	Cu	1.06×10^{-9} (//c), 3.04×10^{-11} (⊥c)	[40]
	Sn	1.81×10^{-16}(//c), 4.43×10^{-16} (⊥c)	[38]
Ni	Ni	8.95×10^{-38}, 1.61×10^{-37}	[38]
Sn	Ni	1.47×10^{-8}(//c), 7.73×10^{-11} (⊥c)	[41]
Cu_3Sn	Cu	5.12×10^{-16}	[24]
	Sn	1.46×10^{-16}	[24]
Cu_6Sn_5	Cu	9.42×10^{-16}	[24]
	Sn	9.44×10^{-16}	[24]
Ni_3Sn_4	Ni	1.36×10^{-17}	[24]
	Sn	6.81×10^{-18}	[24]

//c parallel to *c*-axis; ⊥*c* perpendicular to *c*-axis

Table 8.3 The atomic/molecular volumes of metals and intermetallic compounds [24]

Phase	Atomic/molecular volume (10^{-29} m^3)
Cu	1.18
Sn	2.70
Cu_3Sn	1.44
Cu_6Sn_5	1.77
Ni	1.09
Ni_3Sn_4	1.78

Cu_6Sn_5 and the growth of Cu_3Sn follow diffusion control kinetics [24]. In the meantime, Kirkendall voids were found in the Cu_3Sn layer. The formation of Kirkendall voids was attributed to the difference in diffusion rate of Cu, $D_{cu,}Cu_{3Sn}$, and Sn, $D_{Sn,}Cu_{3Sn}$, in the Cu_3Sn. The vacancy flux in Cu_3Sn, $J_{v,Cu3Sn}$, is given by [24],

$$J_{v,\ Cu_3Sn} = -(J_{Cu,Cu_3Sn} + J_{Sn,Cu_3Sn}) = (D_{Cu,Cu_3Sn} - D_{Sn,Cu_3Sn})\frac{\partial C_{Cu,Cu_3Sn}}{\partial x} \quad (8.1)$$

According to Eq. (8.1), there is a net vacancy flux from the Sn-rich side to the Cu-rich side. Similar behavior of vacancy flux was also found for the annealing of the Ni/Sn/Ni microbump, yet with much smaller vacancy flux as the element diffusivity is smaller in Ni_3Sn_4 compound. The voids were found at the center after long time annealing, indicating the voids may form as a result of the impingement of IMC from top and bottom. In addition to the effect of diffusion rate difference, the atomic volume decrease on the formation of intermetallic compound (Table 8.3) from pure metal will also induce voids or even cracks.

The IMC growth rate during thermal ageing is generally described by,

$$d = kt^n \quad (8.2)$$

where d is the thickness, t is reaction time, n is 0.5 for diffusion-controlled kinetics and is 1 for reaction-controlled process, k is reaction constant. This general expression is valid regardless of the bump dimension. The growth of Ni_3Sn_4 in Ni/SnAg/Ni microbump during annealing at 180 °C was found to have a rate constant of 8.0×10^{-14} cm^2/s and $n = 0.5$ [37]. The consumption of Ni layer is also diffusion controlled with a rate constant of -2.7×10^{-14} cm^2/s.

8.5 The Microstructure and Failure Mechanism of Microbump

The volume of microbump is largely reduced comparing with conventional C4 bump [22, 28]. The microbump solder is generally a cap on Cu pillar after reflow. The volume of the microbump is at least one order of magnitude smaller than that of the C4 bump. The afterward reflow treatments like thermal ageing, electrical current stressing will convert almost the entire volume of the solder to intermetallic compound even with Ni barrier layer. The intermetallic compounds formed will fully occupy the solder joint to give essential intermetallic-like joint. The Cu/solder joint was completely transformed to intermetallic compound after reflow when the solder thickness was less than 20 μm [42]. A micro-tensile test investigation indicates that the as-reflowed solder joint will behave brittle when the joint thickness is 70 μm or less. The mechanical property of the joint of 15 μm thickness dominated by the intermetallic compound current stressing [42] shows potential brittle failure behavior. The Young's modulus of various intermetallic compounds and solder are 133.3 GPa (Ni_3Sn_4), 108.3 GPa (Cu_3Sn), 85.56 GPa (Cu_6Sn_5), and 52.73 GPa (Sn3.5Ag) [25]. These values indicate the potential high brittleness of the intermetallic-like microjoint. The failure mechanism of microbump under current stressing very much depends on the original metallizations that affect the following reaction within the bump. A Cu pillar/SnAg cap/Cu trace structure will become Cu/Cu_3Sn/Cu_6Sn_5/solder/Cu_6Sn_5/Cu_3Sn/Cu after assembly. The structure converted to Cu/Cu_3Sn/Cu_6Sn_5/Cu_3Sn/Cu after current stressing. The Cu_3Sn layer kept growing during current stressing at the consumption of Cu_6Sn_5 and Cu sources on both end. The continuing growth of Cu_3Sn resulted in the extensive formation and coalescence of Kirkendall voids to form crack that leads to failure [12]. On the other hand, when the Ni barrier is introduced between Cu pillar and solder, the Ni barrier retarded the consumption of Cu pillar during current stressing, partly due to the slow reaction rate of the Ni layer. The solder was completely converted to a mixed layer of Cu_6Sn_5 and $(Cu, Ni)_6Sn_5$. Meanwhile, the Cu_3Sn layer grew rapidly at massive consumption of the Cu trace layer and induced Kirkendall voids. The failure at long time current stressing was mainly due to the consumption of most of the Cu trace [12].

Kirkendall void is a typical defect formed at the interface when the diffusivities of counter diffusing elements are different. The diffusivities of Cu and Sn are different in Cu_3Sn while similar in Cu_6Sn_5, Table 8.2. The formation rate of Cu_6Sn_5 is larger than Cu_3Sn. Accordingly, in the case of microbump where Cu_6Sn_5 and Cu_3Sn are formed sequentially, Kirkendall void will more likely form in the Cu_3Sn layer or at the interface between Cu_3Sn and Cu. Void was generally not observed in the Cu_6Sn_5 intermetallic compound. One of the potential methods of avoiding the formation of Kirkendall void is by removing the voids from the diffusing media Cu_3Sn. The manufacturing of nanotwinned Cu pillar by electroplating was able to sink the void formed in the above-mentioned intermetallic transformation reaction in a Cu pillar/Sn/Cu pillar structure [43]. No void was observed in the structure during the transformation of Cu_6Sn_5 to Cu_3Sn under long time thermal ageing. The nanotwinned Cu pillar was produced by appropriate controlling the electroplating conditions [44].

Both Pd and Au form intermetallic compounds with Sn when adopted as metallization. $PdSn_4$ grows faster than $AuSn_4$. A large columnar $(Pd, Ni)Sn_4$ may form in the Cu/Ni/Pd/Au/Sn/Ni/Cu microbump structure, Fig. 8.8a [45]. The thermal cycle test or electromigration test of such joint eventually convert the (Pd, Ni)

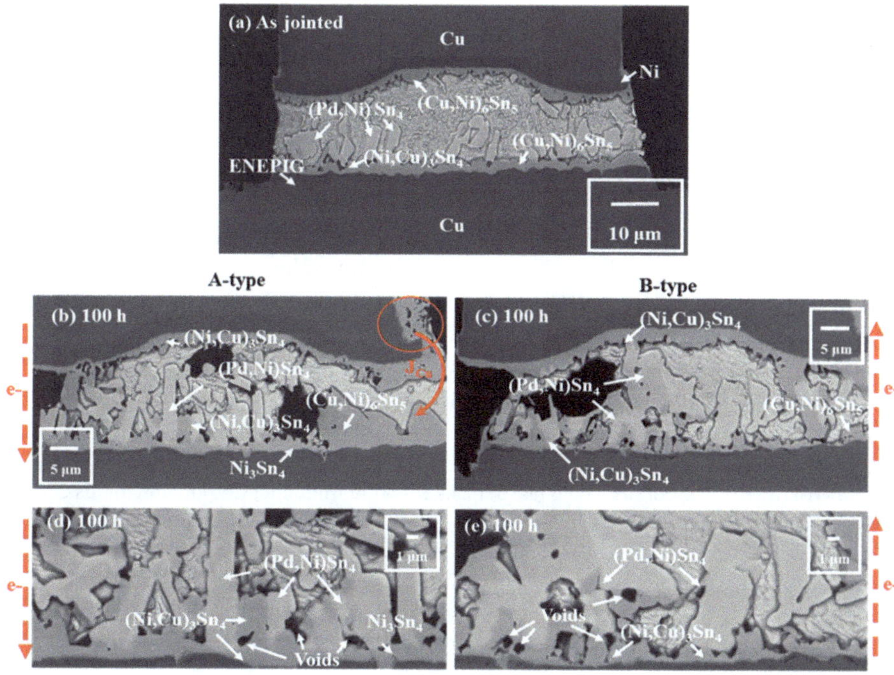

Fig. 8.8 Cross-sectional BSE images (**a**) as-jointed; after stressing with 1.8×10^4 A/cm^2 at 100 °C for 100 h (**b**) current from Cu pillar/Ni to ENEPIG, (**c**) current from ENEPIG to Cu pillar/Ni, (**d**) enlarged view of the micrograph shown in (**b**), (**e**) enlarged view of the micrograph shown in (**c**) [45]

Sn_4 to Ni_3Sn_4. The molar volume difference between the intermetallic compounds will induce void or crack at the interface between the converting intermetallic compounds [22], Fig. 8.8b–e [45] (readers are referred to Chap. 7 for the detail discussion regarding electromigration). The molecular volumes are 71.13 cm^3/mol for $PdSn_4$ and 75.25 cm^3/mol for $NiSn_4$. The compounds conversion actually involved (Pd, Ni)Sn_4 and (Ni, Cu)$_3$$Sn_4$. Thus, a volume difference of 9.65–22.3 % is expected that causes the volume shrinkage during the conversion [45]. The volume shrinkage will result in the formation of voids or cracks at the interface between the intermetallic [45].

8.6 Summary and Future Challenge

The industrial microbumping and bonding process has been mature and promising. In light of the massive shrinkage in bump dimension comparing with conventional flip chip solder bump, the solder materials compositions have been kept as simple as possible. Sn or Sn-Ag have been the predominant solder constituents of the microbump. Electroplating is the major deposition method for the solder bump because of the small dimension of the microbump with Cu pillar design and the productivity concern. The reduced volume of the solder joint results in high volume fraction of intermetallic compound in the as-reflowed and prolonged thermal-aged microbump joint. One of the challenges for reliability concern will be the manipulation of the vacancy formed as a result of the intermetallic compound formation. The molar volume difference between solder and intermetallic compound formed during the bumping process and product performance will induce void, vacancy, and crack at the UBM/solder, substrate metallization/solder, and TSV/solder interfaces. Appropriate adoption of metallization and UBM may help reduce the undesired defects. The ever shrinkage of the electronic products and the idea of the 3D IC are always pushing for finer pitch and reducing joint volume. The microbump technology is facing the challenge for further volume reduction and bump structure simplification.

Acknowledgment The editors would like to thank Pilin Liu from Intel Corporation for his critical review of this chapter.

References

1. C.K. Lee, C.J. Zhan, J.H. Lau et al., Wafer bumping, assembly, and reliability assessment of μbumps with 5μm Pads on 10μm Pitch for 3D IC Integration. In *2012 I.E. 62nd Electronic Components and Technology Conference CPMT*, San Diego, 29 May 2012–1 June 2012, p 636
2. J.W. Peng, Y.S. Chen, Y. Chen et al., Removed organic solderability preservative (OSPs) by Ar/O2 microwave plasma to improve solder joint in thermal compression flip chip bonding. In *2014 I.E. 64th Electronic Components and Technology Conference (ECTC), CPMT*, Lake Buena Vista, FL, 27–30 May 2014, p 1584
3. L. Panchenko, K. Croes, I. De Wolf et al., Degradation of Cu_6Sn_5 intermetallic compound by pore formation in solid-liquid interdiffusion Cu/Sn microbump interconnects. Microelect. Eng. **117**, 26–34 (2014)
4. C. Li, X. Wang, S. Song et al., 21-Layer 3-D chip stacking based on Cu-Sn bump bonding. IEEE Trans. Comp. Pack. Manu. Tech. **5**(5), 627–633 (2013)
5. K.N. Tu, H.Y. Hsiao, C. Chen, Transition from flip chip solder joint to 3D IC microbump: its effect on microstructure anisotropy. Microelect. Rel. **53**, 2–6 (2013)
6. S.Y. Huang, T.C. Chang, R.S. Cheng et al., Failure mechanism of 20 μm pitch microjoint within a chip stacking architecture. In *2011 I.E. 61st Electronic Components and Technology Conference (ECTC), CPMT*, Lake Buena Vista, FL, May 31 2011–June 3 2011, p 886
7. W.H. Chen, C.F. Yu, H.C. Cheng et al., IMC growth reaction and its effects on solder joint thermal cycling reliability of 3D chip stacking packaging. Microelec. Rel. **53**, 30–40 (2013)
8. Y.J. Chen, C.K. Chung, C.R. Yang et al., Single-joint shear strength of micro Cu pillar solder bumps with different amounts of intermetallics. Microelec. Rel. **53**, 47–52 (2013)
9. T. Suzuki, K. Asami, Y. Kitamura, T. Fukushima et al., Challenges of high-robustness self-assembly with Cu/Sn-Ag microbump bonding for die-to-wafer 3D integration. In *2015 I.E. 65th Electronic Components and Technology Conference (ECTC) CPMT*, San Diego, 26–29 May 2015, p 312

10. P. Soussan, B. Majeed, P.L. Boterf et al., Evaluation of Sn-based microbumping technology for hybrid IR detectors. 10μm pitch to 5μm pitch, In *2015 I.E. 65th Electronic Components and Technology Conference (ECTC) ,CPMT*, San Diego, 26–29 May 2015, p. 597
11. H.Y. Chen, C.H. Tung, Y.L. Hsiao et al., Electromigration immortality of purely intermetallic micro-bump for 3D integration. In *2015 I.E. 65th Electronic Components and Technology Conference (ECTC), CPMT*, San Diego, 26–29 May 2015, p. 620
12. K.H. Kuo, C. Mao, K. Wang et al., The impact and performance of electromigration on fine pitch Cu pillar with different bump structure for flip chip packaging. In *2015 I.E. 65th Electronic Components and Technology Conference (ECTC), CPMT*, San Diego, 26–29 May 2015, p. 626
13. J.W. Shin, Y.S. Kim, H.G. Lee et al., Effects of thermos-compression bonding parameters on joint formation of micro-bumps in non-conductive film. In *2015 I.E. 65th Electronic Compo nents and Technology Conference (ECTC), CPMT*, San Diego, 26–29 May 2015, p. 910
14. N. Islam, G. Kim, K.O. Kim, Electromigration for advanced Cu interconnect and the challenges with reduced pitch bumps. In *2014 I.E. 64th Electronic Components and Technology Conference (ECTC) CPMT*, Lake Buena Vista, FL, 27–30 May 2014, p. 50
15. K.H. Kuo, J. Lee, F.L. Chien et al., Electromigration performance of Cu pillar bump for flip chip packaging with bump on trace by using thermal compression bonding. In *2014 I.E. 64th Electronic Components and Technology Conference (ECTC)*, CPMT, Lake Buena Vista, FL, 27–30 May 2014, p. 56
16. Y. Liu, M. Li, D.W. Kim et al., Filler trap and solder extrusion in 3D IC thermo-compression bonded microbumps. In *2014 I.E. 64th Electronic Components and Technology Conference (ECTC), CPMT*, Lake Buena Vista, FL, 27–30 May 2014, p. 609
17. K. Murayama, M. Aizawa, M. Higashi, Study of electro-migration resistivity of microbump using SnBi solder. In *2014 I.E. 64th Electronic Components and Technology Conference (ECTC), CPMT*, Lake Buena Vista, FL, 27–30 May 2014, p. 1166
18. J. De Vos, L. Bogaerts, T. Buisson et al., Microstructural and morphological characterization of SnAgCu micro-bumps for integration in 3D interconnects. In *2013 I.E. 63rd Electronic Components and Technology Conference CPMT*, Las Vegas, NV, 28–31 May 2013, p. 1122
19. J. Bertheau, P. Bleuet, R. Pantel et al., Microstructural and morphological characterization of SnAgCu micro-bumps for Integration in 3D interconnects. In *2013 I.E. 63rd Electronic Components and Technology Conference, CPMT*, Lake Buena Vista, FL, 28–31 May 2013, p. 1127
20. B. Dang, S. Wright, J. Maria et al., NiFe-based ball-limiting-metallurge (BLM) for microbumps at 50 um pitch in 3D chip stacks. In *2013 I.E. 63rd Electronic Components and Technology Conference* CPMT, Lake Buena Vista, FL, 28–31 May 2013, p 1595
21. H.Y. Son, S.K. Noh, H.H. Jung et al., Reliability studies on micro-bump for 3-D TSV integration. In *2013 I.E. 63rd Electronic Components and Technology Conference ,CPMT*, Lake Buena Vista, FL, 28–31 May 2013, p. 29
22. J.Y. Juang, S.Y. Huang, C.J. Zhan et al., Effect of metal finishing fabricated by electro and electro-less plating process on reliability performance of 30 μm-pitch solder microbump interconnection. In *2013 I.E. 63rd Electronic Components and Technology Conference, CPMT*, Lake Buena Vista, FL, 28–31 May 2013, p. 653
23. Y. Ito, T. Fukushima, K.W. Lee et al., Flux-assisted self-assembly with microbump bonding for 3D heterogeneous integration. In *2013 I.E. 63rd Electronic Components and Technology Conference, CPMT*, Lake Buena Vista, FL, 28–31 May 2013, p. 891
24. Y. Wang, S.H. Chae, J. Im et al., Kinetic study of intermetallic growth and its reliability implications in Pb-free Sn-based microbumps in 3D integration. In *2013 I.E. 63rd Electronic Components and Technology Conference, CPMT*, Lake Buena Vista, FL, 28–31 May 2013, p. 1953
25. Y. Wang, S.Y. Chae, R. Dunne et al., Effect of intermetallic formation on electromigration reliability of TSV-microbump joints in 3D interconnect. In *2012 I.E. 62nd Electronic Compo nents and Technology Conference CPMT*, San Diego, CA, 29 May 2012–1 June 2012, p. 319

26. C.K. Lee, C.J. Zhan, J.H. Lau et al., Wafer bumping, assembly, and reliability assessment of μbumps with 5μm pads on 10μm pitch for 3D IC integration. In *2012 I.E. 62nd Electronic Components and Technology Conference, CPMT*, San Diego, CA, 29 May 2012–1 June 2012, p. 636
27. Y.S. Park, J.W. Shin, Y.W. Choi et al., A study on the intermetallic growth of fine-pitch Cu pillar/SnAg solder bump for 3D-TSV interconnection. In *2012 I.E. 62nd Electronic Compo nents and Technology Conference, CPMT*, San Diego, CA, 29 May 2012–1 June 2012, p. 2053
28. R. Dunne, Y. Takahashi, K. Mawatari et al., Development of a stacked WCSP package platform using TSV (through silicon via) technology. In *2012 I.E. 62nd Electronic Compo nents and Technology Conference, CPMT*, San Diego, CA, 29 May 2012–1 June 2012, p. 1062
29. Y. Choi, J. Shin, K.W. Paik, 3D-TSV vertical interconnection method using Cu/SnAg double bumps and B-stage non-conductive adhesives (NCAs). In *2012 I.E. 62nd Electronic Compo nents and Technology Conference, CPMT*, San Diego, CA, 29 May 2012–1 June 2012, p. 1077
30. Y.M. Lin, C.J. Zhan, J.Y. Juang, J.H. Lau et al., Electromigration in Ni/Sn intermetallic microbump joint for 3D IC chip stacking. In *2011 I.E. 61st Electronic Components and Technology Conference (ECTC), CPMT*, Lake Buena Vista, FL, 31 May 2011–3 June 2011, p. 332
31. A. Syed, K. Dhandapani, R. Moody et al., Cu pillar and μ-bump electromigration reliability and comparison with high Pb, SnPb, and SnAg bumps. In *2011 I.E. 61st Electronic Compo nents and Technology Conference (ECTC), CPMT*, Lake Buena Vista, FL, 31 May 2011–3 June 2011, p. 332
32. W. Zhang, B. Dimcic, P. Limaye et al., Ni/Cu/Sn bumping scheme for fine-pitch micro-bump connections. *2011 I.E. 61st Electronic Components and Technology Conference (ECTC), CPMT*, Lake Buena Vista, FL, 31 May 2011–3 June 2011, p. 109
33. Y. Ito, T. Fukushima, K.W. Lee et al., Reductant-assisted self-assembly with Cu/Sn microbump for three-dimensional heterogeneous integration. Jpn. J. Appl. Phys. 52, 04CB09-1-04CB09-6
34. Y.H. Ko, M.S. Kim, J. Bang et al., Properties and reliability of solder microbump joints between Si chips and a flexible substrate. J. Elect. Mater. **44**(7), 2458–2466 (2015)
35. H.Y. Hsiao, A.D. Trigg, T.C. Chai, Failure mechanism for fine pitch microbump in Cu/Sn/Cu system during current stressing. IEEE Trans. Comp. Pack. Manufact. Tech. **5**(3), 314–319 (2015)
36. M. Hansen, K. Anderko, *Constitution of Binary Alloys*, 2nd edn. (McGraw-Hill, New York, 1985)
37. H.Y. You, Y.S. Lee, S.K. Lee et al., Reliability of 20μm microbump interconnects. In *2011 I.E. 61st Electronic Components and Technology Conference (ECTC), CPMT*, Lake Buena Vista, FL, 31 May 2011–3 June 2011, p. 608
38. E.A. Brandes (ed.), *Smithells Metals Reference Book* (Butterworths, London, 1983)
39. K. Hoshino, Y. Iijima, K. Hirano, Interdiffusion and Kirkendall effect in Cu-Sn alloys. Trans. Jpn. Inst. Met. **21**(10), 674–682 (1980)
40. B.F. Dyson, T.R. Anthony, D. Turnbull, Interstitial diffusion of copper in tin. J. Appl. Phys. **38**, 3408 (1967)
41. D.C. Yeh, H.B. Huntington, Extreme fast-diffusion system: nickel in single-crystal tin. Phys. Rev. Lett. **53**, 1469–1472 (1984)
42. Y. Wang, I.M. De Rosa, K.N. Tu, Size effect on ductile-to-brittle transition in Cu-solder-Cu micro-joints. In *2015 I.E. 65th Electronic Components and Technology Conference (ECTC), CPMT*, Lake Buena Vista, FL, 26–29 May 2015, p. 632
43. W.L. Chiu, C.M. Liu, Y.S. Haung et al., Formation of nearly void-free Cu_3Sn intermetallic joints using nanotwinned Cu metallization. Appl. Phys. Lett. **104**, 171902 (2014)
44. H.Y. Hsiao, C.M. Liu, H.W. Lin et al., Unidirectional growth of microbumps on (111)-oriented and nanotwinned copper. Science **336**, 1007–1010 (2012)
45. Y.H. Hsiao, K.L. Lin, The formation and conversion of intermetallic compounds in the Cu pillar Sn-Ag micro-bump with ENEPIG Cu substrate under current stressing. J. Mat. Sci Mat. Elect. **27**, 2201–2205 (2016).

Chapter 9
Fundamentals of Electromigration in Interconnects of 3D Packaging

Pilin Liu

9.1 Introduction

To meet the demands for higher packaging density, better performance, and smaller form factors, microelectronics industry is currently transitioning from traditional flip chip technology to 3D (or 2.5D) integrated circuits (ICs). In 3D ICs, several Si chips are jointed together by micro bumps and through Si vias (TSV). Redistribution layer (RDL) is used in the TSV chip to connect TSV with micro bumps [1, 2]. The diameter of the micro bumps is about 10–20 μm, which is almost one order of magnitude smaller than flip chip first-level interconnect (FLI). The current density in micro bumps will increase dramatically with the same power demand and the joule heating will also be much higher when all the Si chips are stacked together. It can be seen that EM in micro bumps, TSV, and RDL will be a key reliability concern in 3D ICs. High joule heating may also induce large thermal gradient and local high temperature, inducing thermomigration failure or circuit burn [1–5].

Micro bumps in 3D packaging are Pb-free solder joints with Ni or Cu metallization [3–5]. The general mechanisms about the EM for larger FLI solder joints will also apply to the micro bump solder joints. On the other hand, Cu TSV and its connected metal lines are electroplated Cu. Their EM mechanisms should be the very same as those for damascene Cu interconnects in Si die.

In this chapter, the general EM factors in solder joints that are applicable to micro bumps will be briefly summarized in Sect. 9.2, followed by review of the unique EM behaviors in micro bumps in Sect. 9.3. Thermal migration induced failure modes will also be included in this section. EM in Cu TSV and its connected

P. Liu (✉)
Intel Corporation, 5000 W. Chandler Blvd., Chandler, AZ 85226, USA
e-mail: pilin.liu@intel.com

Y. Li, D. Goyal (eds.), *3D Microelectronic Packaging*, Springer Series in Advanced Microelectronics 57, DOI 10.1007/978-3-319-44586-1_9

metal layers will be discussed in Sect. 9.4 with the brief summaries about the key EM factors of Cu damascene interconnects.

9.2 Key Modulators for EM in Solder Joints

EM can cause solder joint failures at IMC/solder interface due to flux divergence induced by Sn self-diffusion, which is a very common EM failure mode and has been well discussed [6–8]. On the other hand, electron wind can also accelerate the dissolution of under-bump metallization (UBM) on cathode side and cause failures due to depletion of UBM [9, 10]. It has been reported that metallization dissolution is closely related to Sn grain orientations [9, 10].

9.2.1 Typical EM Fail Caused by Sn Diffusion

In general, EM can induce failures through a void nucleation and propagation mechanism. Voids nucleate near the entrance of electrons due to moving away of the predominant diffusion species, which is caused by the momentum exchange between electron wind and metal ions. The fundamental force equation governing electromigration is given by Eq. (9.1) [11].

$$F_{EM} = Z^{*}eE = \left(Z_{el}^{*} + Z_{wd}^{*}\right)e\rho j \tag{9.1}$$

where F_{EM} stands for electromigration driving force, Z^{*} is effective charge representing transfer of momentum from electrons to atoms, e is electron charge, ρ is resistivity, j is current density, Z_{el}^{*} is nominal valence of the diffusing ion for the direct force, and Z_{wd}^{*} is the effective valence for the electron wind force.

In Sn-based Pb-free solders, the predominant diffusion specie is Sn atom and the EM failure is between IMC and solder on cathode side. Figure 9.1 shows the typical EM fail on substrate side of a Pb-free solder joint when the electron flow goes from substrate to Cu bump. Sn flux divergence happens at the Sn/IMC interface on substrate side during EM and causes the void nucleation and propagation along that interface.

Solder systems are quite complex metallurgical systems. The dissolution of solutes (Cu, Ni, Pd, Au) from the substrate surface finish and Cu bumps into the solder during reflow makes the solder alloys extremely complex multicomponent thermodynamic system. The solute elements dissolved in the solder are interstitial in nature and hence diffuse substantially fast in the solder matrix. Additionally, solders are multiphase alloys and undergo several intermetallic reactions at the interfaces. This makes the solder joint EM more complex than Al or Cu traces in electronic packaging.

Fig. 9.1 Typical solder joint EM fail induced by Sn diffusion flux divergence between IMC and Sn solder at cathode side [7]

Many factors can impact the EM failure induced by Sn diffusion, including solder alloying, IMC formation, current crowding, metallization dissolution, and solder joint height. All these factors are applicable to micro bumps in 3D packaging also.

A. The impact of IMC reaction and solder alloying

One unique feature for solder joint EM is its very dynamic nature under electron wind and high temperature. Metallization from both cathode and anode will keep dissolving into the solder alloy and react with Sn to form IMCs. This will either coarsen the existing IMC particles in the solder matrix or increase the thickness of the IMC on anode side. The IMC reaction between metallization (typically Cu or Ni) and Sn causes volume shrinkage, which will generate Sn vacancies during the reaction. This will increase the probability of solder voids nucleation and growth at IMC/solder interface on cathode side, where Sn atoms flux divergence happens. From this point of view, IMC reaction and the instability of the solder system can reduce the EM capability. This IMC reaction impact on EM fail will be even more dramatic in smaller 3D packaging micro bumps since the Sn volume is very limited and easy to be exhausted.

Solder alloying can also impact the stability of the solder system. Seo et al. [12] compared Sn-1.8Ag and Sn-0.5Cu with different joining path to control the Ni and Cu dissolution in the solder system. They found Sn-1.8Ag shows much more stable grain structure and stable Ag_3Sn particles, which slow down Cu dissolution into the solder. Stable microstructure can benefit both Sn diffusion and metallization dissolution failure modes. Lu et al. [13] studied the effect of Zn doping on SnAg solder microstructure and EM stability. They found that Zn reacts strongly with alloying elements, such as Cu, Ag, and Ni. Zn-doped solders improve the stability of the interface as well as the bulk microstructure. Stable microstructure in a Zn-doped solder can effectively suppress EM failure induced by Sn atom flux divergence.

Overall, it can be seen that factors that can slow down the IMC reactions or metallization dissolution can increase the stability of the solder joint and improve EM.

B. The impact of solder joint height

Based on Al EM experiments, Blech [14] observed a threshold current density under which no EM damage happens. This threshold is inversely proportional to the length of the strip. This so-called Blech effect has since been well studied in different systems including solder joints [15–17]. It is a result of back stress gradient induced by accumulation of atoms at anode and extra vacancies at cathode, balancing out the electron wind force. Since the current density threshold is inversely proportional to the length of the stressed conductor, the Blech effect states the existence of critical Blech product $(j \times L)_c$ (current density times conductor length), below which the conductor will be never failed by EM. This critical product tells that at given current density, EM damage can be eliminated when the conductor length is short enough. Efforts to measure the critical Blech product in solder joint have been made using either edge displacement method [16, 17] or plot of time to fail (TTF) vs. Blech product [15]. The results are summarized in Table 9.1.

In general, the critical product decreases dramatically with temperature increases. This raises a new challenge to Black's law [18] based reliability model used across industries, in which the current density factor (n) and activation energy (Q) are used as the key reliability projection parameters. Blake's equation doesn't include the $(jL)_c$ term, assuming nonexistence of critical current density. EM damage will happen as long as current is present. This will underestimate the solder joint EM capability and make the reliability risk assessment too conservative. In order to consider the length and critical product impacts on EM lifetime, Black's equation can be modified as follows:

$$\frac{1}{\mathrm{MTTF}} = A\left(j - \frac{(jL)_c}{L}\right)^n \exp\left(-\frac{Q}{\mathrm{RT}}\right)$$

where MTTF is the mean time to failure, j is current density, L is solder joint height, n is current density exponent, and Q is activation energy.

This modified Black's equation includes the solder joint height impact, which is very important for 3D packaging because solder joint height can vary from

Table 9.1 Pb-free solder critical products at different temperatures

Solder	Temp (°C)	L (μm)	Critical product (A/cm)	References
SnAg1.8	145	50–150	30	[15]
SnAg1.8	145	75	26	[15]
SnAg3.8Cu0.7	80	350	1505	[16]
	100		1120	
	120		490	
SnAg3.0Cu0.5	140	100, 200, 500	253	[17]

several hundred microns to several microns in one component. A reliability model that considers the solder joint geometry change is highly needed. The fact that critical product changes with temperature suggests that at higher temperature, for the same height of solder joint, the critical current density will be much smaller than that at low temperature. This makes EM reliability lifetime very hard to be projected based on accelerated high temperature testing. More efforts are needed to fundamentally understand the physics of temperature impact on $(jL)_c$ so that a more accurate reliability model can be established for solder joints in 3D packaging.

9.2.2 EM Fail Caused by Metallization Dissolution

EM can accelerate the cathode metallization dissolution and promote the formation of thicker IMC on anode side. When the cathode metallization is thick enough, the joint can be eventually transformed to a full IMC joint or a joint with very short Sn height and the EM resistant of the joint can be high. However, when the cathode metallization is thin, EM can dissolve the entire metallization layer and induce solder joint open [9, 10]. This is a different failure mode from Sn diffusion induced IMC/solder separation discussed in Sect. 9.2.1. One interesting phenomenon related to this failure mode is its sensitivity to Sn grain orientations [9, 10, 19–21]. It has been reported that [20] the time to dissolve 2 μm Ni UBM can vary from 100 h (when Sn *c*-axis is in parallel to electron flow direction) to 1800 h (when *c*-axis is perpendicular to electron flow) at given temperature and current density. Figure 9.2a

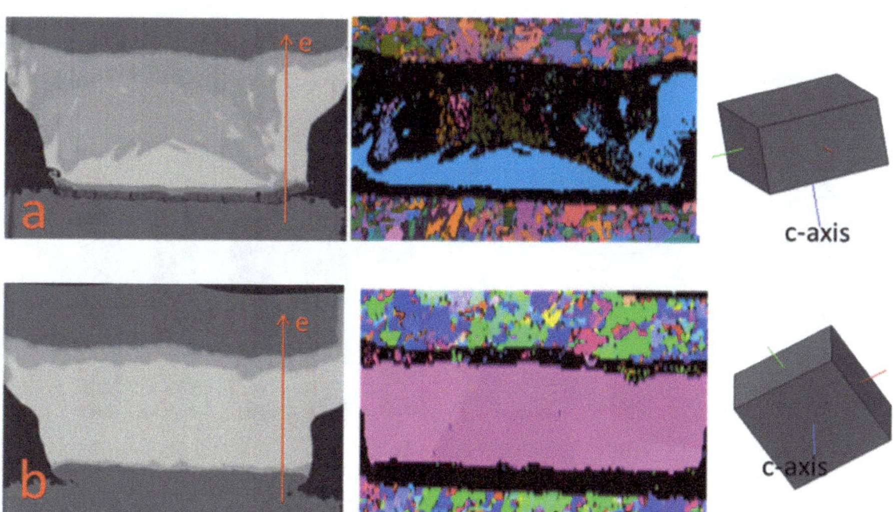

Fig. 9.2 SEM and EBSD show very good correlation between *c*-axis of Sn and cathode side Ni-P EM damage. (**a**) Fast Ni dissolution when *c*-axis is in line with the electron flow direction; (**b**) slow Ni dissolution when *c*-axis is perpendicular to the electron flow direction [21]

[21] shows the EM failure caused by Ni dissolution of the electroless Ni-P surface finish on cathode side of the stressed solder joint. EBSD clearly shows *c*-axis of Sn solder aligns well with the electron flow direction in this joint. Figure 9.2b is another solder joint that stressed together with Fig. 9.2a with exactly the same current and temperature. No any sign of EM damage can be seen at the cathode side and the *c*-axis of Sn is perpendicular to the e-flow direction. It is clear that Sn grain orientation is the key modulator for fast UBM dissolution during EM. It is also found that the key assembly process parameters such as cooling rate and Cu dissolution in the solder can modulate Sn grain orientation distributions.

A. The impact of assembly cooling rate

Solder joints with cooling rate about 1 and 50 °C/s are studied [21]. EM testing shows that MTTF for those with faster cooling rate is about three times shorter than those with slower cooling rate and the cross section on the EM damaged parts shows Ni dissolution is the main cause for fail. This data clearly shows that faster cooling rate increased the chance of Ni dissolution in EM-stressed solder joints. To understand the impact of cooling rate on grain orientation distributions, statistic EBSD analysis has been done to compare solder joints with different cooling rates. The angles between *c*-axis and the substrate normal were calculated and statistically analyzed as shown in Fig. 9.3. It is found that the solder joints with faster cooling rate show higher chance of forming *c*-axis oriented Sn grains along the substrate normal or the electron flow direction and this directly correlates to the worst EM performance in contrast to the joints formed with slower cooling rate.

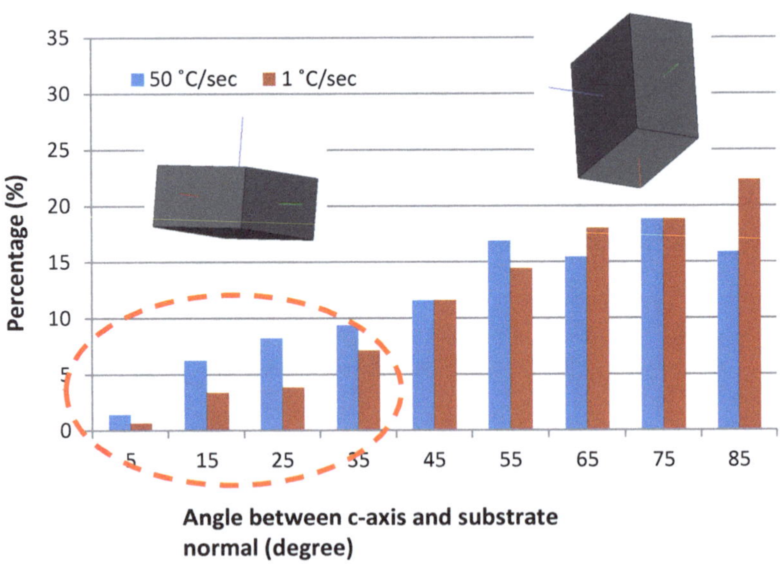

Fig. 9.3 Statistical EBSD results on solder joints before EM shows that faster cooling rate increases the percentage of *c*-axis along with substrate normal [21]

This EBSD data in general explains the result that solder joints with faster cooling rate show higher chance of Ni dissolution. Further studies indicated that the critical angle between the *c*-axis Sn grains and substrate normal for fast Ni dissolution is about 40°. It is found that the angle between 0° and 40° is the range for the fast metal dissolution which results in EM fails. Based on this, the probability for EM fails caused by fast Ni dissolution is measured as ~25 % for fast cooling rate and ~15 % for slow cooling rate.

The reason that fast cooling rate can cause the Sn orientation distribution difference is related to the fact that solder orientations are not randomly distributed. The angles between 0° and 40° are always in small percentages for both cooling rates. Inverse polar results indicate that [110] is the preferred direction aligned with substrate normal. Given the fact that the thermal profiles during solidification are normal to the substrate or Si die, it can be concluded that [110] is the preferred grain growth direction following the thermal flow direction during solidification, as reported in another study [22]. Since *c*-axis of the tetragonal Sn is always perpendicular to its [110] direction oriented along the substrate normal direction, the *c*-axis of the grain remains perpendicular to the normal direction. This explains the unique percentage distribution of the angles of the *c*-axis solder grains for their preferred perpendicular direction to the substrate for both fast and slow cooling rates. During solidification, it can be assumed that slower cooling rate would allow the system to have enough time to grow grains along their favored directions. On the other hand, fast cooling rate would make the grain orientation more random, which means more grains with smaller angles between *c*-axis and substrate normal. This explains the cooling rate impact on Sn grain orientation distributions.

B. Impact of alloying

While solder solidification rate is the key parameter to impact Sn grain orientations, solder alloying is another factor that impacts the Sn grain size and orientation distribution. Sylvestre et al. [23] studied the impact of Ag content on the cyclic twin formation for SAC solder alloys and showed that 2.2 wt% Ag in SAC alloy can dramatically increase the chance of forming cyclic twins, which commonly occurs during solidification of SnAg, SnCu, and SAC alloys [24, 25]. This is a very unique type of twinning and very less frequently observed in other metal systems. The formation of cyclic twins in Pb-free solder can possibly impact the distribution of *c*-axis along with substrate normal direction. We know that cyclic twins share the [100] or [010] common directions, but not in the direction of [001]. If one of the three cyclic twins happens to have its *c*-axis in the direction of electron flow, the *c*-axis of the other two grains cannot align with the electron flow because their common axis is not of [001] orientation. In other words, for cyclic twinning structures, three orientations are bundled together and they cannot have all their *c*-axis align with the electron flow direction. This scenario of Sn crystalline structure will reduce the chance of *c*-axis aligns with substrate normal. Therefore, adding Ag to the SnCu system can improve the EM performance induced by metallization dissolution, which has been experimentally observed [12, 15].

On the other hand, Cu atoms in the liquid solder can possibly play the same role as Ag. It is reported that the formation of the cyclic twin structure initiates from a small hexagonal cluster of Sn atoms centered on a Cu or Ag atom in the molten solder [25]. The quick dissolution of Cu from Cu column into the solder system during assembly can reduce the possibility of *c*-axis of Sn grain align with substrate normal, which can delay the metallization dissolution during EM [21].

C. Impacts of Solder Joint Height

Besides Sn grain orientation and solder alloying, solder joint height can also impact metallization dissolution during EM. However, the effect of solder joint height on metallization dissolution under EM has been seldom reported. Lu et al. [19] observed that the resistance change vs. stress time plot of SnAg solder joint during EM can reach a plateau and they considered that as an evidence of Blech effect in SnAg solder joint controlled by Sn diffusion failure mechanism. They further concluded that solder joint height should not impact metallization dissolution since no such plateau was observed in SnCu solder joints, which were mainly dominated by Ni dissolution failure mode. However, no direct comparison has been done between longer and shorter solder joints in their study.

Zhang et al. [26] studied the solder joint height impact on Cu dissolution of Cu/SAC305/Cu joint with same volume but different height. They observed that the Cu dissolution rate increases with the solder joint height increase, as shown in Fig. 9.4.

The same phenomenon is also observed on Ni dissolution in Cu/Sn0.7Cu/Ni solder joint [21]. In one controlled DOE, the solder joint height was purposely

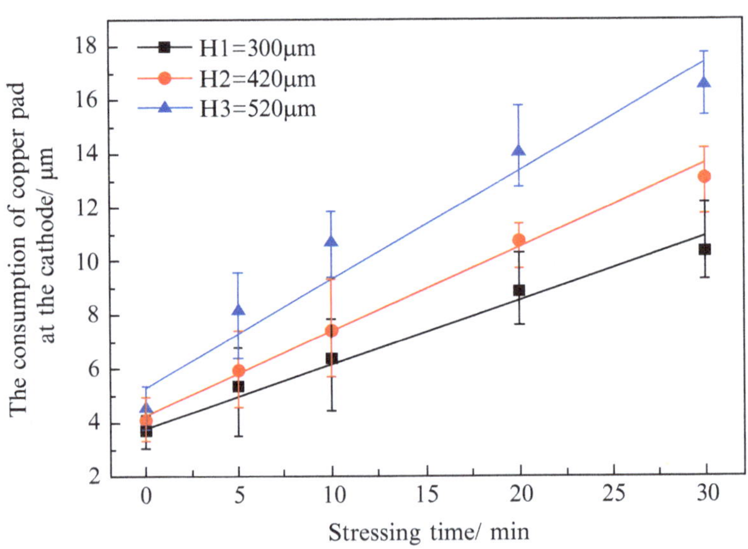

Fig. 9.4 Cu dissolution rate increase with the increase of solder joint height [26]

stretched during reflow while the solder volume was fixed, which ended up with two DOE legs with about 20 μm solder joint height difference. EM testing indicates the MTTF for the shorter solder joint height is about three times longer than that with higher solder joint height. The EM failure mode is mainly caused by Ni dissolution.

Solder joint height impact on EM failure dominated by Sn diffusion flux divergence has been discussed in Sect. 9.2.1. However, the back stress theory in Sn diffusion cannot be applied to explain the observed phenomena that Cu or Ni dissolution is dramatically accelerated during EM in longer solder joint height because both Cu and Ni diffusion path are through the interstitials [27] and the back stress impact should be low in contrast to Sn self-diffusion. Moreover, back stress is hard to be built up with interstitial diffusion because no vacancies will be generated during the diffusion. The impact of the solder joint height on metal dissolution can be discussed from the diffusion point of view instead of back stress.

In order to promote fast metal dissolution mechanism, the dissolved Ni atoms need to be quickly removed from the interface. The schematic for the process of Ni dissolution is shown in Fig. 9.5 [21].

Fast Ni dissolution flux ($J_{dissolution}$) needs quick and effective J_{Ni} (Ni flux that quickly removes Ni from the cathode interface). If the solder joint height is short, when the electron wind blows the Ni atoms away from the cathode interface, they will quickly meet the anode side of the solder joint and will be subjected to back diffusion. In other words, the anode side acts as a barrier for Ni atoms to diffuse further away. In this case, Ni atoms cannot be effectively removed from the interface and the deficiency of metallization at the cathode interface is hard to be established for a quick $J_{dissolution}$, since J_{back} (flux of Ni atoms diffuses back to cathode) diffusion can reach to that interface. On the

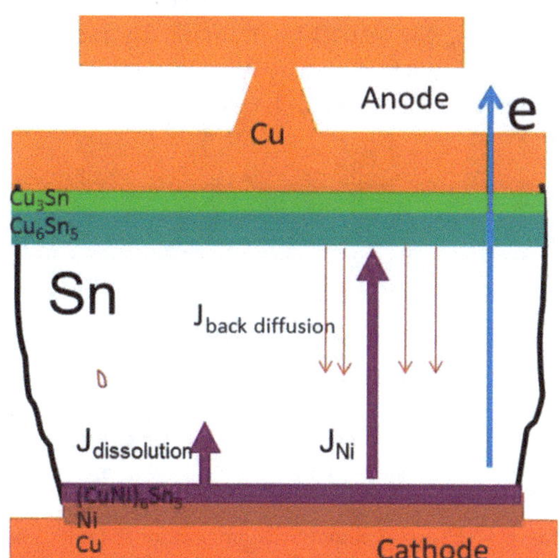

Fig. 9.5 Schematic of a solder joint shows the Ni diffusion fluxes during EM [21]

other hand, if the solder joint is long, the back diffusion atoms will be hard to come back to increase the Ni concentration at the cathode side interface. The Ni deficiency at the interface can be easily established.

9.3 EM in Solder Joints of 3D Packaging

9.3.1 *EM Damage due to Sn Flux Divergence in Micro Bumps*

Compared with larger solder joints in C4 flip chip and BGA packaging, unique EM behaviors happen in micro bumps of 3D packaging due to their smaller dimensions. Failure between solder and IMC as shown in Fig. 9.1 is hard to find in micro bumps. This can be simply because of the Blech effect that the back stress for short micro bumps is high enough to dramatically delay or eliminate the EM damage caused by Sn flux divergence. Take an example of Sn1.8Ag solder, its critical product $(jL)_c$ is about 30 A/cm at 145 °C, as shown in Table 9.1. Based on this, the critical length at 5×10^4 A/cm^2 is 6 μm, which is about the typical solder joint height of micro bumps. This indicates Sn1.8Ag micro bumps with 6 μm solder joint height will not have any EM damage induced by Sn flux divergence when stressed at 5×10^4 A/cm^2 and 145 °C.

It is worthy to note that some studies report the partial failure of micro bumps with almost full transformation to IMC joints, as shown in Fig. 9.6 [28]. The authors comment this partial damage is close to the cathode side and can be the result of Sn flux divergence. Since the remained Sn is very minimal next to the partial cracks, it is less possible that they are induced by Sn flux divergence from the back stress point of view. EM back stress for Sn diffusion is closely related to the height of remaining Sn. It will be dramatically increased and can easily balance the electro wind force when nearly full IMC joint is formed. The partial crack close to the cathode in Fig. 9.6 can be explained as following.

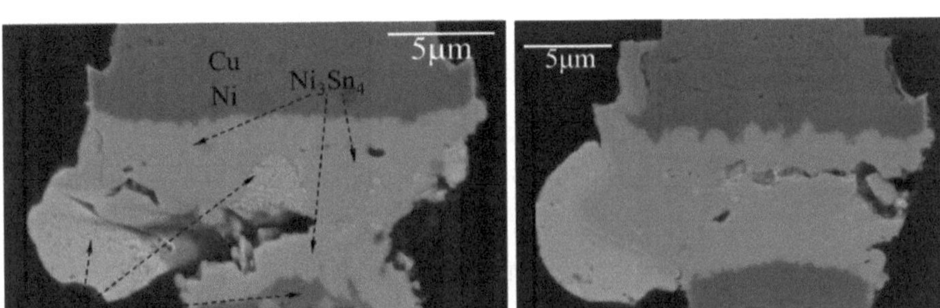

Fig. 9.6 Cross-sectional SEM images for the micro bump stressed at 0.12 A at 150 °C for 350 h (**a**) with upward electron flow and (**b**) with a downward electron flow [28]

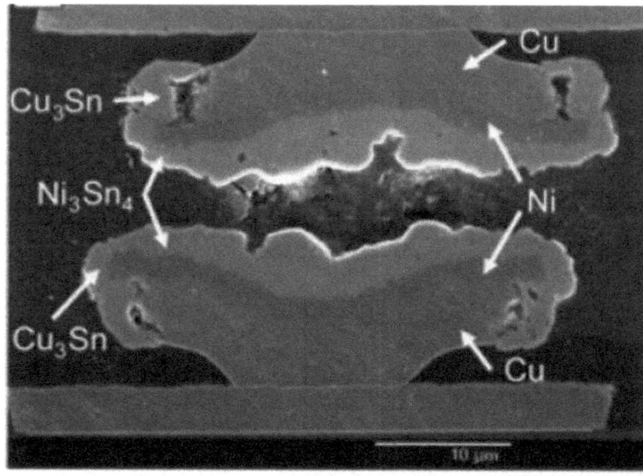

Fig. 9.7 Big gap is formed in the middle of micro bump when Sn is totally consumed by IMC reactions after 20 min reflow at 260 °C [29]

During EM, electron wind can accelerate the metallization dissolution on cathode side. The dissolved metallization (Ni in this case) is quickly removed from the cathode by electron wind and diffused to anode. Thicker IMC can be formed on anode side due to electron wind-assisted quick metallization dissolution. The formation of the thick IMC on anode side can reduce the effective Sn-based solder height and delay or eliminate the EM damage due to Sn flux divergence. However, the IMC reaction is volume shrinkage process. During EM stressing, thicker anode side IMC and thinner cathode side IMC will eventually merge together with the consumption of Sn. When more Sn are exhausted, the gap can be formed close to cathode side because IMC on anode side is thicker.

The similar phenomenon is also observed in micro bumps after long time reflow [29]. Figure 9.7 is a bump after 20 min reflow at 260 °C, Sn is totally exhausted by the IMC reaction, and a gap is formed between IMCs on upper and bottom sides. The gap is in the middle because the IMC thickness is the same on upper and bottom sides.

9.3.2 *The Transformation of Full IMC Joint Under EM*

Micro bump under EM has another unique behavior comparing with larger solder joints. It typically has smaller solder to metallization volume ratio, which can form a full IMC bump before the metallization is fully consumed. This is the reason that metallization dissolution as discussed in Sect. 9.2.2 is not the key reported failure mode in micro bumps.

The threshold volume ratio of metallization to solder to form full IMC instead of induce metallization dissolution failure mode can be simply calculated based on the IMC reaction equation. Take Cu/Sn reaction as an example, Cu_6Sn_5 IMC is always formed first with the dissolution of Cu under EM. It can be calculated that when Cu to Sn volume ratio is larger than 0.52, a full Cu_6Sn_5 IMC bump will be formed. In

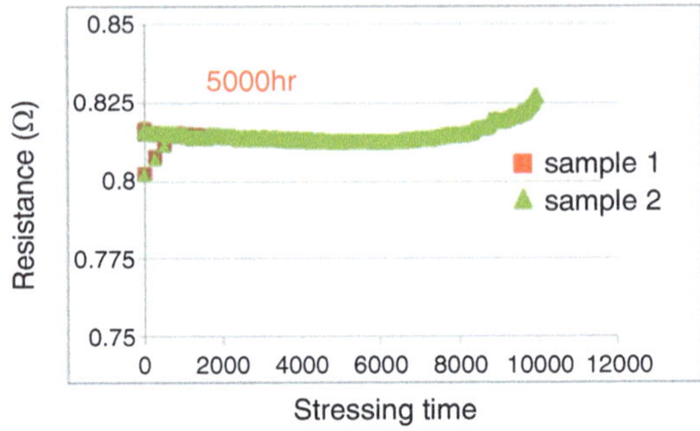

Fig. 9.8 Resistance vs. stressing time plot shows full IMC micro bumps have no EM damage after 5000 h of stressing at 1.4×10^5 A/cm^2 and 170 °C [30]

other words, as long as the Cu height is longer than half of the solder height, a full IMC joint will be formed before it failed as Cu dissolution.

At typical solder joint EM testing conditions, full IMC micro bumps are found to be almost immortal [30, 31]. Chen et al. [30] reported that a full IMC micro bump has no resistance increase after 5000 h of stressing at 1.4×10^5 A/cm^2 and 170 °C, as shown in Fig. 9.8. The later resistance increase in the plot is found to be related to Al trace damage instead of IMC damage. They further estimated that the IMC critical current density in their micro bumps is about 3×10^5 A/cm^2. This means the test current density is below the threshold and no EM damage can be induced no matter how long to test in this condition.

Wei et al. [32] studied the EM behavior in Sn-Cu IMCs using edge displacement method as shown in Fig. 9.9. They found Cu_6Sn_5 is more susceptible to EM than Cu_3Sn, which is related to the lower solidus temperature and higher resistivity of Cu_6Sn_5. They also observed Blech effect in the strips of various length of IMCs. The critical product for Cu_6Sn_5 IMCs is between 2.5 and 5 A/cm at 225 °C. Just as typical EM fails, voids are formed at cathode side interface where the atomic flux divergence is present. However, the predominant diffusion specie could not be determined in their report because no Cu-rich or Sn-rich areas are found under SEM/EDS examinations after EM testing. The critical product for Cu_6Sn_5 IMC cannot be directly compared with solder critical products listed in Table 9.1 since big temperature difference. It is expected IMCs should have much higher critical product than Pb-free solders at the same temperature because all the studies indicate that solder joint EM can be dramatically improved once a full IMC joint is formed [30, 31].

While the intrinsic EM resistance of full IMC bump is very high, in some cases it can fail at Cu_3Sn/Cu interface, as reported in Cu/SnAg/Cu micro bumps [33]. Figure 9.10 shows the failure of the solder joints between Cu_3Sn and Cu under EM test. The separation only happens on one side of the solder joint regardless of the EM polarity. The same fail was also observed in the baked joints in the same study.

The results indicate this failure is not purely resulted from EM but more related to baking. In fact, it has been well reported that Cu/Cu_3Sn separation is mainly caused by the impurities (such as S) from Cu plating [34, 35]. The formation of

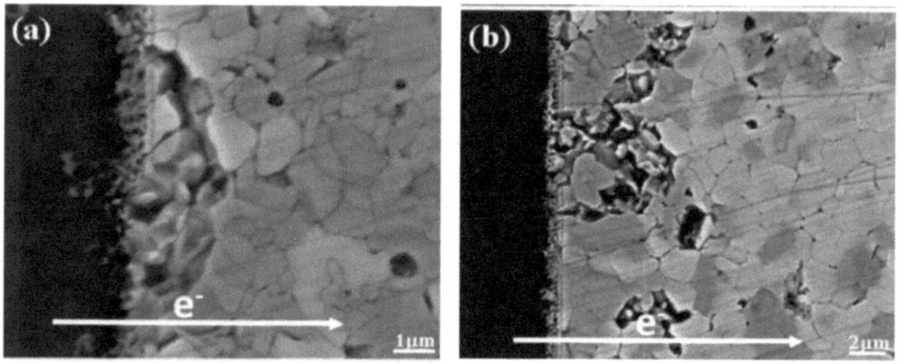

Fig. 9.9 SEM images shows Cu_6Sn_5 EM damages at the cathode side of interface where flux divergence is present [32]

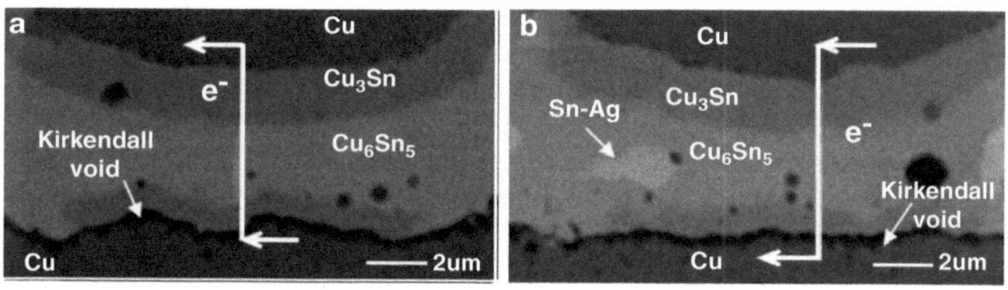

Fig. 9.10 Cu/Cu_3Sn separation can be observed on one side of the full IMC joint regardless of the EM electron flow direction [33]

Cu_3Sn can have two reaction fronts. Cu can diffuse through the IMC and react with Sn, and at the same time, Sn can diffuse through the IMC and react with Cu. Since Cu diffuse through Cu_3Sn is several times faster than Sn through it [36], vacancy flux is formed towards the Cu during the reaction. Vacancy flux towards Cu can condense to form voids and induce Cu_3Sn/Cu separation when impurities are present at the interface. Normally, the interface can be viewed as dislocation networks and are actually very good vacancy sinks. The vacancy condensation follows the classic condensation theory that needs nucleation and growth. If the interface is a good vacancy sink, it is very hard to cumulate enough vacancies for voids nucleation. However, when impurities are present at the interface, they can pin the dislocations at the interface and cause the interface less movable. In this case, the interface will not be an effective vacancy sink anymore and voids will be formed due to vacancy condensation. Only monolayer of impurities at interface can trigger this mechanism [37]. Actually, a more intuitive way to view this can be as following. Interfacial voids cannot be generated when the reaction interface is smoothly moveable and follows well with the reaction front. However, if impurities pin the interface and make it less moveable, the interface can not follow well with the reaction front and voids have to be generated. The reason that the full IMC joint only failed at lower side regardless of the EM polarity can be due to the Cu impurity difference between upper and lower Cu bumps.

In general, EM resistance in micro bumps is more robust because of the transformation of full IMC joint. However, impurities from Cu plating should be avoided to prevent the solder joint from Cu/Cu_3Sn failure. While this is mainly a backing-related phenomenon, EM can definitely increase the failure rate due to its acceleration of cathode side Cu dissolution.

9.3.3 Thermomigration Accompanied by EM

When current is applied to solder joints in electronic packaging, electron wind can induce EM damages as discussed above. Meanwhile, Joule heating due to the high current density can result in thermal gradient and induce thermomigration of the solder joint [38–40]. The thermal gradient is closely related to the geometrical configurations of the electronic packaging. Typically, joule heating on the substrate side of the solder joint is much less than that from the chip side where transistors and finer interconnects will generate more heating. It is reported that a temperature difference of 10 °C across a 100 μm diameter solder joint (1000 °C/cm) is sufficient to induce thermomigration and elements redistribution [38]. The large thermal gradient is expected to be worse in micro bumps of 3D packaging since its diameter is more than one order of magnitude smaller and stacked Si could generate more heat due to bad thermal dissipation.

Similar to EM force, solder joint under thermal gradient force can also result in two impacts on diffusion species. One is the migration of Sn atoms; the other is metallization (Cu, Ni) dissolution.

Ouyang et al. [39] studied the thermomigration behaviors of micro bumps in 3D packaging and compared them with flip chip solder joints. They observed that Sn in solder alloy migrated to the hot side in the flip chip solder joint, but no Sn thermomigration was observed in micro bumps of 3D packaging, as shown in Figs. 9.11 and 9.12. They attribute this to the back stress gradient difference caused by solder joint height. Just as mentioned in Blech effect, shorter solder joint height will result in larger back stress gradient which can balance out the thermal migration force on Sn atoms induced by thermal gradient. They further calculated the back stress (F_{BS}) and the driving force from thermal gradient (F_{TM}) in their tested 3D packaging based on the simulated temperature gradient of 5345 °C/cm across the micro bumps and the working temperature of about 134 °C. It is concluded that F_{BS} is about 3–4 times larger than F_{TM}, which explains why no Sn thermomigration happens in their tested micro bumps. In other words, in order for Sn thermomigration to happen in the micro bumps, the temperature gradient needs to be larger than about 20,000 °C/cm. Compared the fact that 1000 °C/cm thermal gradient will induce thermomigration in flip chip solder joint [38], micro bumps are more resistant to Sn thermomigration due to the back stress effect. It can be seen that both thermomigration and EM risks due to Sn flux divergence are low for micro bumps.

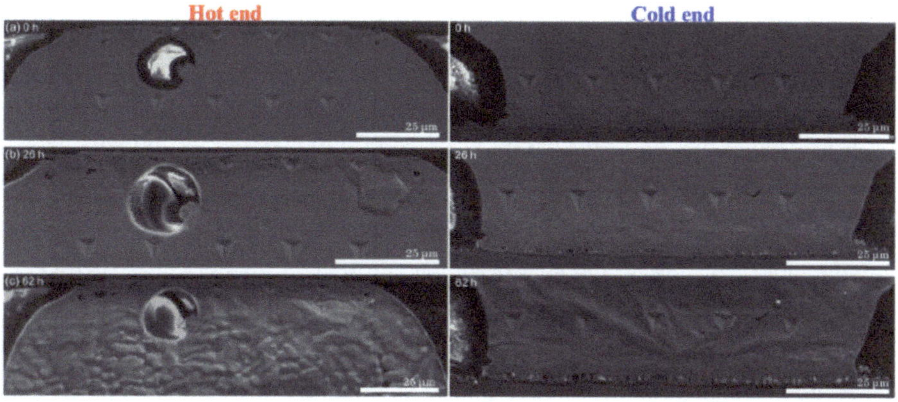

Fig. 9.11 Sn protrusions on hot end and voids on cold end are observed in flip chip solder joint under thermal gradient [39]

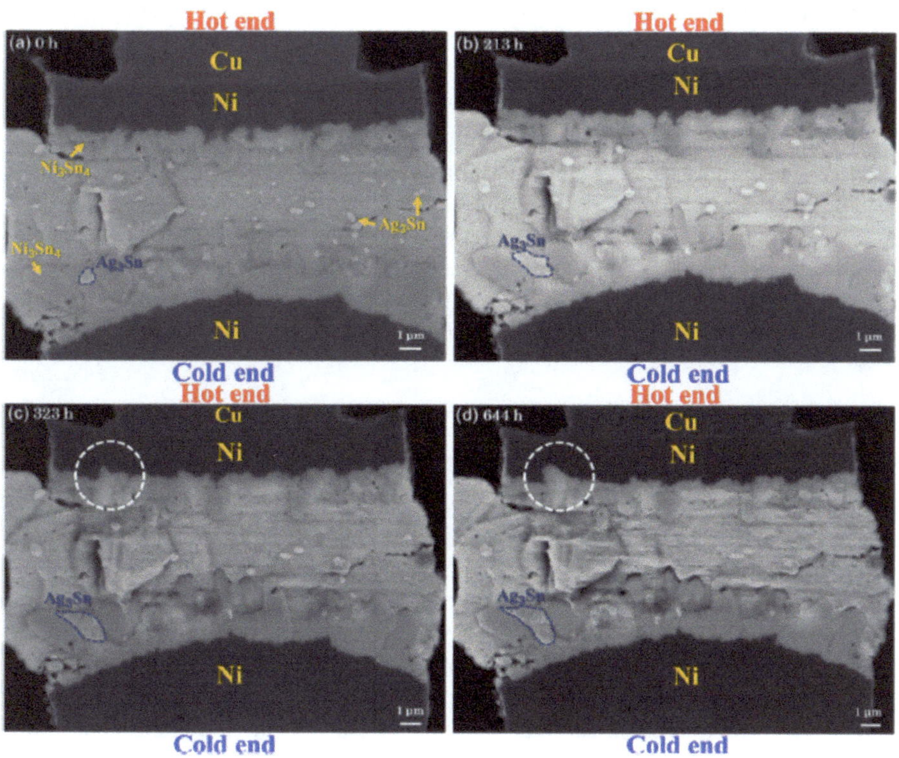

Fig. 9.12 No sign of Sn migration is observed in a micro bump under thermal gradient. Ni and Ag migration can be seen from hot end to cold end [39]

While thermomigration damage due to Sn diffusion can be minimal in micro bumps due to back stress effect, metallization dissolution induced by thermal gradient could be another concern that needs to be addressed. Chen et al. summarized the thermomigration of Ni, Cu, Ag, and Ti in solder joints [38]. It is noted that they all migrate to the opposite direction of Sn atoms under thermal gradient. Sn migrates from cold end to hot end, while Ni, Cu, Ag, and Ti move from hot end to cold end. This correlates to the positive or negative sign of heat of transport Q^*, which is defined by the difference between heat carried by a moving atom per mole to the heat of atoms per mole at the hot end. When Q^* is negative, the diffusion specie moves from cold to hot (i.e., Sn), and when Q^* is positive, atoms move from hot to cold (i.e., Ni, Cu, Ag, Ti). Cu thermomigration in solder joint is much easier than Ni. Part of the reason can be related to the low Ni solubility in solder alloy. It is reported that the solubility of Ni is only 0.28 wt% in Pb-free solder at 250 °C, while the solubility of Cu is 1.54 wt% at 260 °C [41]. Cu migration was observed at thermal gradient of 1000 °C/cm in flip chip solder joint [42], however, no Ni migration at thermal gradient of 1429 °C/cm [43]. When thermal gradient is 5345 °C/cm, Ni thermomigration was observed in micro bumps as shown in Fig. 9.12 [39]. In a recent study, Ouyang et al. [44] reported that Ag_3Sn IMC particles or plates can suppress thermomigration of Cu. They can inhibit the dissolution of Cu at the hot end and abnormal accumulation of IMCs at the cold end. The ability of Ag_3Sn to against thermomigration of Cu is speculated to be related to the less susceptibility of Ag atoms to temperature gradient.

Thermomigration can induce the fast metallization dissolution and cause reliability concern. However, similar to the discussion in EM case, metallization dissolution will eventually transform the solder joint into full IMC joint when the volume ratio of metallization to Sn is optimized. In this case, both thermomigration and EM concerns should be minimal in micro bumps.

9.4 EM in TSV of 3D Packaging

In 3D packaging, micro bumps that we discussed above are interconnects between Si dies. Inside a Si die, TSV, RDL, and metal layers connected to TSV are also key 3D packaging interconnects that may have EM reliability concerns due to their small dimensions. Similar to Cu interconnects in typical Si die, this TSV-related interconnects are also using damascene process. Passivation layers and barrier metals are used as well for the electroplated Cu TSV and its connected metal layers. The EM fundamentals that summarized from Si die Cu damascene interconnects will be also applicable to TSV-related interconnects.

9.4.1 EM for Cu Damascene Interconnects

EM for Cu damascene interconnects in Si die has been well reported [45–47]. Different from Al-based interconnects that are grain boundary dominated failure, EM damage in Cu-based metal lines is interface dominated. The interfaces at via sidewalls or between Cu line and barrier/passivation layer are the fast diffusion paths and hence the failure locations. Several key EM factors summarized from Cu damascene interconnects, which are believed also applicable to TSV-related interconnects, are listed as follows:

1. Current crowding: because of the nonuniformity of interconnect dimensions and complexity of routing, the current density is not uniform along the interconnects. It has been found that higher current density can be achieved at the inner corner of a bend conductor, where the EM failure starts to happen [48]. Current crowding can also induce local self heat increase and generate the thermal gradient and thermal mechanical stress gradient; both are the key driving forces for Cu diffusion hence can impact the EM failure also.
2. Influence of passivation layer: since Cu EM is interfacial diffusion dominated, the passivation layer plays very important role on the EM damage. Ways that can improve the Cu/passivation adhesion and slow down the interfacial Cu diffusion will help to improve the EM lifetime. It has been reported that the EM MTTF for Cu interconnects coated with SiN_x is about 12 times longer than those coated with SiC [46]. This correlates very well with the experimental results that the debond energy for Cu/SiC is much less than Cu/SiN [49].
3. Reservoir effect: Extensions of metal lines next to via connections (also described as overhang region or reservoir) can delay the EM failure of interconnects in Si die because the extensions serve as reservoir for void growth [50]. Lower levels of stress and vacancy concentration in the extension are believed to contribute to the better EM performance. Gan et al. [47] calculated the current density distributions with different length of M2 extensions in dual-damascene Cu interconnects. They observed a low current density zone in the M2 extension corners above the cathode, which retards the void migration into the extensions. Based on this, they proposed that there is a critical extension length beyond which increasing extension lengths will not lead to longer EM lifetimes.
4. Thermal mechanical stress impact: as described in Blech effect, stress gradient can be the driving force for atomic flux [51]. The stress gradient in Blech effect arises from electron wind force induced vacancy density difference between cathode and anode. Besides that, the complexity of design structures, manufacture processes, and dielectric materials for Cu interconnects can also induce thermal mechanical stress gradient due to thermal mismatch at high temperature EM testing. It has been reported that passivation layer thickness [52] and interconnect processing temperature [53] can dramatically modulate the thermal stress gradients in metal lines and change their EM performance.

9.4.2 EM Failure in TSV

In a recent study, Oba et al. [54] reported the EM testing on a typical TSV structure. Figure 9.13 shows the schematic of the EM testing structures. The TSV is connected to the top die back-end-of-line (BEOL) layer and bottom die redistribution layer (RDL). The RDL is bonded to bottom die BEOL through micro bumps. Two patterns are tested as shown in Fig. 9.13. Type 1 is TSV + bump connection and type 2 is TSV + RDL connection.

The EM failure modes in these testing structures are shown in Fig. 9.14. For upstream electron flows, the failures are at the bottom of TSV for both type 1 (Fig. 9.14b) and type 2 (Fig. 9.14d) testing patterns. The micro bump is not EM damaged in Fig. 9.14b even though current goes through it, indicating micro bump in this testing is more EM resistant than TSV. It is mentioned in the report that the bottom part of TSV is separated from RDL by barrier metal layers of PVD Ta and PVD TiW. Under upstream e-flow EM, Cu flux divergence is expected at the interface of bottom TSV and barrier metal layer where EM voids will be generated to cause failure. For downstream e-flow testing, no TSV EM damages are observed (Fig. 9.14a, c). In this case, flux divergence is not expected at the bottom of the TSV, but expected in the RDL next to TSV. However, no RDL EM damage in Fig. 9.14c, which is believed to be related to the reservoir effect because the left RDL portion of TSV in Fig. 9.14c can be viewed as the Cu extension that discussed in Sect. 9.4.1. It is noted that enough redundancy design of upper die BEOL

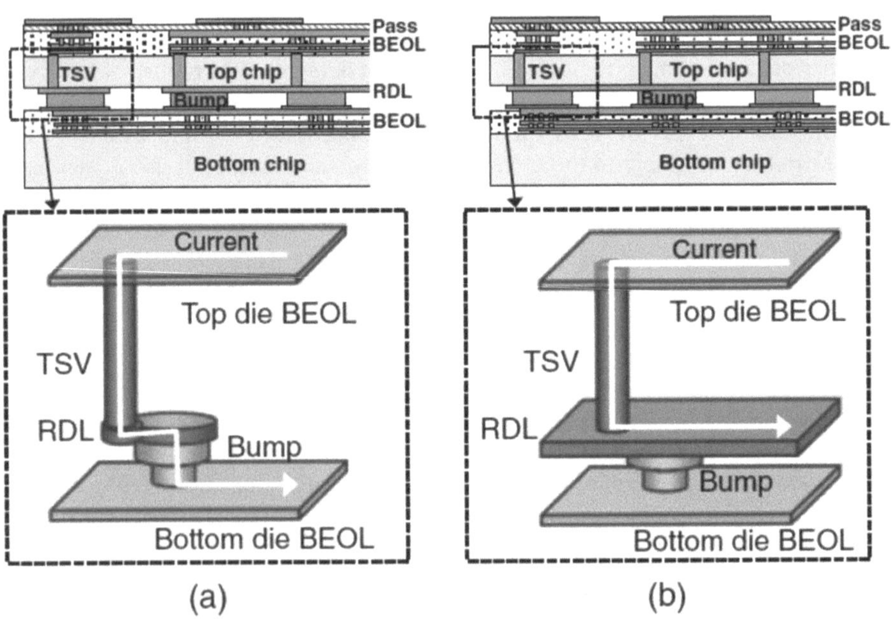

Fig. 9.13 Schematic cross section of EM testing pattern: (**a**) TSV + micro bump connection (type 1) and (**b**) TSV + RDL connection (type 2) [54]

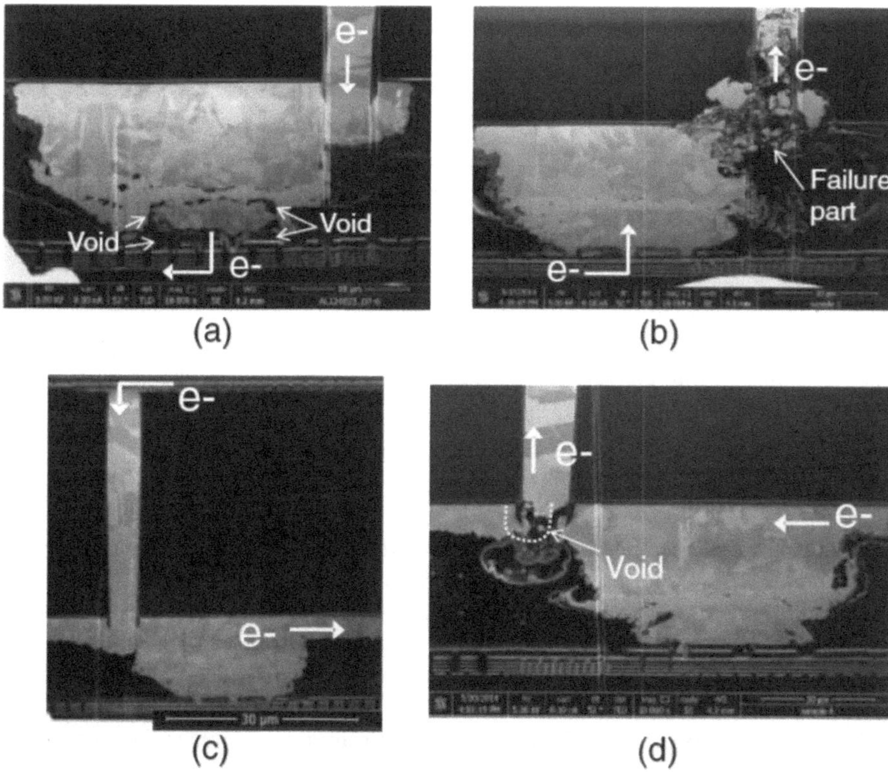

Fig. 9.14 TSV EM failure modes: (**a**) downstream and (**b**) upstream e-flow with TSV + bump connection (type 1) and (**c**) downstream and (**d**) upstream e-flow with TSV + RDL connection (type 2) [54]

purposely makes the upper TSV/BEOL more robust and no EM damages are expected.

While this study shows that EM damages are mainly within the TSV where flux divergence happens, other researchers observed that the EM failure can be also within the metal layers connected to TSV [55–57]. Figure 9.15 shows a tested TSV structure and its failure mode reported by Frank et al. [57]. The voids initiated at the Cu/SiN interface, which is known to be the typical failure interface in Cu damascene interconnects [46]. They further studied the same TSV structures with thicker metal lines and found the failure mainly happens at metal line/TiN interface next to the TSV. Cu/SiN voids are also observed in thick metal line but not dominated. The EM failures in metal lines are also observed in other studies [55, 56].

Only limited studies are reported about the TSV EM failure modes. It can be seen that the failure can happen in the TSV or in the metal layers connected to TSV. The reasons about the failure mode change are not very clear and more systematic studies are needed. Metal line and TSV dimensions definitely play the role. Other than those, all the factors discussed in Sect. 9.4.1 can also modulate the TSV failure modes.

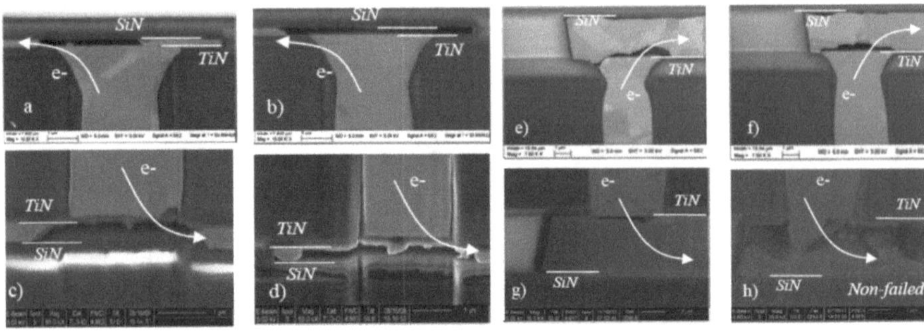

Fig. 9.15 Failure modes show metal line void at Cu/SiN with thin line and thick processes (**a**, **b**) thin line upstream; (**c**, **d**) thin line downstream; (**e**, **f**) thick line upstream; and (**g**, **h**) thick line downstream [57]

9.5 Summary

In 3D packaging, the interconnects that have EM reliability concerns can be the micro bumps, TSV, and its connected metal layers (i.e., RDL and BEOL). Micro bumps are small solder joints that have the same basic EM fundamentals as in large FLI solder joints. On the other hand, TSV and its connected metal layers are electroplated Cu interconnects and they follows the mechanisms for damascene Cu interconnects in Si die.

For solder micro bumps in 3D packaging, their unique features make them more EM robust for failures induced by both Sn flux divergence and metallization dissolution. (1) Short solder joint heights of micro bumps can induce large back stress during EM and balance out the electron wind force on Sn atomic flux. This largely reduced the risk of EM failure due to Sn flux divergence at IMC/solder interface. (2) Typically, solder to metallization volume ratios for micro bumps are large enough that a full IMC joint can be formed before metallization are totally consumed to cause failures. This eliminated the metallization dissolution failure mode that usually causes early EM fails in large FLI solder joints. Full IMC joints are much more EM robust than solder joints. Overall, although higher current density and higher joule heating are expected in 3D packaging micro bumps, their EM risk should be low if they have no quality issues and the metallization layer thickness is not very thin.

For TSV and its connected metal layers, EM experimental studies are limited. Failures can be inside the TSV close to the connection interface where Cu flux divergence happens, or in the metal layers adjacent to TSV. When failed in the metal layers, EM behavior is very the same as that in well-reported Cu damascene interconnects that the voids initiated at the interface between Cu and passivation layer. The key factors that modulate the TSV failure modes are not very clear based on the limited reports and more studies are needed. It is believed that the key parameters that modulate the EM of Cu damascene interconnects, including current crowding, Cu/passivation interface, Cu reservoir, and thermal mechanical stress gradient, will be also the key factors for TSV-related EM failure.

Acknowledgment The editors would like to thank Kwang-Lung Lin from National Cheng Kung University in Taiwan for his critical review of this chapter.

References

1. S.Q. Gu, MRS Bull. **40**, 233 (2015)
2. X. Zhang, J.K. Lin, S. Wickramanayaka, S. Zhang, R. Weerasekera, R. Dutta, K.F. Chang, K.J. Chui, H.Y. Li, D.S.W. Ho, L. Ding, G. Katti, S. Bhattacharya, D.L. Kwong, Appl. Phys. Rev. **2**, 021308 (2005)
3. K.N. Tu, H.Y. Hsiao, C. Chen, Microelectron. Reliab. **53**, 2 (2013)
4. K.N. Tu, T. Tian, Sci. China **56**, 1740 (2013)
5. K.N. Chen, T.N. Tu, MRS Bull. **40**, 219 (2015)
6. L. Xu, J.K. Han, J.J. Liang, K.N. Tu, Y.S. Lai, Appl. Phys. Lett. **92**, 262104 (2006)
7. P. L. Liu, F. Y. Ouyang, S. Ou, Z. Fu, G. Deepak, in *ASME InterPACK Conference*, vol. 1, 2009, pp. 703–709
8. C. Chen, H.M. Tong, K.N. Tu, Ann. Rev. Mater. Res. **40**, 531 (2010)
9. M.H. Lu, D.Y. Shih, P. Lauro, C. Goldsmith, D.W. Henderson, Appl. Phys. Lett. **92**, 211909 (2008)
10. Y. Wang, K.H. Lu, V. Gupta, L. Stiborek, D. Shirley, S.H. Chae, J. Im, P.S. Ho, J. Mater. Res. **27**, 1131 (2012)
11. K.N. Tu, J. Appl. Phys. **94**, 5451 (2003)
12. S.K. Seo, S.K. Kang, D.Y. Shih, H.M. Lee, J. Met. **61**, 22 (2010)
13. M. Lu, D.Y. Shih, S.K. Kang, C. Goldsmith, P. Flaitz, J. Appl. Phys. **106**, 053509 (2009)
14. I.A. Blech, J. Appl. Phys. **47**, 1203 (1976)
15. M.H. Lu, D.Y. Shih, P. Lauro, C. Goldsmith, Appl. Phys. Lett. **94**, 011912 (2009)
16. Y.C. Hsu, C.K. Chou, P.C. Liu, C. Chen, D.J. Yao, T. Chou, K.N. Tu, J. Appl. Phys. **98**, 033523 (2005)
17. M.S. Yoon, M.K. Ko, B.N. Kim, B.J. Kim, Y.B. Park, Y.C. Joo, J. Appl. Phys. **103**, 073701 (2008)
18. J. R. Black, in *Proceedings of the 6th Annual International Reliability Physics Symposium*, 1967, p. 148
19. T.C. Huang, T.L. Yang, J.H. Ke, C.H. Hsueh, C.R. Kao, Scr. Mater. **80**, 37 (2014)
20. Y. Wang, P.S. Ho, Appl. Phys. Lett. **103**, 121909 (2013)
21. P. L. Liu, O. Alan, D. Goyal, in *Electronic Components and Technology Conference (ECTC)*, 2015, p. 95
22. A.U. Telang, T.R. Bieler, Scr. Mater. **52**, 1027 (2005)
23. J. Sylvestre, A. Blander, J. Electron. Mater. **37**, 1618 (2008)
24. L.P. Lehman, Y. Xing, T.R. Bieler, E.J. Cotts, Acta Mater. **58**, 3546 (2010)
25. B. Arfaei, N. Kim, E.J. Cotts, J. Electron. Mater. **41**, 362 (2012)
26. H. Zhang, F. Sun, X. Li, Y. Liu, T. Xin, in *16th International Conference on Electronic Packaging Technology*, 2015, p. 1125
27. D.C. Yeh, H.B. Huntington, Phys. Rev. Lett. **53**, 1469 (1984)
28. Y.W. Chang, C. Chen, T.C. Chang, C.J. Zhan, J.Y. Juang, A.T. Huang, Mater. Lett. **137**, 136 (2014)
29. C. Chen, D. Yu, K.N. Chen, MRS Bull. **40**, 257 (2015)
30. H. Y. Chen, C. H. Tung, Y. L. Hsiao, J. L. Wu, T. C. Yeh, L. L. C. Lin, C. Chen, D. C. H. Yu, in *Electronic Components and Technology Conference (ECTC)*, 2015, p. 620
31. H. Y. Chen, D. Y. Shih, C. C. Wei, C. H. Tung, Y. L. Hsiao, D. C. H. Yu, Y. C. Liang, C. Chen, in *Electronic Components and Technology Conference (ECTC)*, 2015, p. 49
32. C.C. Wei, C.F. Chen, P.C. Liu, C. Chen, J. Appl. Phys. **105**, 023715 (2009)

33. Y. B. Park, S. H. Kim, J. J. Park, J. B. Kim, H. Y. Son, K. W. Han, J. S. Oh, N. S. Kim, S. Yoo, in *Electronic Components and Technology Conference (ECTC)*, 2013, p. 1988
34. J.Y. Kim, J. Yu, Appl. Phys. Lett. **92**, 092109 (2008)
35. S. Kumar, J. Smetana, D. Love, J. Watkowski, R. Parker, C.A. Handwerker, J. Electron. Mater. **40**, 2415 (2011)
36. Y. Wang, S. H. Chae, J. Im, P. S. Ho, in *Electronic Components and Technology Conference (ECTC)*, 2013, p. 1953
37. P.L. Liu, J.K. Shang, Scr. Mater. **53**, 631 (2005)
38. C. Chen, H.Y. Hsiao, Y.W. Chang, F.Y. Ouyang, K.N. Tu, Mater. Sci. Eng. R **73**, 85 (2012)
39. F.Y. Ouyang, W.C. Jhu, J. Appl. Phys. **113**, 043711 (2013)
40. H. Hsu, T.Y. Lin, F.Y. Ouyang, J. Electron. Mater. **43**, 236 (2013)
41. K. Zeng, K.N. Tu, Mater. Sci. Eng. R **38**, 55 (2002)
42. H.Y. Chen, C. Chen, K.N. Tu, Appl. Phys. Lett. **93**, 122103 (2008)
43. H.-Y. Chen, C. Chen, J. Mater. Res. **26**, 983 (2011)
44. W.N. Hsu, F.Y. Ouyang, Mater. Chem. Phys. **165**, 66 (2015)
45. C.M. Tan, A. Roy, Mater. Sci. Eng. R **58**, 1 (2007)
46. C.K. Hu, L. Gignac, E. Liniger, B. Herbst, D.L. Rath, S.T. Chen, S. Kaldor, A. Simon, W.T. Tseng, Appl. Phys. Lett. **83**, 869 (2003)
47. Z.H. Gan, W. Shao, S.G. Mhaisalkar, Z. Chen, H. Li, K.N. Tu, A.M. Gusak, J. Mater. Res. **21**, 2241 (2006)
48. C.M. Tan, A. Roy, A.V. Vairagar, A. Krishnamoorthy, S.G. Mhaisalkar, IEEE Trans. Device Mater. Reliab. **5**, 198 (2005)
49. M.W. Lane, E.G. Liniger, J.R. Lloyd, J. Appl. Phys. **93**, 1417 (2003)
50. K.N. Tu, C.C. Yeh, C.Y. Liu, C. Chen, Appl. Phys. Lett. **76**, 988 (2000)
51. J.R. Lloyd, Microelectron. Eng. **49**, 51 (1999)
52. J.R. Lloyd, P.M. Smith, J. Vac. Sci. Technol. **A1**, 455 (1983)
53. L. Kisselgof, S. P. Baranowski, M. C. Broomfield, T. Spooner, L. Elliott, L. Brooke, J. R. Lloyd, in *Proceedings of the SPIE* 1805, 1992, p. 154
54. Y. Oba, J.D. Messemaeker, A.M. Tyrovouzi, Y. Miyamori, J.D. Vos, T. Wang, G. Beyer, E. Beyne, I.D. Wolf, K. Croes, Jpn. J. Appl. Phys. **54**, 05EE01 (2015)
55. H. J. Choi, S. M. Choi, M. S. Yeo, S. D. Cho, D. C. Baek, J. Park, in *Interconnect Technology Conference (IITC), IEEE International*, 2012, p. 1
56. S. Moreau, D. Bouchu, in *Reliability Physics Symposium (IRPS), IEEE International*, 2013, p. CP.1.1
57. T. Frank, S. Moreau, C. Chappaz, L. Arnaud, P. Leduc, A. Thuaire, L. Anghel, in *Electronic Components and Technology Conference (ECTC)*, 2012, p. 326

Chapter 10
Fundamentals of Heat Dissipation in 3D IC Packaging

Satish G. Kandlikar and Amlan Ganguly

10.1 Introduction

A schematic of a 3D IC stack is shown in Fig. 10.1. It consists of individual chips or chip stacks that are separated by cooling layers. The cooling layer consists of microchannels or finned passages that provide increased surface area and enhancement for heat transfer from the stack surfaces to the coolant flowing in the cooling layers. The design of the cooling layer is restricted by the placement of the through-silicon-vias (TSVs) connecting between adjacent chip stacks. Electrical and thermal issues require a high degree of codesign for efficient electrical as well as thermal performance of the 3D ICs.

Using a heat sink and an air cooled fin stack is the most prevalent and cost-effective option available for cooling 2D ICs. Somewhat less stringent space restrictions above the chip in the third dimension make the placement of a heat sink with air flow passages readily feasible. Such an approach is generally feasible for a two-chip stack, but introduces severe heat dissipation limits for stacks with three or more chips. As the heat fluxes in 2D IC chips become large, air cooling is no longer adequate, and various liquid cooling options such as embedded microchannels [2], spray cooling [3], and jet impingement [4, 5] have been considered in the literature. Among these, the embedded microchannels, proposed by Tuckerman and Pease [2] in early eighties, have received the greatest attention due to their compactness, ability to contain the liquid, and superior thermal performance in terms of higher heat dissipation rates, lower pressure drops, and higher heat

S.G. Kandlikar (✉)
Mechanical Engineering Department, Rochester Institute of Technology, Rochester, NY, USA
e-mail: sgkeme@rit.edu

A. Ganguly
Computer Engineering Department, Rochester Institute of Technology, Rochester, NY, USA
e-mail: axgeec@rit.edu

Y. Li, D. Goyal (eds.), *3D Microelectronic Packaging*, Springer Series in Advanced Microelectronics 57, DOI 10.1007/978-3-319-44586-1_10

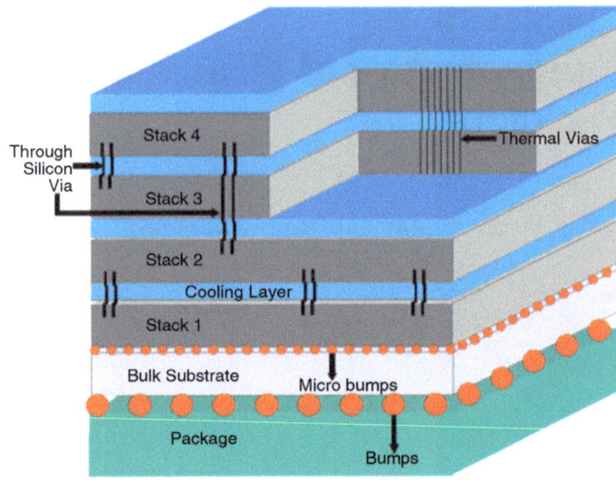

Fig. 10.1 A schematic of a three layer 3D IC structure with an external heat sink. Redrawn from [1]

transfer coefficients [6]. A practical implementation of an IBM silicon chip cooler was demonstrated by Colgan et al. [7] that is capable of addressing the >1 kW/cm^2 cooling needs. Their chip cooler utilized offset strip fins instead of straight microchannels to provide lower thermal resistance (determined to be ~1×10^{-6} m ^{2}°C/W) [8]. Kandlikar et al. [9] discuss various cooling options available for 2D and 3D ICs.

As the chips are stacked in a 3D IC package, the placement of the heat sink becomes possible only above the outermost layer of the 3D IC stack. This approach, often referred to as conduction cooling, introduces additional thermal resistance for the interior chips placed in the second layer and further below the directly cooled chip. The alignment of heat generating devices needs to be carefully evaluated to avoid localized hot spots created by this stacking effect. Conduction cooling can be combined with aggressive air cooling designs with compact heat exchangers, pool boiling in vapor space or liquid cooled microchannels as shown in Fig. 10.2 [9].

10.2 Thermal Performance Parameters for 3D ICs

In an effort to address the effect of device placement on different layers, two factors are introduced in converting the heat flux levels in a 3D chip to the corresponding levels for the heat sink design [9]. The Thermal Intensification Factor, TIF, takes into account the overlapping of heat generating devices from different layers along the heat flow path.

$$\mathrm{TIF} = \frac{q''_{\mathrm{3D}-H}}{q''_{\mathrm{3D}-U}} \tag{10.1}$$

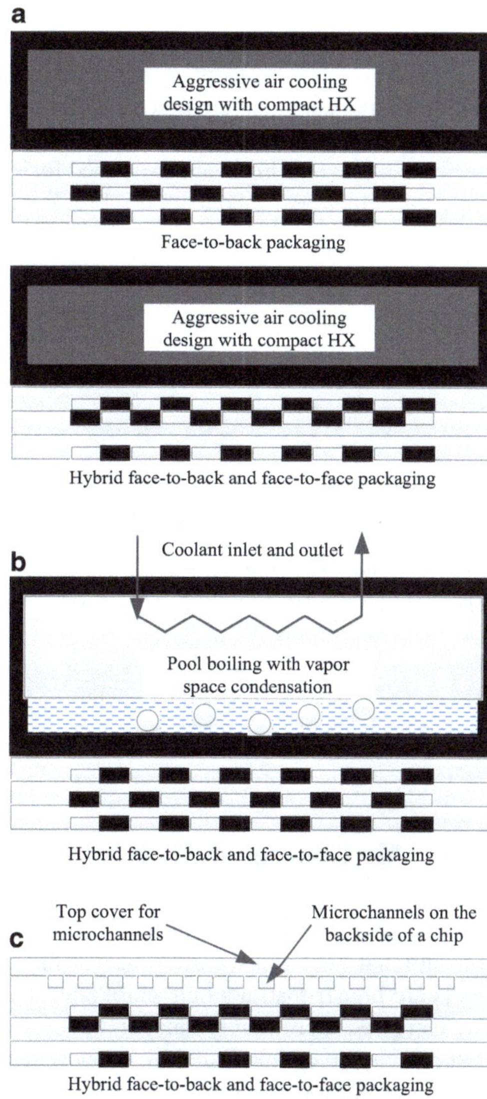

Fig. 10.2 Integrated conduction cooling of 3D ICs with other cooling systems: (**a**) aggressive air cooling with compact HX, (**b**) pool boiling with vapor space condensation, (**c**) microchannel cooling, [9]. Redrawn from [9]

where q''_{3D-U} is the uniform heat flux based on the total power being dissipated at the active cooling area and q''_{3D-H} is the local value of heat flux apparent at the heat sink at any given location. A TIF map can thus be generated at the heat sink level to estimate the localized cooling needs over the chip surface adjacent to the heat sink. A detailed three-dimensional conduction analysis would be needed to estimate the actual values of TIF with multiple IC layers and distributed devices. This mapping will enable generating a heat flux map of a 3D stack to its corresponding planar chip profile, which can be used for designing or analyzing the cooling performance.

The Thermal Derating Factor, TDF, accounts for the additional resistance introduced by the interface between the heat generating device on a chip in a 3D stack and the chip adjacent to the heat sink. This becomes especially important when the adjacent chips are mechanically bonded together. The additional thermal resistance is accounted for by increasing the heat flux that is apparent on the chip at the heat sink level. This heat flux is used for estimating the device temperature map, while the actual heat fluxes are used for coolant heat balance calculations.

$$\mathrm{TDF} = \frac{q''_{\mathrm{3D}-U}}{q''_{\mathrm{1D}-U}} \tag{10.2}$$

Alternatively,

$$q''_{\mathrm{1D}-U} = \frac{q''_{\mathrm{3D}-U}}{\mathrm{TDF}} \tag{10.3}$$

where $q''_{\mathrm{1D}-U}$ is the equivalent 1D heat flux that will result in the same device level temperature profile corresponding to a uniform heat flux of $q''_{\mathrm{3D}-U}$. Combining Eqs. (10.1) and (10.3), the maximum heat flux equivalent is given by:

$$q''_{\mathrm{1D}-H} = \frac{q''_{\mathrm{3D}-H}}{\mathrm{TDF}} = q''_{\mathrm{3D}-U}\,\frac{\mathrm{TIF}}{\mathrm{TDF}} \tag{10.4}$$

10.3 Air Cooling of 3D ICs

Stacking two or more IC chips under conduction cooled configuration allows using conventional heat sinks on the face of the 3D IC chip stack. Since the heat flux levels are higher than a single chip, advanced air cooling options may be needed. The increased thermal resistance in the heat flow pathway also needs to be considered and can be compensated by a lower thermal resistance in the heat sink.

Advanced air cooling options utilize forced air cooling with air delivered over the finned heat sink using a fan. As the heat flux (heat dissipation per unit heat sink footprint area) increases, higher velocities may be needed. Utilizing ducted air inlets and outlets reduces the noise level as well as the pressure losses at the entrance and exit from the heat sink. Employing compact heat exchanger surfaces (originally developed for automotive radiators) in the heat sink matrix results in significantly higher heat dissipation rates as the heat transfer surface area is significantly increased due to the secondary fins. Although the pressure drop and fan power are increased, significantly higher heat dissipation rates are achieved as compared to the natural convection cooled heat sinks. A comprehensive coverage of compact heat exchanger design and performance is given by Kays and London [10].

10.4 Jet Impingement and Spray Cooling

High speed jets from a nozzle impinging on a target surface provide an effective technique to dissipate high heat fluxes. A single jet provides cooling in the immediate vicinity of the jet impingement area, while an array of jets could be used to cover a larger area. Air or liquid can be used as the cooling medium, while liquids are most suitable for higher heat flux dissipation levels.

Although the jets provide a high heat transfer performance, in general, they require high pressure drops in the range of 50–500 kPa. Apart from the fabrication and leakage issues, the high velocity jets may lead to vibrations and erosion of the heat transfer surfaces.

The jet impingement technique cannot be readily applied to conduction cooled 3D ICs. Placing jets in the interlayers can also lead to very complex fluid distribution system design from the fabrication perspective. Leakage remains a critical concern due to high pressures employed in delivering the high velocity jets.

Spray cooling involves spraying liquid droplets from a nozzle on a target surface where they evaporate. This technique employs evaporation as against jet impingement cooling which is based on either on single-phase heat transfer or in some cases forced convection boiling. The target surface may be coated with porous coatings or microstructures to enhance the evaporation rate. The spray distribution pattern, the effect of returning vapor on the droplet velocity, and controlling the flow rate to avoid flooding are some of the issues that require careful design considerations. Although the pressures required for generating sprays are considerably lower than the jets used in impingement cooling, the fabrication issues still remain because of the complex liquid distribution network required in 3D ICs. Additionally, the vapor generated needs to be transported back to a condenser and resulting two-phase flow due to condensation in the return lines may pose some instability concerns. A recent review of spray cooling techniques provides an in-depth information about different spray cooling techniques and their performance comparisons [11]. These systems require higher pressures (from 100 to 300 kPa) and spray nozzle plates that require periodic maintenance. This precludes their applicability to electronics cooling where a high degree of reliability and simplicity in cooling system configuration are important. The small passage dimensions in the 3D ICs also present challenges for spray cooling systems.

10.5 Microchannel Cooling

Microchannels are generally classified as channels with their minimum dimension in the range of 10–200 μm [12]. The microchannels are either etched directly into the silicon substrate or etched or machined in a separate silicon or copper chip that is bonded on the back side of the IC chip in case of conduction cooling. These passages can also be etched between the interlayers in 3D ICs.

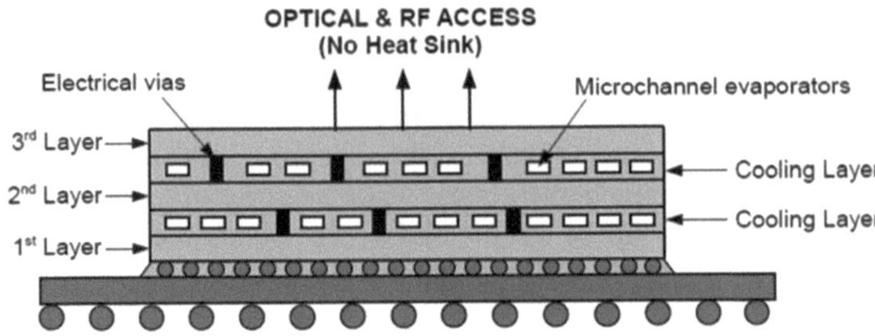

Fig. 10.3 A schematic representation of a microchannel cooled three layer 3D IC structure. Microchannels can be employed in both single-phase and two-phase (evaporative) modes. Redrawn from [13]

Figure 10.3 shows a schematic of a microchannel cooled three layer 3D IC structure by Koo et al. [13]. This schematic shows how the flow passages and TSVs (electrical vias) need to be configured in the cooling layer between adjacent IC layers.

Microchannel cooling offers the most promising cooling method for 3D ICs. Careful considerations are however needed to account for the placement of the TSVs, and control the pressure drop and temperature nonuniformity. One of the most important considerations is the height of the cooling layer, which is severely restricted by the length of the TSV that is permissible from the electrical design viewpoint. These issues will be addressed in the following sections in this chapter.

10.6 Thermal Design Considerations in 3D IC Architectures

Three-dimensional Integrated Circuits (3D ICs) have emerged as a feasible solution to overcome the performance limitations of 2D planar ICs [14]. However, utilizing the third dimension to provide additional device layers poses thermal challenges as stacking vertical layers significantly increases the power dissipation density and the thermal footprint per unit area [15]. One of the major issues in the implementation of 3D ICs is the excessive heat flux generated by stacking multiple microprocessors, giving rise to an increase in the power generated per unit surface area as well as in the peak temperature [16, 17]. While 3D multicore SoCs reduce some of the power and performance bottlenecks, the increase in power density due to reduced footprint results in severe thermal issues. These issues lead to higher temperatures requiring sophisticated cooling technologies. The power density numbers can easily exceed 50 W/cm^2 in 3D ICs which is the limit for forced air cooling [18]. Figure 10.4 shows the current power density representative of various categories of 3D ICs with data compiled from the International Technology Roadmap for Semiconductors (ITRS), 2012. Power dissipation densities exceeding 100 W/cm^2 up to 500 W/cm^2

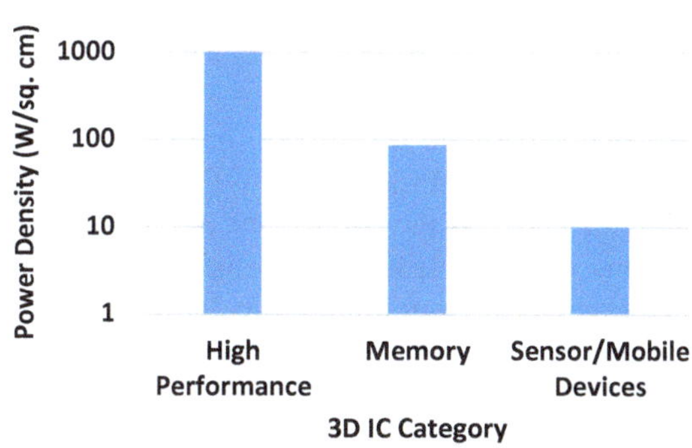

Fig. 10.4 Trends in power density for various 3D IC categories

or higher are also reported and expected in near future [1]. Such high power dissipation densities can result in quick temperature rises on the chip and cause thermal emergencies. Novel and aggressive cooling techniques are therefore necessary for the heat extraction in 3D ICs.

Dynamic Thermal Management (DTM) [19–23] is a widely adopted technique used to avoid thermal emergencies. One of the most common DTM techniques used in planar multicore chips is dynamic temperature-aware task migration reallocation [24]. Several heuristics are proposed over the last several years for achieving optimal task migration from the perspective of best possible thermal profile of the chip. Removal of hot spots and maintaining low average chip temperature are among the chief goals of these heuristics. The underlying principle for all the task migration heuristics is to move the most high power dissipating processes from hot cores to relatively cooler cores. However, DTM techniques like task migrations work best to avoid thermal hot spots when core utilization is sparse with room for task migration. Moreover, consistently high utilization of all or most cores in a multicore chip results in high heat dissipation leaving little room for task migrations.

Task migration in 3D multicore environments presents its own new set of challenges due to the smaller footprint of the IC. It is significantly more challenging to find cooler cores to migrate high power consuming processes given the limited floorplanning of the 3D ICs. Comprehensive thermal management techniques are developed in [16] where a combined approach utilizing dynamic voltage frequency scaling (DVFS), temperature-aware task allocation and liquid cooling is proposed. They also have overheads for task migrations, context switching and synchronization issues, which may negatively impact performance. These conventional DTM mechanisms designed for 2D planar ICs are often not suitable for 3D environments nor are tailored for the new constraints of monolithic 3D ICs. Consequently, we argue that task migration-based DTM mechanisms are not the most suitable for 3D multicore ICs.

10.6.1 Thermal Considerations in TSV Placements

Another technique to remove the high heat fluxes encountered in 3D ICs is by deploying Through-Silicon-Vias (TSVs) referred to as thermal vias. The concept of thermal vias was envisioned for 3D Multichip Modules (MCM) in several publications [25, 26]. It is noted in ref. [27] that while thermal vias can be useful in extracting heat from monolithic 3D ICs there is a trade-off between the size of the thermal via islands and the area left for routing. This limits the number and placement options for the thermal via islands which consist of several individual vias each. Consequently, placement and optimization techniques, fabrication and packaging methodologies, and Computer Aided Design (CAD) tools that support the deployment of thermal vias are essential for successfully inserting thermal vias in a monolithic 3D IC to provide paths for heat extraction.

The overheads of traditional DTM techniques and the challenges of place and route for thermal vias can be mitigated while providing heat extraction paths using microfluidic cooling layers for 3D ICs [28, 29]. Cooling 3D ICs has been mainly addressed in literature using microchannels between two adjacent 3D IC layers. Single-phase cooling utilizes water or a dielectric fluid that transports heat without undergoing phase change [30].

10.6.2 Thermal Analysis Tools for 3D ICs

For thermal analysis of 3D ICs, one can adopt either simulation-based environments or real experimental beds. A few thermal estimation simulators based on finite element method exist today. The commonly used ones are HotSpot 3D [31] and 3D-ICE [32]. These simulators require power dissipation profiles of the IC layers and then perform meshing using heat flow equations to model the propagation of heat through the ICs.

10.6.3 Performance Considerations

3D integration has been explored to alleviate the limitations of 2D chips for SoC and multicore architectures. In [32], the authors discuss the advantages and challenges of designing and manufacturing 3D ICs. It is shown that 3D ICs can improve power, noise, logical span, density, and performance. It even enables integration of heterogeneous technologies in a single die. Processor-memory stacks with very wide TSV-based vertical busses where memory caches are directly stacked vertically above respective processor cores are shown in [34]. Short vertical distances coupled with ultrawide TSV-based busses improve the memory access latency several times ushering in new paradigms in computer design. In [35] a coupling of 3D IC technology and multicore architectures is proposed. However, because of the thermal issues encountered in 3D ICs the number of vertical layers possible to realistically stack has remained low to only about two to four layers. Moreover, architectural innovations are also limited to only vertical processor-memory stacks to reduce power dissipation at the same location which would otherwise result in thermal hot spots.

Conventional DTM techniques have overheads, whereas thermal vias require complex place and route optimizations and CAD tool support. Microfluidic interlayer coolers provide a solution to the thermal issues in 3D ICs without impacting the performance of the IC as they do not require interference with the functionality of the IC or the design of the active layers.

10.7 Liquid Cooling with Integrated Microchannels

Microchannels can be embedded as a cooling interlayer in a 3D IC chip stack. These layers can be placed between each of the IC chips, or after each two or more chips. The challenge faced by the microchannel cooling is its integration with the TSVs. The TSVs can be located only at the channel walls, and hence large number of closely spaced microchannels are desired. Combining this with the requirement of short TSV distances, the desirable channel dimensions tend to be short and narrow microchannels. Such channels will provide high heat transfer performance due to small hydraulic diameters, but the coolant flow pressure drop through these channels becomes prohibitively excessive.

Brunschweiler et al. [36] conducted an exhaustive study on the interlayer cooling using water as the coolant in plain and pin fin populated microchannels of various geometries. They showed that the application of this technique is severely challenged due to the temperature rise as the channel hydraulic diameters are below 200 μm. Effectively, high coolant flow rate is required resulting in excessive pressure drop. An interlayer microchannel cooler with 200 μm hydraulic diameter resulted in a heat flux dissipation of 150 W/cm^2 at a coolant volumetric flow rate of 0.8 L/min. Introducing pin fins through which the TSVs were connected resulted in significantly higher pressure drops. The authors provide extensive plots for 1–4 cm^2 chip areas depicting thermal resistance and pressure drops performance with different microchannel and pin fin configurations.

10.7.1 Variable Fin Density in Microchannel Passages

The height of the TSV is an extremely important consideration in the cooling interlayer design. As shorter heights are employed to improve the TSV performance, the flow resistance increases due to resulting smaller hydraulic diameters. Although the heat transfer coefficient improves with the smaller diameters, the flow

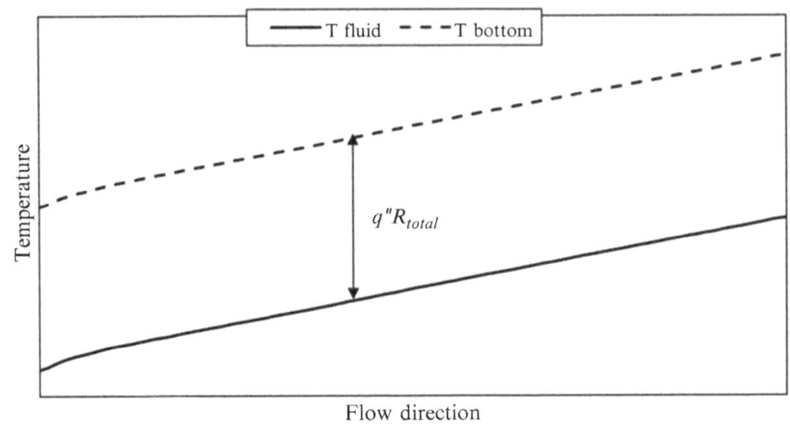

Fig. 10.5 Temperature profile of the coolant and the substrate under constant heat flux boundary condition. Redrawn from [9]

rate decreases under a given pressure drop limit and the resultant heat dissipation decreases. This makes it critical to utilize the available pressure drop in the most efficient manner. Rubio-Jimenez et al. [9] proposed a variable fin density concept to accomplish this goal as described below.

As the coolant traverses through the interlayer passages, its temperature rises linearly assuming a constant heat flux and a uniform heat transfer coefficient (except for the higher value at the entrance due to entrance effects). Figure 10.5 shows the coolant and the substrate temperature profiles. The coolant as well as the substrate reaches their respective highest temperatures at the outlet section. However, the entire chip surface remains well below this maximum value, which dictates the coolant flow rate.

Figure 10.6 shows a configuration in which the heat transfer coefficient increases, or the thermal resistance decreases, progressively toward the outlet [9]. This variation is adjusted to provide a uniform substrate temperature throughout, while the fluid temperature increases along the flow direction due to heat gain.

This approach requires the thermal resistance to decrease along the flow. Since the height of the microchannel in an interlayer is uniform, Rubio-Jimenez et al. [9] proposed varying the pin fin density along the flow length. Figure 10.7 shows a design in which the fin density was varied in different zones along the flow length. They numerically analyzed several such configurations and presented the substrate temperature along the flow length. The surface temperature nonuniformity was reduced and a significant pressure drop reduction was also obtained. The effects of different fin shapes, as suggested by Brunschwiler et al. [36], were also studied. It was noted that elliptical cross-section fins resulted in the lowest pressure drop, but the rectangular offset strip fins provided a better configuration with their higher thermal performance and comparable pressure drops.

As seen from Figs. 10.8 and 10.9, the temperature variation of the substrate is reduced considerable for the variable density fins while reducing the pressure drop

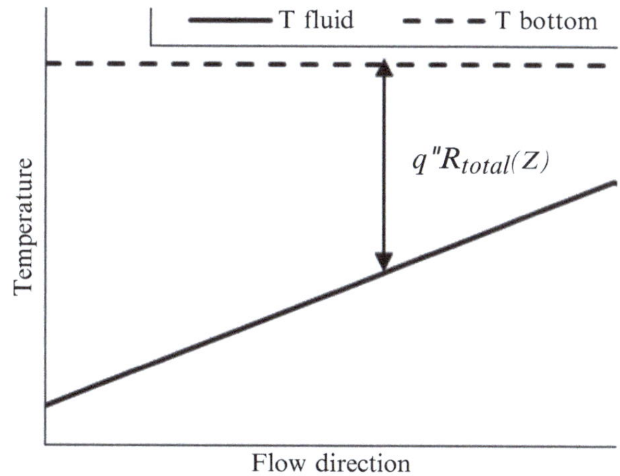

Fig. 10.6 Uniform substrate temperature obtained by introducing a variable total thermal resistance between the fluid and the substrate

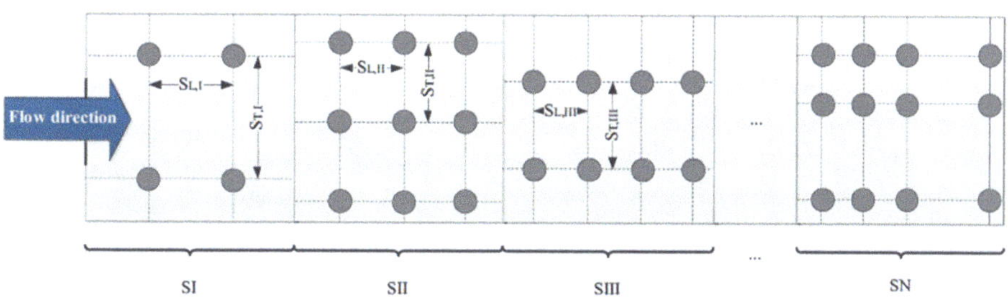

Fig. 10.7 Increasing the fin density in several discrete zones along the flow length

at the same time. As summarized by Kandlikar et al. [37], the temperature nonuniformity in a 3D IC affects the reliability of the 3D ICs through: (1) diffusion effects, (2) dielectric breakdown, (3) ion movement, (4) electromigration, (5) thermal cycling, and (6) performance drift. Several follow-on works in this area provide further details of the effect of spacing, fin geometry, and interlayer height on the temperature nonuniformity and associated pressure drops.

Different fin designs, interlayer heights, and other enhancement techniques such as short pin fins were analyzed for their heat transfer and pressure drop performance [37–39]. Although the variable fin density technique reduces the temperature nonuniformity, it may present challenges while implementing it in conjunction with the TSV placements. Since the TSVs are built within the fins, the fin density and their placement do not remain only a thermal issue. Nevertheless, this approach may provide an additional pathway in the interlayer design.

Integration of TSVs and interlayer cooling requires codesign with electrical and thermal engineers. Zhang et al. [40] discuss the design issues related to thermal requirements as well as need to minimize TSV parasitics that impact latency, bandwidth density, and power consumption. Alfieri et al. [41] studied the microchannel interlayers with cylindrical pin fins using a porous medium approach

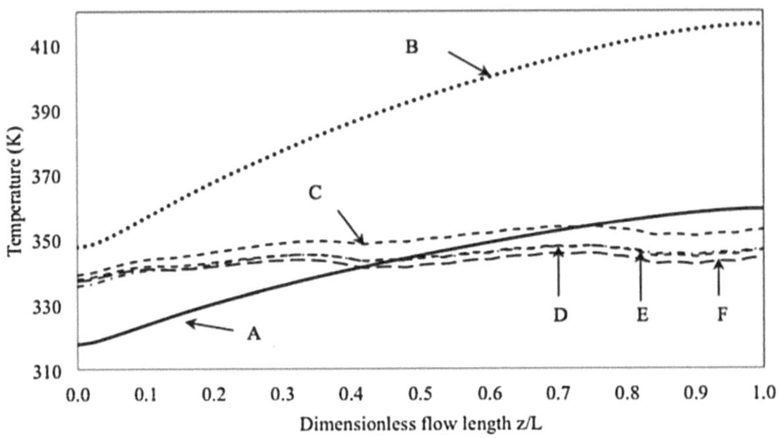

Fig. 10.8 Variation of substrate temperature along the flow direction. (**a**) microchannel heat sink, (**b**) plain gap (without fins, shown for the comparison purpose only, not a practical design as TSVs cannot be routed), (**c–f**) different pin fin pitch and configurations. Chip area of 1 cm^2, 100 W/cm^2 heat flux from each top and bottom surfaces of an interlayer, and a coolant flow rate of 1 mL/s and inlet temperature of 293 K. Redrawn from [9]

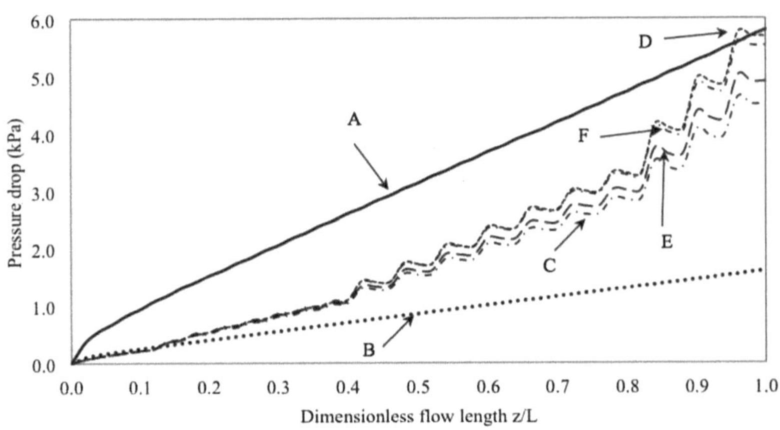

Fig. 10.9 Pressure drop variation along the flow length for interlayer configurations shown in Fig. 10.4. Redrawn from [9]

considering the conjugate conduction and convection heat transfer. This allowed them to analyze the hot spots as well. They found that it is necessary to consider the property variations and local thermal nonequilibrium as well as orthotropic heat conduction.

10.7.2 Two-Phase Cooling

Two-phase cooling option is attractive because of the low flow rates as compared to single-phase cooling, relative temperature uniformity, and ability to remove large amounts of heat. For higher heat fluxes, the conventional microchannel designs with two-phase flow are not satisfactory due to their instability issues, and lower performance. However, recent developments have shown that it is possible to dissipate heat fluxes greater than 1 kW/cm^2 using novel designs such as tapered microchannels or microchannels with embedded pin fins operated at high flow rates [42, 43]. The header complexity with the tapered gap microchannels and high pressure drops with micro pin fins still remain outstanding issues. A recent review of two-phase cooling for 3D IC by Green et al. [44] considers high heat flux dissipation as well as hot spot cooling potential of several options, including localized microgaps over hot spots to dissipate heat fluxes of 2 kW/cm^2 in the hot spot region. In spite of these research efforts, a comprehensive demonstration of 3D IC stack that integrates electric and thermal integration for TSV placement, interlayer and coolant selection to improve both electrical and thermohydraulic performance is warranted.

10.8 Future Directions

Monolithic 3D IC technology has been envisioned to solve the global interconnect problem by integrating multiple active layers within a short vertical distance. This processing and memory blocks are within a few tens of microns along the vertical dimension resulting in ultralow latency interconnects. Vertical processor-memory stacks interconnected by dense bundles of TSVs provide super high bandwidth memory interconnects creating new frontiers of computer design removing the classical constraints on memory bandwidth. However, in spite of these advantages, one of the primary challenges of 3D IC technology is an increased power density as noted earlier. Multiple blocks on the same vertical axes contribute to the power dissipation at a particular location. This fact makes effective heat extraction from layers away from the heat sink very difficult, a fact that is exacerbated by the low thermal conductivity of silicon. Only the layer closest to the conduction cooler such as the heat sink and fan ensemble can lose the heat quickly. These factors result in trapping the generated heat within the layers of the 3D ICs intensifying the problem as the number of layers increases. This trapped heat results in sharp increases in temperature and affects the performance, reliability, and durability of the ICs. Consequently, due to the thermal concerns in 3D ICs being more severe than traditional planar chips, the number of layers possible to integrate is practically limited. Additionally, architectural limitations are imposed so that the most practical 3D ICs are limited to only a single layer of processors and multiple layers of memory stacks.

To alleviate the heat dissipation problem particularly from the internal layers of the 3D ICs, researchers have proposed using interlayer cooling infrastructures built using microfluidic channels. Microchannels through the silicon substrates can circulate coolant fluids to effectively extract heat from the layers of the 3D ICs that are separated from the conventional heat sink. However, there is an inherent conflict in the considerations about 3D IC performance and the ability of the microfluidic interlayer coolers to extract heat. For optimal hydraulic and thermal performance the height of the microchannels needs to be of the order of hundred microns. This results in a significant increase in the TSV length negatively affecting the performance of the 3D interconnections which was the primary goal of using 3D IC technology. Hence, careful codesign of interlayer cooling infrastructure, keeping the performance constraints of the target 3D IC, is necessary to achieve best thermal characteristics without compromising performance.

The design of microfluidic channels for cooling 3D ICs has received a lot of attention in the recent years. Besides simply analyzing the cooling performance as a function of channel dimensions and coolant characteristics, researchers have introduced sophisticated and advanced features like fins, variable fin densities, and localized cooling to help alleviate high heat flux dissipation and thermal hot spot issues in 3D ICs. We envision that in the near future non-TSV type 3D interconnection technology using wireless, inductive, or capacitive coupling methods can help us overcome the electrical and thermal challenges in the 3D IC technology.

Acknowledgement The editors would like to thank Songhua Shi from Medtronic and Ravi Mahajan from Intel Corporation for their critical review of this chapter.

References

1. S.G. Kandlikar, Review and projections of integrated cooling systems for 3D ICs. J. Electron. Packag. **136**(2), 024001. doi:10.1115/1.4027175, 11 p
2. D.B. Tuckerman, R.F. Pease, High performance heat sinking for VLSI. IEEE Electron. Dev. Lett. **2**(5), 126–129 (1981)
3. S.G. Kandlikar, A.V. Bapat, Evaluation of Jet impingement, spray, and microchannel chip cooling options for high heat flux removal. Heat Transfer. Eng. **28**(11), 911–923 (2007)
4. A.C. Cotler, E.R. Brown, V. Dhir, M.C. Shaw, Chip-level spray cooling of an LD-MOSFET RF power amplifier. IEEE Trans. Comp. Packag. Technol. **27**(2), 411–416 (2004)
5. C.J. Chang, H.T. Chen, C. Gau, Flow and heat transfer of a Microjet Impinging on a Heated Chip: Part I—micro free and impinging Jet. Nanosc. Microsc. Thermophys. Eng. **17**(1), 50–68 (2013)
6. C.J. Chang, H.T. Chen, C. Gau, Flow and Heat transfer of a Microjet Impinging on a Heated Chip: Part II—heat transfer. Nanosc. Microsc. Thermophys. Eng. **17**(2), 92–111 (2013)
7. E.G. Colgan, B. Furman, M. Gaynes, N. LaBianca, J.H. Magerlein, R. Polastre, R. Bezama, K. Marston, R. Schmidt, High performance and subambient silicon microchannel cooling. J. Heat. Transfer. **129**(8), 1046–1051 (2007)
8. M. E. Steinke, S.G. Kandlikar, Single-Phase Liquid Heat Transfer in Plain and Enhanced Microchannels. In *ASME 4th International Conference on Nanochannels, Microchannels and Minichannels,* Limerick, June 19–21, ASME Paper No. ICNMM2006-96227

9. S.G. Kandlikar, D. Kudithipudi, C.A. Rubio-Jimenez, Cooling mechanisms in 3D ICs: thermo-mechanical perspective, In *IEEE International Green Computing Conference and Workshops (IGCC)*, Orlando, 25–28 July 2011.
10. W.M. Kays, A.L. London, *Compact Heat Exchangers* (McGraw Hills, New York, 1984)
11. W.-L. Cheng, W.-W. Zhang, H. Chen, L. Hu, Spray cooling and flash evaporation cooling: the current development and application. Renew. Sustain. Energy Rev. **55**, 614–628 (2016)
12. S.G. Kandlikar, W.J. Grande, Evolution of microchannel flow passages—thermohydraulic performance and fabrication technology. Heat Trans. Eng. **24**(1), 3–17 (2003)
13. J.-M. Koo, S. Im, L. Jiang, K.E. Goodson, Integrated microchannel cooling for three-dimensional electronic circuit architecture. J. Heat Transfer. **127**(1), 49–58 (2005)
14. A.W. Topol, D.C. La Tulipe, L. Shi, D.J. Frank, Three-dimensional integrated circuit. IBM J. Res. Dev. **50**(4.5), 491—506 (2010)
15. H. Mizunuma, L.Y. Chang, Y. Chia-Lin, Thermal modeling and analysis for 3-D ICs with integrated microchannel cooling. IEEE Trans. Comput. Aided. Design Integ. Circuits Syst. **30** (9), 1293–306 (2011). doi:10.1109/TCAD.2011.2144596
16. J.H. Lau, T.G. Yue, Thermal management of 3D IC integration with TSV (through silicon via), In *59th Electronic Components and Technology Conference*, San Diego, 26–29 May 2009, pp. 635–640
17. K. Puttaswamy, G.H. Loh, Thermal analysis of a 3D die-stacked high-performance microprocessor. In *Proceedings of the 16th ACM Great Lakes Symposium on VLSI*, New York, pp. 19–24
18. W. Huang, M.R. Stan, S. Gurumurthi, R.J. Ribando, K. Skadron, Interaction of scaling trends in processor architecture and cooling. In *26th Annual IEEE SEMI-THERM*, Santa Clara, 2010, pp. 198–204
19. P. Chaparro, J. González, G. Magklis, C. Qiong, A. González, Understanding the thermal implications of multi-core architectures. IEEE Trans. Parallel Distrib. Syst. **18**(8), 1055–1065 (2007)
20. I. Yeo, C.C. Liu, E.J. Kim, Predictive dynamic thermal management for multicore systems. In *Proceedings of DAC*, Anaheim, 2008, pp. 734–739
21. H.F. Sheikh, H. Tan, I. Ahmad, S. Ranka, B. Phanisekhar. Energy-and performance-aware scheduling of tasks on parallel and distributed systems. J. Emerg. Technol. Comput. Syst. **8**(4), 32 (2012)
22. D. Cuesta, J. Ayala, J. Hidalgo, D. Atienza, A. Acquaviva, E. Macii, Adaptive task migration policies for thermal control in MPSoCs. In *VLSI 2010 Annual Symposium*, Lixouri, 5–7 July 2010, pp. 83–115
23. T. Ge, P. Malani, Q. Qiu, Distributed task migration for thermal management in many-core systems. In *Proceedings of DAC*, Anaheim, 8–13 June 2008, pp. 579–584
24. J. Cui, D. Maskell, A fast high-level event-driven thermal estimator for dynamic thermal aware scheduling. IEEE Trans. Comput. Aided Design Integr. Circuits Syst. **31**(6), 904–917 (2012)
25. D. Yinon, T.D. Dudderar, B.J. Han, A.M. Lyons, Thin packaging of multi-chip modules with enhanced thermal/power management, U.S. Patent 5,646,828, 8 July 1997
26. M. Qing, H. Fujimoto, Silicon interposer and multi-chip-module (MCM) with through substrate vias, U.S. Patent 6,229,216, 2001
27. G. Brent, S.S. Sapatnekar, Placement of thermal vias in 3-D ICs using various thermal objectives. IEEE Trans. Comput. Aided Design. Integr. Circuits Syst. **25**(4), 692–709 (2006)
28. D.B. Tuckerman, R.F.W. Pease, High-performance heat sinking for VLSI. Electron Dev. Lett. **2**(5), 126–129 (1981)
29. B. Dang, M.S. Bakir, D.C. Sekar, C.R. King Jr., J.D. Meindl, Integrated microfluidic cooling and interconnects for 2D and 3D chips. IEEE Trans. Adv. Packag. **33**(1), 79–87 (2010)
30. S.G. Kandlikar, Fundamental issues related to flow boiling in minichannels and microchannels. Exp. Therm. Fluid Sci. **26**(2–4), 389–407 (2002)
31. HotSpot. http://lava.cs.virginia.edu/HotSpot/links.htm
32. 3D-ICE. http://esl.epfl.ch/3d-ice.html

33. A.W. Topol, D.C. La Tulipe, L. Shi, D.J. Frank, K. Bernstein, S.E. Steen, A. Kumar et al.. Three-dimensional integrated circuits. IBM J. Res. Dev. **50**(4.5), 491–506 (2006)
34. P. Jacob, O. Erdogan, A. Zia, P.M. Belemjian, R.P. Kraft, J.F. McDonald, Predicting the performance of a 3D processor-memory chip stack. Design Test Comput. **22**(6), 540–547 (2005)
35. L. Feihui, C. Nicopoulos, T. Richardson, Y. Xie, V. Narayanan, M. Kandemir, Design and management of 3D chip multiprocessors using network-in-memory. IEEE Comput. Soc. **34**(2), 130–141 (2006)
36. T. Brunschwiler, B. Michel, H. Rothuizen, U. Kloter, B. Wunderle, H. Oppermann, H. Reichl, Forced convective interlayer cooling in vertically integrated packages. In *Proceedings of the 11th IEEE Intersociety Conference on Thermal and Thermomechanical Phenomena in Electronic Systems*, vols. 1–3, Orlando, 28–31 May 2008, pp. 1114–1125
37. D. Lorenzini-Gutierrez, S.G. Kandlikar, Variable Fin density flow channels for effective cooling and mitigation of temperature nonuniformity in three-dimensional integrated circuits. J. Electron. Packag. **136**, 021007–021011 (2014)
38. J.-L. Hernandez-Gonzalez, S.G. Kandlikar, Performance assessment comparison of variable fin density microchannels with offset configurations. Heat Transfer Eng. **37**(2), 1369–1381 (2016). doi:10.1080/01457632.2015.1136146
39. S.G. Kandlikar, D. Kudithipudi, A. Rubio-Jimenez, Cooling mechanisms in 3D ICs: thermo-Mechanical Perspective. In *Proceedings of Green Computing Conference and Workshops (IGCC)*, Orlando, 25–28 July 2011. doi:10.1109/IGCC.2011.6008573
40. Y. Zhang, C.R. King, J. Zaveri, Y.J. Kim, V. Sahu, Y. Joshi, M.S. Bakir, Coupled electrical and thermal 3D IC centric microfluidic heat sink design and technology. In *61st Electronic Components and Technology Conference (ECTC)*, Lake Buena Vista, 2011, pp. 2037–2044. doi:10.1109/ECTC.2011.5898797
41. F. Alfieri, M.K. Tiwari, I. Zinovik, D. Poulikakos, T. Brunschwiler, B. Michel, 3D integrated water cooling of a composite multilayer stack of chips. J. Heat Transfer **132**(12), 121402 (2010). doi:10.1115/1.4002287. 9 p
42. A. Kalani, S.G. Kandlikar, Combining liquid inertia with pressure recovery from bubble expansion for enhanced flow boiling. Appl. Phys. Lett. **107**, 181601 (2015). doi:10.1063/1.4935211
43. Y. Zhu, D.S. Antao, K.H. Chu, T.J. Hendricks, E.N. Wang, Enhanced flow boiling heat transfer in microchannels with structures surfaces. In *15th International Heat Transfer Conference*, Kyoto, 2014
44. C. Green, P. Kottke, X. Han, C. Woodrum, T. Sarvey, P. Asrar, X. Zhang, Y. Joshi, A. Fedorov, S. Sitaraman, M. Bakir, A review of two-phase forced cooling in three-dimensional stacked electronics: technology integration, J. Electron. Packag. **137**, 040802, 9 p

Chapter 11
Fundamentals of Advanced Materials and Processes in Organic Substrate Technology

Songhua Shi and Peter Tortorici

11.1 Introduction

Over the past several decades, Moore's law has successfully predicted integrated circuit (IC) technology advancement. However, IC technology began hitting both technology and cost barriers. Conventional die shrinkage and advanced deep-submicron semiconductor technology is no longer able to meet the cost-to-performance ratio that the world desires in the near future. Three-dimensional (3D) packaging has caught broad attention and is poised to help continue the Moore's law by vertically integrating multiple IC chips into same footprint [1]. Although not all 3D package substrates require higher wiring density or smaller feature size than non-3D packages, 3D package does accelerate the substrate material and fabrication technology advancements to enable overall smaller, faster, and cheaper substrate demands. From the substrate material and fabrication technology level, there are no fundamental differences between 3D package substrate and non-3D package substrate. For example, substrate warpage control and finer feature sizes are the trends for both the 3D and non-3D package substrates. The information discussed in the following paragraphs is applicable to all organic substrates, including 3D package applications.

In this chapter, an overview of substrate technology evolution over the past several decades will be discussed. The materials used in substrates will be reviewed, with a concentration on organic substrate materials. The discussion will cover key consideration points in material selection and application. A general review of the substrate fabrication technology in addition to the process technologies on the fabrication of cores, build-up dielectric layers, metal layers and traces, plated through holes (PTH) and vias, contact pads, solder mask, in addition to

S. Shi (✉) • P. Tortorici
Medtronic, 2343 W. Medtronic Way, Tempe, AZ 85281, USA
e-mail: shawn.shi@medtronic.com; peter.c.tortorici@medtronic.com

Y. Li, D. Goyal (eds.), *3D Microelectronic Packaging*, Springer Series in Advanced Microelectronics 57, DOI 10.1007/978-3-319-44586-1_11

surface finishes will be covered. The general recommendation in selecting and applying the appropriate process technologies will be recommended. The purpose of this chapter is to provide consideration points in selecting and making substrates and PWBs as well as the information leads to help the readers dig deeper for their particular applications. This chapter is not written for material and fabrication solutions for a particular substrate or PWB application.

11.2 Overview of Substrate Technology Evolution

The role of substrate and PWB is to transmit electrical power and signals among electronic components of electronic systems and provide the protection and form factor needs to meet final use requirements. Since the commercialization of transistors in 1950s, semiconductor technology advancement has been driving the advancement of substrate technology in the past half century. Most recently, customer preference for higher performance and product miniaturization has become another strong driving force for PWB and package technology advancements. The PWB line/space (L/S) width (in micrometers) has evolved from 250/250 L/S in 1955 to 200/200 L/S in 1960s–1980s to 100/100 L/S in 1990s and 2000s to less than 25/25 L/S recently. The package substrate line/space width in microns has evolved from 30/30 L/S in early 1990s to 10/10 L/S in late 1990s to 5/5 L/S most recently to 3/3 L/S expected resolution in the next couple of years [2]. How to economically and reliably embedding electrical and optical components into substrates is the emerging area of development. In the foreseeable future, organic PWB and package substrates will continue to be mainstreamed to enable faster, thinner, and smaller portable electronics and 3D packaging. The main challenges will be substrate warpage control, and finer line and space width enablers. The heavy burden of addressing these challenges will rest upon the packaging materials. New high T_g and low CTE materials, through design of molecules with higher order structure, nanoparticles, super-fine photosensitivity technology, and combination of electrical and optical technologies are expected to be some of the main possible solutions moving forward [2].

11.3 Organic Substrate Materials

11.3.1 Materials Employed in Organic Substrate Production

For any substrate used for microelectronic packaging, its performance is determined by two factors: bulk material properties and interface properties. The interface properties often result from the chemical nature of the bulk materials (including additives) and the process of creating the interface. Over the past half a century, the materials for organic printed wiring board (PWB) and substrate applications have gone through significant advancement, both in available material varieties and material properties to meet different application needs. However, the basic structural components of PWB or substrate have remained almost unchanged. Generally, substrate and PWBs consist of three structural components: reinforcements (i.e., mechanical reinforcement structure), an organic resin system, and a conductor. In the case of coreless substrates, reinforcement is omitted.

Reinforcements are utilized to achieve desired mechanical and physical properties such as substrate mechanical rigidity and strength, dimensional target and stability, and coefficient of thermal expansion (CTE) matching. The most commonly employed reinforcement is woven fiberglass. Other reinforcements include glass matte, linear continuous-filament fiberglass, paper, aramid fiber, Teflon, and metals such as copper-invar-copper, aluminum, and fillers [3].

The resin system (including its additives) is used to build up layers and glue all the layers together. The commercially available resin systems are epoxy resins, epoxy blends, bismaleimide triazine (BT)/epoxy mixtures, cyanate ester, polyimide, polyester, polytetrafluoroethylene (PTFE), and alkylated polyphenylene ether (APPE). Most resin systems contain additives to achieve desired properties. These additives include curing agents and catalysts, adhesion promoters, flame retardants, ultraviolet (UV) blockers, fluorescing agents, colorants, and photosensitizers. Inorganic fillers are sometimes added to the resin system for CTE and stiffness control of the build-up layers [3].

The conductors are used to establish the electron movement path to fulfill the electrical design goal. The conductors are located in-between dielectric layers, on the wall of pin-through holes (PTHs) and via holes. Copper is the single most dominant conductor material in substrate and PWB applications because of its excellent combination of conductivity, ductility, processability, and affordability. Other conductors such as gold, nickel, palladium, vanadium, silver, tin, and conductive adhesives can also find their use in some area of substrate and PWB such as on exposed pad and in filled vias [3].

The selection of the above material components strongly depends on application and cost targets. Although the material technical data sheet are readily available from material suppliers, designing and making a successful substrate and PWB for high density interconnection applications is often a strenuous job. Raw materials are first screened through the lenses of application requirements, existing equipment, and cost targets. The screened material candidates are down-selected through processability tests, material characterization, and substrate and PWB performance and reliability tests. These tests are often aided by modeling and simulation.

To facilitate the detail discussion of material considerations for 3D substrate applications, the materials are categorized into substrate and PWB cores, dielectric materials, PTH and via filling materials, solder masks, and surface finishes. The proper integration of these materials yields all kinds of thick core, thin core, and coreless substrates as well as rigid, flexible, and rigid-flex PWBs.

11.3.2 General Considerations

3D substrate material selection is primarily driven by performance and cost considerations. The substrate cost is mainly determined by both base material cost and substrate manufacturability. The substrate manufacturability is not only impacted by the intrinsic difficulty level and the number of process steps of processing base materials and yield, but also by the existing equipment capability. Adding new manufacturing capability can trigger major capital investments and results in higher cost to a substrate manufacturer. A detailed discussion of material cost and substrate manufacturability is beyond the scope of this chapter. Readers can refer to microelectronics packaging handbooks and printed circuits handbooks to get the desired information.

Substrate performance is correlated to the properties of base material and a variety of interfaces generated during the manufacturing process. Although the property of an interface is often impacted by the process, it is significantly related to the surface properties of involved base materials. For a particular application, the thermal, physical, mechanical, and electrical properties of base materials must be considered. In many cases, some of these properties are correlated to each other. Substrate manufacturers must study and understand all these properties and their interactions in an integrated way. The most common bulk material properties relevant to microelectronic packaging substrates and PWB applications include glass transition temperature (T_g), coefficient of thermal expansion (CTE), modulus, toughness, flexural strength, density, decomposition temperature (T_d), flammability, moisture absorption, chemical resistance, dielectric constant or permittivity, dissipation factor or loss tangent, volume resistivity, electrical strength, and dielectric breakdown. The common surface properties of the base materials include surface roughness, surface energy or surface tension, surface resistivity, and arc resistance. The measurement of these properties is beyond the scope of this chapter. However, readers can refer to IPC-TM-650, ASTM, and JEDEC standards for testing procedures of most of the mentioned properties.

Glass transition temperature (T_g) is used to describe the response of a molecular structure movement as a whole to the temperature change for solid amorphous materials. For polymeric materials used in microelectronic substrates, glass transition refers to the material change from rigid glassy state to soft rubbery state. T_g is a temperature above which the free volume becomes large enough for macromolecular chain segments to move and change their conformation. Below T_g, the chain segments do not have sufficient free volume and they are locked in place. Although T_g can be measured by multiple methods and often expressed by a single temperature point, the glass transition occurs across a temperature range. T_g is a critical polymeric material property as most of the material properties mentioned above change greatly during the glass transition region. Among them, the CTE and modulus changes during the glass transition region often catch most attention in base material selection for microelectronic substrate applications.

The coefficient of thermal expansion (CTE) is an intrinsic physical property of a material. Like most of materials, the base materials used in substrates expand when they are heated and contract when they are cooled. The CTE can be calculated by $\Delta l/(l \cdot \Delta T)$, where Δl is the linear dimensional change in a given direction, l is the original dimension in the direction, and ΔT is the change of temperature. Starting with the same dimensional length, materials with larger CTE expand or contract more over same range of temperature change. CTE is one of key physical parameters in the selection of base materials because the large CTE differences among the substrate base materials generate large internal stresses and can result in interface layer delamination and substrate warpage.

Modulus is a mechanical property of a solid material, measuring a material's resistance to being deformed to an applied force, often expressed by Young's modulus and dynamic modulus. The Young's modulus is also called tensile modulus, or elastic modulus. It is the slope of stress–strain curve in the elastic deformation region of a material. The Young's modulus is commonly used for elastic materials such as glasses, ceramics, and metals. The dynamic modulus is the ratio of stress to strain under vibratory conditions. It is more commonly used for viscoelastic materials such as polymeric materials. The dynamic modulus consists of two components: storage modulus and loss modulus. The storage modulus measures the stored energy representing elastic portion at the given cyclic loading frequency and strain level. The loss modulus measures the energy dissipated as heat representing the viscous portion (i.e., plastic deformation portion). Since the CTE of all materials used in a substrate is not the same, material modulus is one of the key parameters in determining internal stress level in the PWB or substrate. High internal stress level is detrimental to substrate quality and reliability and often causes substrate warpage, delamination, and cracks.

Toughness is the ability of a material to absorb energy through deformation without fracturing. It is a parameter used to measure the ability of a material that can deform before its rupture under a given strain rate. Since material fracture is one of the major failure modes in microelectronic packaging, toughness is an important material mechanical property to consider in material selection, especially for the applications where the substrates need to go through large bending or cyclic deformation in the later assembly process or field use. Solder mask material selection for flex substrate applications is a good example. Because material toughness strongly depends on particular application, measurement of material toughness is often performed in-house.

Flexural strength represents the maximum stress that a material can sustain before its fracture and is often measured by the so-called three-point bending method, where a material sample is supported at the ends and loaded in the center. The flexural strength sets maximum total stress level (internal stress plus external stress) limit that a material can bear from manufacturing through product life time. In real applications, the maximum designed stress level on any material in a substrate or PWB is much lower than its flexural strength. The material flexural strength is readily available from substrate or PWB material handbooks. IPC-4101 specifies the minimum flexural strength of various materials for substrate applications.

Material density is considered when substrate weight becomes a factor such as in aerospace applications. For cored substrates, the core weight is the main portion of the total weight. Therefore, the selection of core material is the main path to meet the weight target.

The decomposition temperature (T_d) is an intrinsic chemical property of a material. In the context of organic substrates, at a predefined heating rate the decomposition temperature refers to the temperature at which the organic resin system starts to break down. There are multiple ways to characterize the decomposition temperature of a material. Thermo-gravimetric analysis (TGA) is the most commonly used technique by which the material weight loss is measured over a period of time at a constant temperature or over a range of temperature at a constant heating rate. T_d is a very important chemical property for a base material as it determines the maximum process or use temperature of a fabricated substrate. Above T_d, the organic material starts to lose its strength, toughness, and adhesion. This property did not cause much concern before the RoHS requirements came into effect because most of organic base materials can sustain eutectic tin/lead solder reflow temperatures. However, significant attention is now paid to this property during material selection, as lead-free solder reflow temperatures begin to adversely impact the substrate reliability for some organic base materials.

Flammability is another chemical property of a base material. This property is significant when high power density is needed for a substrate or PWB to meet the application needs. The Underwriters Laboratories (UL) classifies the flammability properties as 94V-0, 94V-1, or 94V-2. 94V-0 represents the lowest level of flammability category of a material. The definition of these classifications can be found in UL web site.

Moisture absorption primarily results from the chemical nature of cured organic resins. Most of the reinforcement and conductor materials in a PWB or substrate do not absorb noticeable moisture. For resin systems commonly used in substrates, the moisture absorption ranges from 0.1 % of FR-4 epoxy to ~0.5 % of BT/epoxy blend and cyanate ester [3]. Moisture absorption is an important property of base materials for substrate applications, especially for high temperature lead-free substrate applications. The absorbed moisture in substrate mainly has two effects on substrate reliability during package assembly process. One is that the absorbed moisture can weaken the interfacial adhesion between resin to reinforcement or conductor through hydrolysis of chemical bonds or intermolecular attraction reduction by water molecule like plasticization effect. The other is the so-called the pop-corning effect during solder reflow. The absorbed moisture molecules can be accumulated in very tiny pores in the cured resin or on the interfaces. The moisture vaporizes and generates high vapor pressure during solder reflow. Both of the effects can result in material cracks and interfacial delamination.

Chemical resistance is the ability of a material to retain its original property from solvent attack or chemical reaction when a material contacts solvents and chemicals during the substrate manufacturing process and the subsequent microelectronic assembly processes. These chemicals and solvents include plating and etching baths, liquid dielectric materials, and defluxing solvents. The methylene chloride absorption test is most commonly used for laminate chemical resistance assessments [3].

The dielectric constant or permittivity is a property of an electrical insulating material (a dielectric) and is equal to the ratio of the capacitance of a capacitor filled with the given material to the capacitance of an identical capacitor in a vacuum without the dielectric material. Dielectric constant is used to measure the ability of a material to store an electric charge. When an electric field is applied onto a dielectric material, dielectric polarization occurs. Positive charges are displaced from its equilibrium positions toward the negative side and negative charges are displaced toward the positive side. This process generates the electrical potential energy and stored in the material. The stronger the dielectric polarization, results in higher dielectric constant. Dielectric polarization is closely related to the molecular polarity of a material. Under the same electric field, the higher the molecular polarity is, the higher the dielectric polarization is. Polymers with symmetric structure such as polytetrafluoroethylene (PTFE) and polyethylene (PE) have lower molecular polarity. Dielectric constant varies with frequency, humidity, and temperature and requires careful control in its measurement. Since dielectric constant can distort electrical signal or voltage drops, it is a critical material parameter in selecting base materials, especially for applications that requires high frequency, such as telecommunication and avionics.

Dissipation factor or loss tangent is another important material property of an insulating material. It is expressed as the ratio of the total power loss in a material to the product of the voltage and current in a capacitor being made of the material between two electrode plates at a given frequency. Dissipation factor measures the efficiency of a material interacting with electrical signals. When a material is placed in an oscillating electrical field, the molecules inside the material oscillate as well. This dielectric polarization oscillation is accompanied by the motion of the atoms with partially electrical charges which result in energy loss. Since high dissipation factor means higher energy loss, and results in larger signal or voltage attenuation, and undesired heating, materials with lower dissipation factor are desired for substrate applications. Materials with less electrical field-induced atom and molecule motion usually have lower dissipation factor.

Volume resistivity is the ratio of potential drop over a unit length of material to the current flowing through a unit cross-sectional area. It is an intrinsic property that quantifies how strongly a given material opposes the flow of electric current. Since the volume resistivity of organic base materials is usually higher than $10^7\,\Omega$ cm [3]. The leakage current is not a concern even for very current-sensitive medical device applications including pacemakers and defibrillators. However, volume resistivity is usually a good measurement to detect process or reliability issues.

Electrical strength refers to the maximum electric field that a dielectric material can withstand without breaking down in the direction perpendicular to the plane of the material. It is an intrinsic property of the bulk material. When a dielectric material is placed in a sufficiently high electrical field, free electrons from ambient radiation may become accelerated to velocities that can liberate bound electrons in the material during collisions with neutral atoms or molecules. This process is termed avalanche breakdown and can happen abruptly. Once avalanche breakdown happens, an electrically conductive path was formed and the material loses its insulation ability. Since common fiberglass-reinforced materials usually have electric strength above 1200 V/mil at ambient condition, electrical strength is not a main concern for most applications [3]. However, some medical devices such as defibrillators can have voltage being pumped to more than 1000 V and the electrical strength of the selected material becomes important. The experimentally measured electrical strength often impacted by the moisture content and preconditioning conditions.

Dielectric breakdown is similar to the electrical strength, but the electrical breakdown happens in the direction parallel to the plane of material. Its measured value is impacted by moisture content and preconditioning conditions. With the increase of substrate wiring density, the pitch between two adjacent traces in the same build up layer is shrinking. Dielectric breakdown consideration is needed for high voltage applications and implementation of electrical static discharge (ESD) controls.

Material surface properties mainly impact the interfacial property of the multiple interfaces in substrate, and the interfacial properties impacts both substrate quality and reliability. Surface roughness, surface free energy, surface resistivity, and arc resistance are the following four properties introduced.

Surface roughness refers to the deviations in the direction of normal vector of a real surface from its imaginary ideal surface (i.e., perfectly flat and smooth) and is a parameter describing surface texture of a real surface. Surface roughness is often expressed by both deviations amplitude and spatial frequency. The larger the deviation is, the rougher the surface is. Surface roughness is an important parameter as it impacts adhesion between build-up layers in the substrates. For thin substrate applications, high surface roughness can result in short circuits or electrical breakdown during substrate manufacturing, testing, or field use. For high frequency RF applications, high surface roughness generates longer signal travel distance with varied capacitance from location to location. This results in signal distortion and attenuation.

Surface energy or surface tension quantifies the amount of energy required to break down intermolecular and interatomic bonds to create a new surface. Compared to the molecules or atoms inside a bulk material, the molecules or atoms at the surface are normally at higher energy state and relatively less stable. Surface energy is a key parameter affecting liquid wetting on solid surface and ultimately impacting adhesion strength of two materials. Higher surface energy of a solid material usually results in better wetting. For a surface that has been exposed to the air for long time, the surface often has absorbed many molecules from the ambient and the surface energy has been reduced noticeably. Different kinds of cleaning methods are used to regenerate the fresh surface to increase the surface energy and help the increase of adhesion strength. Some of the cleaning methods include solvent cleaning, plasma, etching, and blasting.

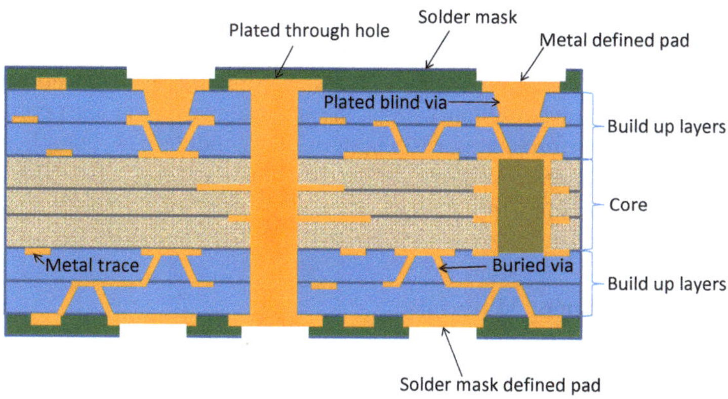

Fig. 11.1 Schematic drawing of a typical cored substrate

Surface resistivity is determined by the ratio of DC voltage drop per unit length to the surface current per unit width. It is an intrinsic material property describing how well the insulating material opposes the current flow on the material surface. The measured surface resistivity of a material strongly depends on the environmental conditions such as humidity and temperature as well as the conductor configuration and contact wellness on surface. Therefore, the measurement procedure needs to be well controlled. Standards, such as ASTM Standard D 257-99 or ESD STM 11.11-2001, can be followed. This parameter is often considered when small pitch exposed conductor traces are required.

Per ASTM D495 Plastic Test Standard, arc resistance is measured by the number of seconds that a material resists the formation of a surface conducting path when subjected to an intermittently occurring arc of high voltage, low current characteristics. The results at 3 mm thickness are considered representative of the material's performance in any thickness. Performance Level Categories (PLC) was introduced by UL to avoid excessive implied precision and bias.

Cored substrates are the most popular substrate utilized in microelectronic packaging. A typical cored substrate is schematically presented in Fig. 11.1. The substrate is composed of a core, build up layer(s), via, PTH, expose metal pad, solder mask, and surface finish.

11.3.3 Substrate and PWB Cores

Core material selection largely impacts the target rigidity, thickness, wiring density, and electrical properties. Based on the thickness level, people often classify core into thick core or thin core, but there is no clear divider between them, and their definition evolves with time. For some people, thin core is designated if the core thickness ranges from 0.05 to 0.80 mm. For other people, thin core is designated if the core thickness ranges from 0.05 to 0.20 mm.

Substrate cores can be simple or complex depending on the application and cost target. Regardless of the complexity level, the substrate core is at least composed of three components: reinforcement, resin system, and conductor. A complicated core can combine different resin systems, different reinforcements, and even different conductors for very unique applications.

11.3.3.1 Reinforcement Materials

Commercially available reinforcements include woven fiberglass cloth, nonwoven fiberglass, aramid fiber, paper, expanded Teflon, and other constraining materials. The predominant reinforcement in substrate and PWB applications is woven fiberglass cloth. Woven fiberglass cloth is usually made of one type of glass material or a combination of different glass materials. At least five glass materials are available for substrate and PWB applications as discussed in the Printed Circuits Handbook by Coombs et al.; they are E-glass, NE-glass, S-glass, D-glass, and quartz [3]. The detail composition of each glass can be found in printed circuits handbooks. Among these materials, E-glass is the most popularly used fiberglass for substrate and PWB applications thanks to its excellent combination of lower cost, electrical, mechanical, and chemical properties. NE-glass found its use in some niche applications where low CTE, low dielectric constant, and low loss factor is preferred. S-glass has higher mechanical strength but wore out drill bit much faster in making PTH or via holes. D-glass and quartz have not gain much attention in substrate and PWB applications. The fiberglass filament is first twisted into a yarn, and the fiberglass yarn is then woven into different cloth styles. The filament diameter usually ranges from 3.8 to 13.0 μm. The woven fiberglass cloth weight ranges from 0.019 to 0.234 kg/m^2 [3]. Considering the considerable numbers of glass compositions, filament diameters, yarn types, and the number of different weave patterns, there are abundant number of fiberglass cloth combinations that can be manufactured. However, plain woven fiberglass cloth is most commonly used due to its good resistance to slippage and fabric distortion during laminate core processing. Coupling agents such as epoxy silane and amino silane are commonly applied to the surface of fiberglass to improve the adhesion between the fiberglass and resin systems. Other types of surface finishes may be applied to the surface of fiberglass for the filament production, abrasion and ESD prevention, and aiding the holding of the filaments together.

Nonwoven fiberglass can be subcategorized into two types based on the orientation randomness of glass fibers: glass matte and linear continuous filament fiberglass. Glass matte either consists of randomly oriented 1–2 in. long chopped fiberglass strand matte distributed evenly and bonded together or consists of continuous strands of fiberglass in a random, spiral orientation. As a result of the fiberglass loading, orientation randomness, and lack of the restriction from fiber to fiber, it has poor slippage resistance and larger localized property variation.

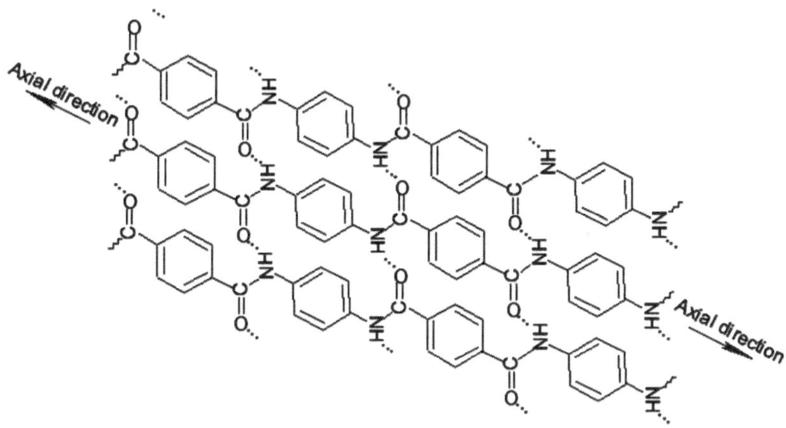

Fig. 11.2 Molecular structure of one aramid material

Therefore, glass matte is only used for simple applications such as CEM-3 and G-10 cores. Linear continuous filament fiberglass consists of three layers: bottom, middle, and top. The fiberglass filaments in bottom and top layers are in parallel but perpendicular to the filaments in the middle layer, and the number of linear filaments in the middle layer is double of that for the top and bottom layers. This type of filament layout improves the laminate core dimensional stability, but its usage is very limited [3].

The full name of aramid is aromatic polyamide, which is commonly used in bullet-proof vests. One typical molecular structure of aramid material is presented in Fig. 11.2. It contains amide functional group and rigid phenylene rings. The polar amide group forms hydrogen bond among the molecular chains and form crystals, and the phenylene rings limits the chain folding. The two effects provide aramid fiber high strength, high thermal resistance, high dimensional stability, high chemical resistivity, and light weight (compared to fiberglass) [4]. Since the rigid polymer chain is so tightly bonded to many other chains by the hydrogen bonding the chain is stretched. When the material is heated, the free volume to each chain segment increases and the chain segment curls back. Therefore, aramid fiber has negative CTE along the axial direction. This negative CTE property is often leveraged to control the substrate core CTE. Since aramid molecule heavily contains carbon atoms, a charcoal layer is formed at the surface during burning and prevents continuous burning. Therefore, the aramid material has a self-retarding property. Aramids can easily be ablated by lasers, so they can be easily developed to enable high density interconnections. This material has extended its uses in military applications such as in MIL-S-13949/15-AF/AFN/AFG boards [3]. This material can be found in substrates made by different suppliers. CoreEZ® HDI substrate made by i3 Electronics, Inc. is just one example [5].

Paper is a cellulose material (also referred to as polysaccharide) consisting of several hundred to many thousands of $\beta(1 \rightarrow 4)$ linearly linked D-glucose units. Polysaccharide chains contain many hydroxyl groups; therefore, cellulose is hydrophilic. The hydroxyl groups on the glucose structure form intramolecular and intermolecular hydrogen bonds and hold polysaccharide chains firmly together and form crystalline domains in amorphous phase. Therefore, cellulose fiber has relatively high strength and does not dissolve in water. Since the cellulose fiber is randomly arranged in paper and has lower rigidity, paper is often used to reinforce substrate cores, with other core reinforcements such as E-glass or other materials that allow punching to form through holes. This reduces cost for low-end consumer electronics applications such as FR-1, FR-2, FR-4, and CEM-1 cores [3]. Additionally, with the increase of environmental protection requirements, paper is also considered for biodegradable substrate applications.

Teflon is the brand name made by DuPont Corp and is a material of PTFE, an abbreviation of polytetrafluoroethylene. Due to its symmetric structure of repeating units in the polymer chain and high electronegativity of fluorine, this material has a superior low dielectric constant of 2.1, a loss factor of 0.0005, and low moisture uptake. It is a semicrystalline polymer with melting point of about 327 °C [3]. Teflon has both a very low surface tension and friction coefficient. For substrate applications, Teflon is usually used in its expanded form, similar to a sponge structure on a microscopic scale. Since the conventional use temperature of a substrate is above its T_g, Teflon is easy to creep and seldom considered as a true reinforcement in substrate core. It is used mainly to improve electric performance for high frequency applications. Since PTFE is radiation reactive and it can be cross-linked above its melting point in an oxygen-free environment, PTFE is not radiation hard material and not proper for applications involving high or frequent radiation.

Constraining materials are sometimes used as substrate cores to achieve required thermal, CTE, and stiffness requirements. Substrates containing constraining materials are often called metal core substrates. The commercially available constraining materials include Copper-Invar-Copper (CIC), Copper-Molybdenum-Copper (CMC), Copper (Cu), Aluminum (Al), Carbon Copper Core (CCC), and Titanium (Ti). Copper-Invar-Copper is a sandwich of Invar, a nickel containing iron alloy, bonded between two layers of copper. The typical relevant properties of these core materials are included in Table 11.1. As the material properties vary with compositions of CCC materials, CCC is not included in the table [6].

Table 11.1 Mechanical and thermal properties of constraining materials

Material name	Unit	CIC	CMC	Cu	Al	Ti
Mechanical properties						
Density	g/cm^3	8.33	9.93	8.9	2.71	4.5
Young's modulus	GPa	119	263	130	70	116
Thermal properties						
CTE (x–y direction)	ppm/K	5.36	6.35	17	23.2	8.9
Thermal conductivity	W/mK	112	204	394	237	17
Heat capability	J/kgK	479	286	385	900	528

11.3.3.2 Resin Systems

The commercially available resin systems for substrate core applications are epoxy, epoxy blend, bismaleimide triazine (BT)/epoxy, cyanate ester, polyimide, and allylated polyphenylene ether (APPE). Fillers are often added to resin systems in order to achieve desired properties.

Among the resin systems mentioned above, the most commonly used resin systems are epoxy resin systems and their blends. Epoxy resin systems possess excellent balanced chemical, physical, mechanical, and electrical properties and find wide use in all kinds of applications. A typical epoxy resin system often consists of epoxy resin and curing agent besides other additives such as catalysts, coupling and toughening agents, flame retardant, and fillers. Over the past more than half a century, many epoxy resins have been developed for choices to meet all kinds of applications. This is one of the main reasons why epoxy resin systems dominate organic core applications. Epoxy resin or epoxide is generally defined as any molecule containing one or more of the 1,2-epoxy group. The most commonly used epoxy resins in microelectronic packaging substrate core applications are di-epoxides or multifunctional epoxides. They can be classified into two main categories: glycidyl-ether type epoxy resins and cycloaliphatic type epoxy resins. The glycidyl ether type epoxy resins are the reaction products of hydroxyl-containing compounds and 1-chloro-2,3-epoxypropane (epichlorohydrin) as presented in Fig. 11.3a. The cycloaliphatic type epoxy resins are mainly the epoxidation products of alkenes, presented in Fig. 11.3b. Based on the varieties of the hydroxyl-containing compounds and alkenes, the two categories of epoxy resins can be further classified into subgroups as shown in Fig. 11.4. Many of the recently developed epoxy resins are multifunctional epoxy resins for lower moisture uptake and higher heat resistance. There are wide choices of curing agents for epoxy resins. The most commonly used curing agents include phenolic resin, amine, anhydride, and dicyandiamide (DICY). By properly mixing the different epoxy resins and curing agents, enormous number of formulations can be made to meet particular application needs. Multifunctional epoxy resins find their use for heat resistance applications [7].

Epoxy blends are mixtures of epoxy resin with other type of resins. The commercially available epoxy blends include epoxy-polyphenylene oxide (PPO), epoxy-cyanate ester, and epoxy isocyanurate. The main purpose is to improve T_g,

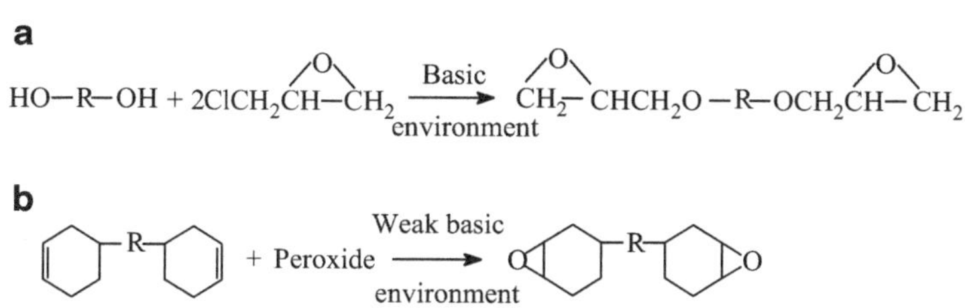

Fig. 11.3 Glycidyl ether type epoxy resin and cycloaliphatic type epoxy resin

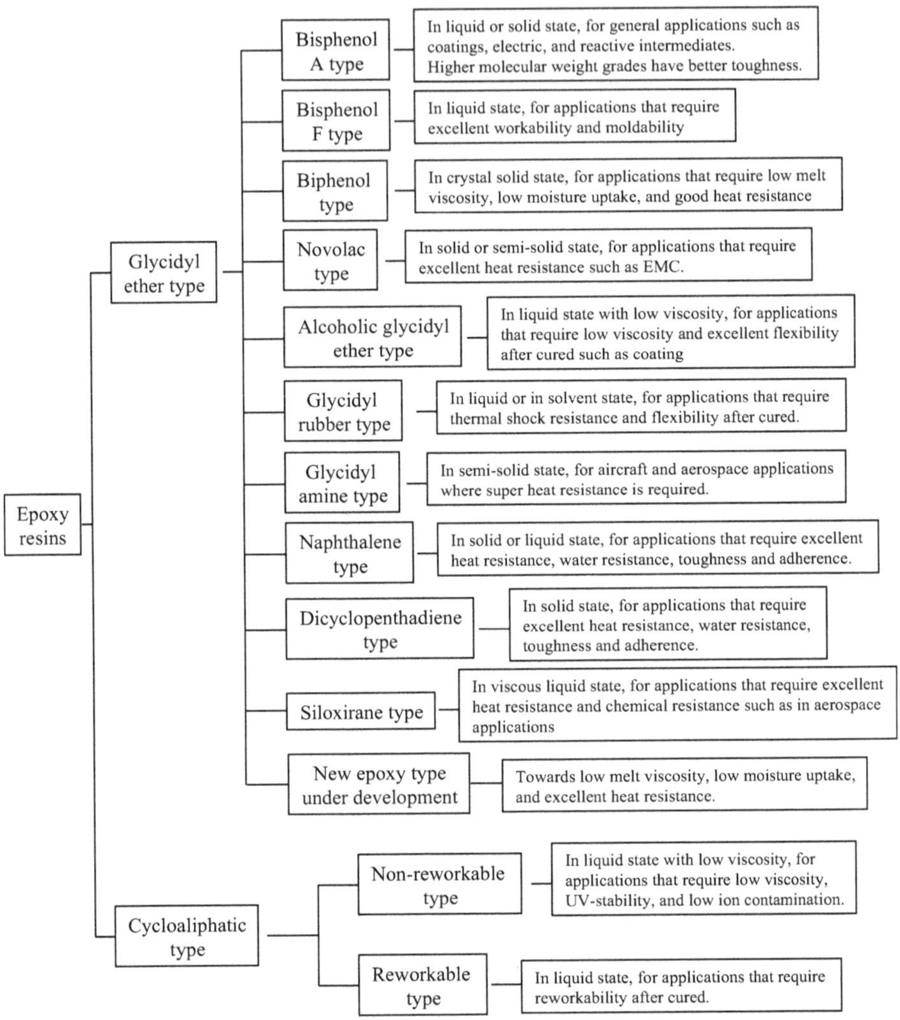

Fig. 11.4 Flowchart of epoxy resin classifications [7]

T_d, and electrical performance of epoxy resins. However, epoxy blends are usually more expensive and may need some tuning of existing substrate fabrication processes.

Bismaleimide triazine (BT)/epoxy resin systems are a mixture of bismaleimide, cyanate ester, and epoxy. The reaction among them is relatively complex. Bismaleimide and cyanate ester can self-polymerize by itself and bismaleimide can react with cyanate ester. Cyanate ester can also react with epoxy. Therefore, this resin system can form a rigid and complex thermosetting resin after curing. The main advantage of BT resin over epoxy is the higher T_g, ranging from 180 to 220 °C, and a better combination of thermal, chemical, and electrical properties

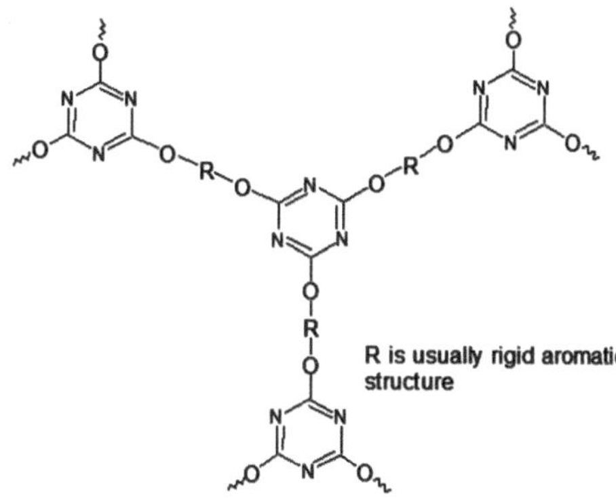

Fig. 11.5 The molecular structure of cured cyanate ester network

[3]. This resin system is widely used for ball grid array (BGA) and chip scale packaging where high density interconnection (HDI) is needed. However, BT resin is usually more expensive. Since BT resins contain more percentage of oxygen and nitrogen atoms in its backbone, it can absorb more moisture, and it is also more brittle due to its rigid cross-linking network.

Cyanate ester resin systems cross-link during curing through the ring forming cyclotrimerization reaction. Since the reaction creates a very high concentration of triazine rings throughout the polymer matrix as illustrated in Fig. 11.5, cured cyanate ester has high T_g, around 250 °C [3]. Due to the low cross-linking density and high concentration of ether linkage, cured cyanate resin usually has a high toughness. The symmetrical arrangement of oxygen and nitrogen atoms around the carbon atoms in cyanate ester result in an absence of strong dipoles, therefore resulting in a low dielectric constant and dielectric loss. However, cyanate ester is relatively expensive and requires special processing, which limits its use to only some niche applications that required high temperature and high speed applications.

Polyimides are condensation polymers derived from bifunctional carboxylic acid anhydrides and primary diamines. With the proper selection of dianhydride and diamine components used as starting materials, aromatic polyimides can achieve outstanding mechanical strength and flexibility, thermal and oxidative stability, high T_g (as high as 260 °C) and low CTE. However, polyimide is a relatively expensive material. Its use in substrate core is limited to niche applications [3, 8].

APPE is a suitable material for high frequency high performance applications, such as RF, wireless, and high speed computing applications as a result of its very low dielectric constant and excellent thermal performance. Early version of APPE materials experienced process difficulties, but recent formulation improvements have made the material compatible with conventional substrate manufacturing processes.

Small particle fillers added to resin systems include talc, silica, kaolin clay powders, tiny hollow glass spheres, and other inorganic materials. Talc is a mineral composed of hydrated magnesium silicate, often used as a lubricant to facilitate material handling. Silica particles are also used to reduce CTE, especially *Z*-axis CTE. Palladium-coated kaolin clay powders are used as catalysts for electroless copper plating. Tiny hollow glass spheres are sometimes added to reduce dielectric constant and loss factor [3].

11.3.3.3 Conductors

The conductor employed in substrates and PWBs is exclusively copper (Cu), in the form of Cu foils and electroplated Cu. However, there are many grades of copper foils to choose from and many aspects about the copper foils must be considered before finalizing their selection. These aspects include thickness, surface roughness, surface treatment, mechanical properties, purity and resistivity, and integration of other metal alloy. The specification of each copper foil grade is documented in IPC-4562. Among the copper foils specified IPC-4562, they can be classified into two types based on manufacturing method: electrodeposited copper foil and rolled-wrought copper foil.

Copper foil thickness is usually expressed in ounce, which is the weight of 1 ft^2 copper foil assuming uniform thickness. The typical thicknesses of copper foils used in substrates is 0.5 oz, 1 oz, 2 oz, and 3 oz, corresponding to thickness of approximately 18 μm, 36 μm, 70 μm, and 140 μm, respectively. Use of thin Cu foils of 0.25 and 0.375 oz thickness is emerging [3]. Since the copper foil thickness directly impacts current density, the total substrate thickness, substrate warpage, and etching productivity, it is usually the first parameter to be considered in substrate manufacturing.

The surface roughness of copper foil impacts the copper foil adhesion to the resin system, electrical performance, and manufacturing capability of trace width. Higher surface roughness foils generate a larger surface area and more mechanical bonding sites. Therefore, it usually results in higher adhesion strength. However, higher frequency electrical signals propagate near the surface of the conductor. An electrical signal needs to travel farther on a rougher surface and results in more signal attenuation. With very thin dielectric layer sandwiched in-between two copper foils, the surface roughness of the copper foil causes capacitance and resistance variation from location to location, and this can distort the signal as well. Copper foil surface roughness also impacts copper etching time. In order to ensure the etching completeness, the etch time must be set up long enough to ensure the thickest location is etched away. This results in trapezoidal circuit traces and reduces the capability of making fine-line circuits in the substrate.

Surface treating copper foil is done for the purpose of achieving specific needs. Coupling agent treatments are used to increase the adhesion between copper foil and resin systems and help prevent copper surface oxidation and contamination. Bonding treatments or nodularization is done by plating copper or copper oxide nodules to the surface of copper foil to increase the surface area. The purpose of this treatment is mainly to achieve good adhesion to high performance resin systems such as BT, cyanate ester, and polyimide resin systems. Thermal barrier treatment is conducted by coating a several hundred angstroms thick of zinc, nickel, or a brass layer over copper foil nodules. Their purpose is to prevent or retard the bond degradation between copper foil and resin systems during laminate, substrate, and microelectronic package manufacturing. Passivation and antioxidant coatings are very thin layers of chromium-based or organic-based coatings on the both sides of the copper foil, preventing oxidation during storage and lamination. The coating is usually removed prior to substrate and PWB manufacturing [3].

The mechanical properties of copper foil that are distinguishable among different grade of copper foil are tensile strength and elongation. These two properties are mainly determined by grain size, orientation, and structure from the manufacturing process. Readers may refer to IPC-4562 specification or get detail technical data from suppliers. Usually, electrodeposited copper foil has higher tensile strength but lower elongation percentage than rolled-wrought copper foil.

The purity and resistivity of copper foils is also specified in IPC-4562. The minimum purity of electrodeposited copper foil without treatment but with any silver counted as copper is 99.8 %. The minimum purity for rolled-wrought copper foils is 99.9 %.

The main purpose of integrating other metal alloys onto copper foil is to make buried resistors. A resistive metal alloy is first coated onto the base copper foil. The prepared integrated foil and a resin layer are then made into a laminate. The integrated foil is then imaged and etched per defined pattern to produce desired circuit pattern and resistive components [9].

11.3.4 Dielectric Materials

The dielectric material, also called the layer build-up material, is mainly used to route the electric power and signal as well as make the package meet application form factor requirements. A successful dielectric material must have compatible chemistry with core material, good adhesion to the core and copper, good dielectric properties to meet electrical performance requirements, good thermal stability for lead-free solder processing, good microvia manufacturability, and satisfactory reliability.

Theoretically, all the core laminates can be used as dielectric materials. However, glass cloth reinforced laminates are not suitable for build-up layers for high density substrate applications as a result of via size and wiring density limitation of mechanical drilling. With the exception of very thin glass cloths, glass cloth reinforced laminates are not practically compatible with other microvia-making methods such as laser drilling, plasma etch, and photoimaging. The commonly used layer build-up materials for high density substrate applications, such as 3D-packaging, are epoxy and epoxy/BT resin, polyimide, aramid reinforced epoxy laminate, APPE, and liquid crystal polymers (LCP) [3, 8–12].

Epoxy and epoxy/BT resin is one of the major dielectric materials employed in high density substrates. The generic chemical structures and properties of epoxy and epoxy/BT resins have been previously discussed. When these resins are used for dielectric build-up layers, one of key considerations is the ability to fabricate via holes, and their quality. Generally speaking, there are three ways to form via holes for the epoxy and epoxy/BT resin dielectric materials: laser drilling, UV photo-defining, and etching. Laser drilling and etching (both plasma dry etching and chemical wet etching) support more choices of resin systems. There are two types of photo-definable epoxy resins [8]. The first type of photo-definable epoxy resin is made by mixing the base epoxy resin with photosensitizing agents. The second type is intrinsically photo-definable epoxy. There are more choices of the first type of photo-definable epoxy than the second type. These resins can be used in either liquid form or B-staged dry film form. Screen printing, curtain coating, and spin coating are used to apply liquid epoxy/BT resin, while vacuum laminating is used for B-staged dry film. Fillers are sometimes added to the resin system in order to gain desirable properties such as lower CTE and easier copper plating. Since most of the epoxy resins used for dielectric layers contains multifunctional epoxy groups with aromatic backbone structures, they are rigid and brittle after curing.

Polyimide is another major material used for substrate build-up layers. This is due to its outstanding integrated properties and availability in different formulations. There are three types of polyimides used for dielectric build-up layers: non-photosensitive polyimide, negative photosensitive polyimide, and positive photosensitive polyimide. Since polyimide is often used in precursor form in both liquid solution or film form, cure shrinkage must be sufficiently considered in processing polyimide dielectric material. Multiple methods can be used to generate microvias in non-photosensitive polyimide layers. Laser drilling is a commonly used technique of making microvias in fully cured polyimide dielectric layers. The via diameter is limited by the laser beam spot size. Eximer laser drilling can generate vias with diameters of around 10 μm. Dry and wet etching, with the help of photoresist, is also a commonly used method to form microvias in polyimide dielectric layers. However, etching processes require many steps, and the achieved via hole diameter depends on the dielectric layer thickness because of etching isotropicity issues. The diameter of via made by etching methods often ranges from 75 μm to several hundreds of microns. Photosensitive polyimides have gained lots of attention because they allow simplified process and have higher resolution compared to eximer laser generated via. Much more material choices are available for negative photosensitive polyimide than positive photosensitive polyimide. Negative photosensitive polyimides are primarily derived from polyamic acid precursors which have either covalently bonded ester or ionic-bonded photosensitive groups attached to the backbone polymer chain. Positive photosensitive polyimides are derived either from incorporating a photosensitive group into the backbone as a side branch of the base polyamic acid, or by blending a photosensitive compound with the polyamic acid precursor. A detailed discussion of the properties and application processes of different types of photosensitive polyimides can be found in Microelectronics Packaging Handbook edited by Rao Tummala et al. [8]. Polyimide dielectric build-up layers are utilized in wide varieties of substrates and PWBs such as thick core, thin core, and coreless substrates, as well as rigid, flex, and rigid-flex PWBs.

LCP is a thermoplastic polymer whose molecular chains have one or two-dimensional spatial orders. LCP gains its attention in substrate applications because of many of its outstanding properties. LCP is "near" hermetic with low 0.02–0.04 % moisture absorption, and its CTE can be custom tuned from 3 to 30 ppm/°C to match copper CTE of 17 ppm/°C. LCP is halogen free, has a similar dielectric constant and loss factor to PTFE and is claimed to be capable of enabling 110 GHz frequency applications [11]. The desirability for aerospace applications is its flexibility and light weight, with its density of 1.4 g/cm^3, which is 40 % lighter than silicon, 65 % lighter than alumina, and almost 30 % lighter than FR4. LCP is a good material candidate for embedded substrate applications because it can be made as thin as 25 μm and as thick as desired. It is radiation hard and biocompatible which is suitable for medical applications, and it is thermoplastic and easy to process for lower costs. Coreless multilayer substrates have been built only by LCP dielectric material and copper. Laser drilling is used to make 50 μm diameter through vias and buried vias. Some substrate base material suppliers have commercialized LCP dielectric material such as Rogers Ultralam 3850 and 3908, and Nippon Steel Espanex L. However, it is worth of mentioning some limitations of LCP material. The material is relatively new and lacks of industrial use database. It is dissolvable in base solution at slightly evaluated temperature and has poor thermal conductivity. The surface roughness is as high as microns in its natural form. LCP requires polishing to reduce its surface roughness [12]. Thorough testing is recommended in applying this material for substrate applications.

11.3.5 PTH and Via Filling Materials

There are multiple purposes of filling PTH and via holes. For high density package applications, filled via enables stacked via and via-in-pad designs, which significantly reduces wasted area for routing wires and sometimes reduces required build-up layer counts. Filled via can eliminate the concern of plating acid residuals, therefore eliminating the concern of acid corrosion to the plated copper in the via hole. Other benefits of filled via include avoiding flux or solder paste blowout during packaging assembly, keeping solder mask ink from migrating into through holes during screen printing, maintaining better printed solder paste volume for via-in-pad design, and enhancing the solder mask planarity. For PTH or through hole vias, filling can prevent vacuum-loss caused mishandling during substrate build up and flux or underfill dripping through the holes during subsequent assembly processes. Filled vias and PTHs are also required for the high power HDI substrate applications.

PTH and via filling materials can be resin based or pure copper. Resin-based materials include B-staged resin coated copper (RCC) laminate, filled nonconductive resins, silver and copper filled resins, photoimageable dielectrics, and solder mask ink. Most of these resins are epoxy, epoxy/BT resin, cyanate ester resin, and epoxy acrylic resins. Via filling with B-staged resin laminates is done by laminating the layer over via holes, where the liquid resin under elevated temperature and pressure flows into via holes and fills them. Although the insulation displacement via formation process is used for some applications, other type of filled resin systems are mainly processed using screen printing since they are in liquid or paste form. Therefore, they need to be easily printable and planarizable. All the filling materials must have good adhesion to plated copper inside vias and PTHs. Since solder mask ink contains solvent, the process stability is impacted by the vaporization of the solvent and film shrinkage. If the via diameter is small, the surface tension may result in solvent residues near the via openings. The adhesion strength of cured solder mask ink to copper is usually not as good as other via filling materials.

Copper filled vias can be made by different ways. One way is called copper pillar technology, similar to insulation displacement via formation technology, but the resin bump is now substituted by a copper pillar. Leveling is typically required after the dielectric layer is applied. Another method is the so-called every layer interconnection (ELIC) by NCAB Group or Any Layer Inner Via Hole (ALIVH), where laser drilled via holes are completely plated by copper followed by copper thickness reduction if needed [13]. Another layer of dielectric material is then applied and the build-up process repeats itself. Copper filled vias have high reliability, high density, and higher heat dissipation. This process is relatively expensive and often used for high performance applications such as smartphone applications.

11.3.6 Solder Mask Materials

The advancement of solder mask materials and process technology has resulted in the utilization of solder mask being used more than just preventing solder bridging and short circuit. Solder mask is also used as a protection layer against copper oxidation of fine traces, chemical attack to dielectric material during surface finish plating, via deterioration under assembly and application.

In addition to all the additives such as adhesion promoters, colorants, catalysts, and fillers, solder mask has two basic chemistries: epoxy and acrylate. When soldermask was originally developed, epoxies were thermally cured and acrylates were UV cured. The thermally cured epoxies require a lot of energy, and curing is typically 30–60 min at 120–150 °C, while UV cure uses very low energy and last only several seconds [9]. Since UV cure is sensitive to surface contamination, the surface cleanliness is very important for UV-cured acrylate solder masks. Since all of these solder mask materials are not photoimageable, they usually require patterned screen printing methods to apply the solder mask material onto the areas that require solder mask. As a result of this, the resolution is low and accurate registration is difficult. Hence, they typically cannot meet feature size requirements for the high density substrate applications.

Photoimageable solder mask development opened the door for finer pitch and features [14, 15]. Photoimageable solder masks can be either pure epoxy resin systems or epoxy-acrylate mixtures. These two photoimageable solder mask materials can be in both liquid form and dry film form. Many application techniques are available for liquid photoimageable solder mask (LPSM) materials. These techniques include screen printing, spray coating, and curtain coating. In the context of dry film solder mask (DFSM), vacuum lamination is the dominant method of application of DFSM onto the substrate or PWB surface. Since DFSM usually does not contain solvent, it nicely conforms to the circuits without trapping voids.

Many factors need to be considered during solder mask selection. Some of these factors include processability, compatibility with surface finish, RoHS compliance, resolution, thickness, ionic cleanliness, and compatibility with later assembly material and process such as soldering fluxes and pastes, de-fluxing agent, and underfill adhesives. The trend of new solder mask material development is toward finer pitch, thinner layers, higher crack resistance, Higher T_g and lower CTE, higher reliability, and direct imaging. The increase of resin purity, use of nano-filler, development of new high T_g and low CTE epoxy resin are some of the approaches in formulating higher performance solder mask materials [15].

11.3.7 Surface Finishes

The surface finish serves as a critical electrically conductive interface between the circuitry of the substrate to chip, electronic component, or system. The surface finish plays an important role in determining interconnect solderability and system quality and reliability. With the lead-free transition, almost every aspect of surface finish technology is under constant evolution, such as surface finish material compatibility, lead-free SMT process compatibility, and surface finish shelf-life.

Detailed discussion of all kinds of surface finishes can be found in many printed circuits handbooks such as the one edited by Clyde Coombs. The properties of same type of surface finish can be somewhat different among different suppliers. The specifications of different surface finishes in Table 11.2 from DYCONEX® are just examples of commonly available finishes in the market. Some vertically integrated companies sometimes have their in-house developed proprietary surface finish compositions.

11.3.8 Summary

In this subchapter, wide varieties of base materials used for today's organic substrates and PWBs were discussed. The discussion has two focuses. One focus is on

Table 11.2 Some of typical surface finish specifications [5]

Process abbreviation	ENIG	E-NI/E-Au	Ni/reductive Au	OSP	Imm. Ag	Imm. Sn	E-Ni/Pd/I-Au—ENEPIG	HASL SnPb	ASIG
Full name	Eletroless nickel immersion gold	Electroplated nickel and gold	Electroless nickel electroless thick gold	Organic solderability preservative	Immersion silver	Immersion tin	Electroless nickel and palladium immersion gold	Hot air solder leveling	Autocatalytic silver immersion gold
Layer thickness (µm)	Ni: 3–5	Ni: 3–8	Ni: 3–6	0.2–0.5	0.2–0.5	0.8–1.2	Ni: 4–7	10	Ag: 0.13–0.25
	Au: 0.05–0.1	Au: 1–2	Au: 0.2–1.0				Pd: 0.05–0.2 Au: 0.03–0.08		Au: 0.02–0.06
Composition	Ni: ~90 % Ni +~7–10 % P	Ni: Pure Ni	Ni: pure NI	Benzimidazoles	Ag	Sn	Ni, Pd, Au	63 % Sn	Ag, Cu
	Au: 99.99 % Au	Au: Au with 0 % or 0.2 % Co	Au: 99.99 %					37 % Pb	
Max. process temp. (°C)	90	90	90	40	50	70	70	250	61

Multiple reflows	4	4	4	4+	5	5	5	5	5
Shelf life (month)	12	12	12	6	12	12	12	12	12
Coplanarity	Very good	Very good	Very good	Very good	Very good	Very good	Very good	Poor	Very good
Minimum space/line (μm)	40/40	20/20	40/40	20/20	40/40	40/40	40/40	100/100	40/40
Cost	High	Very high	Very high	Low	Medium	Medium	Very high	Very high	Medium
US-Wire bonding (Al wire)	Yes	No	No	No	Yes	No	Yes	No	Yes
TS-Wire bonding/TAB (Au wire)	No	Yes	Yes	No	No	No	Yes	No	Yes

the discussion of the definition and implication of base material properties. The purpose of this discussion is to help reader to generate a proper set of material properties for their application-oriented study and assessment of their selected material candidates. The other focus is on the discussion of chemical fundamentals and their effects of many base materials commonly used in substrates and PWBs. The purpose of this discussion is to help reader select proper material candidates from the commercially available material pools to meet particular application needs. The substrate material advancement enables the overall substrate and PWB technology toward smaller, faster, and cheaper feature sets. This trend is for the entire industry level substrate need as a whole, but not for 3D packages only. The need from 3D package somewhat help the acceleration of substrate material advancement. To enable smaller packages, all the materials from cores, to dielectric build-up layers, to solder mask to photoresist must be high resolution definable. In order to make faster signals, low dielectric constant and low loss factor materials together with shorter electrical path such as stacked vias are enablers. The materials must be easy to fabricate and enable easy and high throughput substrate fabrication process for cost considerations. RoHS requirements add another layer of challenge in new material developments. High heat resistance and high T_g materials such as new multifunctional epoxy resins are the materials for these current applications. The materials must be compatible with different types of surface finish plating processes.

11.4 Organic Substrate Fabrication

Organic PWB substrates have gradually evolved to become the platform of electronics assembly for many of the most important industries in the world from medical to aerospace to consumer applications. The fabrication of organic PWB substrates is an industry requiring the skills of many different engineering and science disciplines. As silicon IC fabrication geometries have pushed well into the nanometer feature range, and are part of the "sexier" chip industry, there is often an under-appreciation of the symphony of efforts required to successfully manufacture circuit boards with the fine features required in conventional packaging in order to marry the functionality required of the modern electronic devices that are now required in our daily lives.

The many details of the organic substrate production process cannot be dealt justice in the space allocated in this particular text. In this context, a general high level overview of the traditional organic substrate manufacturing process will be reviewed. The reader is directed at several excellent resources that have become the standards in the PWB industry, in addition to the electronics industry's professional societies for a wealth of knowledge and information on the varying topics and subtopics of organic PWB fabrication, in addition to the newer advancements in materials and process technologies [3, 10, 16–19].

This chapter will highlight the most important information during the substrate manufacturing process for end users and design engineers hoping to obtain a high level overview on the manufacturing process of organic substrates. This general flow is presented in Fig. 11.6.

11.4.1 Substrate Raw Material Selection and Preparation

Fabrication of a printed circuit board begins with selecting the proper materials for the final desired stackup thickness These consists of an inner layer core material and sequentially laminated prepreg materials of the desired thickness. Both the core material and the prepreg materials comprise woven glass/epoxy dielectrics with a thin sheet of Cu foil are laminated to the dielectric material. In the context of the core material, the Cu is laminated to both sides of the fully cured dielectric, whereas the prepreg material only has the Cu laminated to one side of the B-stage epoxy glass dielectric. A schematic of the core material is presented in Fig. 11.7. There exist many different raw material manufacturers and choices as to the types of raw material depending on the type of application (e.g., RoHS, high speed, high reliability, high power/voltage) [10]. The reader is recommended to consult the appropriate resources for guidance on raw material selection, as the number of new materials and applications is changing rapidly.

Quality testing is normally required of these materials as part of the incoming/receiving procedure of many printed circuit board fabricators. Typically, Cu is stripped and the dielectric is examined for embedded particles, which could compromise the electrical performance of the material. Additionally, dielectric breakdown strength testing is typically conducted on a sample of core material from the production lot to understand if there exist raw material defects. In order to have an acceptable bonding strength, the Cu foil will typically have a "tooth" that facilitates the bonding with the base dielectric material. However, this can result in electrically sensitive field intensification that can result in compromising the dielectric strength of the core material. Recently, higher I/O density drives the lower profile copper foil.

The copper thickness of the cores and prepregs is typically designated in terminology such as ½ oz., 1 oz foil. This is the mass of the foil per 1 ft^2 area. Conventionally, ½ oz. Cu foil is equivalent to a thickness of 18 μm and 1 oz Cu foil equal to 36 μm thickness. Typically, the thinner the starting Cu thickness, the finer the lines and spaces features possible on the finished multilayer organic substrate design. The raw materials are typically sized into $16'' \times 20''$, $24'' \times 36''$ or 400 mm × 500 mm sized panels for processing through the manufacturing process. While smaller development runs of five panels are possible, a typically manufacturing lot consists of 20–30 panels processed through the full manufacturing flow.

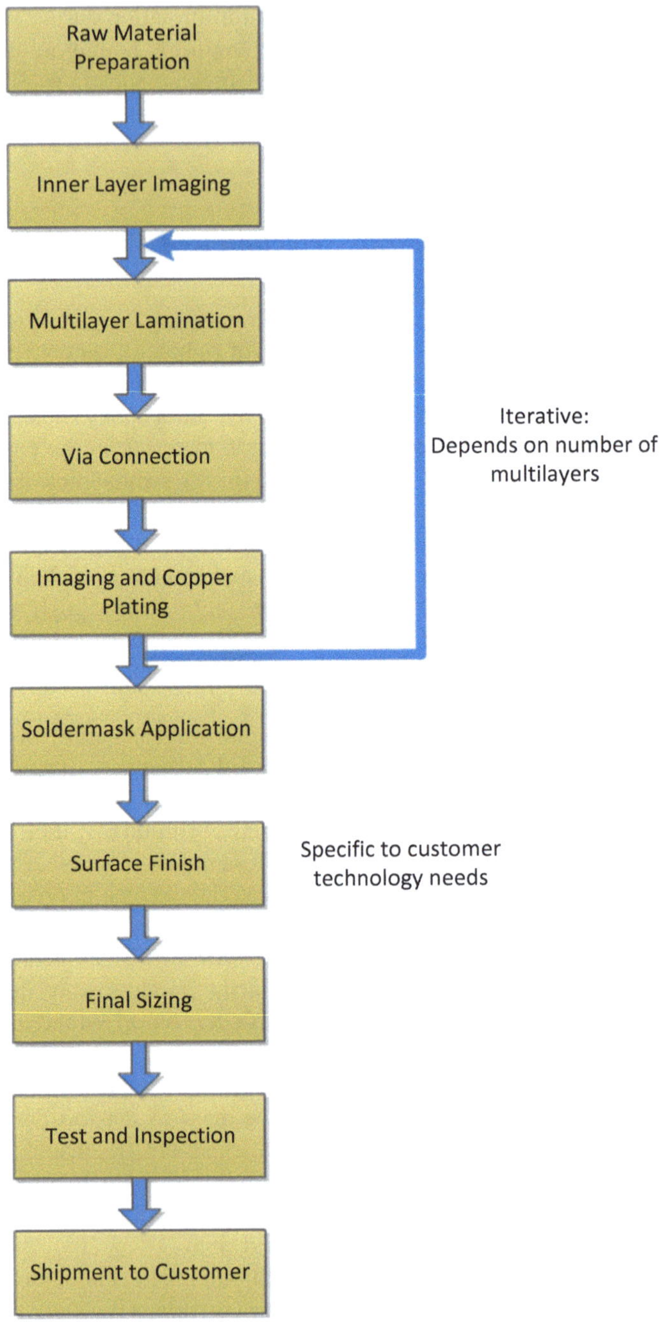

Fig. 11.6 General flow: organic substrate manufacturing

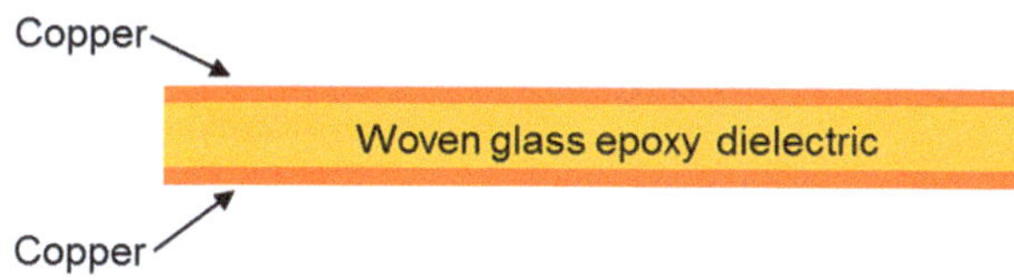

Fig. 11.7 Schematic cross section: organic substrate core material

11.4.2 Inner Layer Imaging

After the raw materials have been selected, the cores are precleaned and moved into a clean room to image the pattern on the core material. Figure 11.8a–e presents a representation of the inner layer core imaging sequence. A photosensitive resist is laminated on the core material. An artwork film which is an image of the line pattern acts as a mask. The inner layer image is exposed by ultraviolet light, cross-linking the exposed resist material. The unexposed resist is rinsed away via a conveyorized horizontal developer. The partially exposed Cu signal layer is etched away. The Cu signal areas covered with the remaining resist are protected from the etchant and remain unetched. The desired signal pattern is generated and finally, the remaining resist is stripped. After the inner layer core fabrication, optical and laser automated optical inspection (AOI) systems scan the etched circuit pattern and compare it with the base artwork. Depending on the customer requirements, certain features may be repaired or reworked at this step. After the completion of the inspection procedure, the remaining Cu is prepared chemically for the multilayer buildup process in the proceeding manufacturing sequence.

11.4.3 Multilayer Buildup

The multilayer organic substrate is built up by adding successive layers of prepregs on each side of the core material via a hydraulic high temperature vacuum press. Prepregs can have the same or different dielectric material as the core material although they only have Cu on one side as the uncured resin bonds to the already imaged core. The press cures and adheres the prepregs to the imaged core. An image of this is presented in Fig. 11.9.

Following the lamination sequence, the connection between the one or more of the signal layers is done via a mechanical or laser drill in the laminate material. If the final laminated structure is not complete, blind vias or micro vias are drilled in the laminated stackup. The holes can be drilled by a combination of high-speed drilling machines, running over 125,000 rpm, laser drills, which drill via laser ablation of the Cu and dielectric material, or a combination thereof. Hole connecting signal layers drilled with lasers are typically termed microvias because of the small nature of the holed, as compared to ones drilled with a conventional drill bit. The drilled holes are then cleaned from the drilling debris and prepared for signal layer imaging. The signal layers are imaged with the same process that was

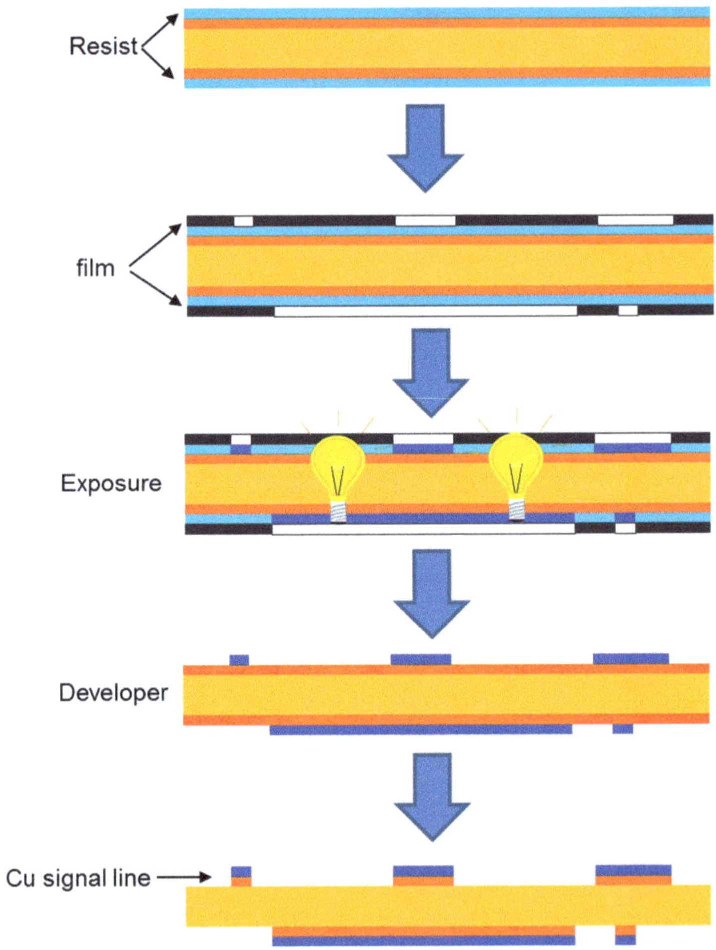

Fig. 11.8 Inner layer core imaging sequence

Fig. 11.9 Schematic of the multilayer lamination process

used during the core fabrication process. The only difference is that the Cu signal layers are electroplated up from the base Cu foil to the desired thickness via a pattern plating or equivalent process.

Depending on the number of layers of the multilayer board, the process repeats until the final outer layers are laminated and etched. This is done by successively laminating pairs of dielectrics with thin Cu foil added on one side to each side of the core material via the hot-press cycle. The primary difference between the successive lamination and signal lamination steps and that of the core imaging is that the exposed dielectric for the drilled vias needs to be catalyzed with a palladium activator solution, and seeded with a thin Cu seed layer that is deposited on to the via walls prior to electroplating the remaining Cu on the wall structure.

The imaged structures are then processed through the AOI and inspection sequence prior to lamination of the next layers in the buildup. This process is repeated for each of the sequential buildup layers, after each lamination step as represented by the manufacturing multilayer loop presented in Fig. 11.6. Once the last multilayer is applied, the final plated through holes and tooling holes are drilled through the completed multilayer stackup. This is done via a high speed mechanical drill process. The plated through holes are plated with the last plating and imaging process with the final outer signal layers. The completed board assembly is ready for the application of a soldermask and the final surface finish.

11.4.4 Soldermask and Surface Finish Application

Soldermask is applied to the outer layers prior to the application of the final surface finish and any printing on the outer layers (if desired). Soldermasks have evolved over the years. Traditional liquid soldermask inks are still widely used throughout the industry. They are screened or roll coated onto the panel sides, tack dried, and photo-imaged, to expose only the required signal layers for the surface finish, and then finally developed, similar to the signal layer imaging process [14]. Newly developed soldermasks consist of dry-film resist sheets which are laminated on the finished multilayer substrate and imaged prior to the final surface finish. The dry-film resists provide the opportunity for more consistent thickness control, thinner overall soldermask thicknesses, and reduced debris and mess compared to the traditionally applied liquid soldermask process.

After the soldermask has been applied, imaged and cured, the surface finish is applied to the exposed pads. This is the part of the process that is dependent on the assembly needs of the organic substrate customer. A schematic of a completed multilayer board with the exposed pads ready for surface finish application is presented in Fig. 11.10. A wide variety of surface finishes are available to accommodate the technology needs of the customer. Table 11.2 lists some of the common surface finishes that are available and used today in the organic substrate industry [20, 21]. Additionally, some of the key comparison characteristics of the surface finishes are listed in the table.

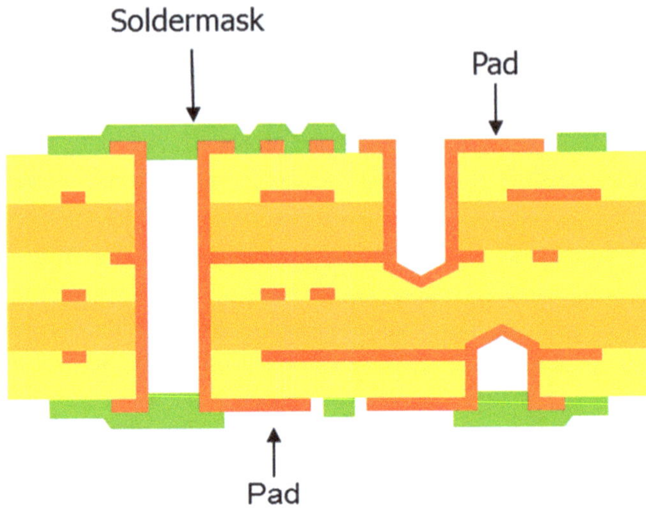

Fig. 11.10 Soldermask application on the completed multilayer assembly with the exposed pads for final surface finish preparation

11.4.5 Final Sizing, Testing, Inspection, and Shipment

Upon the completion of the finished circuit board, the individual circuits are sized out from the larger panels and electrically tested. The electrical testing can range from very simple clamshell fixtures to flying probe test systems. The comprehensiveness of the test requirements can add a significant cost to the overall final substrate cost and should be considered when reviewing with the designers and manufacturers. Upon completion of the electrical testing, the substrates are processed through a final cleaning before a visual inspection process prior to packing and shipping. Visual inspection processes are typically manual in nature and customer requirements can be as little as following simple IPC class I, II, or III inspection requirements to custom-designed quality requirements that can approach 100 pages in length. Upon completion of the inspection procedure, the substrates are packed and then shipped to the customer for assembly into the desired electronic module.

Acknowledgment The editors would like to thank Sangil Lee from Invensas Corporation and Kyu-oh Lee from Intel Corporation for their critical review of this chapter.

References

1. I. Szendiuch, Development in electronic packaging—moving to 3D system configuration. Radioengineering **20**(1), 214–220 (2011)
2. Y. Nakamura, S. Katogi, Technology trends and future history of semiconductor packaging substrate material, Hitachi Chemical Technical Report No. 55, Hitachi Chemical, 2013
3. C.F. Coombs Jr., *Printed Circuits Handbook*, 6th edn. (McGraw-Hill, New York, 2008)

4. M. Jassal, S. Ghosh, Aramid fibres—an overview. Indian J. Fibre Text. Res. **27**, 290–306 (2002)
5. S. Bagen, D. Alcoe, F. D. Egitto, R. N. Das, G. Thomas, Advanced organic substrate technologies to enable extreme electronics miniaturization. Presented on IEEE Components, Packaging and Manufacturing Technology Chapter, Santa Clara Valley, Endicott Interconnect Technologies, New York, 13 Feb 2013
6. *Design Guidelines 2013 (External)* (DYCONEX, MST Company, Bassersdorf, Switzerland, 2013)
7. S. Shi, Study on no-flow underfill materials for low-cost flip-chip applications, Ph.D. Thesis, Georgia Institute of Technology, 28 Mar 2000
8. R.R. Tummala, E.J. Rymaszewski, A.G. Klopfenstein, *Microelectronics Packaging Handbook—Semiconductor Packaging, Part II*, 2nd edn. (Kluwer Academic, Berlin, 1997)
9. M.L. Minges, *Electronic Materials Handbook: Packaging*, vol 1 (ASM International, Novelty, OH, 1989)
10. M. Jawitz, in *Printed Circuit Board Materials Handbook*, ed. by M. Jawitz (McGraw-Hill, New York, 1997)
11. D.C. Thompson, O. Tantot, H. Jallageas, G.E. Ponchak, M.M. Tentzeris, J. Papapolymerou, Characterization of liquid crystal polymer (LCP) material and transmission lines on LCP substrates from 30 to 110 GHz. IEEE Trans. Microwave Theory Tech. **52**(4), 1343–1352 (2004)
12. N. Kingsley, Liquid crystal polymer: enabling next-generation conformal and multilayer electronics. Microwave J. **51**(5), 188–200 (2008)
13. T. Shiraishi, K. Amami, Y. Bessho, K. Sakamaoto, K. Eda, T. Ishida, Flip chip MPU module using high performance printed circuit board "ALIVH". Int. J. Microcircuits Electron. Packag. **21**(2), 205–211 (1998)
14. IPC-HDBK-840, *Solder Mask Handbook.* (Association Connecting Electronics Industries (IPC), Bannockburn, IL, 2006)
15. T. Nagoshi, S. Tanaka, K. Yoshizako, S. Fukuzumi, K. Kurafuchi, Photosensitive solder resist film for semiconductor package "FZ Series", Hitachi Chemical Technical Report No. 54, Hitachi Chemical, 2012
16. R.S. Khandpur, *Printed Circuit Boards, Design Fabrication and Assembly and Testing* (McGraw-Hill, New York, 2006)
17. K. Gileo, *Printed Circuit Board* (ET-Trends LLC, Port Orange, FL, 2014)
18. Recent Japanese Developments in Printed Wiring Boards for SMT, IEEE Electr. Insul. Mag. **7** (2), 9–16 (1991)
19. Advanced Full Additive Process for High Density Printed Wiring Boards, in *Japan IEMT Symposium*, 1989, pp. 141–146
20. D. Walsh, G. Milad, D. Gudeczauskas, Know your final finish options. Printed Circuit Des. Manuf. **23**(2), 38 (2006)
21. The PWB Magazine, February 2015

Chapter 12
Die and Package Level Thermal and Thermal/Moisture Stresses in 3D Packaging: Modeling and Characterization

Liangbiao Chen, Tengfei Jiang, and Xuejun Fan

Nomenclature

α_{Cu}	CTE of copper
α_{si}	CTE of silicon
β	Coefficient of hygroscopic swelling
σ_θ	Circumferential stress
σ_r	Radial stress
σ_{rz}	Shear stress in *r*–*z* plane
σ_z	Normal stress in *z* direction
σ_{T}	Mismatch thermal stress
$\Delta\omega_3$	Frequency shift of the longitudinal Raman mode
ΔH_r	Residual via extrusion
ΔT	Thermal loading
ΔT_{y}	Critical thermal load for plastic yielding
ε_{c}	Hygroscopic strain
ε_{p}	Vapor pressure-induced strain
ε_{T}	Thermal strain
$\overline{\varepsilon}_{\mathrm{T}}$	Average thermal strain
ϕ	Porosity
$\dot{\phi}$	Porosity growth rate
γ_{e}	The rate of elastic extrusion

L. Chen • X. Fan (✉)
Department of Mechanical Engineering, Lamar University, Beaumont, TX 77710, USA
e-mail: liangbiao.chen@lamar.edu; xuejun.fan@lamar.edu

T. Jiang
Department of Material Science and Engineering, University of Central Florida, Orlando, FL 32816, USA
e-mail: Tengfei.Jiang@ucf.edu

Y. Li, D. Goyal (eds.), *3D Microelectronic Packaging*, Springer Series in Advanced Microelectronics 57, DOI 10.1007/978-3-319-44586-1_12

γ_p	The rate of plastic extrusion
η	Generalized Poisson's ratio
φ	Normalized concentration or activity
κ	Curvature
κ_{app}	Applied external curvature
κ_{nat}	Natural bending-induced curvature
ν	Poisson's ratio
π_{11}, π_{12}, π_{44}	Piezoresistance coefficients
ρ_a	Apparent moisture density
ρ_g	Density of saturated water vapor
a	Via radius
A	Geometry factor for calculating thermal stress in multilayered structure
C_{sat}	Saturated moisture concentration
d	Via diameter
D	Moisture diffusivity
D_T	Thermal diffusivity
E	Young's modulus
e	Total volumetric strain
G and λ	Lame's elastic constants
H	Via depth
h_b	Bending axis
h_i	Top surface coordinate
h_m	Midpoint coordinate
k_P	Factor for hydrostatic pressure gradient-driven diffusion
k_T	Factor for thermal gradient-driven diffusion
P	Hydrostatic pressure
p	Partial water vapor pressure
p_{amb}	Ambient partial water vapor pressure
p_g	Saturated water vapor pressure
Q	Activation energy for moisture diffusivity
S	Solubility
T_m	Maximum temperature during thermal cycling
T_R	Room temperature
T_{ref}	Stress-free temperature
u_i	Displacement vector
w	Wetness
W	Warpage
X_i	Body force vector

12.1 Introduction

3D packaging with through-silicon vias (TSVs) is one of the key technologies implemented in semiconductor manufacturing in order to increase bandwidth, enhance performance, shrink form factor, and reduce power [1–3]. Due to the introduction of new features like TSV for providing short vertical interconnections, there exist unique reliability requirements for 3D technology [4, 5]. Compared to conventional 2D packages, particular attention must be given to the new failure mechanisms of TSV components and 3D chip stacks under various stress conditions [5–8].

Thermal stress in 3D package is generated due to the mismatch in coefficients of thermal expansion (CTEs) between dissimilar materials when there is temperature change. At die level, high thermal stress in TSV chips may cause many reliability issues, including extrusion of TSV, cracking of silicon chip, and unintentional carrier mobility change around TSVs. The issues in TSV chips will eventually affect the reliability of the whole 3D system. At package level, reliability issues caused by thermal stresses may include fatigue fracture of solder ball joints and other interconnects, the warping of dies, substrates and package, and the delamination and cracks initiated at the interface between dissimilar materials [8–12]. The warping or nonuniformity at package level can be especially critical because larger single die or multiple dies are often used in 3D structure. The overall warpage of a chip-substrate structure will affect the bonding process at board level [13].

In addition to thermal stress, particular attention is also given to moisture-induced stress. A typical 3D package will contain moisture-sensitive materials, i.e., polymeric materials at different levels, which can absorb a certain amount of moisture from the ambient. Excessive moisture absorption may increase the corrosion rate of the metal parts and aggravate the electrochemical migration [14, 15]. - Moisture-induced stress, such as hygroscopic stress, may also cause leakage to bulk chips, metal delamination, cracking and extrusion in TVS chains [5], and the hygroscopic strain could be comparable or even higher than thermal strain [16–18]. More significantly, during reflow soldering or the highly accelerated temperature and humidity stress test (HAST), devastating material ruptures or interface delamination may take place [19–27].

This chapter aims to provide fundamental theories, modeling and characterization methods for analyzing the effects of thermal and moisture-induced stresses in 3D packaging. First, the analysis and measurement of thermal stress in and around the TSVs are presented in Sect. 12.2. The results reveal the near-surface characteristics of thermal stress and the important role of Cu microstructure in controlling stress. Stress-induced reliability issues, including the keep-out zone and via extrusion, are discussed. In Sect. 12.3, thermal stress-induced warpage at package level is studied. Analytical modeling of an idealized multilayered structure is described first. Then, several warpage control methods with experimental testing and numerical modeling are introduced. A new warpage control method is presented to achieve warpage-free flip-chip package. In Sect. 12.4, moisture-induced stress is calculated and coupled with thermal stress to form an integrated stress modeling. A theoretical framework based on damage micromechanics and effective stress concept is presented to achieve stress integration, which is able to combine multiple field variables such as temperature, moisture concentration, vapor pressure, and void volume fraction. An example of integrated stress analysis is then performed based on the theory. Section 12.5 summarizes the findings of thermo-mechanical and moisture-induced stresses in 3D packaging.

12.2 Thermal Stress and Its Effects on TSV Structures

12.2.1 Introduction

Due to the large thermal expansion mismatch between Cu TSVs and Si ($\alpha_{Cu} = 17$ ppm/°C, $\alpha_{Si} = 2.3\,\text{ppm/°C}$), significant thermal stresses can be induced in and around TSVs to cause serious reliability issues [28–31]. To successfully implement TSVs in 3D integration, it is essential to analyze and understand the characteristics of thermal stresses at die level, as well as their impact on TSV structures.

The confinement by the surrounding Si results in complex stress state in and around the TSV. The microstructure of Cu also plays an important role in controlling the stress of the TSV structure. To capture these unique characteristics, this section will first present stress analysis by both a semi-analytic method and numerical analysis, followed by stress measurement using two complementary techniques. Finally, two reliability issues induced by stress, the device keep-out zone (KOZ) and via extrusion will be discussed.

12.2.2 Characteristics of TSV Stress by Semi-analytic and Numerical Solutions

2D plane-strain solution to the classical Lamé problem has been employed to analyze the thermal stresses of TSVs, but the 2D solution inherently could not capture the 3D nature of the stress field near the wafer surface [32]. To overcome this limitation, a semi-analytic solution is derived for an isolated TSV embedded in silicon using a linear superposition method [33]. The stress contours calculated from the semi-analytic solution are plotted in Fig. 12.1. At the wafer surface ($z = 0$), the normal stress σ_z is zero as required by the traction-free boundary condition and is nonuniform away from the surface (Fig. 12.1a). At the wafer surface, the shear stress (σ_{rz}) has a singular concentration at the via/Si interface (Fig. 12.1b), and the radial stress (σ_r) and circumferential stress (σ_θ) are non-zero (Fig. 12.1c, d). Both the radial and shear stresses contribute to the driving force for interfacial delamination. Because the radial stress (σ_r) is tensile during cooling and compression

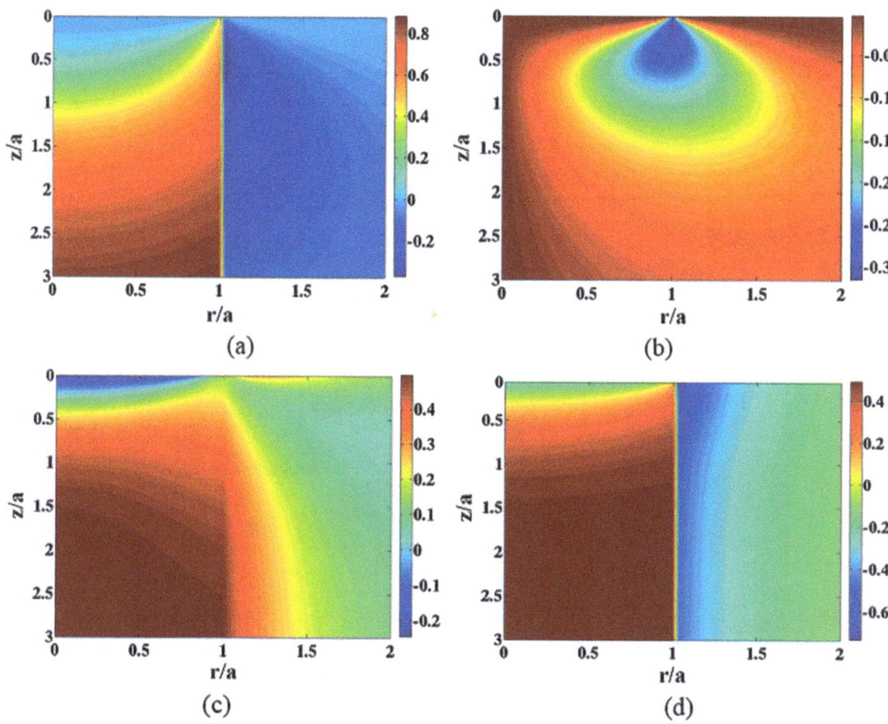

Fig. 12.1 Near-surface stress distributions predicted by the semi-analytical solution for a thermal load of ΔT. The stress magnitudes are normalized by $\sigma_T = -E\varepsilon_T/(1-\nu)$, and the radial and depth coordinates (r and z) are normalized by the via radius $a = d/2$: (**a**) out-of-plane stress (σ_z); (**b**) shear stress (σ_{rz}); (**c**) radial stress (σ_r); (**d**) circumferential stress (σ_θ) [33]

during heating, the driving force for interfacial delamination under cooling is about twice of that under heating. Based on the semi-analytic solution, the steady-state energy release rate of interfacial delamination is found to be linearly proportional to the via diameter, but to the square of the thermal mismatch strain, $\varepsilon_T = \Delta\alpha\Delta T$. This result suggests that decreasing the TSV diameter and the thermal load, or selecting via materials with smaller thermal expansion mismatch with Si are potential solutions to reduce the delamination driving force.

Finite element analysis (FEA) has been used to verify the semi-analytic solution and to further examine the thermal stress in the TSV. Three-dimensional (3D) FEA models are constructed for blind via structures. The via dimension is 10 μm × 55 μm (diameter × depth), and the Si thickness is 780 μm. There is an oxide barrier layer at the TSV/Si interface of 0.4 μm thick, and the thin Ta barrier layer is neglected in the model. The material properties used in the model are listed in Table 12.1.

For a thermal load of $\Delta T = 200$ °C, the von-Mises stress is plotted in Fig. 12.2 [34]. The von-Mises stress is much lower than the yield strength of Cu in most of the via, except for a small region near the via/Si interface at the top of the via. Since von-Mises stress is the effective shear stress driving plastic deformation, the FEA

Table 12.1 Thermo-mechanical properties used in finite element analysis

Material	CTE (ppm/°C)	Young's modulus (GPa)	Poisson's ratio
Cu	17	110	0.35
Si	2.3	130	0.28
Oxide	0.55	72	0.16

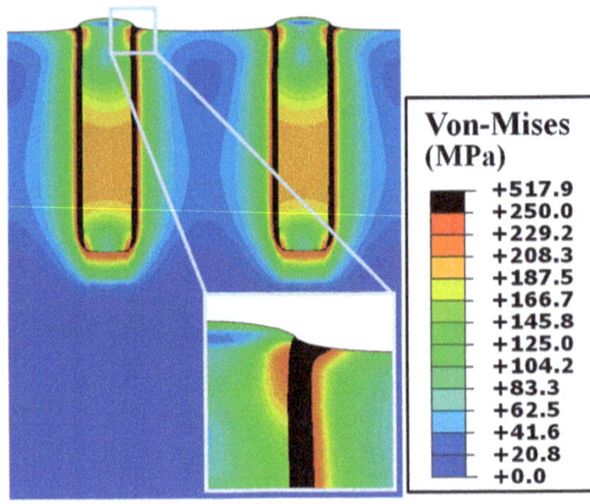

Fig. 12.2 Von-Mises stress for a thermal load $\Delta T = 200$ °C. The via dimension is 10 μm × 55 μm (diameter × depth), and Si thickness is 780 μm thick [34]

result suggests that the Cu via will be mostly elastic under temperature excursion, and plasticity will be localized near the junction between the via/Si interface and the wafer surface. The localized plasticity in TSVs can be attributed to the triaxial stress state in the TSV structure, which is distinctively different from the biaxial stress state of Cu thin film. As a result of the triaxial stress state, the effective shear stress causing plastic deformation is relatively low in TSV structures. This has important implications on TSV reliability which will be discussed in the rest of this section.

12.2.3 Measurement of Thermal Stress

Several experimental techniques have been developed to measure the stress behavior of TSV structures, including wafer curvature and micro-Raman spectroscopy [34–38]. While individual technique has its limitations, when combined and by correlating to the microstructure evolution and FEA, the results can provide basic understanding of the stress characteristics of the TSV structures.

The wafer curvature method utilizes an optical lever setup to measure the bending curvature of the sample during thermal cycling. For a blind TSV structure illustrated in Fig. 12.3, this method is used to obtain the overall thermo-mechanical behavior of the TSV structure [35]. For TSV specimens going through single

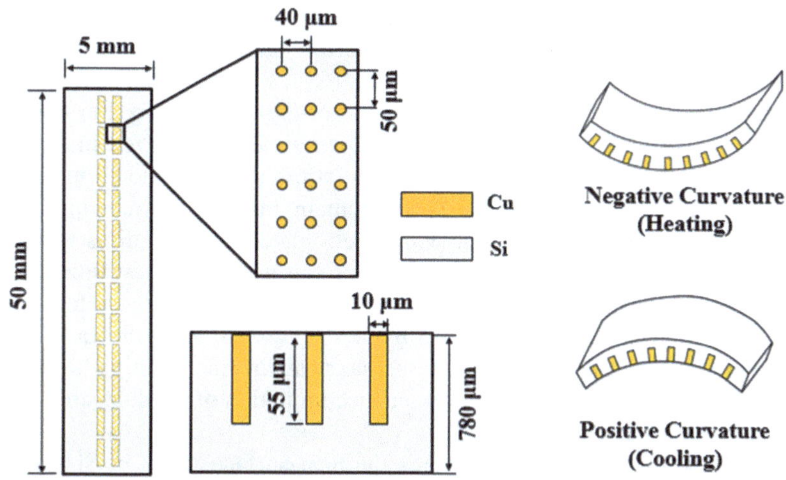

Fig. 12.3 Illustration of the TSV specimen for wafer curvature test. The specimen size is 5 mm × 50 mm × 780 μm (width × length × thickness) and contains periodic arrays of vias along the centerline of the specimen. The dimension of the TSV is 10 μm × 55 μm (diameter × depth), and the pitch distances are 50 μm and 40 μm in the longitudinal and transverse directions, respectively. The top surface of Si is covered with an oxide layer of 0.8 μm thick which was removed by mechanical polishing before measurement. The thickness of oxide barrier layer at the via/Si interface is about 0.4 μm. Adapted from [35]

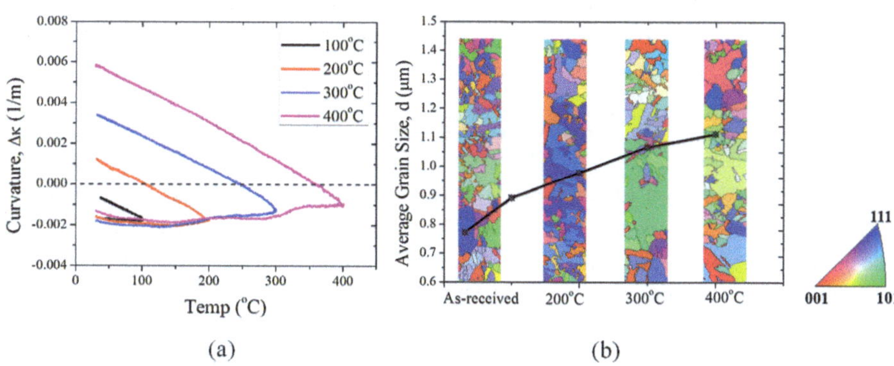

Fig. 12.4 (**a**) Curvature measurement of four TSV specimens subjected to single thermal cycling to 100 °C, 200 °C, 300 °C, and 400 °C, respectively. (**b**) Grain growth in TSVs as shown by the crystal orientation maps and grain sizes (*black line*) obtained by electron backscatter diffraction (EBSD) after the TSVs have been thermally cycled to different temperatures (adapted from [41])

thermal cycles to peak temperatures of 100 °C, 200 °C, 300 °C, and 400 °C, the curvature change, Δk, as a function of temperature, T, are plotted in Fig. 12.4a.

Some distinct features of the curvature-temperature behavior of TSV structures as seen in Fig. 12.4 are: stress relaxation during heating; linear curvature-temperature relation during cooling; and the absence of hysteresis loops. The stress relaxation observed in the heating portion of each test is correlated to Cu grain growth in the via, which is measured by electron backscatter diffraction (EBSD) (Fig. 12.4b). A decrease of additive elements (Cl^-, F^-, S^-, and CN^-) is found to accompany grain growth as obtained by time-of-flight secondary ion mass spectroscopy (TOF-SIMS) [39]. The linear cooling curvature observed in the experiment can be directly correlated to the triaxial stress state (as shown in Fig. 12.2) obtained by FEA, as the Cu via would remain mostly elastic with localized plasticity. This is in sharp contrast with electroplated Cu thin films which have biaxial stress state and show hysteresis loops during thermal cycling [40]. The temperature at which the cooling curvature crosses zero gives the reference temperature at which the average stress in the TSVs is zero. With increased thermal cycling temperature, the amount of stress relaxation increases during heating. This leads to higher stress-free temperatures and accumulation of residual stress at room temperature after each thermal cycle.

Micro-Raman spectroscopy measures the near-surface stress in Si around the TSV and complements the measurement by wafer curvature method. TSV specimen as shown in Fig. 12.3 is used for the micro-Raman measurement [35]. Two specimens, referred to as A and B, are subjected to different thermal treatment and their stress-free temperatures are determined to be $T_{\text{ref}}^{\text{A}} = 100\,°\text{C}$ and $T_{\text{ref}}^{\text{B}} = 300\,°\text{C}$, from wafer curvature measurement. The micro-Raman measurement is performed using [001] backscattering configuration, and the signals are collected by scanning the laser across two neighboring vias along the [110] direction in each sample.

In the test configuration, longitudinal Raman frequency shift, $\Delta\omega_3$, is measured and can be correlated to the sum of the in-plane normal stresses, $\sigma_r + \sigma_\theta$, through Eq. (12.1) [42, 43]:

$$\sigma_r + \sigma_\theta(\text{MPa}) = -470\Delta\omega_3\left(\text{cm}^{-1}\right) \tag{12.1}$$

As the measurement obtains only the sum of the two individual stress components in Si, FEA is used to delineate the stress components. The FEA model considers the anisotropic elastic properties of Si while treating Cu as linear elastic as deduced from the linear cooling curve in the wafer curvature measurement. The material properties in Table 12.1 are used in FEA. The thermal load is $\Delta T_{\text{A}} = -70$ °C and $\Delta T_{\text{B}} = -270$ °C, as obtained from the reference temperatures. Since the Raman signal penetrates up to 0.2 μm from the wafer surface, the stress components are extracted from 0.2 μm below the wafer surface in the model [35]. The stress sums from both experimental measurement and FEA are plotted in Fig. 12.5 and show reasonable agreement. Away from the Cu/Si interface, the sum of the stresses first increases sharply, and then gradually decreases. Between the two adjacent vias, the stress depends on the pitch distance as a result of stress interaction. Because of the larger negative thermal load for specimen B ($|\Delta T_{\text{B}}| > |\Delta T_{\text{A}}|$), specimen B (Fig. 12.5b) shows a higher stress level in Si than specimen A (Fig. 12.5a). This establishes the correlation between the Raman measurement and wafer curvature measurement, which together show the dependence of residual stress on the thermal history of TSVs.

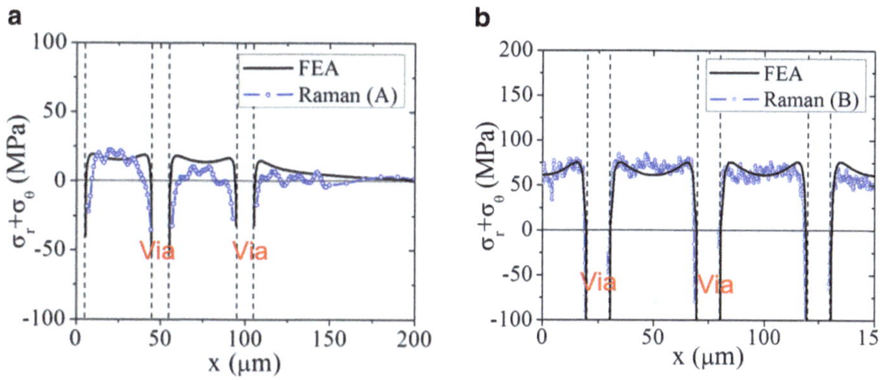

Fig. 12.5 Comparison of the near-surface stress distribution between Raman measurements and FEA: (**a**) specimen A ($\Delta T_A = -70$ °C), (**b**) specimen B ($\Delta T_B = -270$ °C)

Table 12.2 Piezoresistance coefficients for n- and p-Type Si (in unit of 10^{-11} Pa^{-1})

	π_{11}	π_{12}	π_{44}
n-type Si	−102.2	53.7	−13.6
p-type Si	6.6	−1.1	138.1

12.2.4 Effect of Thermal Stress on Carrier Mobility and Keep-Out Zone

The thermal stress in Si can degrade the performance of the MOSFET devices through the piezoresistivity effect [44]. This results in a keep-out zone (KOZ) around the TSV, where the placement of active devices should be avoided [45–47]. For [001] Si with current flows in the [100] direction, the piezoresistance coefficients for n- and p-Type Si are listed in Table 12.2 [44].

With the piezoresistance coefficients, the carrier mobility change around the TSVs can be calculated from the thermal stress obtained in FEA and shown in Fig. 12.6 for an isolated via embedded in Si [47]. The via dimension is 10 μm × 200 μm (diameter × height), the Si thickness is 780 μm, and the thermal load is $\Delta T = -250$ °C. The dashed lines in the figures mark the boundaries of 5 % change in mobility, which defines the KOZ. Since both the elastic properties and the piezoresistance of Si are anisotropic, the mobility change is highly directional and is very different for n- and p-type Si. For transistors with [100] channel, a sizable KOZ is developed in n-type silicon (Fig. 12.6a), but no KOZ exists in p-type Si. In contrast, for [110] channel, considerable KOZ exists for p-type silicon (Fig. 12.6b) while no KOZ exist in n-type Si. Further analyses by FEA show that the KOZ increases monotonically with TSV diameter [47]. The plasticity of Cu can relax stress close to the via/Si interface to reduce the size of KOZ, but this effect is rather localized. When the TSVs are placed closer together, stress interaction between neighboring vias can increase the overall stresses in Si and thus the size of KOZ, but stress interaction is negligible when the ratio of via pitch to diameter is larger than 5.

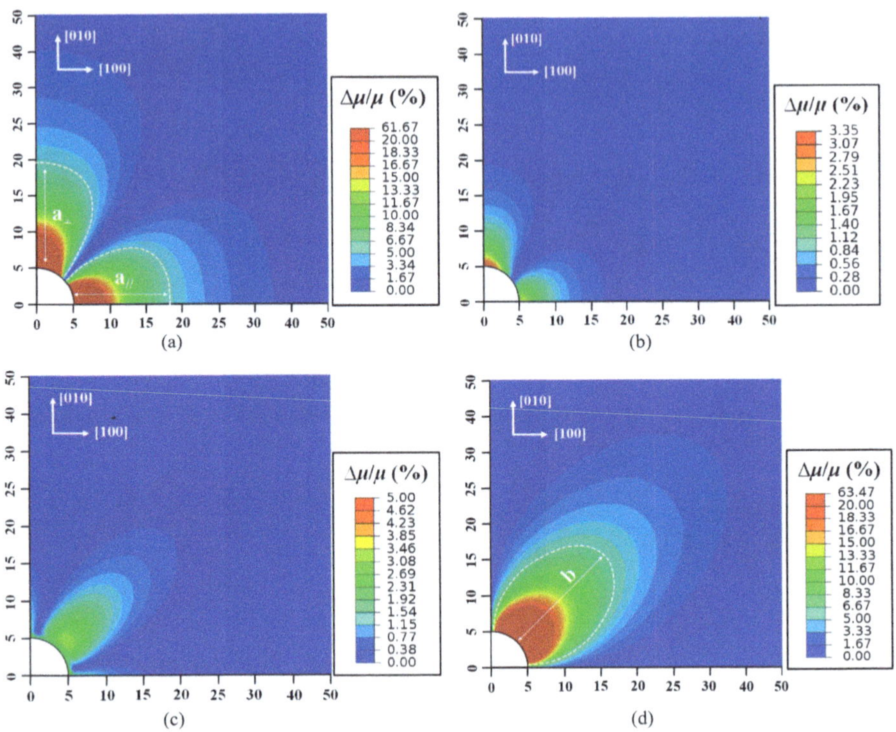

Fig. 12.6 Distribution of mobility changes in (**a**) n-type MOSFETS with the electric field and current density in [100] direction and (**b**) mobility change for p-type MOSFET with [100] channel. (**c**) Mobility change for n-type MOSFET with current density in [110] direction. (**d**) p-type MOSFETS with the electric field and current density in [110] direction. The *dashed lines* indicate the 5 % mobility change ($d = 10$ μm, $H = 200$ μm, and $\Delta T = -250$ °C) [40]

12.2.5 Thermal Stress Induced via Extrusion

Via extrusion is the non-recoverable deformation of Cu near the top of the via after thermal cycles. Grain growth, plastic yielding by dislocation glide, and diffusional creep along grain boundaries and/or the via/liner interface are possible relaxation processes responsible for via extrusion. A high-temperature post-plating annealing has been adopted in TSV fabrication to stabilize the grain structure to reduce via extrusion during subsequent thermal processing [28, 48, 49]. In general, grain growth could lower the yield strength of Cu due to the Hall-Petch effect, which in turn would cause more plastic deformation and Cu extrusion [50].

The local plasticity in Cu is observed experimentally by synchrotron x-ray micro-diffraction and correlated to the magnitude of via extrusion (Fig. 12.7) [51]. To elucidate the effects of Cu plasticity on via extrusion, a simple analytic model is formulated as follows. When the TSV is subjected to a thermal cycle from

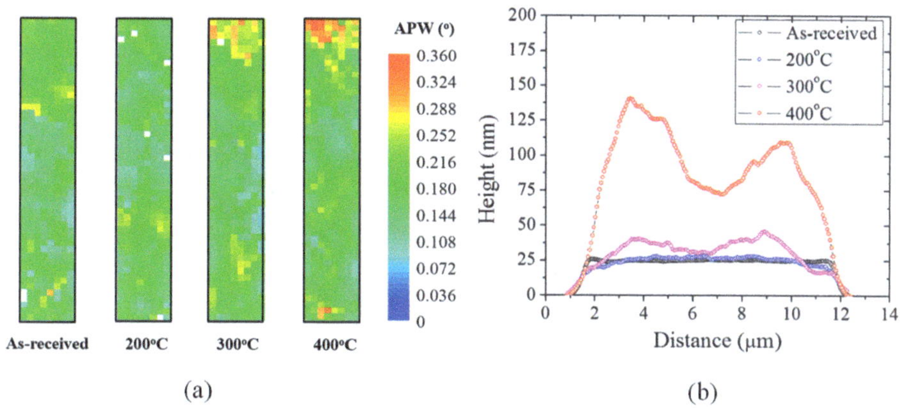

Fig. 12.7 (**a**) Average peak width (APW) of Cu for the as-received via and vias after thermal cycling to 200 °C, 300 °C, and 400 °C. Increased APW indicates increased plasticity in the Cu vias. (**b**) Via extrusion profiles obtained from AFM measurements [51]

room temperature (T_R) to an elevated temperature T, the mismatch of thermal expansion between the Cu via and Si induces a compressive stress in Cu upon heating ($\Delta T = T - T_R$). Assuming perfect plasticity with a yield strength σ_y for the Cu via, plastic yielding of Cu is predicted when heating above a critical temperature, $\Delta T_y = \frac{\sigma_y}{\alpha_{Cu}-\alpha_{Si}}\left(\frac{1-\nu_{Cu}}{E_{Cu}} + \frac{1+\nu_{Si}}{E_{Si}}\right)$, which is proportional to the yield strength of Cu. Beyond the critical temperature $\left(\Delta T > \Delta T_y\right)$, the Cu via deforms plastically, leading to more via extrusion at higher temperatures. The plastic extrusion does not vanish after cooling, resulting in a non-zero residual extrusion after a full thermal cycle: $\Delta H_r = H\left(\gamma_p - \gamma_e\right)\left(\Delta T_m - \Delta T_y\right)$, where H is the via height, $\gamma_e = 20.64\,\text{ppm}/°\text{C}$ and $\gamma_p = 46.4\,\text{ppm}/°\text{C}$ are the rate of extrusion as a function of temperature calculated from the thermo-mechanical properties in Table 12.1, and $\Delta T_m = T_m - T_R$ where T_m is the maximum temperature during thermal cycling [51]. This shows that the magnitude of the residual extrusion depends on the peak temperature during thermal cycling and the plastic yield strength of the Cu via. Increasing the yield strength of Cu would increase the yield temperature ΔT_y and thus decrease the residual extrusion for the same thermal load [50].

The effect of interfacial properties on via extrusion has been studied by numerical simulations using FEA models (Fig. 12.8) [50–52]. FEA shows that a perfectly bonded interface between the Cu via and Si reduces the residual extrusion at room temperature by about three times. Using a cohesive model for the via/Si interface, the via extrusion is found to be bound by two limits: the frictionless sliding in the analytical model as the upper bound and the perfectly bonded interface (no sliding) as the lower bound. This analysis implies that via extrusion can be suppressed by improving interfacial adhesion to resist sliding.

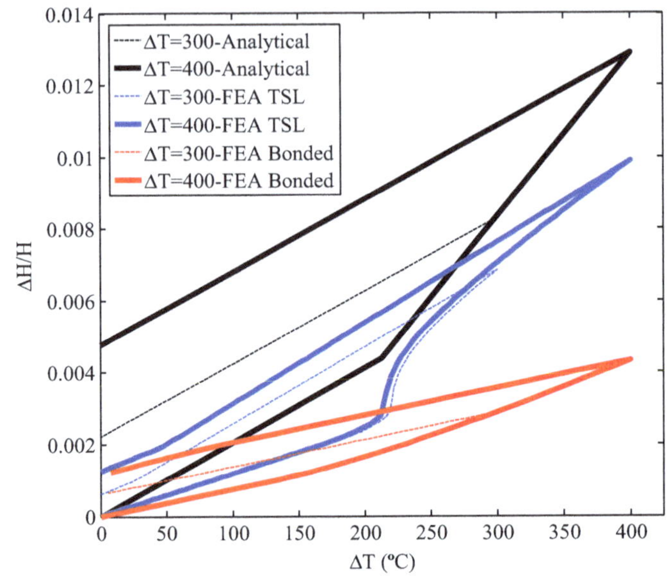

Fig. 12.8 Comparison of extrusion ratio, $\Delta H/H$, between analytical and FEA results assuming interfaces with perfect bonding and bilinear traction separation relationship. Assume zero initial residual stress [51]

For a large ensemble of vias, via extrusion is found to follow a lognormal distribution [53, 54]. For two via sizes of 5 × 50 μm and 10 × 100 μm with different post-plating annealing temperatures, the average via extrusion at 50 % is about half for the smaller TSVs (Fig. 12.9a), but the largest extrusions at 99.9 % of the lognormal distribution is about the same for both sets of TSVs, independent of via size and anneal conditions (Fig. 12.9b). These results highlight the statistical nature of via extrusion, and indicated that the TSV reliability will not be improved by simply reducing via diameter, since 3DICs contain a large number of TSVs. Rather, the reliability is determined by the largest extrusions, i.e., the weakest link, in spite of their small percentage (about 0.1 %). The statistical spread in via extrusion can be correlated to the grain size and orientation near the top of the vias and smaller extrusion is observed for vias with a larger portion of twin boundaries [54]. It is therefore important to optimize the electroplating chemistry and annealing process to achieve grain structures with a large portion of twin boundaries and large yield strength in order to minimize via extrusion. However, it is not clear how such grain structures could be so prevalent to completely eliminate the tail distribution of the largest extrusions.

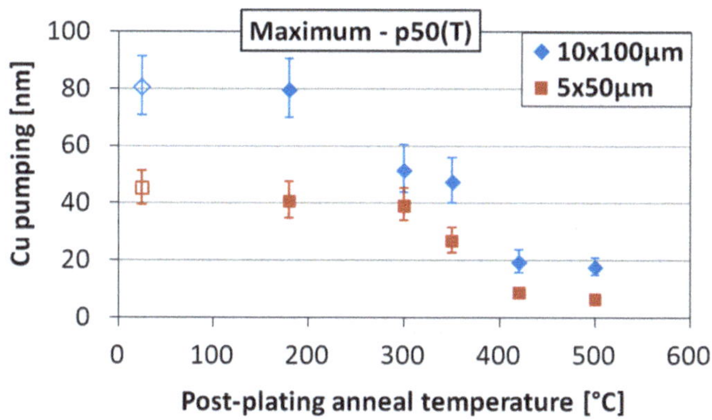

(a)

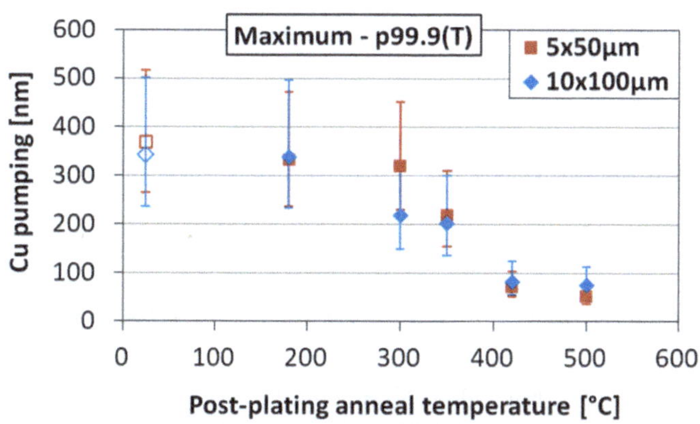

(b)

Fig. 12.9 (**a**) Maximum average (50 %) residual extrusion and (**b**) maximum tail distribution (99.9 %) of residual extrusion for two sets of TSVs of 5 × 50 μm and 10 × 100 μm dimensions as a function of post-plating anneal temperature [53]

12.3 Thermal Stresses and Warpage Control at Package Level

12.3.1 Introduction

Warpage by thermal stress is one of the most common but critical issues for electronic packaging [55–64], which could pose a great challenge in the manufacture, application, and reliability testing of 3D packages, especially for the case when large, thin chip stacks are connected to organic laminates or substrates [8–10, 65]. Excessive warpage will give rise to difficulty in chip interconnecting and bonding to the substrate, usually leaving opens. The warping in chips can also cause compressive forces on molten solder balls, resulting in lateral expansion and unintentional bridging. Operations at board level are also affected by excessive package-level warpage. Examples include solder ball bridging or opening during surface mount process and package failures during package functional test [60, 61].

To reduce thermal stress-induced warpage, traditional control methods, such as adding a stiffener ring or a lid, may not be efficient. These methods are typically substrate-control methods, which could result in additional stress into the package. Therefore, it is necessary to develop more effective control methods based on deep understanding of warpage control mechanism [66].

To address the thermal stress-induced reliability issues (e.g., excessive warpage and solder joint failures) at the package level and provide potential solutions, this section will first introduce analytical theories for the calculation of thermal stress and warpage for multilayered structures under temperature loadings. Then, the warpage mechanism during the packaging process is demonstrated, followed by a discussion on the effectiveness of traditional warpage control methods in Sect. 12.3.3. In Sect. 12.3.4, a new control approach is presented for achieving potential warpage-free package. Numerical modeling, as well as experimental testing, is performed to verify the new method. Different cases are studied and the results are discussed in details.

12.3.2 Thermal Stresses in a Multilayered Structure

As 3D chip stacks and other microelectronic devices can be conceived as multi-layered structures, analytical modeling based on the existing theories may be adopted. The main advantage of analytical modeling is that one can perform a quick calculation of thermal stress for any multilayered systems with regular geometry, especially for those including very thin layers.

There are several analytical methods available for calculating the thermal stress in laminated structures or multilayered composite used in electronic packaging. One theory introduced here is called the natural bending theory [67], which is based on the assumptions of linear elasticity and small deformation. The basic idea of the natural bending theory is illustrated in Fig. 12.10, where a two-layer system is used for simplicity. Due to different CTEs of layer 1 and layer 2 ($\alpha_2 < \alpha_1$), thermal mismatch is produced in the cooling process from high temperature (T_H) to room temperature (T_R). The theory first considers a stress-free case, so that each material will contract accordingly to their CTEs, resulting in mismatches in stress-free thermal strain $\left(\varepsilon_{T1}^0 = \alpha_1 \Delta T > \varepsilon_{T2}^0 = \alpha_2 \Delta T\right)$. To accommodate the mismatches, layer 1 will be constrained and then bonded to layer 2. After the constraint is released and the stack will be bent. The process of natural bending theory can be used to explain how warpage is developed by thermal stress in dissimilar materials. Note that uniform temperature is considered in the theory.

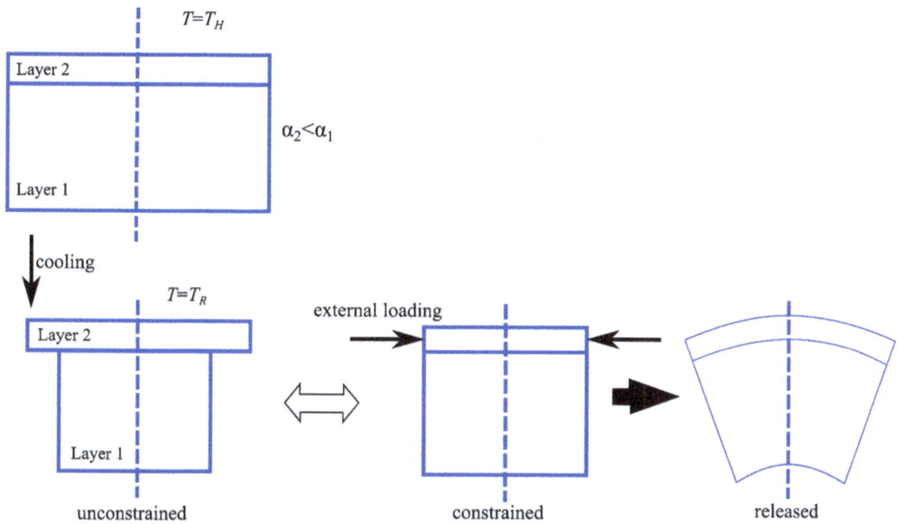

Fig. 12.10 Illustration of natural bending theory for warpage in a multilayered device [67]

Based on the natural bending theory, the warpage of a multilayered device can result from both thermal loading and external applied curvature. The normal stress in x direction in each layer i of a composite can be expressed as a function of thermal strain and applied curvature, as [67]

$$\sigma_{xi} = \overline{E}_i\left[\left(\kappa_{\text{nat}} + \kappa_{\text{app}}\right)(y - h_{\text{b}}) + \overline{\varepsilon}_{\text{T}} + \nu_i A - \eta_i \varepsilon_i^0\right] \quad (12.2)$$

where $\bar{E}_i$ is elastic modulus based on geometry configuration; k_{nat} the curvature caused by natural bending; k_{app} is applied external curvature; y is the position in thickness direction; h_{b} is the bending axis; $\overline{\varepsilon}_{\text{T}}$ is average thermal strain; ν_i is Poisson's ratio; A is a geometry factor; η_i is generalized Poisson's ratio; and ε_i^0 is the stress-free thermal strain. The expressions for h_{b}, $\overline{\varepsilon}_{\text{T}}$, and k_{nat} are given in Eqs. (12.3), (12.4), and (12.5), respectively:

$$h_{\text{b}} = \sum \overline{E}_i H_i h_{mi} \Big/ \sum \overline{E}_i h_i \quad (12.3)$$

$$\overline{\varepsilon}_{\text{T}} = \sum \overline{E}_i H_i \varepsilon_i^0 \Big/ \sum \overline{E}_i H_i \quad (12.4)$$

$$\kappa_{\text{nat}} = \frac{\sum \overline{E}_i H_i (h_{mi} - h_b)\left(\eta_i \varepsilon_i^0 - \overline{\varepsilon}_{\text{T}} - \nu_i A\right)}{\sum \overline{E}_i D_i \left[h_i^2 + H_i^2/3 - h_i H_i + h_{\text{b}}(h_{\text{b}} - 2h_{mi})\right]} \quad (12.5)$$

where h_{mi} and h_i are the coordinates of the midpoint and the top surface of each layer, respectively; and H_i is the thickness of each layer. The equations from natural bending theory can be used for different geometry configurations, such as uniaxial plain stress, uniaxial plain strain, and biaxial model, by choosing corresponding $\bar{E}_i$, geometry factor A, and η_i (see p. 48 in [67]). The theory provides a convenient tool for a rapid and easy modeling and monitoring the warpage of multilayer structures due to thermal stresses and applied external bending or curvature [67].

Another analytical method for calculating thermal stress and warpage in multi-layered structures is the beam-type plate theory proposed by Wen and Basaran [68], which was mainly developed to consider the stress behavior along the interfaces between layers due to thermal loading. The analytical model follows Timoshenko's bi-thermostat beam theory [69] (which was improved by Suhir [70] later), as well as the model proposed by Valisetty and Rehfield [71]. The model treats each layer of an N-layer laminated structure as a beam-type plate with orthotropic material properties. The thermal stress for isothermal loading can be obtained by solving the differential equations for the classical plate theory with beam-like behavior assumptions, and the displacements of each layer can be solved by the introduction of thermal strain terms into Valisetty model [71]. Despite the derived formulation being a little complicated, the model could yield very accurate results in a rapid manner.

12.3.3 Warpage Mechanism and Control Methods

In reality, an IC package structure is much more complicated than a simplified multilayered structure. In this section, a flip-chip package is adopted to demonstrate the development of warpage during different steps of packaging assembly. The structure, primarily comprising a chip and a substrate, is shown in Fig. 12.11, which is considered as one of the fundamental forms that can be used to study the warpage mechanism.

As shown in Fig. 12.11, the chip having electrically conductive bumps on its active surface is flipped and attached to the top surface of the substrate. An underfill material is usually dispensed into the gap between the chip and the substrate through a capillary force to protect solder bumps. The degree of warpage will vary during the assembly process of the package. Figure 12.11 shows the basic steps for the assembly process, including: step (a) for die attach process; step (b) for underfill dispensing and curing process; and step (c) or (d) for stiffener/lid attachment process for warpage control. After the die attach step (a), the electrical connection between the die and substrate has been completed, and the warpage at this point is small due to the quick relaxation of the solder bump stress that is caused by the visco-plastic property of the solder material. At this stage, the mechanical connection between the die and substrate is still weak. In order to enhance the mechanical connection and to protect the solder bumps, the underfill is filled into the gap between the die and substrate. The underfill is usually cured at a high temperature, such as 165 °C. During underfill curing, the flip-chip package is still very flat, i.e., the warpage at this point is very small due to the same reason as the visco-plastic property of the solder material. However, after the underfill is cured, the die and substrate will be strongly connected, and large warpage can be

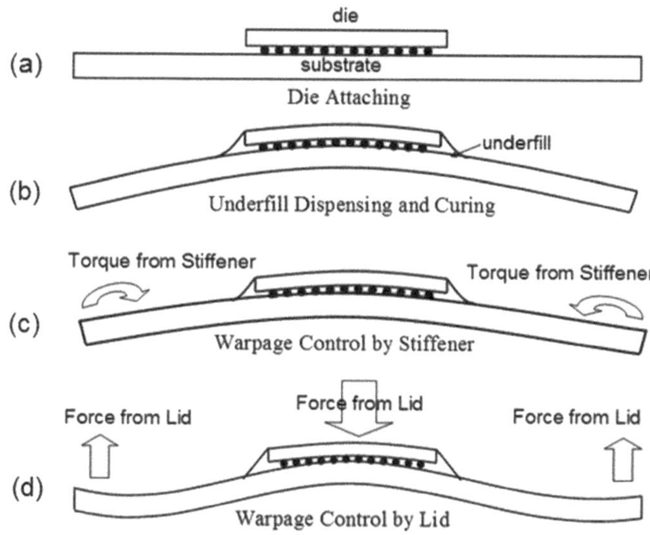

Fig. 12.11 Assembly processes of a flip-chip structure and mechanism of conventional ways for warpage control: (**a**) die attach, (**b**) underfill dispensing and curing, and (**c**, **d**) stiffener or lid attach [66]

developed when cooling down to the room temperature or heating up to the reflow temperature. This can also be seen by the natural bending theory, and the reason for the warpage is due to the big CTE difference between the die and substrate. An example of downwards warpage after cooling down to room temperature (also called coplanarity issue) is shown in Fig. 12.11b. JEDEC specification has defined 8 mil or 200 μm for large flip-chip packages as the tolerance limit. However, for a 45 mm × 45 mm flip-chip package with a bare die, a warpage over 12 mil or 300 μm could exist.

To reduce the warpage, a stiffener ring or a lid is conventionally applied in the next processing step after underfill curing process [66, 72]. The mechanism of using a stiffener or lid to reduce the warpage is illustrated in Fig. 12.11c, d, where the stiffener or lid applies a force or torque at the edge of the substrate, forcing it to deform upwards. So, it is seen that the conventional ways using a stiffener or lid to reduce the warpage will re-deform the substrate, or in other words, they are substrate control methods. As a result, the stress level in the flip-chip package rises when the substrate is re-deformed by the stiffener or lid. For example, for the flip-chip package with 45 mm × 45 mm substrate size and 23 mm × 23 mm die size, the warpage at room temperature is about 12 mil, 8 mil, and 5 mil for the package types of bare die, stiffener, and lid, respectively. Though the lid method gives the lowest warpage, it causes the highest stress in the package, as compared to the bare die and stiffener packages.

Three main points about the warpage control by using stiffener or lid can be summarized as followings: (1) warpage has been developed after underfill

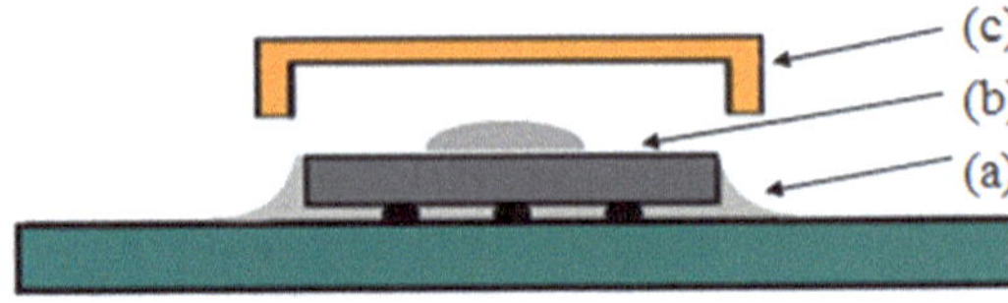

Fig. 12.12 Assembly process for capped-die flip-chip package [66]

dispensing and curing process and has been frozen in the package; (2) stiffener or lid attach process is done after underfill curing process; and (3) the conventional ways using stiffener or lid to control warpage is only to re-deform the substrate. Therefore, the above methods are substrate control methods, which will introduce additional stress into the package.

12.3.4 A Capped-Die Approach for Warpage Control

A capped-die approach is introduced here, which is illustrated in Fig. 12.12. The approach is based on die-control principle and is fundamentally different from substrate-control approaches.

As shown in Fig. 12.12, there are four basic steps to assemble a capped-die package (after die-attach process): (a) dispensing underfill material between the die and the substrate; (b) dispensing adhesive material on top of the die or the stacking dies in 3D; (c) covering the cap onto the die; and (d) concurrently curing the underfill and adhesive materials. Two key differences can be identified between the capped-die flip-chip package design and the conventional flip-chip packages using stiffener or lid. The first one is that the die cap mainly constrain the die, forming a capped-die; therefore, it is a die control method. The second difference is that the underfill material between the die and the substrate is concurrently cured with the adhesive material between the die cap and die. As a result, the CTE between the capped die and the substrate may get matched. Therefore, the warpage after the underfill dispensing and curing process as showed in Fig. 12.11b may be avoided before it is frozen in the package if a die cap with a proper thickness is selected. The same base resin can give a good joint of both underfill and adhesive materials at the die edge region.

One potential failure mode of capped-die flip-chip packages is the shear failure of cap edge, which is caused by a high stress around the die edge region when the die cap constrains the deformation of the die during temperature change (Fig. 12.13a). Based on this consideration, an improved die cap design can be used to lessen the risk of this failure mode (Fig. 12.13b). Besides that, using the adhesive material which has the same T_g as that of the underfill material may also help because both materials consistently become softer or harder when temperature passes T_g. It is preferred that both materials use the same resin as their base materials, and their fillers may be different. For example, for the thermal purpose,

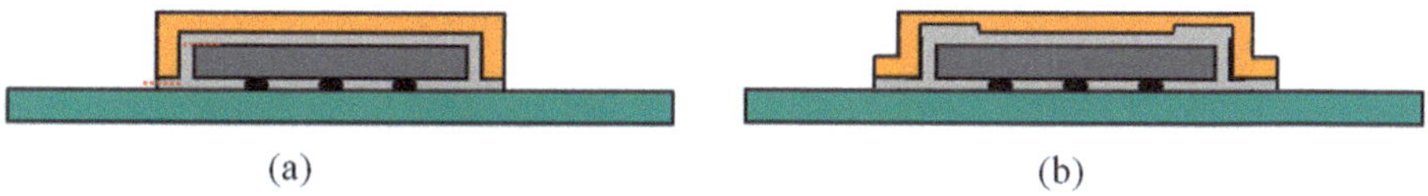

Fig. 12.13 (a) Potential failure modes for capped-die flip-chip packages, (b) a capped-die flip-chip package using an improved die cap design [66]

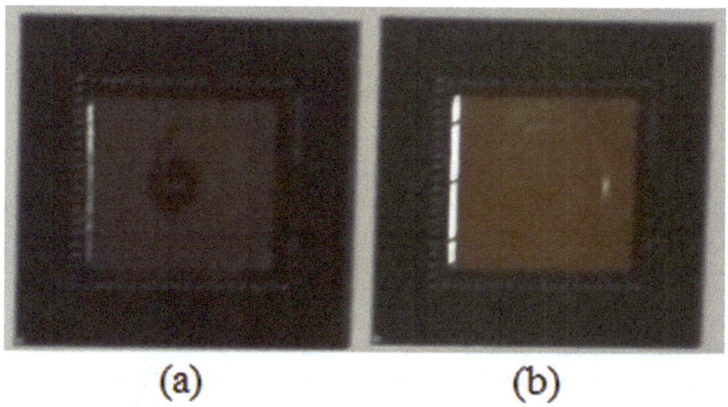

Fig. 12.14 Photos of test vehicles for verifying the capped-die concept: (a) bare die package structure, (b) capped-die package structure [66]

the fillers of the adhesive material for bonding the die cap with the die may use high conductive fillers, such as silver flakes. Generally, the capped-die approach can be used to protect 3D chip stacks from large warpage, where a cap can be made to constrain the deformation of the stacking dies on a large organic substrate.

12.3.5 Warpage Characterization by Experimental Testing

Experimental testing by shadow moiré technique can be used to characterize the warpage deformation and evaluate the warpage control methods. Here, we compare two different dies: one is a bare die and the other is a capped die, as shown in Fig. 12.14. The test vehicle is a flip-chip package with 45 mm × 45 mm substrate size, 23 mm × 23 mm die size and 0.78 mm die thickness. The substrate of the flip-chip package has eight metal layers and 0.8 mm core thickness. The CTE and Young's modulus of the core are 11.3 ppm/°C and 32 GPa. In the test, the same underfill material is used as the adhesive material for bonding the die cap and the die. Two underfill materials, U6 and U2 are selected. The T_g of U6 and U2 are 95 °C and 135 °C, respectively.

Two different thickness, 0.2 and 0.4 mm, were tested. It was found that the warpage control using 0.2 mm thickness of die cap is only a little better than that by

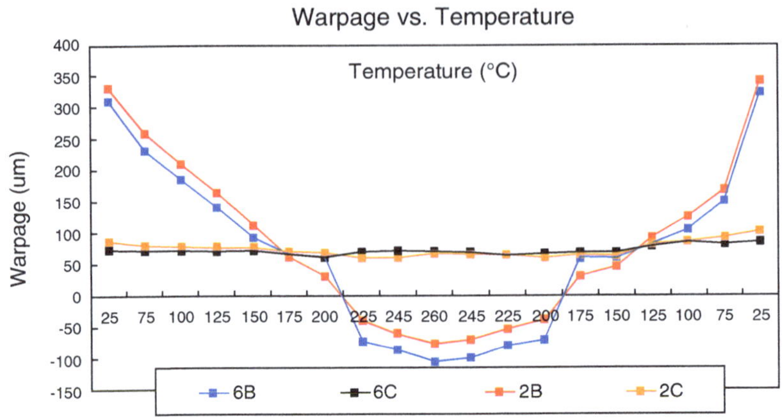

Fig. 12.15 Shadow Moiré data for bare die and capped-die flip-chip packages, where the symbol 6B and 2B stand for bare die package structure using underfill material U6 and U2, and 6C and 2C stand for capped-die package structure using underfill material U6 and U2 [66]

a stiffener ring, whereas much better warpage control is achieved by 0.4 mm thickness of die cap. In Fig. 12.15, the test data of warpage vs. temperature are plotted for both bare dies and capped dies with 0.4 mm thickness. It is seen that when temperature changes, the capped-die package does not warp up and down, and the warpage curve is almost flat and at a warpage value of about 65 μm. This is an important feature that can enhance board-level reliability when mounting the package on a board and under thermal cycling. Also, this feature means that the stress variation inside the package during thermal cycling loading under component level is low, thus giving high component-level reliability.

Another feature of the capped-die package is that the warpage is independent of underfill T_g, which is different from conventional flip-chip package where higher T_g of underfill causes higher warpage at room temperature. It is known that there are two issues for conventional flip-chip package to use very high T_g of underfill material: one is higher warpage and the other is higher stress at die corner. The two issues may be avoided by using capped-die flip-chip packages. So, for capped-die flip-chip package, underfill materials with higher T_g than 120 °C may be applicable. As a result, the underfill material will not transition from hard to soft under application or test loading conditions, improving the reliability of bumps and low-k layers of the package.

It is also interesting to know why the capped-die package has stable warpage at a small value (about 65 μm). Actually, the small warpage is from the warpage of the flip-chip package at curing temperature 165 °C. As can be seen in Fig. 12.15, the bare die package has similar warpage value around that temperature. The warpage value of the capped die package no longer varies with temperature because the CTE of capped die and substrate gets matched. Two factors may cause the small warpage value. One is the residual stress from solder bumps, and the other is the initial warpage of bare substrate due to unsymmetrical metal layers above and below the core.

According to the experimental testing, the capped die concept may give a promising package structure with the advantages of warpage-free control and low stress or high reliability in component as well as board levels for 3D packages. In order to determine the optimized thickness for the cap, FEA analysis can be performed with a parametric study. In the following section, FEM simulation is presented as a preliminary investigation for capped die package design.

12.3.6 Numerical Modeling for Optimizing Warpage Control Design

Numerical simulation is used here to verify the effectiveness of the warpage control method described above and also to optimize the cap design. FEM software ANSYS is utilized to carry out the thermo-mechanical simulation. Several cases are investigated and the results are given below.

12.3.6.1 Comparison of Different Control Methods

The same flip-chip package used in the preceding test vehicle is used as an example for FEM simulation. The stress-free temperature is assumed as the curing temperature of 165 °C, at which the small warpage is not considered based on the test data. Four package structures are compared, including bare die, stiffener, lid, and capped-die packages. The stiffener dimension is 0.6 mm thick and 7 mm wide, the lid dimension is 1.5 mm thick and 4 mm foot width, and the die cap is 0.4 mm thick.

The simulation results for the contour plot of the deformation of the four packages at room temperature are shown in Fig. 12.16, where W_{25C} stands for the

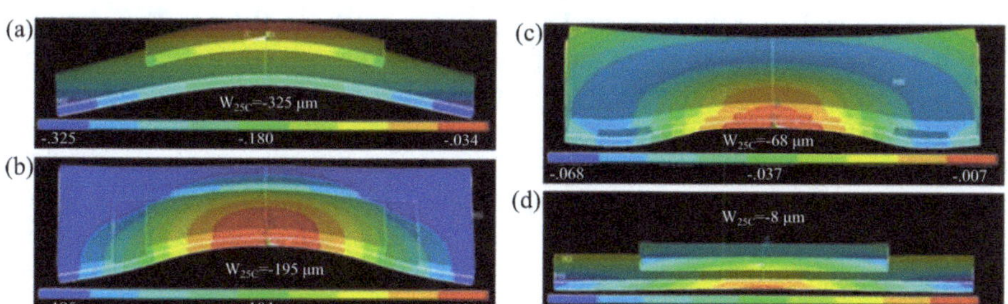

Fig. 12.16 FEM results of warpage for comparing various package structures: (**a**) bare die, (**b**) die with stiffener, (**c**) die with lid, (**d**) die with a cap [66]

warpage at room temperature which is defined as the difference between the substrate bottom center and the lowest point. It is seen from Fig. 12.16 that the capped-die package gives an ideal warpage control. Note that the simulation result shown in Fig. 12.16a for the room temperature warpage of the bare die package is in agreement with the test data showed in Fig. 12.15, validating the FEM model.

12.3.6.2 Optimization of Cap Thickness to Achieve Warpage-Free Packages

There are several variables to be considered when designing a capped die package: the size of die cap, the gap size between the die edge and the cap, and the thickness of die cap. The size of a die cap usually follows the size of the corresponding die, being a little bigger than the die. For the gap size between the die edge and the side of the die cap, a proper value needs to be determined by experimental test for each case. It is suggested that the gap size should be in the range of 100–300 μm. The smaller the gap is the stronger constraints that the die cap applies to the die, but the higher risk the failure between the die and the die cap becomes.

Die cap thickness is the third major design parameter, and a proper thickness may vary with different size of flip-chip packages. For demonstration, a large size, a medium size, and a small size of packages are considered in the FEA simulation. The large package is the same as the test vehicle, i.e., substrate size is 45 mm × 45 mm, die size and thickness are 23 mm × 23 mm and 0.78 mm. The substrate has eight metal layers and 0.8 mm thickness of core. The CTE and Young's modulus of the core are 11.3 ppm/°C and 32 GPa. For the medium-sized package, its substrate size is 30 mm × 30 mm, die size and thickness are 17 mm × 17 mm and 0.3 mm. For the small package, its substrate size is 20 mm × 20 mm, die size and thickness are 10 mm × 10 mm and 0.1 mm. The substrate of the medium-sized package has six metal layers and 0.4 mm thickness of core. The substrate of the small package has six metal layers and 0.15 mm thickness of core. The core materials of the medium and small packages are the same as the big package.

With trial and error method through multiple rounds of simulations, the proper die cap thicknesses for the three packages are determined as 0.4 mm, 0.3 mm, and 0.25 mm, respectively. The simulation results of the contour plot of the deformation of the packages at room and high temperatures are showed in Fig. 12.17, where W_{25C} and W_{260C} stand for the warpage at 25 °C and 260 °C, respectively. It is seen that a warpage-free packaging for the different size of packages are theoretically achieved by a proper thickness of die cap. In practice, it is difficult to achieve such a small warpage because there is an initial warpage when curing the underfill material. However, it is ideal to achieve a flat warpage curve vs. temperature, as shown by the test data in Fig. 12.15.

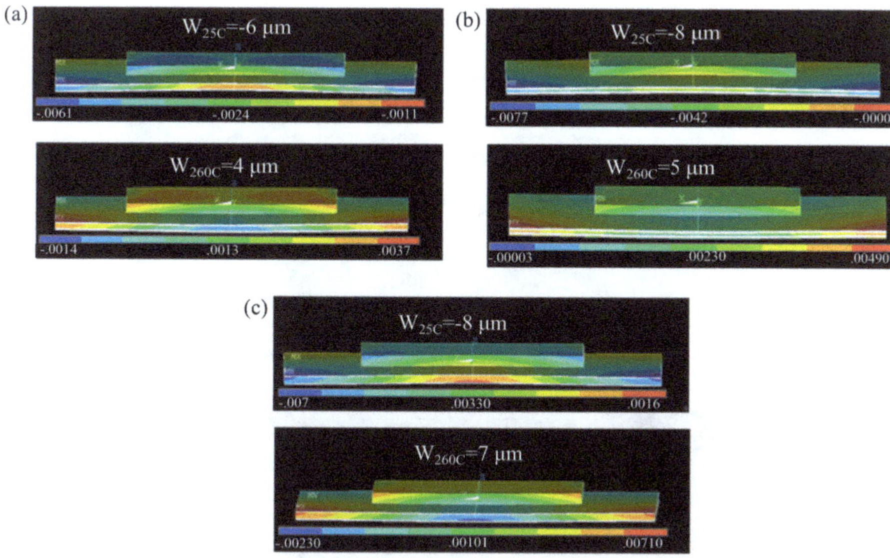

Fig. 12.17 Contour plot of the deformation of different sizes of package at room and high temperatures with optimized cap thickness [66]. (**a**) Small size, (**b**) medium size, (**c**) large size

12.3.6.3 Overcontrolled Warpage

In the preceding section for achieving warpage-free packaging, the die cap thickness is selected by trial and error method through multiple rounds of simulations for different package sizes. Here, an interesting phenomenon, called overcontrolled warpage is shown for the capped-die method by selecting a thicker die cap. In the simulation example, the same large package is used, but a thinner substrate core of 0.4 mm thickness is considered. It has been seen that for 0.8 mm thickness of core, 0.4 mm thickness of die cap gives an ideal warpage control. By using the same 0.4 mm thickness of die cap for the package with the thinner core, the phenomenon of overcontrolled warpage can be simulated, as shown in Fig. 12.18b. For the thin core, a thinner die cap, i.e., a 0.3 mm thickness of die cap is proper for an ideal warpage control, as shown in Fig. 12.18c. Note that the conventional stiffener and lid for warpage control never give the phenomenon of overcontrolled warpage even though a very thick stiffener or lid is used.

12.3.6.4 Warpage-Free Control for Coreless Substrate

Compared to substrates with core, coreless substrates are desired for packaging application because it can shrink substrate size as well as improve the functional performance of flip-chip package. However, flip-chip packages using coreless substrates have very excessive warpage due to the low rigidity of coreless

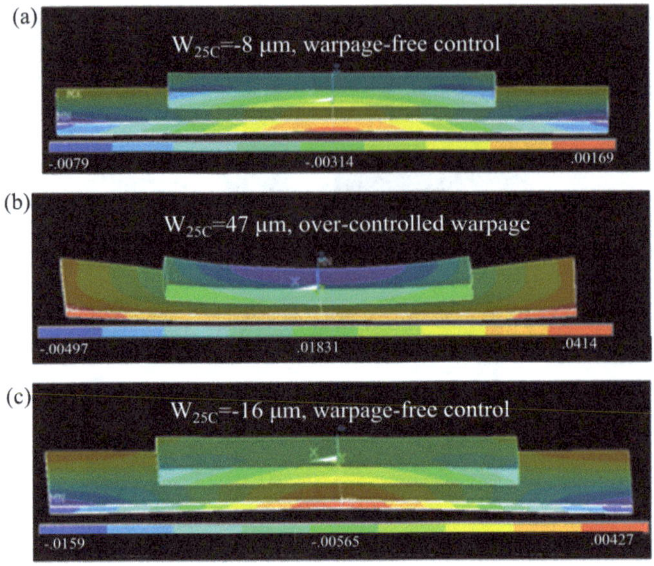

Fig. 12.18 FEM results for proper die cap thickness vs. substrate core thickness [66]. (**a**) 0.4 mm thick of die cap for 0.8 mm substrate core, (**b**) 0.4 mm thick of die cap for 0.4 mm substrate core, (**c**) 0.3 mm thick of die cap for 0.4 mm substrate core

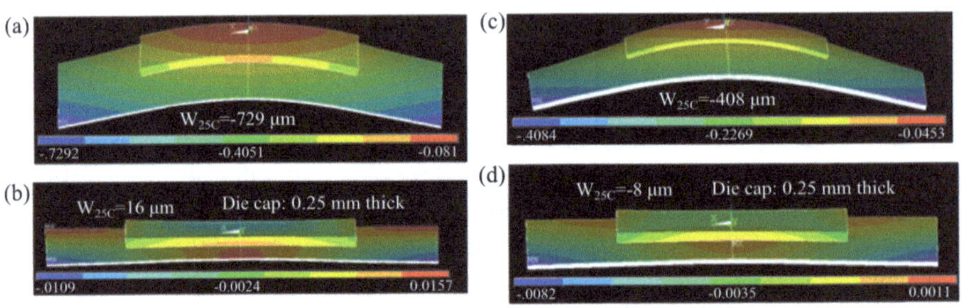

Fig. 12.19 Contour plot of the deformation of large package using eight layers of coreless substrate (**a**, **b**) and six layers of coreless substrate (**c**, **d**) at room temperatures [66]

substrates. And it is not easy to control the large warpage using a stiffener or lid for big package size. It is known that coreless substrates exhibit a severe W-shape when its substrate warpage is controlled by stiffener or lid, limiting its applications.

FEM simulation is used to investigate the effect of capped die method for controlling the warpage of packages using coreless substrates. Two package sizes are simulated, and their dimensions and substrate metal layers are the same as the large and small packages done in the preceding section except the substrate type. For comparison, bare die packages are also simulated. Simulation results are shown in Fig. 12.19 for both the large and small flip-chip packages. It is seen that the huge room temperature warpage is ideally controlled by a 0.25 mm thickness of die cap.

One interesting thing is that for both large and small packages, the warpage-free control happens to be achieved by the same thickness of die cap. Note that the substrate has eight metal layers and die thickness is 0.78 mm in the large package (45 mm × 45 mm), and the substrate has six metal layers and die thickness is 0.1 mm in the small package (10 mm × 10 mm).

12.4 Integrated Stress Analysis for Combining Moisture and Thermal Effects

12.4.1 Introduction

In addition to reliability issues caused by thermal stress, many failures could also occur due to moisture-induced hygroscopic stress and vapor pressure at elevated temperature [14, 15, 25, 73–76]. The problem of moisture absorption and subsequent package failure at elevated temperature has become an important issue concerning the reliability of microelectronic devices. The adverse effects of moisture may include adhesion reduction, hygroscopic swelling, interface delamination, and other material degradation problems [16–18, 25, 26, 77]. One extreme case is reflow soldering, in which temperature ramps up quickly to a peak temperature of 260 °C and the encapsulated moisture in packaging material may evaporate to generate high internal vapor pressure, causing interface delamination and even pop-corning [14, 27, 77–79]. During the highly accelerated temperature and humidity stress test (HAST), moisture-induced stress also becomes very critical to cause package failures as high temperature, high humidity, and high atmospheric pressure are applied.

Due to the deleterious effects caused by moisture, extensive efforts have been made to investigate the moisture diffusion, desorption, and the ensuing mechanical stresses in encapsulated microelectronics devices [16, 26, 80–84]. More importantly, since moisture-induced stress is often accompanied with thermal stress (e.g., during reflow or HAST condition), it is necessary to perform integrated stress analysis to combine all these effects including thermal expansion, hygroscopic swelling, and internal vapor pressure development [17, 63, 77, 85].

To investigate the combined effects of thermal stress and moisture-induced stress, this section starts with the examination of moisture diffusion equation with varying temperature and relative humidity (RH) conditions in Sect. 12.4.2. Then, the theory for incorporating moisture-induced stress is given in Sect. 12.4.3, where the concept of effective stress is introduced. The model for calculation of vapor pressure is described in Sect. 12.4.4, in which the void volume fraction and vapor pressure are introduced as two internal field variables. Section 12.4.5 then summarizes the complete set of governing equations including temperature, moisture, displacement, void volume fraction, and vapor pressure based on the literature [17]. Commercial finite element analysis software ANSYS is used to solve the equations. With a representative bi-material configuration, the solutions for moisture diffusion, vapor pressure, and integrated stress effects are presented in Sect. 12.4.6.

12.4.2 Moisture Diffusion

The transient moisture diffusion equation based on the Fick's law [86] is the most widely used in semiconductor engineering, which has a form as:

$$\frac{\partial C}{\partial t} = -\nabla \cdot (-D\nabla C) \tag{12.6}$$

where C is the local moisture concentration and D is the diffusivity (the rate of diffusion), and t is time. However due to dissimilar materials are used in 3D packaging, the moisture concentration C is discontinuous at the interface of different materials, creating numerical difficulty. As a result, a normalized concentration, or the activity of the diffusing material (denoted as φ here), is introduced [87]

$$\varphi = C/S(T) \tag{12.7}$$

where S is the solubility of each material in a multilayered system at temperature T. When two dissimilar materials are joined at an interface, the normalized concentration is continuous across the interface between the two different materials. Substitute Eq. (12.7) into Eq. (12.6) and the transient moisture diffusion in terms of the normalized concentration becomes

$$S\frac{\partial \varphi}{\partial t} + \varphi\frac{\partial S}{\partial T}\frac{\partial T}{\partial t} = \nabla \cdot \left[D\left(S\nabla\varphi + \varphi\frac{\partial S}{\partial T}\nabla T\right)\right] \tag{12.8}$$

To describe non-Fickian behaviors, a general form of moisture diffusion constitutive model can be written as [88]

$$\mathbf{J} = -SD(\nabla\varphi + k_{\mathrm{T}}\nabla T + k_{\mathrm{P}}\nabla P) \tag{12.9}$$

where $\mathbf{J}$ is the total moisture flux, k_{T} and k_{P} are the factors for temperature gradient-driven and hydrostatic pressure-driven diffusion, respectively. Here, the hydrostatic pressure P is defined as $-\sigma_{ii}/3$.

A comparison between Eqs. (12.8) and (12.9) yields

$$k_{\mathrm{T}} = \frac{\varphi}{S}\frac{\partial S}{\partial T} \tag{12.10}$$

which indicates that the normalized diffusion equation in (12.8) is actually a special case of the general moisture diffusion equation without considering the pressure-driven diffusion.

Another normalized variable, wetness w, was also proposed [24], which is:

$$w = C/C_{\text{sat}} \tag{12.11}$$

where C_{sat} is the saturated moisture concentration of each absorbent material for a given relative humidity and temperature. Since C_{sat} is not exactly a material property due to its dependence on the relative humidity, it is ambiguous to extend the above approach to a varying humidity condition, such as in a process from soaking to reflow, where relative humidity changes. Another general form for wetness can be used:

$$w = C/C_{\text{sat}}(\text{RH}_0, T) \tag{12.12}$$

where RH_0 is the reference relative humidity. Compared to Eq. (12.11), the wetness defined in Eq. (12.12) has an unambiguous definition and is continuous at the interface of dissimilar materials. For a varying relative humidity process, the denominator in Eq. (12.12) always refers to a "reference" saturated moisture concentration at RH_0 and T, even though RH_0 may be different from the current relative humidity conditions. With Henry's law [86]:

$$C_{\text{sat}}(\text{RH}_0, T) = p_{\text{amb}}(\text{RH}_0)S(T) = p_{\text{g}}\text{RH}_0 S(T) \tag{12.13}$$

where p_{amb} and p_{g} are partial pressure at RH_0 and saturated vapor pressure of moisture at temperature T.

It has been observed that the saturated concentration is linearly proportional to the ambient relative humidity but not dependent on the temperature, as long as it is far away from the glass transition temperature [14, 80]. In this case, the $C_{\text{sat}}(\text{RH}_0)$ can be considered as a constant within the interested temperature range. Substituting Eq. (12.13) into Eqs. (12.12) and (12.6), we obtain

$$\frac{\partial w}{\partial t} = \nabla \cdot (D \nabla w) \tag{12.14}$$

which indicates that the thermal-moisture analogy holds true for a temperature-dependent moisture diffusion process when w in Eq. (12.12) is used. The Dirichlet boundary condition for Eq. (12.14) then becomes

$$w\big|_{\text{bc}} = RH_{\text{bc}}/RH_0 \tag{12.15}$$

where RH_{bc} is the relative humidity at the boundary.

Other methods like direct concentration approach (DCA), convection-diffusion models [78, 79], have also been developed, where normalization is not required.

Particularly, the convection-diffusion model might be useful in the case of rapid heating where vapor flow is important. Details of these models can be referred to the literature [78, 79] and are not covered in this chapter.

12.4.3 Moisture-Induced Strain and Effective Stress Theory

Moisture-induced stress may include both hygroscopic stress and vapor pressure p. To characterize hygroscopic swelling, thermal-moisture analogy can be used:

$$\varepsilon_{\mathrm{c}} = \beta C \tag{12.16}$$

where β is the coefficient of hygroscopic swelling, which is a material property that can be determined by experimental testing with TGA and TMA [14].

To consider vapor pressure-induced strain, the effective stress concept is introduced by assuming polymeric material as porous media. As moisture is vaporized, the generated vapor pressure is exerted on the solid phase of the porous material. The coupling phenomena between vapor pressure and matrix material deformation can be described by the effective stress concept, as

$$\sigma_{ij} = \sigma_{ij}^{'} - p\delta_{ij} \tag{12.17}$$

where σ_{ij} is the total stress tensor component, p is the partial water vapor pressure, and $\sigma_{ij}^{'}$ is the effective stress. Note the vapor pressure p in Eq. (12.17) has a different meaning from the pressure stress (which is denoted by capital P in Eq. (12.9)). Usually, the vapor pressure p in Eq. (12.17) has a positive value.

According to the definition of effective stress, the deformation of porous skeleton is governed by the effective stress only. Therefore, the constitutive model for linear, isotropic elastic material can be described by the effective stress as follows:

$$\varepsilon_{ij} = \frac{1+v}{E}\sigma_{ij}^{'} - \frac{v}{E}\sigma_{kk}^{'}\delta_{ij} + (\alpha\Delta T + \beta C)\delta_{ij} \tag{12.18}$$

where E is Young's modulus, v is Poisson's ratio, $\Delta T = T - T_{\mathrm{o}}$ with T_{o} as the reference temperature. It can be seen that both thermal strain and hygroscopic strain are included. The effect of vapor pressure can be revealed by substituting Eq. (12.17) into Eq. (12.18), which leads to

$$\varepsilon_{ij} = \frac{1+v}{E}\sigma_{ij} - \frac{v}{E}\sigma_{kk}\delta_{ij} + \left(\alpha\Delta T + \beta C + \frac{1-2v}{E}p\right)\delta_{ij} \tag{12.19}$$

Equation (12.19) indicates that the vapor pressure contributes to the matrix deformation as an additional volumetric strain term. To summarize, the total thermal/moisture induced volumetric strain can be written as

$$\varepsilon^{\text{volu}} = \varepsilon_{\text{T}} + \varepsilon_{\text{h}} + \varepsilon_{\text{p}} = \alpha \Delta T + \beta C + \frac{1 - 2\upsilon}{E} p \tag{12.20}$$

With Eq. (12.20), Eq. (12.17) can be rewritten as

$$\sigma_{ij} = 2G\varepsilon_{ij} + \lambda \varepsilon_{kk}\delta_{ij} - \left(\frac{E}{1 - 2v}(\alpha \Delta T + \beta C) + p \right) \delta_{ij} \tag{12.21}$$

where G and λ are Lamé's elastic constants. Based on the constitutive law in Eq. (12.21), the deformation equilibrium now becomes [17]

$$G \nabla^2 u_i + (\lambda + G) e_{,i} - \left[\frac{E}{1 - 2v}(\alpha \Delta T + \beta C) + p \right]_{,i} + X_i = 0 \tag{12.22}$$

where u_i is the component of displacement vector, e is the total volumetric strain, X_i is the component of body force vector, and ",i" represents the comma derivative. Equation (12.21) or (12.22) implies that all applied loads, thermal expansion, hygroscopic swelling, and vapor pressure, are in the forms of body loads. The resulting stress/deformation response caused by temperature excursion and moisture diffusion and evaporation can be obtained by solving the force equilibrium equation with the above strain definitions and the governing equations of heat conduction and moisture diffusion. Since they are virtually independent of the stress state (if stress gradient-driven diffusion mechanism can be neglected), the temperature and moisture concentration fields can be determined, respectively, prior to a stress analysis, and then they can be incorporated into a subsequent stress analysis.

12.4.4 Vapor Pressure Modeling

To evaluate the total strain, vapor pressure must be determined first. Generally, vapor pressure is related to the local moisture concentration and free volume fraction as well (according to the porous media assumption). Although moisture diffusion is analyzed at a macroscopic level, the vapor pressure model should be considered at a microscopic level. Based on this principle, a micromechanics-based vapor pressure model has been proposed [89] and is well accepted in semiconductor packaging. Other vapor pressure models, such as those based on Henry's law [19, 20, 78, 79], do not consider the microscopic effect. Therefore, the micromechanics-based model is adopted and described in this chapter.

The micromechanics-based model considers a representative elementary volume (REV) as shown in Fig. 12.20. The RVE exists at any considered point in porous medium. All moisture absorbed is either in liquid water or vapor form. Moisture collects at the micro-/nano-pores, in free volumes, at the interfaces, and/or in micro/macro-voids. Unbound moisture will evaporate during the reflow process. The total

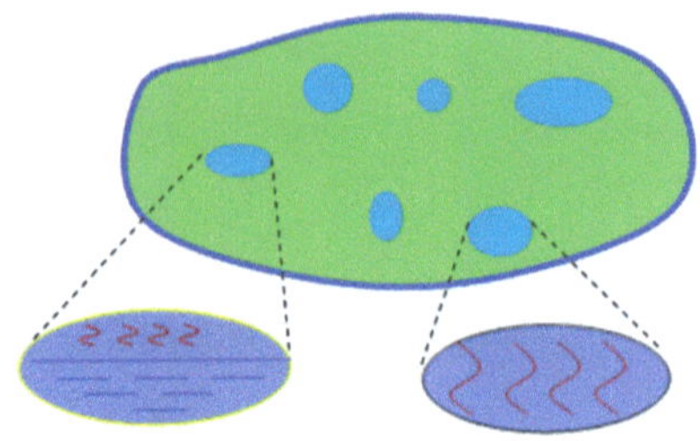

Fig. 12.20 Representative elementary volume (REV) describing two distinct states of moisture in pores in polymer materials [89]

moisture content in an REV is obtained from the local moisture concentration C at a macroscopic level. If the void volume fraction ϕ (or interstitial space fraction) is known, then the "apparent" moisture density (ρ_a) in pores can be defined as [89]

$$\rho_a = C/\phi \tag{12.23}$$

The calculation of vapor pressure starts with determining the state of moisture in the voids. If the voids contain only vapor and the vapor pressure reaches the saturation point, the density of saturated water vapor, ρ_g, can be expressed according to the ideal gas law:

$$\rho_g(T) = \frac{p_g(T)}{RM(H_2O)T} \tag{12.24}$$

where $p_g(T)$ is the saturated vapor pressure at a given temperature; the gas constant $R = 8.314$ J(mol K)$^{-1}$, and the water molecular mass $M(H_2O) = 18$ g mol^{-1}. The vapor pressure is then evaluated by comparing ρ_a with ρ_g:

$$p(T) = \begin{cases} \dfrac{RT}{M(H_2O)\phi} \cdot C, & \rho_a < \rho_g \\ P_{sat}(T), & \rho_a \geq \rho_g \end{cases} \tag{12.25}$$

Equation (12.25) provides a model to calculate vapor pressure. It indicates that vapor pressure is related to moisture diffusion (moisture concentration C), and temperature field (T), as well as the void volume fraction ϕ. The void volume fraction may be considered as a damage field variable. The growth of the void must follow the continuity equation, as follows [17, 90]:

$$\dot{\phi} = (1 - \phi)\dot{u}_{i,j} \tag{12.26}$$

which is coupled with the continuum's deformation in Eq. (12.22).

12.4.5 Governing Equation for Integrated Stress Analysis

From the analysis in the previous sections, the basic field variables of the problem include: temperature T, normalized moisture concentration φ, displacement vector u_i, internal vapor pressure p, and void volume fraction ϕ. The vapor pressure p and void volume fraction ϕ can be considered as internal field variables. Void volume fraction ϕ is a damage parameter to describe the progression of damage in a system of interest. The governing equations for solving the problem are summarized as follows [17]:

$$\frac{\partial T}{\partial t} = D_{\mathrm{T}} \nabla^2 T \tag{12.27}$$

$$S\frac{\partial \varphi}{\partial t} + \varphi \frac{\partial S}{\partial t} = \nabla \cdot [D(S\nabla\varphi + \varphi\nabla S)] \tag{12.28}$$

$$G\nabla^2 u_i + (\lambda + G)e,_i - \left[\frac{E}{1-2\upsilon}(\alpha\Delta T + \beta C) + p\right],_i + X_i = 0 \tag{12.29}$$

$$\dot{\phi} = (1-\phi)\dot{u}_{i,j} \tag{12.30}$$

where D_{T} is the thermal diffusivity. As normalized concentration is used, the equations can be solved for multi-material system such as 3D chip stacks. It is seen that an integrated stress modeling during reflow soldering requires five types of modeling, namely (1) moisture diffusion during moisture preconditioning and reflow, (2) thermal modeling, (3) hygro-mechanical modeling, (4) thermo-mechanical modeling, and (5) vapor pressure modeling.

12.4.6 Case Studies

To demonstrate the application of the integrated stress theory, a bi-material system is used to perform the analysis as shown in Fig. 12.21. The bi-material system can be viewed as the most fundamental structure in a 3D device. In the example, copper is chosen for material 1 and a mold compound is chosen for material 2. The geometry and material properties are given in Tables 12.3, 12.4, and 12.5, respectively.

The moisture diffusivity of the mold compound is dependent on temperature, which can be described by the Arrhenius equation:

$$D = D_{\mathrm{o}} \mathrm{e}^{-\frac{Q}{RT}} \tag{12.31}$$

where D_{o} is a pre-factor, and Q is the activation energy (ev). Copper does not absorb moisture and thus its diffusivity and saturated moisture concentration are assigned

Fig. 12.21 A bi-material assembly for integrated stress analysis

Table 12.3 Bi-material assembly geometry [17]

Geometry	Copper	Mold compound
Thickness (mm)	0.2	0.3
Length (mm)	2	2

Table 12.4 Temperature-dependent material properties for mold compound [17]

Temperature (°C)	CTE (ppm/°C)	*E* (GPa)
85	20	20.0
125	20	20.0
155	40	5.0
195	40	1.0
265	40	1.0

Table 12.5 Material properties of bi-material assembly [17]

	Copper	Mold compound
C_{sat} at 85 °C/85%RH (μg/mm^3)	N/A	4.0
Diffusivity pre-factor D_o (mm^2/s)	N/A	5.0
Diffusivity activation energy Q (ev)	N/A	0.4
Initial void volume fraction	N/A	0.05
CTE (ppm/°C)	17	see Table 12.4
Young's modulus (GPa)	130	see Table 12.4
Poisson's ratio	0.34	0.3
Coefficient of hygroscopic swelling (mm^3/μg)	N/A	0.5×10^{-3}

to very small (e.g., 10^{-6} lower) values. The values of D_o and Q are shown in Table 12.5.

The assembly initially experiences preconditioning at 85 °C/85%RH for 196 h, followed immediately by a reflow process. The time history of the reflow process is depicted by Fig. 12.22. According to the literature, the reflow temperature profile greatly affects the moisture distribution [22].

Sequentially coupled moisture diffusion, vapor pressure, and integrated stress analysis is performed based on the governing equations described in Sect. 12.4.3. To solve the multi-field equations, we apply the thermal-moisture analogy and built-in user-defined swelling functions in ANSYS. The results are given and discussed in details below.

Figure 12.23 shows the absolute moisture concentration contour plots at three different temperatures during reflow. It can be seen that the maximum moisture

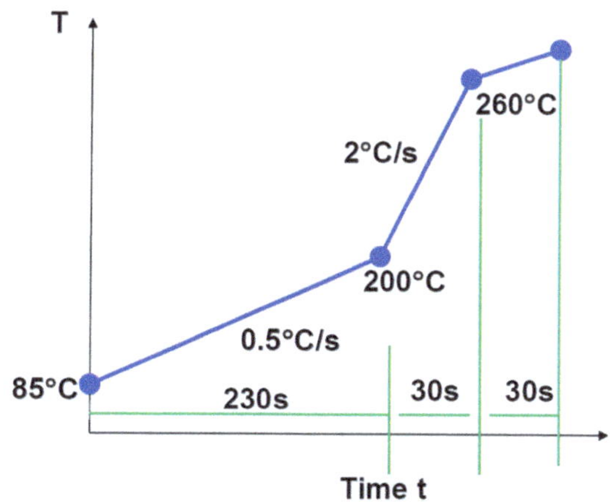

Fig. 12.22 Reflow temperature profile used for the integrated stress analysis [17]

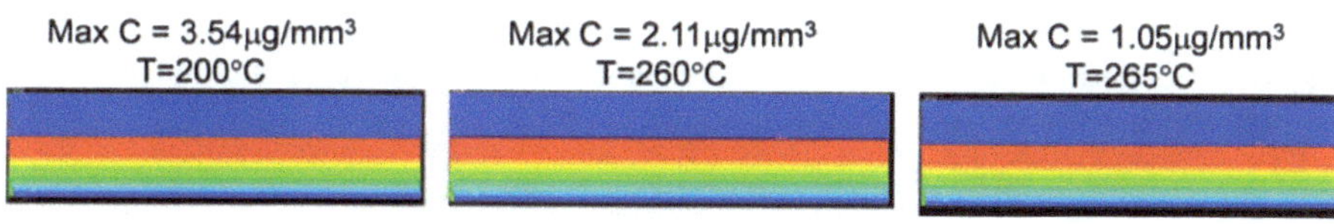

Fig. 12.23 Moisture contours at different temperatures [17]

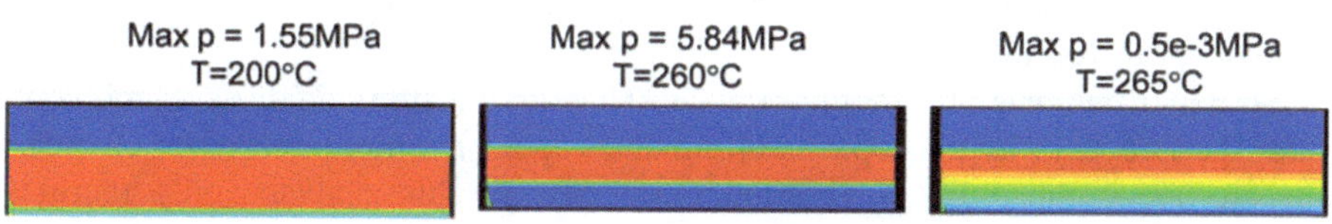

Fig. 12.24 Vapor pressure contours at different temperatures [17]

concentration at copper/mold compound interface decreases significantly with the increase of reflow temperature.

Figure 12.24 gives the vapor pressure contour plots at the same three temperatures based on the micromechanics approach. It can be seen that the vapor pressure increases with the increase of temperature from 200 to 260 °C, despite the decrease in the moisture concentration. However, with more moisture lost, vapor pressure starts to decrease even the temperature continues to rise to 265 °C. This indicates that there is insufficient residual moisture to keep the high vapor pressure as the desorption process goes on.

After the moisture concentration and vapor pressure fields are determined, they can be incorporated into a subsequent stress analysis using Eq. (12.22). Figure 12.25 shows the comparison of moisture distributions at 1 h and 18 h in preconditioning under 85 °C/85%RH, and the corresponding deformed shapes of the assembly and the maximum von Mises stresses in mold compound. In the beginning of preconditioning, the assembly is in convex shape due to thermal mismatch from

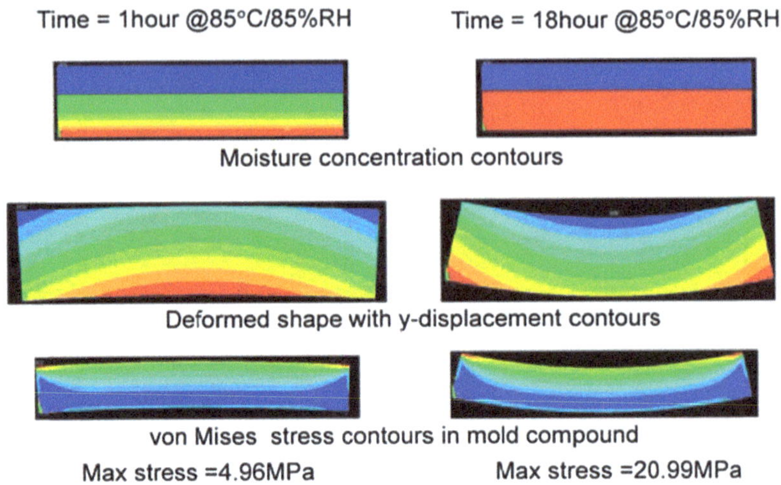

Fig. 12.25 Comparison of moisture concentration, deformation, and stress at different time in preconditioning process [17]

stress free temperature of 160–85 °C. Afterwards, mold compound swells with moisture absorption, and the assembly deforms from convex to concave shape. At the same time, stress level in mold compound increases significantly due to hygroscopic swelling.

To better demonstrate the importance of integrated stress analysis which consider the effects of hygroscopic swelling and vapor pressure, three types of stress analysis are compared: (1) Thermal load only. In this analysis, only thermal mismatch is considered at the reflow. Since this is a linear analysis, the stress state depends on the final temperature state only. (2) Combined thermal and hygroscopic swelling loads. At different times in the studied process (either in preconditioning or reflow stage), moisture distributions can be applied as body loads. The stress state depends on the temperature state and instantaneous moisture distributions. (3) Combined thermal, hygroscopic swelling, and vapor pressure loads. In this case, Eq. (12.22) is used to apply a body load which comprises both hygroscopic swelling and vapor pressure.

Numerical results from the above three types of analysis are summarized in Tables 12.6, 12.7, and 12.8. Table 12.6 shows the maximum warpage, maximum von Mises stresses in copper and mold compound, respectively, at three different times during reflow: 200 °C, 260 °C, and 260 °C after 30 s hold. At 200 °C, hygroscopic swelling and vapor pressure-induced stresses play significant roles in final stress state (Table 12.6). Results indicate that the effects of hygroscopic swelling and vapor pressure cannot be overlooked. Comparing Tables 12.7 and 12.8, it is apparent that the effects of hygroscopic swelling and vapor pressure fade when more moisture is lost at the same temperature 260 °C. The maximum stress state can be determined at a certain temperature with given reflow profile and the

Table 12.6 Maximum warpage (W) and von Mises stresses (σ_e) in copper and mold compound at 200 °C [17]

	W (mm)	σ_e, MC (MPa)	σ_e, Cu (MPa)
Thermal stress only	0.4e-3	2.4	8.7
Thermal + hygroscopic	0.7e-3	4.9	17.3
Thermal + hygroscopic + vapor pressure effect	1.0e-3	6.3	22.9

Table 12.7 Maximum warpage (w) and von Mises stresses (σ_e) in copper and mold compound at 260 °C [17]

	W (mm)	σ_e, MC (MPa)	σ_e, Cu (MPa)
Thermal stress only	1.3e-3	6.0	21.9
Thermal + hygroscopic	1.5e-3	7.2	27.5
Thermal + hygroscopic + vapor pressure effect	2.1e-3	11.2	38.9

Table 12.8 Maximum warpage (w) and von Mises stresses (σ_e) in copper and mold compound at 260 °C after 6 s hold [17]

	W (mm)	σ_e, MC (MPa)	σ_e, Cu (MPa)
Thermal stress only	1.3e-3	6.0	21.9
Thermal + hygroscopic	1.4e-3	6.3	24.7
Thermal + hygroscopic + vapor pressure effect	1.4e-3	6.3	24.7

maximum stress may not occur at the peak reflow temperature since moisture continues to escape during temperature rise.

12.5 Summary

In this chapter, thermal-/moisture-induced stresses and their effects on 3D packaging have been discussed. Techniques for thermo–mechanical and moisture modeling are introduced to characterize the stresses and the corresponding issues in 3D packaging under both thermal and moisture loadings. Evaluation based on these modeling techniques can help enhance the understanding of the failure mechanisms, provide guidance for package design, and improve the performance and reliability of 3D products.

In Sect. 12.2, the effects of die-level thermal stress on TSV structures are investigated. The thermal stress in TSV is induced by the CTE mismatch between Cu via and Si and has triaxial stress state due to the confinement of the Si substrate. To describe the distinct nature of near-surface stress in TSV structures, a semi-analytic 3D solution has been developed for an isolated TSV embedded in the silicon wafer. As a numerical approach, finite element method has been used to analyze the thermal stresses in 3D integrated structures and to evaluate stress-induced keep-out zone. Experimentally, wafer curvature method and micro-Raman spectroscopy are combined to measure the thermal stress in and around TSVs. The microstructure of Cu plays an important role in affecting the stress relaxation and residual stress after thermal cycling. The microstructure also has significant implications on via extrusion reliability, as grain growth, plasticity, and diffusional creep are underlying relaxation mechanisms causing via extrusion. It is important to optimize the processing conditions of TSVs, including the electroplating chemistry and annealing, to control the thermal stress and improve the reliability of TSV structures.

In Sect. 12.3, the effects of thermal stress at package level of 3D packaging are discussed, with a focus on the reliability issue associated with warpage. Thermal stress-induced warpage in multilayered structures can be evaluated by analytical modeling. Two theories, the natural bending theory and the beam-like plate theory, are introduced. To identify the warpage mechanism during the assembly process, a flip-chip package is studied, showing that warpage usually occurs after underfill is cured followed by a sudden temperature change. The conventional control methods, such as applying a stiffer ring or a lid, are based on substrate control principle and thus may introduce additional stress into the package. On the contrary, the capped-die approach, which is based on die control principle, is considered more efficient and is able to achieve warpage-free design in 3D packaging. The key control mechanism of the capped-die approach is to reduce the mismatch of CTEs between the chip and substrate. The capped-die concept proves more effective than the conventional control methods through both experimental testing and numerical modeling. Ideally, it is possible to achieve warpage-free packaging in the full range of temperature variation with the approach. With warpage being controlled at a small and stable level, the corresponding reliability issues caused by thermal stress at package level can be effectively mitigated.

In Sect. 12.4, moisture-induced stresses, which include both hygroscopic stress and vapor pressure, have been considered together with thermal stress in order to take account of the circumstance with both high temperature and high humidity (e.g., reflow process or HAST environment). To couple moisture and thermal effects, a damage micromechanics continuum theory is presented, which allows the coupling of moisture diffusion, heat conduction, matrix material deformation, as well as the growth of void volume fraction. The total strain is formulated as the combination of hygroscopic swelling, thermal expansion, and vapor pressure-induced strain. A new constitutive law is then established according to the effective stress concept. A complete set of governing equations is provided for integrated stress analysis, which can be solved by thermal-moisture analogy and built-in user-defined swelling functions in ANSYS. A bi-material assembly is used to illustrate the integrated analysis of moisture diffusion, vapor pressure evolution, and stress determinations. The results show that the moisture-induced stresses play an important role in final stress state during reflow, and the moisture-induced deformation (which includes both hygroscopic swelling and vapor pressure-induced strain) can be as significant as thermal expansion. It is concluded that the theory for integrated stress analysis can serve as a useful tool to analyze, understand, and improve the reliability of a 3D IC package in circumstances like reflow soldering or HAST conditions, where moisture-induced stress cannot be overlooked. With the normalized concentration used in diffusion equation, the theory can be applied to any 3D assembly that has many dissimilar materials. It is worth noting that there is still no consensus of vapor pressure model to be used in microelectronics packaging. However, the theoretical framework provided in this chapter does not depend on the vapor pressure model adopted.

Acknowledgment The editors would like to thank Zhiheng Huang from Sun Yat-sen University in China for his critical review of this chapter.

References

1. Z. Wang, J. Microelectromech. Syst. **24**, 1211 (2015)
2. J.U. Knickerbocker et al., IBM J. Res. Dev. **52**, 553 (2008)
3. P. Garrou, C. Bower, P. Ramm, *Handbook of 3D Integration* (Wiley-VCH, Germany, 2008)
4. M. Jung, J. Mitra, D.Z. Pan, S.K. Lim, Commun. ACM **57**, 107 (2014)
5. JEDEC-JEP158, (JEDEC Solid State Technology Association, 2009)
6. S.W. Yoon, J.H. Ku, F. Carson, in *SEMICON Korea 2010* (Seoul, Korean, 2010)
7. K.N. Tu, Microelectron. Reliab. **51**, 517 (2011)
8. C.-H. Liu, J.-L. Tsai, C. Hung-Hsien, C.-L. Lu, S.-C. Chen, in *IEEE 64th Electronic Components and Technology Conference*, 2014, p. 1628
9. K. Sakuma et al., in *IEEE 64th Electronic Components and Technology Conference (ECTC)*, 2014, p. 647
10. L. Mirkarimi, R. Zhang, A. Agrawal, H. Shaba, B.-S. Lee, R. Katkar, E. Chau, S. Arkalgud, *Analytical and Experimental Studies of 2.5D Silicon Interposer Warpage: Impact of Assembly Sequences, Materials Selection and Process Parameters*. Paper presented at 11th International Wafer Level Packaging Conference, San Jose, CA, 2014
11. C.-H. Liu, Y.-H. Liao, W.-T. Chen, C.-L. Lu, S.-C. Chen, in *IEEE 65th Electronic Components and Technology Conference (ECTC)*, 2015, p. 1502
12. K. Sakuma et al., in *IEEE 65th Electronic Components and Technology Conference (ECTC)*, 2015, p. 318
13. S.Y. Yang, W.-S. Kwon, S.-B. Lee, Microelectron. Reliab. **52**, 718 (2012)
14. X.J. Fan, E. Suhir, *Moisture Sensitivity of Plastic Packages of IC Devices* (Springer, New York, 2010)
15. X.J. Fan, T.B. Lim, in *ASME International Mechanical Engineering Congress and Exposition, IMECE/EPE-14*, 1999
16. J. Zhou, T.Y. Tee, X.J. Fan, in *Moisture Sensitivity of Plastic Packages of IC Devices* (Springer, Berlin, 2010), p. 153
17. X.J. Fan, J.-H. Zhao, in *12th International Conference on Thermal, Mechanical and Multi-Physics Simulation and Experiments in Microelectronics and Microsystems (EuroSimE)*, 2011, p. 1
18. X.Q. Shi, Y.L. Zhang, Z. Wei, X.J. Fan, IEEE Trans. Compon. Packag. Technol. **31**, 94 (2008)
19. M.N. Kitano, A. Nishimura, S. Kawai, in *Proc. IRPS*, 1988, p. 90
20. C.G. Shirley, IEEE Trans. Device Mater. Reliab. **14**, 426 (2014)
21. B. Xie, X.Q. Shi, X.J. Fan, in *Proceedings of the 57th Electronic Components and Technology Conference*, 2007, p. 242

22. B. Xie, X.J. Fan, X. Shi, H. Ding, J. Electron. Packag. **131**, 031011 (2009)
23. B. Xie, X.J. Fan, X. Shi, H. Ding, J. Electron. Packag. **131**, 031010 (2009)
24. E.H. Wong, Y.C. Teo, T.B. Lim, in *Proceedings of the Electronic Components and Technology Conference*, 1998, p. 1372
25. S. Liu, Y. Mei, IEEE Trans. Compon. Packag. Manuf. Technol. Part A **18**, 634 (1995)
26. X.J. Fan, J. Zhou, A. Chandra, in *58th Electronic Components and Technology Conference*, 2008, p. 1054
27. X.J. Fan, G.Q. Zhang, W.D. van Driel, L.J. Ernst, IEEE Trans. Compon. Packag. Technol. **31**, 252 (2008)
28. J.V. Olmen et al., Microelectron. Eng. **88**, 745 (2011)
29. K. Sinwoo et al., in *IEEE International 3D Systems Integration Conference (3DIC)*, 2011, p. 1
30. A. Mercha et al., in *IEEE International Electron Devices Meeting (IEDM)*, 2010, p. 2.2.1
31. A.P. Karmarkar, X. Xu, V. Moroz, in *IEEE International Reliability Physics Symposium*, 2009, p. 682
32. K.H. Lu, X. Zhang, S.-K. Ryu, J. Im, R. Huang, P.S. Ho, in *59th Electronic Components and Technology Conference (ECTC)*, 2009, p. 630
33. S.-K. Ryu, K.-H. Lu, X. Zhang, J.-H. Im, P.S. Ho, R. Huang, IEEE Trans. Device Mater. Reliab. **11**, 35 (2011)
34. S.-K. Ryu, T. Jiang, K.H. Lu, J. Im, H.-Y. Son, K.-Y. Byun, R. Huang, P.S. Ho, Appl. Phys. Lett. **100**, 041901 (2012)
35. T. Jiang, S.-K. Ryu, Q. Zhao, J. Im, R. Huang, P.S. Ho, Microelectron. Reliab. **53**, 53 (2013)
36. I. De Wolf, in *AIP Conference Proceedings*, 2011, p. 138
37. S.-K. Ryu, Q. Zhao, M. Hecker, H.-Y. Son, K.-Y. Byun, J. Im, P.S. Ho, R. Huang, J. Appl. Phys. **111**, 063513 (2012)
38. W.S. Kwon, D.T. Alastair, K.H. Teo, S. Gao, T. Ueda, T. Ishigaki, K.T. Kang, W.S. Yoo, Appl. Phys. Lett. **98**, 232106 (2011)
39. T. Jiang, S.-K. Ryu, J. Im, H.-Y. Son, N.-S. Kim, R. Huang, P.S. Ho, in *IEEE International Interconnect Technology Conference (IITC)*, 2013, p. 1
40. D. Gan, P.S. Ho, R. Huang, J. Leu, J. Maiz, T. Scherban, J. Appl. Phys. **97**, 103531 (2005)
41. T. Jiang, J. Im, R. Huang, P.S. Ho, MRS Bull. **40**, 248 (2015)
42. W. De Ingrid, Semicond. Sci. Technol. **11**, 139 (1996)
43. D.J. Gardiner, *Practical Raman Spectroscopy* (Springer, New York, 1989)
44. Y. Sun, S. Thompson, T. Nishida, *Strain Effect in Semiconductors* (Springer, New York, 2010)
45. A. Mercha et al., in *2010 Symposium on VLSI Technology (VLSIT)*, 2010, p. 109
46. S. Cho et al., in *International Interconnect Technology Conference and 2011 Materials for Advanced Metallization (IITC/MAM)*, 2011, p. 1
47. R. Suk-Kyu, L. Kuan-Hsun, J. Tengfei, J.-H. Im, H. Rui, P.S. Ho, IEEE Trans. Device Mater. Reliab. **12**, 255 (2012)
48. J.C. Lin et al., in *International Electron Devices Meeting (IEDM)*, 2010, p. 2.1.1
49. T.C. Tsai, W.C. Tsao, W. Lin, C.L. Hsu, C.L. Lin, C.M. Hsu, J.F. Lin, C.C. Huang, J.Y. Wu, Microelectron. Eng. **92**, 29 (2012)
50. T. Jiang, C. Wu, J. Im, R. Huang, P.S. Ho, J. Microelectron. Electron. Packag. **12**, 118 (2015)
51. T. Jiang et al., Appl. Phys. Lett. **103**, 211906 (2013)
52. S.-K. Ryu, T. Jiang, J. Im, P.S. Ho, R. Huang, IEEE Trans. Device Mater. Reliab. **14**, 318 (2014)
53. J. De Messemaeker, O.V. Pedreira, B. Vandevelde, H. Philipsen, I. De Wolf, E. Beyne, K. Croes, in *IEEE 63rd Electronic Components and Technology Conference (ECTC)*, 2013, p. 586
54. J. De Messemaeker, O.V. Pedreira, H. Philipsen, E. Beyne, I. De Wolf, T. Van der Donck, K. Croes, in *IEEE 64th Electronic Components and Technology Conference (ECTC)*, 2014, p. 613

55. D. Campos et al., in *2013 European Microelectronics Packaging Conference (EMPC)*, 2013, p. 1
56. R.L. Hubbard, P. Zappella, IEEE Transactions on Components. Packag. Manuf. Technol. **1**, 1957 (2011)
57. K.M. Jansen, B. Öztürk, IEEE Transactions on Components. Packag. Manuf. Technol. **3**, 459 (2013)
58. W. Lin, B. Baloglu, K. Stratton, in *IEEE 64th Electronic Components and Technology Conference (ECTC)*, 2014, p. 1401
59. W. Lin, S. Wen, A. Yoshida, J. Shin, in *IEEE 62nd Electronic Components and Technology Conference (ECTC)*, 2012, p. 1406
60. A.-H. Liu, D.W. Wang, H.-M. Huang, M. Sun, M.-R. Lin, C. Zhong, S.-J. Hwang, H.-H. Lu, in *IEEE 61st Electronic Components and Technology Conference (ECTC)*, 2011, p. 431
61. T.H. Wang, C.-I. Tsai, C.-C. Lee, Y.-S. Lai, Microelectron. Reliab. **53**, 297 (2013)
62. D.G. Yang, K.M.B. Jansen, L.J. Ernst, G.Q. Zhang, W.D. van Driel, H.J.L. Bressers, X.J. Fan, in *Proceedings of the 54th Electronic Components and Technology Conference*, 2004, p. 98
63. X.J. Fan, J. Zhou, G.Q. Zhang, Microelectron. Reliab. **44**, 1967 (2004)
64. K. Sakuma, E. Blackshear, K. Tunga, C. Lian, S. Li, M. Interrante, O. Mantilla, J.-W. Nah, in *IEEE 63rd Electronic Components and Technology Conference (ECTC)*, 2013, p. 667
65. P.S. Huang, M.Y. Tsai, C.Y. Huang, P.C. Lin, L. Huang, M. Chang, S. Shih, J.P. Lin, in *2012 14th International Conference on Electronic Materials and Packaging (EMAP)*, 2012, p. 1
66. Y. Shen, L. Zhang, W.H. Zhu, J. Zhou, X.J. Fan, *IEEE Transactions on Components, Packaging and Manufacturing Technology*, doi:10.1109/TCPMT.2016.2592947, 2016
67. A.N.V. da Silva, Dissertation, Universidade de Lisboa, 2010
68. Y. Wen, C. Basaran, J. Electron. Packag. **125**, 134 (2003)
69. S. Timoshenko, J. Opt. Soc. Am. **11**, 233 (1925)
70. E. Suhir, J. Appl. Mech. **53**, 657 (1986)
71. R.R. Valisetty, L.W. Rehfield, A theory for stress analysis of composite laminates. Paper presented at 4thAIAA/ASME/ASCE/AHS structures, structural dynamics and materials conference, Lake Tahoe, NV, 1983
72. X. Liu, M. Li, D.R. Mullen, J. Cline, S.K. Sitaraman, IEEE Trans. Device Mater. Reliab. **14**, 512 (2014)
73. A.A.O. Tay, T.Y. Lin, in *Inter-Society Conference on Thermal Phenomena in Electronic Systems. I-THERM V*, 1996, p. 67
74. A.A.O. Tay, T.Y. Lin, in *The Sixth Intersociety Conference on Thermal and Thermomechanical Phenomena in Electronic Systems*, 1998, p. 179
75. J.E. Galloway, B.M. Miles, IEEE Trans. Compon. Packag. Manuf. Technol. Part A **20**, 274 (1997)
76. X.J. Fan, Mechanics of moisture for polymers: fundamental concepts and model study. Paper presented at EuroSimE 2008—international conference on thermal, mechanical and multiphysics simulation and experiments in microelectronics and micro-systems, 2008
77. T.Y. Tee, Z. Zhong, Microelectron. Reliab. **44**, 105 (2004)
78. L. Chen, J. Adams, H.-W. Chu, X.J. Fan, J. Mater, Sci. Mater. Electron. **27**(1), 481–488 (2015)
79. L. Chen, H.-W. Chu, X.J. Fan, J. Polym. Sci. B **53**, 1440 (2015)
80. X.J. Fan, S.W.R. Lee, Q. Han, Microelectron. Reliab. **49**, 861 (2009)
81. X.J. Fan, J. Zhou, G.Q. Zhang, A. Chandra, in *Moisture Sensitivity of Plastic Packages of IC Devices*, ed. by X.J. Fan, E. Suhir (Springer, Berlin, 2010), p. 279
82. M.D. Placette, X. Fan, J.-H. Zhao, D. Edwards, Microelectron. Reliab. **52**, 1401 (2012)
83. W.D. van Driel, M. van Gils, X.J. Fan, G.Q. Zhang, L.J. Ernst, IEEE Trans. Compon. Packag. Technol. **31**, 260 (2008)
84. X.J. Fan, G.Q. Zhang, W.D. van Driel, L.J. Ernst, in *Proceedings of the Electronic Components and Technology Conference*, 2003, p. 733
85. F. Le, S.W.R. Lee, K.M. Lau, C.P. Yue, J.K.O. Sin, P.K.T. Mok, K. Wing-Hung, C. Hoi Wai, in *IEEE 64th Electronic Components and Technology Conference (ECTC)*, 2014, p. 919

86. J. Crank, *The Mathematics of Diffusion* (Oxford University Press, Oxford, 1956)
87. S.R. DeGroot, P. Mazur, *Non Equilibrium Thermodynamics* (North Holland Publishing, North Holland, Amsterdam, 1962)
88. P. Sofronis, R.M. McMeeking, J. Mech. Phys. Solids **37**, 317 (1989)
89. X.J. Fan, J. Zhou, G.Q. Zhang, L.J. Ernst, J. Electron. Packag. **127**, 262 (2005)
90. Y.C. Fung, *Foundations of Solid Mechanics* (Prentice Hall, Upper Saddle River, NJ, 1965)

Chapter 13
Processing and Reliability of Solder Interconnections in Stacked Packaging

Paul Vianco

13.1 Introduction

This chapter provides a synopsis of stacked packaging technology. The discussion will highlight key aspects of materials and assembly processes, which have rapidly matured to meet the implementation of this technology in the consumer electronics market place. Added focus will be placed on long-term reliability because the military, space, and other high-reliability electronics sectors have yet to fully embrace stacked package technology across its market place.

There is a wealth of information available to the reader, which addresses the materials, processes, and reliability aspects of stacked packages. Those resources range from magazine feature articles to conference proceedings and peer-reviewed journal articles. Search engines allow the reader easy access to nearly all of these publications. Therefore, in order to avoid filling the available pages with reference citations, which could easily be the case, the author provides a limited sampling of citations within each topical area. The reader is recommended to further search for additional resources that can provide additional details.

13.1.1 Miniaturization and Functionality Trends

The miniaturization of electronic packages, whether passive devices such as capacitors and resistors or active devices ranging from diodes to microprocessors, has allowed a greater number of components to be assembled to printed circuit boards

P. Vianco (✉)
Sandia National Laboratories, Albuquerque, NM, USA
e-mail: ptvianc@sandia.gov

Y. Li, D. Goyal (eds.), *3D Microelectronic Packaging*, Springer Series in Advanced Microelectronics 57, DOI 10.1007/978-3-319-44586-1_13

(PCBs). The benefit to the consumer electronics as well as high reliability electronics (military, satellite, and space) communities has been not only smaller, lighter products, but also a significant increase in their functionality.

This trend has proven to be particularly advantageous for the larger active devices such as microprocessors. These components traditionally have the largest area footprint on the PCB. Over the past 40 years, as the silicon (Si) die have reduced in size and the accompanying package has followed suit. The size reduction has been accompanied by a change of the package input/output (I/O) configuration in order to compliment the decrease of package size as well as maintain yields in assembly processes. The pin-in-hole geometry of early generation, dual in-line packages (DIPs) gave way to J-lead and gull-wing lead configurations. As package functionality increased, more leads were required for power, ground, and signal transmission. However, their size had to decrease along with the reduced package dimensions. As pitches reduced to 0.4 mm and less (so-called "fine pitch" in the 1990s), the thin, flimsy leads became susceptible to handling damage, which negatively impacted manufacturing yields.

Therefore, the peripheral-leaded package geometry was replaced with an area-array of I/O placed on the bottom of the package. First, the area-array I/O configuration used the pin-grid array that required pin-in-hole solder joints or sockets. The ball-grid array (BGA) package eliminated the need for through-hole interconnections, which improved PCB producibility because the interconnections are entirely surface mount. The BGA has solder ball interconnections placed in an array on the bottom of the injected molded, plastic package. The counterpart to the plastic ball-grid array (PBGA) package was the ceramic column-grid array (CCGA) package that meets the need to package the die in a hermetic environment, yet can meet reliability requirements when assembled to a laminate PCB.

The consumer products industry and, in particular, the explosive growth in hand-held products, from the cell phone and to the iPhone (iPhone is a trademark of Apple Corporation, Cupertino, CA), has led the way to further miniaturization *and* increased functionality of electronic assemblies. The designers were forced to decrease the PCB footprint to meet size limitations. Therefore, the only remaining option to increase functionality was to add capability in the vertical ("*z*") dimension, which has led to the development of three-dimensional (3D) packaging.

It is important to realize that miniaturization was not the only driver for designers to consider 3D packaging. At the same, product performance required that packages, and even the dice in the packages, be in closer proximity to one-another so as to reduce the travel time of signals between them. Moreover, shorter conductor lengths reduced parasitics in the circuit, which reduced power consumption as well as improved functional performance.

13.1.2 3D Packaging Variations

There are two primary 3D packaging methodologies: *stacked die* or *stacked package*. As the terms imply, stacked die describes the mounting of individual die upon one-another within a single, molded package. Interconnections were wire bonds between individual die. Today, through-silicon vias (TSVs) are replacing wire bonds. Stacked packaging refers to the mounting of molded packages upon one another. The earliest renditions of stacked packages used peripherally leaded components [1]. This practice began by stacking gull-wing leaded packages atop one another to increase the memory die available on an assembly [2]. Today, stacked packages are based upon the BGA format. Increased integration is being achieved in stacked packaging technology by stacking dice in each of the stacked packages.

Faced with market demands for miniaturization and increased product functionality, the consumer electronics industry has quickly moved from stacked packaging to stacked die technology. However, stacked packaging remains more attractive to the high-reliability electronics community. The particular advantage scenario is based upon the known good die (KGD) concept, except that the latter is extended to "known good package." Stacked packaging technology allows for verifying the performance and reliability of each individual package through burn-in, qualification, and acceptance test regimens. Therefore, known good packages are established prior to their assembly together and then to the PCB.

On the other hand, in the case of stacked die, wafer burn-in precedes die stacking (mounting and wire bonding) and subsequent molding into the package. The packaged, stacked die component may be placed through a second series of burn-in, qualification, and acceptance testing. However, with multiple die in each device, there is an increased likelihood that die yield, when multiplied by three, four, or however many dice are present, will reduce the yield of each stacked die package, which also implies statistically the potential for a reduced long-term reliability in the field. This chapter will focus on the stacked packing technology.

Although theoretically many packages can be stacked upon one another, there are practical limitations. First, there are design considerations to keep products, especially those for the consumer market, as thin as possible. Secondly, an increased count of individual packages requires additional signal routing to address the added functional complexity, which means higher I/O counts that increase the risk-reduced assembly yields. Third, an increased number of packages in the stack raises concern with thermal management, in particular, removing the heat generated by a microprocessor die as well as from multiple memory dice. Fourth, manufacturability challenges add up with each package addition to the point they can interfere with the clearances available in pick-and-place or reflow equipment. Lastly, taller stacks increase their susceptibility to damage during handling and transportation as well as reduced reliability caused by shock and vibration environments experienced in service. Therefore, the two common variants are: *package-on-package (PoP)* that stacks two (2) packages atop each other, and

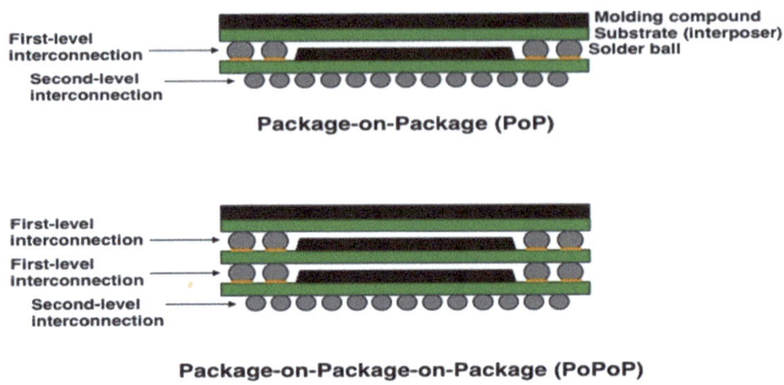

Fig. 13.1 Schematic diagrams illustrate the (**a**) PoP and (**b**) PoPoP components

package-on-package-on-package (PoPoP) that is a three-package stack. Any needs for further miniaturization cause the stacked packaging option to give way to the stacked die technology.

Schematic diagrams are shown in Fig. 13.1 that illustrates the general configurations of PoP and PoPoP components. (The features are not to scale.) The most important distinction to be made for either configuration is the solder joints. There are the *first-level* interconnections that route signals between the individual packages. Then, there are the *second-level* interconnections, which attach the PoP or PoPoP to the PCB. Although exposed to the same environments, the responses are entirely different between the first-level and second-level interconnections due to the different materials and structures between them.

The solder joint arrays differ between the first-level and second-level interconnections. The first-level joints are typically arranged in a perimeter array around the molding compound on the package below it. The second-level joints can be patterned to a full array that covers the entire bottom-side of the package footprint, or similarly arranged in a perimeter configuration, often three or four rows to a side. The added capacity is required of the second-level joints since all of the signals, ground, and power are routed between the stack and the PCB through these interconnections.

13.1.3 Applications Drive PoP and PoPoP Component Requirements

The performance and reliability requirements that are placed on PoP and PoPoP are determined by the application. Applications have been categorized by two product regimes: *consumer electronics* and *high-reliability electronics*. Consumer electronics includes hand-held devices (cell phones, laptops, tablets, etc.) as well as desktop computers, digital storage media, and entertainment systems. The high-reliability sector has products such as automotive electronics, telecommunications, high-end servers, avionics as well as military, space, and satellite electronics. In the case of consumer electronics, short product cycles and relatively benign use environments (generally offices and homes) allow performance and low-cost to drive the requirements placed on stacked packaging technology. The primary reliability concern is mechanical shock caused by dropping hand-held devices to the ground.

High-reliability applications place different requirements on electronic packaging and associated interconnections. Shock and vibration are less severe than a drop, but have longer durations such as experienced in launch cycles or the constant flow of forced air in a cooling system. The latter environments generate *high-cycle fatigue* conditions for the interconnections on the printed wiring assembly (PWA). The particular implication for PoP and PoPoP components is not so much degradation to the stacked package itself, but rather bending of the PWA that exposes the second-level solder joints to the fatigue loads.

The primary concern in high reliability applications is *temperature cycling*. Aside from potential thermal degradation of materials at the elevated temperature portion of the cycle, there is the concern for *low-cycle fatigue* damage to the solder joints that caused the mismatch of coefficients of thermal expansion (CTE) between the various materials that comprise the package, solder joints structures, and the PCB. The degradation scenario is referred to as *thermal mechanical fatigue (TMF)*, which can lead to a crack solder joint and thus, an electrical open.

The TMF condition presents a particularly complex problem for PoP and PoPoP due to the multiple materials and construction of the components. Besides the molding compound and substrate (interposer) structures of the individual packages, there is also the PCB—laminate or ceramic—to which is attached the component. The introduction of encapsulants, underfills, and conformal coating further complicate the overall stacked package structure. The variety of materials sets and geometries present in PoP and PoPoP components limits the practicality of empirical programs to obtain accelerated aging data to predict the TMF of the solder joints for every variation of these contributing factors. Computational modeling, together with validation tests of more limited scope, are finding greater use in predicting the reliability of PoP and PoPoP solder joints that is required for high-reliability applications.

13.2 Soldering Assembly Processes

Even today, probably the predominant contributing factor to yield loss within stacked package technology are solder joint defects that are caused by *warpage* [3, 4]. The study described in [4] examined, specifically, the role of the top package's material properties—CTE, modulus, etc.—on the extent of its warpage during the reflow process. Unfortunately, the process engineer rarely has the opportunity to select PoP or PoPoP packages having material properties that are optimized for an assembly process. Therefore, the topic of warpage is highlighted in the Soldering Assembly Process section because its effect is felt most strongly at assembly. As such, optimizing the assembly process parameters provides the greatest likelihood of minimizing warpage behavior.

This section will provide an overview of these factors that comprise the soldering assembly process: solder alloys, fluxes and pastes, assembly methodologies, conformal coatings and underfills, as well as inspection techniques as they pertain to stacked packaging technology. Also, as noted above, the discussion will address warpage.

13.2.1 Solder Alloys

13.2.1.1 Sn-Pb Solders

The advent of stacked packages, whether early versions that combined peripherally leaded packages or the current area-array geometries, used the eutectic Sn-Pb solder. This alloy has the composition of 63Sn-37Pb (wt.%) and equivalent solidus and liquidus temperatures of T_{solidus} (T_{s}) and T_{liquidus} (T_{l}) = 183 °C. The soldering processes, which are well-established in the electronics community, use the lowest temperatures in a reflow profile that minimizes warpage-related defects. Most high-reliability electronics are still assembled with Sn-Pb solder, albeit with greater challenges in a supply chain that is moving away from this technology.

13.2.1.2 Pb-Free Solders: "High Ag" Alloys

The time frame having the most significant growth of stacked package has coincided with the consumer electronics industry rapid change over to Pb-free solutions. Eutectic Sn-Pb solder balls destined for both first- and second-level interconnections were replaced with Sn-Ag-Cu alloy balls. The predominant Sn-Ag-Cu alloy used in Pb-free PoP and PoPoP components was the 96.5Sn-3.0Ag-0.5Cu (SAC305) solder. This composition is near eutectic, having T_{s} and T_{l} that are nearly equivalent at 217 °C. This melting temperature is considerably higher than that of Sn-Pb solder. Therefore, the SAC305 has peak reflow temperatures in the range of 235–245 °C (as opposed to 220–235 °C for Sn-Pb solder). Of course, in the case of PoP and PoPoP components, the immediate concern with the higher process temperatures is an increased warpage experienced by each individual package.

There are also companion alloys that have slightly different compositions, but with the SAC305 solder, are categorized together as the "high Ag" solders: 95.5Sn-4.0Ag-0.5Cu (SAC405), 95.5Sn3.9Ag0.6Cu (SAC396), and 95.5Sn-3.8Ag-0.7Cu (SAC387). These solders perform very similarly to the SAC305 alloy. They are only slightly off the ternary eutectic composition, which adds from one (1) to four (4) Celsius degrees to their respective liquidus temperatures.

13.2.1.3 Pb-Free Solders: "Low Ag" Alloys

The second set of Pb-free solders are the "low Ag" alloys that were developed, initially, to improve the resistance of BGA solder joints to drop-shock failure. These alloys have now gained relevance in the stacked packaging arena for the same reasons. The basic compositions are 98.3Sn-1.2Ag-0.5Cu (SAC125) and 98.5Sn-1.0Ag-0.5Cu (SAC105). There has been a rapid proliferation of other quaternary as well as higher-order alloys with elemental additions such as Ni, Mn, etc. that allow the assembled component to meet or exceed the JEDEC requirements [5, 6]. A number of studies have shown the improved resistance to failure under drop-shock conditions provided by the low-Ag alloys. For example, Wang, et al. showed that the SAC105 solder ball provided better performances than did underfilled, SAC305 interconnections [7]. A particular drawback of all of the low-Ag alloys is a higher liquidus temperature. In some cases, the liquidus temperature can be as high as 228 °C. The higher liquidus temperature drives up the peak reflow temperature in the assembly process, which in turn increases the risk of package warpage.

13.2.1.4 Mixed Solder Joints

Mixed solder interconnections are defined as those joints that were formed by the combination of two different solder compositions, one for the solder ball and different composition for the solder paste. This scenario, which is referred to as *backwards compatibility*, is most likely to take place with the introduction of stacked packages into high-reliability applications (military, space, etc.). The PoP and PoPoP components likely come from the commercial supply chain, and thus, have Pb-free solder balls to meet the environmental restrictions of that market. On the other hand, high-reliability electronics are still assembled (albeit, to a declining degree) with Sn-Pb solder.

A mixed solder scenario warrants that consideration be given to the solder joint microstructure generated by the reflow process. That microstructure may be homogeneous such as that shown in Fig. 13.2a. Or, if the reflow processes is not sufficiently "severe," that is, the peak temperature is too low or the time above liquidus temperature is too short, a segregated microstructure develops due to poor intermixing by the ball and paste alloys as shown in Fig. 13.2b. The different microstructures can have different responses to the shock, vibration, and TMF loads placed on the interconnections. Moreover, the change to those properties may not always be intuitive. Shown in Fig. 13.3 is the Weibull statistics-based, failure plot for the solder joint illustrated in Fig. 13.2a (−55 °C/125 °C; 10 min hold times; ramp rates of 6–8 °C/min). The mixed solder interconnection has a longer fatigue life as benchmarked by a characteristic lifetime of 2660 ± 90 cycles versus the 100 % SAC305 solder joint, which exhibited 2030 ± 90 cycles [8]. However, as will be illustrated later on in this chapter, this trend was *not* repeated when the solder joint in Fig. 13.2a was exposed to the same temperature cycle, but is in the presence of an underfill.

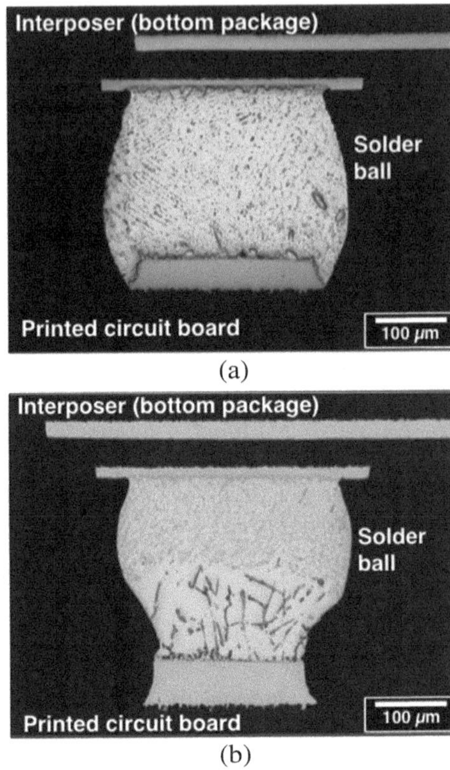

Fig. 13.2 Optical micrographs illustrate the two variations of microstructure that result from a backwards compatibility scenario based upon SAC305 solder ball and Sn-Pb solder paste used to construct the second-level interconnections: (**a**) a homogeneous microstructure and (**b**) an inhomogeneous microstructure interconnection

13.2.2 Fluxes and Pastes

Stacked package technology has led to several variations to the typical flux and paste technologies used to solder assemble PWAs. While the second-level assembly of the PoP or PoPoP components to the PCB uses those mainstream fluxes and pastes, the assembly of the package stack requires alternative materials because it is not possible to print paste on the top of the packages to form the first-level interconnections. Fortunately, those first-level interconnections do not require a significant addition of solder metal to form the joint because of their small size and fine pitch. Although dispensing flux or paste by either a typical dispenser equipment or jetting is a viable process, it can create a bottleneck in high-volume manufacturing lines.

The first-level interconnections are made with either flux, only, or with the use of a paste (flux plus solder metal). The two materials are referred to as *dip flux* and *dip paste* [9–11]. As the terms imply, flux or paste is added to the balls of the upper packages by dipping the solder spheres into baths of either material. Dip flux is also referred to as *tacky* flux, which signifies its added function to prevent inadvertent movement between the two packages, once they have been stacked upon one another. Such movement can lead to solder joint defects. As described in the

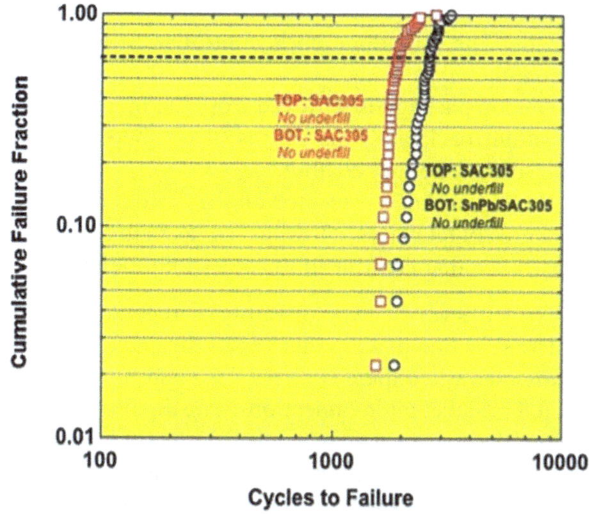

Fig. 13.3 Weibull failure plot shows the results for two PoP assemblies [8]. Both assemblies used SAC305 alloy for the top solder joints. One test vehicle had 100 % SAC305 solder for the bottom joints and the other had backwards compatibility interconnections formed from SAC305 balls and Sn-Pb paste. The thermal cycle had these parameters: lower temperature limit of −55 °C; upper temperature limit of 125 °C; 10 min hold times; and 6–8 °C/min ramp rates

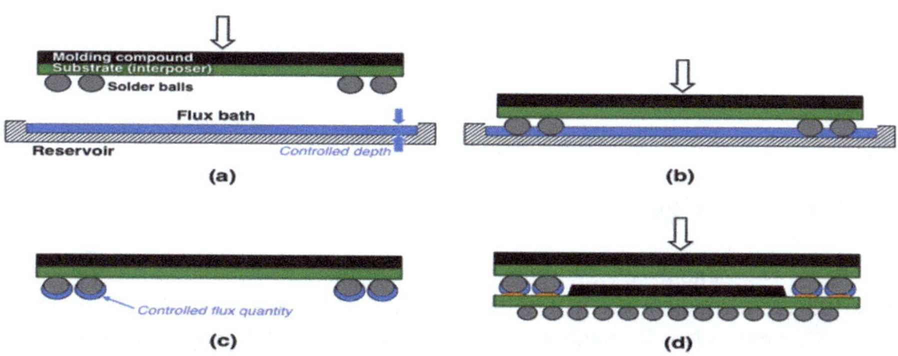

Fig. 13.4 Schematic diagram shows the flux dipping process used to make the first-level interconnections, in this case, for a PoP: (**a**) a flux bath is created with a tightly controlled depth to deliver a consistent quantity of flux to each solder ball, (**b**) top package is immersed in the flux bath, (**c**) the package is lifted from the bath and transferred to the bottom package, and (**d**) the package is placed on top of the bottom package

cited references, these materials are precisely engineered to control the amount and consistency of either flux or paste that is transferred to the solder balls in order to optimize assembly yields.

The schematic diagram in Fig. 13.4 illustrates the dipping process steps:

- A flux reservoir is created having a specific depth of flux. The depth of the bath is controlled, precisely, in order to deliver a consistent quantity of flux to each solder ball.

- The top package has its solder balls dipped into the flux. The flux wets the solder balls. The time duration is controlled to further ensure consistency of the flux coating.
- The top package is removed from the flux bath and transported to the site of the bottom package. This step is also timed, precisely, in to prevent inadvertent drying of the flux coating.
- The top package is located on the bottom package. Although the flux has a high degree of tackiness, care must be taken to prevent excessive movement of the top package that could lead to solder joint defects.

The steps (a)–(c) would simply be repeated one more time for the PoPoP case.

The selection of a dip flux versus dip paste is based upon several factors. The dip flux is used when there is no need for additional solder metal as is the case for finer pitch, first-level interconnections. The dip flux is preferred in the event that warpage causes the two packages to displace closer together, which increases the risk of short-circuits due to excess solder causing bridge defects. The dip paste provides added solder metal to the joints and is used to compensate for significant warpage in the event that the latter causes the two packages to separate relative to one-another, which can lead to solder joint open.

13.2.3 Assembly Methodologies

13.2.3.1 Stacked Packages

There are two general assembly processes for stacked packaging technology. The first process assembles the individual packages into a PoP or PoPoP component, first, using multiple solder reflow steps. This method is often referred to as the *two-pass process* for PoP components, or *three-pass process* for PoPoP components. Returning to Fig. 13.4, the unit illustrated in Fig. 13.4d would next be placed into a reflow process to form the second-level interconnections. Steps (a)–(d) would be repeated, *including another reflow step*, to create the PoPoP component. Then, the now-assembled PoP or PoPoP components are pick-and-placed on to the PCB with other surface mount components and passed through a final reflow process to create the second-level interconnections.

The two- or three-pass process is not the preferred approach for the following reasons: First of all, the multiple exposures to soldering temperatures increase the likelihood of not only damage to the internal structures (e.g., die attach, passivation layers, etc.), but also warpage that can cause solder joint open or bridging defects. Secondly, the stacked package is more susceptible to handling damage in the tray as well as during the pick-and-place process. Third, the multiple reflow steps lengthen the manufacturing time.

The second and preferred process is illustrated in Fig. 13.5. It is referred to as the *single-pass process* and is represented here with dip flux. The same procedure

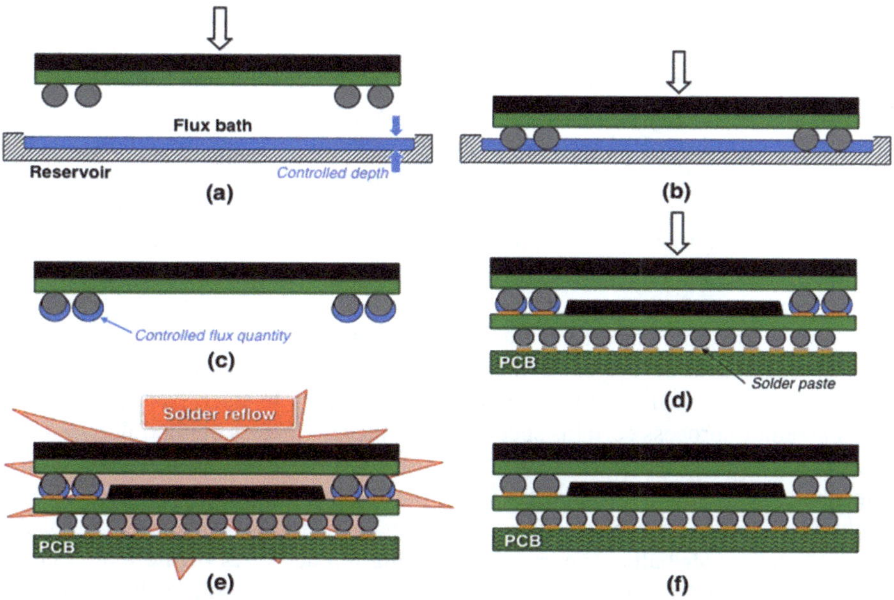

Fig. 13.5 (**a–d**) The flux dipping steps are repeated, here, as described in Fig. 13.4. (**e**) The assembly is placed in the solder reflow furnace to melt the first-level interconnection balls, now covered with flux, and the second-level solder balls together with the printed paste. (**f**) The completed PoP component is shown soldered to the PCB

applies to a dip paste. The steps (a) through (d) are the same as in Fig. 13.4. The exception is that the bottom package has already been placed on the PCB with printed solder paste to create the second-level interconnections. The remaining steps are described below:

- The PoP assembly is placed into the reflow furnace to make both the first-level (top package-to-bottom package) and second-level (bottom package-to-PCB) solder joints—hence, the term, single-pass process.
- The diagram illustrates the final PoP assembly.

It is important to recognize that, although the single-pass process is the preferred approach, there may be circumstances that warrant the two- or three-pass process, either due to the characteristics of the overall PWA or limited equipment availability. Nevertheless, it was demonstrated by Srinivas, et al. that, under accelerated aging by means of temperature cycling, the two assembly processes did not result in a statistically significant difference in long-term reliability of the interconnections [12].

The assembly of stack packages is performed almost exclusively by automated pick-and-place equipment. The fine-pitch of the first- and second-level interconnections causes hand placement to be very difficult. Also, when using dip flux, there is little molten solder metal to provide self-alignment by the *surface tension effect*, which places further challenges on the precision of manual placement.

13.2.3.2 Soldering Assembly (Second-Level Interconnections)

There are not any special measures that must be exercised in the reflow (second-level assembly) process for a PWA that has stacked packages on it. Given the complexity of today's PWAs, it often proves difficult to adjust the reflow process solely to account for PoP or PoPoP components—e.g., to minimize warpage—while also maintaining a satisfactory yield for the solder joints of all other devices. In the event that there is a suspicion that the PoP or PoPoP materials set and geometry causes the component to be susceptible to excessive warpage or damage due to process peak temperatures, hold times, and ramp rates, it is necessary to perform the "up-front" development studies to understand these potential degradation mechanisms as part of establishing a PWA soldering process.

The one aspect that must be considered with respect to stack packages and the PWA assembly process pertains specifically to the single-pass methodology. In this case, the PWA process must also reflow, for the first time, the first-level solder joints between individual packages. At the present time, alternative Pb-free solders may be used for the first-level interconnections, which may differ from that of the second-level solder balls. The range of melting temperatures of these alternative solders (Sect. 13.4.1) must be understood in order to be assured that first-level solder balls, or solder balls-plus-(dip) paste, will reflow during the second-level assembly process. This point becomes particularly important in the case of high reliability PWAs being assembled with the traditional eutectic Sn-Pb solder. The reduced temperature profile may not be sufficient to melt the higher Sn solder balls. Running a Sn-Pb second-level assembly process "hotter" than normal to alleviate this discrepancy poses a risk of thermal damage to the other components on the PWA.

13.2.3.3 Cleaning Considerations

The consumer electronics industry uses processes that are based upon *no-clean* or *low-residue* tacky fluxes or similar fluxes contained in the solder pastes. These flux technologies are utilized for the assembly of first-level interconnections as well as the final soldering step used to make all second-level interconnections to the PCB. Therefore, cleaning processes are not always required in this manufacturing environment.

On the other hand, PWAs for high-reliability systems are required to have flux residues removed by a suitable cleaning process. This action avoids potential reliability concerns posed by corrosion mechanisms. Cleaning processes are also recommended to establish pristine surfaces that will optimize both the flow and adhesion of conformal coatings, encapsulants, and underfills.

By-and-large, stacked packages can accommodate cleaning systems—defined as the cleaning materials and the equipment. The cleaning materials, which are selected based on the type of flux residues, have no greater propensity to attack

PoP and PoPoP materials than they degrade the more traditional plastic encapsulated microcircuits. The gaps between individual packages do not present a particular challenge to current cleaning systems. Because stacked packages have a higher profile than do BGAs, passive devices, etc., it is prudent to confirm that high-pressure jets do not damage the first- or second-level interconnections.

However, post-soldering cleanliness requirements may affect the choice of first-pass versus two-pass (PoP) or three-pass (PoPoP) methods as a part of the overall soldering process. The first-pass approach is preferred because both first- and second-level solder joints are made in the same process step. The tacky flux or tacky paste has a short residence time on the parts prior to reflow, and residues after reflow, which increases the efficacy of the subsequent cleaning step. Conversely, the multiple pass process requires that flux residues be cleaned from the *individual stacked components* shortly after making each set of first-level interconnections. Otherwise, there is the possibility of a long-time interval developing between making the first-level interconnections and the second-level reflow soldering step. The flux residues can become more tenacious, which increases the difficulty of removing them in the post-reflow (second-level assembly) cleaning step.

A second challenge to removing flux residues at the stacked package level is the need for special fixtures to hold the PoP or PoPoP units. Trays or other fixture methodologies must secure the components against movement under the pressures jets in the cleaning machine. Otherwise, the package structures and, in particular, the bottom package solder balls become susceptible to damage, which reduces assembly yields of the PWA.

13.2.3.4 Rework

The difficulty of performing a manual assembly of PoP and PoPoP parts predisposes rework operations to only high-reliability assemblies. The rework of stacked package components was discussed by McCormick et al. [13]. A critical finding that was drawn by this author from that work is that the existing package cannot be readily salvaged, even if the defect can be traced to the first- or second-level interconnections. Overall miniaturization, together with the fine pitch of both first- and second-level interconnections, simply adds too much risk of ancillary degradation to the individual package structures to warrant their re-use.

The premise that rework leads to the scrapping of the pre-existing package (s) places a significant cost burden on stacked package use on high-reliability electronics because piece-parts typically have very high unit values. High unit costs are due to short-run (low-volume) manufacturing contracts and the need for extensive testing, particularly when up-screening test regimens are applied to COTS components. Therefore, the best approach to this scenario is to optimize the process through development up to first article assembly. This method assures the process engineer of acceptable yields in order to enable stacked packages for high-reliability applications.

13.2.4 Inspection Techniques

The post-assembly inspection of PoP and PoPoP components on a PWA is generally not unlike that of other BGA components. The acceptance criteria are placed on both first-level and second-level interconnections. The same defects are observed on stacked packages: damaged to the molding compound and/or substrate (interposer); missing solder balls; open solder joints; and solder joint bridges that lead to short-circuits.

The inspection methods are also the same as those used for area-array packages. A visual inspection can be made of the perimeter joints. A boroscope can be used to examine the internal connections. However, the latter method may have limited application with PoP and PoPoP first- and second-level interconnections. The trend for these packages is towards fine pitches that limit the space around the solder joints to maneuver the light source and camera.

X-ray inspection provides the means to examine the internal interconnections. However, there is an added complexity due to the multiple rows of solder joints. This point is illustrated by the x-radiographs in Fig. 13.6, which shows the top-down views of (a) a PoP and (b) a PoPoP components. The PoP component radiograph (Fig. 13.6a) can be interpreted for defects because the two levels of solder joints can be distinguished from one another by number of perimeter rows and solder ball size. The two rows of large solder balls comprise the first-level interconnections and four rows of smaller solder balls are the second-level solder joints.

However, the challenge increases significantly to interpret x-ray images of PoPoP components as is evident in Fig. 13.6b. Defects can become hidden by the shadows of interconnections above *and* below the suspect solder joint. Also, there is uncertainty as to which of the overlapping interconnections actually contains the defect.

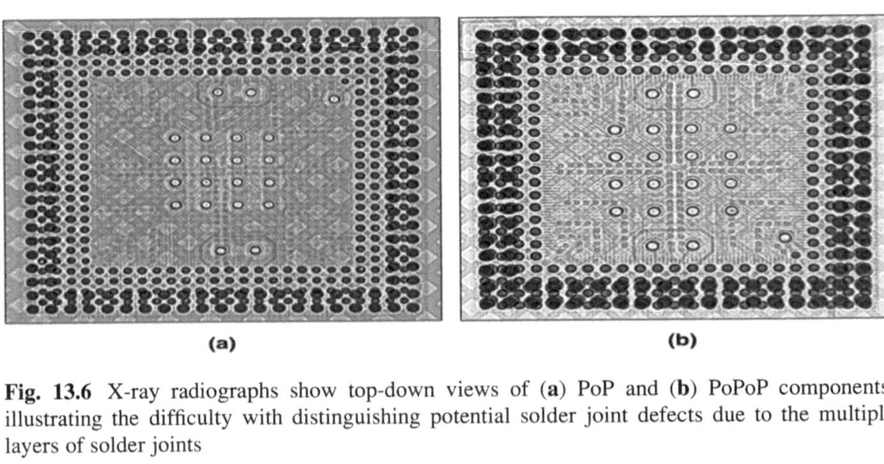

(a) (b)

Fig. 13.6 X-ray radiographs show top-down views of (**a**) PoP and (**b**) PoPoP components, illustrating the difficulty with distinguishing potential solder joint defects due to the multiple layers of solder joints

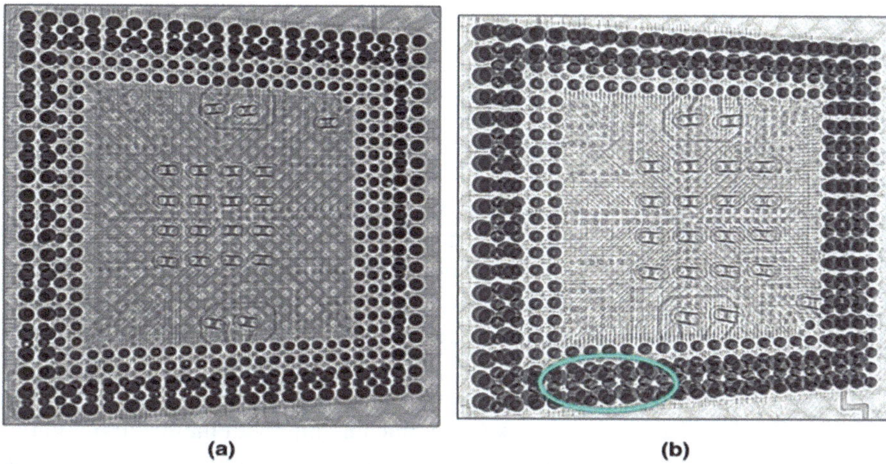

Fig. 13.7 The assemblies were tilted in an effort to view all of the rows of solder joints in the (**a**) PoP and (**b**) PoPoP component; the effort was left with minimal success, especially of the PoPoP component as highlighted by the *cyan oval*

At first glance, the approach to mitigate this shadowing effect is to carefully tilt the component in order to project a separate view of each level of solder joints. Figure 13.7 shows 30° views, which is the optimum angle for the same PoP and PoPoP components. Clearly, the radiograph interpretations were made even *more* difficult for identifying solder joint defects in either package but, particularly so, for the PoPoP unit—e.g., the region within the cyan oval.

The risk of a misinterpretation of x-ray images can be remedied through the use of the *computer tomography* (*CT*) technique. This method produces image "slices" through the device. Those image slices are then stitched together by a computational algorithm to create a three-dimensional (3D) construct of all of the interconnections—and their defects. However, the time required to perform the scans and construct the image (upwards of 2–4 h) is prohibitive, certainly for high-volume assembly processes, but also for the low-volume production runs that characterize the manufacturing of high-reliability electronics.

It is clear that the difficulties associated with visual and x-ray inspection of stacked packages are fundamental to the construction of these components, regardless of the equipment and manpower resources. Therefore, from the Design for Manufacturing (DfM) aspect of assuring that the product is made with the required reliability, it is more advantageous to optimize (a) the PWA design to achieve the desired functionality and reliability as well as (b) the assembly process steps for PWAs having PoP and PoPoP components. This approach is far more cost-effective than is "inspecting-in" performance and reliability at the quality assurance stage.

13.2.5 Underfill, Conformal Coatings, and Encapsulants

13.2.5.1 Underfill

Underfills provide added assurance that a stacked package will withstand the mechanical loads caused by shock and vibration conditions. This mitigation is particularly attractive to prevent the "drop-shock" failure of hand-held consumer electronics [14]. In particular, underfills are an attractive alternative to the so-called shock-resistant, (low-Ag) Pb-free solders when the latter are not viable because acceleration forces are particularly high, or when the higher processing temperatures of the low-Ag, Pb-free solders degrade the manufacturing yields of PoP and PoPoP assemblies.

The use of underfills is not simply a drop-in step for the assembly process. High-volume manufacturing processes take a significant "hit" with the incorporation of an underfill step. There are the capital and recurring costs associated with the dispensing equipment. Secondly, although the actual application of the underfill to a stacked package can be performed in less than a few seconds, ancillary factors such as curing time and limited pot life of the underfill on the manufacturing floor add to the overall process flow time [15].

High-reliability applications place less emphasis on drop-shock resistance. Rather, PWAs and the stacked packages on them must withstand the vibration and shock environments associated with military, space, and satellite applications. Those conditions include vibration during launch and shock loads resulting from staging sequences. It is anticipated that these environments will necessitate the use of underfills with PoP and PoPoP components.

However, these applications also require that consideration be given to the effects of underfills on solder joint TMF when exposed to *temperature cycling conditions*. In particular, reliability under temperature cycling can be a controlling factor that determines, outright, whether or not underfills can be used between some, or all, of the packages. More details are provided in Sect. II that addresses, specifically, solder joint reliability and the effects of underfills and encapsulants.

The use of an underfill begins by identifying the appropriate material. Adapting an underfill step to an assembly process will establish requirements such as pot life, viscosity, and curing schedule details. Then, the underfill physical and mechanical properties must accommodate the reliability requirements. Underfill materials that do not contain filler particles (e.g., silica) are generally better suited for drop shock resistance. On the other hand, those underfills with filler components provide better resistance to solder joint TMF under temperature cycling conditions [15]. The filler particles are used to tailor the underfill CTE to better match the CTEs of the packages (molding compounds), PCB, and the solder joints. Attention should also be given to the glass transition temperature (T_g) of the underfill material. Although it is not possible to avoid the glass transition point and the abrupt change in materials properties associated with crossing it during the reflow process, it is preferred to avoid T_g during accelerated aging and/or in service to minimize solder

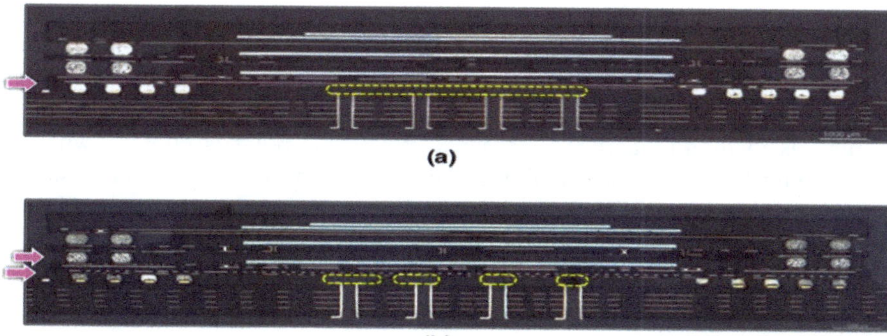

(a)

(b)

Fig. 13.8 Optical micrographs show PoPoP components in cross sections: (**a**) underfill is present only in the bottom gap (*magenta arrow*) and (**b**) underfill is present in the middle and bottom gaps. The *yellow, dashed ovals* highlight voids in the underfill and, in particular, the different morphologies that they can take in the gap

joint degradation. The wide variety of underfill materials on the market, today, allows design and process engineers the opportunity to optimize this technology over a range of performance and reliability requirements.

The second step is to determine which stacked package gaps require an underfill to meet the assembly reliability requirements. Photographs are shown in Fig. 13.8 of two PoPoP components. Figure 13.8a has underfill only between the bottom package and the PCB, that is, around the second-level interconnections (magenta arrow). The photograph in Fig. 13.8b shows a unit that has underfill in the middle gap (first-level interconnections) and in the bottom gap. The dashed, yellow ovals accentuate voids in the underfill. There is one, long void over the pattern of through-vias in Fig. 13.8a and four small voids over each of four through-vias in Fig. 13.8b. These images demonstrate the sensitivity of underfill flow behavior to discontinuities in the surfaces because of a driving force to minimize its liquid surface energy.

Underfill is applied after the assembly of the stacked package to the PCB. When multiple gaps are to be filled, one approach is to perform an “L” pattern pass at each gap, beginning with the second-level interconnection gap. The underfill flows through the gap by capillary action. A fillet is formed between the bottom-most package and the PCB. This fillet is important as it reduces the residual stresses that can develop at the edge of the packages after curing as well as under shock or temperature cycling environments.

A second approach is to fill all of the gaps by a *single pass* of the dispensing needle. The advantage is reduced process time. However, this technique requires a greater control of the underfill material’s flow properties (e.g., temperature, flow rate, etc.) because a larger quantity of material is being dispensed at one time. The failure in either regard can quickly lead to voids and a “mess” on the PCB. Also, the fact that the volume of underfill must fill multiple gaps at the same time increases the likelihood of voids and poor fillet development.

13.2.5.2 Conformal Coatings

The use of conformal coatings on stacked packages has not been extensively explored in the electronics community. The primary application of conformal coatings is in high-reliability electronics. When this point is coupled with the slow introduction of PoP and PoPoP technology into military, space, and satellite system, there has been only a limited need for such studies, that is, until very recently. Several resources provide excellent reviews of the four basic types of conformal coatings: (a) acrylics, (b) urethanes, (c) silicones, and (d) parylene, including attributes versus drawbacks, material properties, and application methods [16, 17].

The primary concern is the extent to which the conformal coating fills the gaps between the first- and second-level interconnections. Reliability concerns arise if the coating completely bridges the gap between the two packages, or between the bottom package and the PCB. The high CTE value of conformal coatings vis-à-vis the solder alloy, coupled with the confined geometry created by the bridged gap, can result in significant residual stresses in the package and interconnections, alike. At elevated temperatures, expansion of the conformal coating can push apart the interconnections due to very high tensile loads. If the conformal coating simply flows into the gap, but does not bridge it, the solder joints are exposed to significantly reduced residual stresses.

Fortunately, the gaps are sufficiently large that conformal coating generally does not completely fill them. This point is illustrated in Fig. 13.9, which is a photograph of a PoP component having a urethane conformal coating over it. The orange arrow shows the first-level solder joint gap; it is not filled with coating material.

The selection of a conformal coating material is not particularly sensitive to a presence of PoP or PoPoP components on the PWA from the assembly perspective, unless there is the risk of bridging a gap. Then, parylene coatings are preferred because their thickness is limited to less than 125 μm. Otherwise, the introduction of underfills can mitigate the potential risks associated with presence of the other conformal coatings into the gaps.

Fig. 13.9 Stereo image of a PoP component having underfill in the *bottom* (second-level solder joints) gap and covered by a conformal coating. The latter did not bridge the first-level interconnection gap (*orange arrow*)

The method of application must consider the higher profile of stack packages. For example, the dipping and hand brushing methods can result in puddling of conformal coating around some, but not all, of the package sides. Puddles lead to asymmetries in the residual stresses arising from the mismatch of CTE values. Asymmetric residual stresses are more damaging to the first- or second-level solder joints than symmetric stresses of higher magnitudes. The better approach is to spray the coating over the PWA for a more even layer.

13.2.5.3 Encapsulants

As is the case with conformal coatings, encapsulations are typically used on high-reliability PWAs to prevent damage when exposed to shock and vibration environments. Therefore, there is not a large amount of empirical data, be it performance or reliability, that considers stack packages on an encapsulated assembly. Although the increased introduction of PoP and PoPoP components into high-reliability PWAs will certainly generate such data, it is more likely that design, process, and reliability engineers will rely more heavily upon computational modeling to guide their respective efforts.

The encapsulation process that is used for a PWA is not expected to depend explicitly on the presence of stack packages. Nevertheless, there are several factors that should be addressed to optimize the encapsulation technology in order to reduce any risk to the long-term reliability of the first- or second-level interconnections. First of all, the higher profiles of PoP and PoPoP components increase the chance of voids forming next to them due to flow disturbances around the package. These unfilled volumes can lead to asymmetrical shear and tensile loading placed on the stacked package and PCB, which will accelerate TMF of the solder joints.

Secondly, the need to mitigate residual stresses caused by the curing cycle or CTE mismatches may require that a compliant or "slush" layer be placed over the component prior to applying the encapsulant. Polysulfide layers are often used to fulfill this role; however, as a pseudo-conformal coating, flow behavior into the gaps must be fully characterized as well as the ensuing effect on solder joint reliability. The best approach to minimize an over-stress or excessive TMF of the solder joints is the development of a controlled application process that dispenses the slush layer in a repeatable manner.

13.2.6 Warpage Effects

Probably the most significant challenge that has faced the execution of stacked package technology is *warpage*. Warpage can pose a challenge to both the assembly process as well as the assurance of solder joint reliability in service conditions that include temperature cycling. The latter scenario is discussed in the following section.

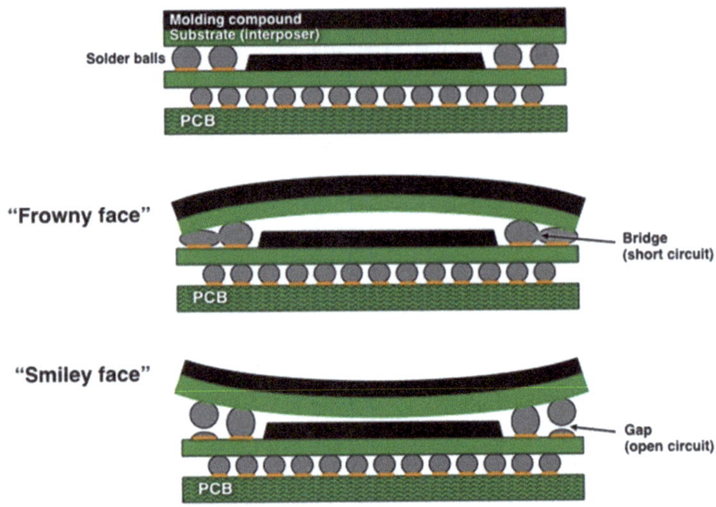

Fig. 13.10 The top schematic diagram shows baseline condition (no warpage) of the PoP device. During assembly reflow, when the solder balls are *liquid*, the middle picture shows the "frowny face" warpage that can lead to a solder bridge (*short circuit*). The bottom image illustrates "smiley face" warpage that leads to a gap (*open circuit*)

Between assembly yield and reliability, warpage has its greatest impact on the former. This premise stems from the fact that the solder joints, being in the molten condition, have zero load-bearing capacity. As a result, their shapes that are taken up at the time solidification are controlled by (a) the molten solder surface tension and (b) the warped geometries taken by the individual packages and PCB. Package warpage determines the yield of the first-level interconnections, while warpage of the bottom package *and* the underlying PCB control the second-level solder joint yield.

The schematic diagrams in Fig. 13.10 illustrate the two types of warpage experienced by packages and the potential defects. The top illustrate shows the baseline condition. Below it is the "frowny face" warpage, which can push together the outer rows of molten solder so as to create a bridge (short circuit). The bottom image is that of the "smiley face" warpage, which pulls apart the outer solder joints, potentially leading to a gap (open circuit).

There are two caveats that accompany these diagrams: First of all, only the top package has been allowed to warp. The solder joint defects become compounded when all of the packages, and the PCB, show a measure of warpage. Secondly, the illustrated warpage occurred in only two-dimensions. There is also the third dimension into/out-of the image plane, which can further complicate any ability to predict defects in the solder joints.

Numerous studies—in fact, too many to describe each in detail here—have examined various factors that affect warpage. For example, Smith, et al. recognized that the top package in a PoP stack has the greater priority in terms of managing warpage [18]. Moreover, the authors recognized that warpage is *dynamic*; it changes with temperature through the reflow profile. Relative displacements can reach upwards of 100–150 μm at the solder balls, which is sufficient to cause the defects illustrated in Fig. 13.10. Zhao and co-workers explored the effects of two factors, substrate design and top package design, on the latter's warpage behavior [3]. Copper density and balance in the substrate, as well as the relative volumes and footprints of the dice and molding compound, are critical factors with respect to package warpage. Similar trends were corroborated by Vianco and co-workers, using a computational modeling routine [8].

The roles of molding compound and substrate *materials properties* were examined as factors controlling top package warpage by Yim et al. [19]. An important finding from that work was that assigning "rules-of-thumb" to warpage versus either CTE or moduli values was not particularly fruitful as design tools that will minimize warpage. In fact, the authors observed that controlling the warpage through optimizing molding compound properties becomes less effective as packages become thinner, which is the current trend in stack package technology.

At present, both empirical and modeling approaches are providing the means to anticipate warpage behavior in stacked package technology during the reflow process. Chiavone explored three-dimensional profile tools for understanding the separation between warped packages as the resulting gap affects the solder joint geometry [20]. A comprehensive modeling approach was described by Lall and co-workers for predicting PoP warpage behavior [21]. Their work began by compiling a range of relevant materials properties, geometries, and reflow parameters into a finite element model that provided a deterministic prediction of warpage behavior. Next, the authors developed a statistical model to predict the probability distribution of warpage as a function of input parameters, over the range of those parameters. The model predictions were validated against experimental data.

An important take-away is that it is not always possible to control warpage by intuition or general qualitative precepts. Rather, computational modeling provides a better method to predict warpage because it can take into account multiple variables—materials properties, geometries, and the process temperature profile. Then, more limited empirical studies can be used to validate the model predictions.

Lastly, most of the several hundreds of investigations addressing stacked packages have, as exemplified by even the limited list of citations in the reference section of this article, considered largely PoP components. Computational modeling, together with validation experiments, will be absolutely necessary to develop assembly processes and reliability prediction for PoPoP components, especially as individual packages grow thinner, and the solder interconnections become smaller in size.

13.3 Solder Joint Reliability

The content of this section examines the long-term reliability of solder interconnections associated with stacked packages. Those interconnections include the first-level solder joints between the individual packages and the second-level interconnections that attach the bottom package to the PCB. A description is given of the environments that affect PoP and PoPoP reliability, followed by a brief synopsis of some empirical findings. The chapter is rounded out with a look at a reliability study that included both computational modeling and empirical testing. Those studies have formed the groundwork to support the implementation of stacked package technology in high-reliability applications.

The fact that this chapter places a focus on solder joint reliability should not be construed to imply that the reliability of the individual package components—molding compound, wire bonds, die, and substrate (interposer)—is inconsequential or even of a lesser importance to PoP and PoPoP technology. The fact is that these structures are sensitive to different specific conditions—e.g., moisture and corrosive contaminants—that do not necessarily impose a significant risk to the solder interconnections. The current trend towards increasingly thinner packages implies that the molding compound provides a reduced barrier between the external environments and the die or wire bonds.

13.3.1 Environments

13.3.1.1 Use Conditions

The stacked package can be exposed to a considerable range of use conditions. The discussion, below, will consider two general categories. A further breakdown of use conditions is described in the current revision of the standard: "Performance Test Methods and Qualification Requirements for Surface Mount Solder Attachments," *IPC*-9701 (IPC; Bannockburn, IL). First, there are the use environments of *consumer electronics*, which include household appliances, office equipment, and personal electronics such as cell phones, tablet computers, etc. The second category is *high-reliability electronics*, which encompasses products used in telecommunications equipment, commercial aircraft avionics, military weapons, satellite hardware, and space flight systems. This category also includes automobile electronics as well as monitoring systems used in the exploration of fossil fuel reserves (oil wells) and the development of renewable energy industry (geothermal wells, solar electric components, etc.). Automotive and energy system electronics are considered high-reliability because they are exposed to very harsh environments and/or pose one or more high-consequence, failure scenarios.

Each of the two use categories can be described according to the details of the environment that affect the reliability of stacked package solder interconnections.

The three primary environments are: (a) temperature fluctuations; (b) mechanical vibration; and (c) mechanical shock. A *conservative* approximation, which is not without its detractors, is to assume that the electronics experience the equivalent temperature cycle, mechanical shock loads, etc. that are present in the external environment. This assumption is conservative because, in the case of temperature cycling, there can be a significant time delay between a component buried into the interior of the system and the external temperature profile. The magnitude of the time delay depends upon the location of the component inside of the product and the thermal conduction of materials located between the two locations.

Vibration conditions are characterized by frequency, power spectral density (a measure of amplitude versus frequency), and direction of the load [22]. The loading direction is described by Cartesian coordinates; or, the vibration is *random* and occurs equally in all directions.

Mechanical shock refers to the loads experienced over a very short time duration, typically 1–5 ms [23]. Mechanical shock is experienced as a "drop shock" event when personal electronics are dropped to a hard surface. On the other hand, shock conditions are also important in high-reliability military, space, and satellite hardware that must survive the rigors of a launch and/or impact environment.

An important factor when considering (mechanical) shock and vibration environments, which is similar to thermal conduction in temperature cycling, is *coupling* between the external condition and the environment at the site of the stacked package. Internal structures can dampen the outside shock and vibration load amplitudes. However, it is equally possible that, in the absence of suitable engineering controls, shock and vibration loads can be amplified by activating characteristic frequencies of the PWA or its mounting structures, leading rapidly to a damaged assembly.

13.3.1.2 Consumer Electronics

The maximum temperature excursions are typically in the range of 0–55 °C [24]. The ramp rates are relatively slow so that consumer electronics rarely encounter conditions that would be categorized as thermal shock ($\Delta T > 20$ °C/min). Exposure time durations can be relatively long, on the order of several hours. Therefore, temperature cycling is, intrinsically, of a lesser concern for stacked package interconnections in consumer products.

By and large, consumer products that contain stacked packages are not exposed to significant vibration conditions.

Probably the most critical environment encountered by consumer electronics is the mechanical shock of a device being dropped to a hard surface—so-called "drop shock" resistance. Electronic assemblies can experience accelerations of 400–4000 Gs [24]. The resulting forces peak, and then diminish, in time frames of 1–5 ms. For example, a study described in [15] used a drop "tower" measuring 1.4 m in height. In spite of the relatively short distance, the impact acceleration was measured at 3700 Gs and duration of 0.4 ms. The extent of damage expected from the impact depends strongly upon the orientation of the assembly with respect to the falling direction. Lastly, land transportation (truck or train) and handling (loading/unloading) can introduce shock conditions on consumer electronics. However, these loads are typically of lesser magnitudes than those caused by dropping the product onto a hard surface.

Because the results of the drop test are strongly dependent upon the test parameters, there is the specification, JEDEC Standard No. 22-B111, Board Level Drop Test Method of Components for Handheld Electronic Products (JEDEC, Arlington, VA, 2003). This document sets controls on the test procedures in order to provide for a measure of commonality for comparison purposes. A number of stacked packages are soldered to the PCB test vehicle. The test vehicles are dropped multiple times. Failures, which are typically electrical opens in a daisy chain loop, are recorded by the test equipment. The failure data are reported in the 2P Weibull distribution format as percent of packages failed as a function of number of drops. The characteristic lifetime is the number of drops to reach a failure of 63.2 % of the sample population.

For example, in [14], the drop test was used to conclude that placing underfill in both gaps of a PoP component improved drop shock performance versus those units receiving underfill only in the bottom gap. Dreiza and co-workers examined the drop test survivability of a PoP component by separating failures of the first-level interconnections from those of the second-level interconnections [25]. The important findings of this early work were: (a) the solder alloy composition has a significant effect on board level reliability and (b) package bending moments are responsible for a majority of the solder joint failures.

Stacked packages are more prone to failure under shock loads than are most other surface mount devices. First of all, PoP and PoPoP components have relatively high profiles, which introduces bending moments into the solder joints. In particular, these bending moments cause peel loads at the solder joints, a loading mode in which solder joints are particularly weak. Secondly, stacked packages have relatively large masses owing to the multiple molding compounds, dice, substrates, and solder balls. The increased mass places higher stresses on the interconnections during shock and vibration events.

13.3.1.3 High-Reliability Electronics

The service temperature conditions for all but automotive under hood and energy sector electronics have temperature minimums and maximums of −40 °C and 85 °C, respectively [24]. Automotive under hood conditions are more severe due to engine heat. The temperature range is typically −40 to 125 °C. The harshest environments are those experienced by down hole oil, gas, and geothermal well-monitoring electronics where maximum operating temperatures can reach 150–200 °C.

High-reliability electronics can be subjected to a wide range of vibration conditions. Automobile under hood electronics and, in particular, modules mounted directly on the engine block or transmission case, experience significant vibration loads [26]. Automotive applications, as well as space vehicles when undergoing launch, have had their vibration power spectra characterized through laboratory and field test experiments. The resulting guidelines, together with finite element modeling software, have allowed designers to estimate vibration conditions down to the level of the PWA and individual components.

Mechanical shock poses a particularly challenging environment for high-reliability electronics. Several conditions and the corresponding shock loads generated by them are shown below:

- Automotive: bumps in the road (10–40 Gs [27]);
- Military weapons: launch, penetrator warheads, (1000–100,000 Gs [23])
- Space and satellite vehicles: launch and staging sequences (100–400 Gs [22]).

Finite element models have been developed to predict the shock loads experienced by electronics assemblies down to the component level. The present challenge is to develop the high loading rate material properties that are necessary to execute the computational models.

13.3.1.4 Accelerated Aging

Printed wiring assemblies, which have stacked packages on them, are exposed to the same accelerated aging regimens that have been used on other surface mount (and through-hole) assemblies. Numerous consensus standards have been developed (e.g., [24, 27]) to identify the appropriate test parameters, test vehicles geometries, sample size, and failure assessments with which to characterize the reliability or the interconnections. The important step is to calculate the appropriate *acceleration factor* that correlates the accelerated aging test results to the predicted behavior under the use conditions. The use conditions should include any next-assembly, transportation, and handling environments.

The complex construction of stacked packages and the multiple materials (including underfills) used in them requires that a careful survey be made of their properties and thus, possible limitations, vis-a-vis the test parameters. This precaution will prevent an over-testing event, which generates failure modes that are irrelevant to the actual wear-out mode. The consequence can be a lower-than-actual reliability of the solder joints due to a high rate of premature failures.

Otherwise, in general, current aging test practices, whether for temperature cycling, vibration, or mechanical shock assessment, are suitable for PWAs having PoP and PoPoP components. In the case of temperature cycling and the use of daisy chain links, there is an added complexity of monitoring some, or all, of the first- and second-level interconnections in a stacked package. Although the current suppliers of "dummy" components offer several varieties of daisy chain patterns, all signals must eventually pass through the second-level interconnections. If the latter fail first, there is no information gained on the longevity of the first-level interconnections. This scenario exemplifies the need for careful, post-test failure analysis to determine the source of the electrical open.

13.3.2 Underfill, Conformal Coatings, and Encapsulants

When considering long-term reliability, conformal coatings, encapsulants, and underfills can have a role that is *active* or *passive* in nature. Encapsulating foams and underfills actively mitigate against the effects of severe vibration and shock by mechanically anchoring the individual packages to one another and to the PCB. Conformal coatings perform this role, but to a lesser degree because they are present only as thin layers that do not completely cover the sides of, or fill the gaps between, the individual packages.

A photograph is shown in Fig. 13.9 in which a PoP component had been coated with a 125 μm thick urethane layer. The method of application was spraying. The coating was present on the top surface of the upper package and on the bottom fillet of the underfill. The coating was discontinuous between these two locations. In fact, the coating did not cross the gap between the top and bottom packages (orange arrow). Therefore, in the case of high-profile stacked packages, conformal coating provides minimal, if any, additional mechanical attachment function.

13.3.2.1 Materials Properties

On the other hand, under temperature cycling conditions, all three of these materials have a passive effect on solder joint reliability. The material properties that determine the magnitude of this effect are:

- Coefficient of thermal expansion (CTE)
- Elastic modulus (E), and
- Glass transition temperature (T_g).

The CTE mismatch between the encapsulant, conformal coating, or underfill generates *additional* shear and tensile loads on the solder joints, depending upon the extent to which they couple to the stacked package and the PCB. The preferred situation is that the material CTE closely matches that of the package or PCB to minimize the additional loads. Materials that fill the gap between packages should have a CTE that matches that of the solder.

The elastic modulus is an important factor. A more compliant material, that is, one of lower elastic modulus, can reduce residual stresses by elastically deforming under the applied loads. Unfortunately, the tradeoff of having a reduced elastic modulus is that the material generally has a higher CTE value. Therefore, the design engineer must select a material, the properties of which minimize solder joint TMF. Computational modeling is a valuable tool to make a "first-cut" in choosing the appropriate encapsulant, underfill, or conformal coating for an assembly.

The glass transition temperature is a key parameter for encapsulants, conformal coatings, and underfills vis-à-vis solder joint reliability. It is preferred that temperature fluctuations, whether those experienced in service or, those used in accelerated aging, avoid crossing T_g. When the temperature excursion crosses T_g, there is a significant *change* to the physical properties of the material. For example, there are *increases of CTE* and *decreases of elastic modulus*, with a positive temperature ramp. The effects are reversible with the negative temperature change. In either circumstance, the consequence is potentially sharp increases in the loads put on the solder interconnections. However, a greater discrepancy can take place in reliability predictions, when T_g is crossed in accelerated aging; but, it is not crossed in the use environment. In effect, the aging tests are potentially causing a different failure mechanism under this scenario.

13.3.2.2 Geometry

Besides material properties, geometry poses two contributing factors to the solder joint reliability of PoP and PoPoP components. First, the stacked package has a higher profile than other devices. The taller height is a particular concern in the case of encapsulants, because it allows the latter a greater area to "grab" the package and exert more loads on the PoP or PoPoP component that could potentially damage the solder interconnections.

The second geometry factor is the size of the gaps between two packages as well as that between the bottom package and the PCB. The gap height determines whether the encapsulant or conformal coating will partially, or completely, fill the gap, top-to-bottom. When either material bridges the gap, as opposed to simply wetting one or both surfaces, the confined geometry causes the material to exert very high tensile stresses on the solder joints when the expanding material pushes the two surfaces apart. Solder joints are particularly weak in tension, with premature failures typically occurring in the IMC layer.

Underfill materials are expected to fill the gap. Therefore, they are designed to have a CTE value that closely matches that of the solder alloy so that the two expand/contract to the same degree upon heating/cooling temperature cycles. In order to achieve this goal, underfills combine a *filler material*, typically alumina or silica particles, with the organic matrix. The effect is to lower the CTE value to one similar to the solder. The tradeoff is an increase of elastic modulus, which causes the underfill to be less compliant, and thus, less able to relieve residual stresses.

13.3.3 Reliability Studies

13.3.3.1 Mechanical Shock and Vibration

As noted above, the early implementation of stack packages took place in the consumer electronics community. The PoP and PoPoP components provided a means to further miniaturize hand-held products while, at the same time, increasing their functionality. As a "lesson-learned" from the introduction of BGA packages, drop-shock was the foremost reliability concern for stacked components in these products. The evidence to this effect came in the form of the multiple investigations, two of which are cited in the reference section [28, 29].

The solder interconnections of stacked packages are more susceptible to damage in mechanical shock and vibration environments than are those of other surface mount devices. Stacked packages have a greater mass and higher center-of-gravity than those latter packages, which places more loads on particularly the bottom solder joints made to the PCB. Encapsulation may be necessary to mitigate the detrimental loads caused by shock and vibration. Interestingly, PoP and PoPoP components have relatively modest footprints so that board bending under these environments has a lesser effect on the solder joints than it would impact, for example, the larger area-array packages.

Solder alloys and surface finishes are the two materials whose physical and mechanical properties generally apply across the wide range of assembly geometries, attachment methodologies, as well as shock and vibration environments. Their effects are discussed, below.

Solder Alloy Effects

As use of stacked packages and, primarily PoP, grew with time, so too were the changes in supporting technologies that affect stacked package reliability. New Pb-free solders were developed to replace the SAC305, SAC396, etc. alloys (which had replaced eutectic Sn-Pb solder) in order to improve drop shock performance of area-array packages, including stacked packages. Those new solders were the low-Ag alloys such as the SAC105 (98.5Sn-1.0Ag-0.5Cu) and SAC 125 (98.3Sn-1.2Ag-0.5Cu) as well as later ternary and quaternary compositions. McCormick and co-workers examined the TMF performance of the high- and low-Ag, Sn-Ag-Cu alloys under the temperature range of −40–125 °C [30]. Their findings confirmed the superior performance of the high-Ag SAC305 and SAC405 alloys in temperature cycling versus the low-Ag SAC105 and SAC125 solders. Other studies have examined the drop-shock behavior of these solders, including test matrices that illustrated the synergistic roles of the underfill materials [14, 31].

Surface Finish Effects

The second supporting technology that has undergone rapid changes concurrent with PoP and PoPoP technology is surface finishes. A variety of surface finishes are available for the substrates (interposers) of the individual packages as well as on the PCB. The surface finish affects the drop-shock reliability of stack packages because the high-strain rate nature of the deformation "forces" the failure path to the solder/pad interface. Fracture usually occurs in the IMC layer, the development of which is determined directly by *both* the surface finish and the solder alloy [18].

The authors, Ejiri et al., used a rapid shear test to simulate the loads placed on solder balls under high strain rate events such as the drop test [32]. They concluded that the combined effects of IMC thickness, which depends upon the solder composition, surface finish, and aging conditions, as well as the "bulk" solder strength, had contributing roles in the failure behavior of the joints. The results of these experiments are critical towards deciphering the extent to which the solder microstructure-strength relationship and the interface physical metallurgy control the fracture behavior of interconnections in the considerably more complicated configuration of stacked packages.

Importance of Test Standards

Mechanical shock and vibration effects are also strong functions of the PWA design, the manner in which it is anchored in the next-assembly, and the use environment. These properties vary considerably between applications, causing significant challenges when an attempt is made to compare relative performance attributes. The JEDEC standard for drop shock testing provides for a consistent test methodology with which to compare data across different product assemblies. Several authors have examined the effect of location on the test vehicle on the failure behavior [31, 33]. The authors observed that different failure rates occurred at different locations on the test PWA location—that observation was anticipated because a higher degree of board bending occurs in the PWA center versus near the support structures. However, the important finding was that the *failure mode* also changed with location on the PWA. In other words, the failure mode of PoP components, and likely also PoPoP components, is very sensitive to the orientation and magnitude of the shock load, per se.

Clearly, the complexity of stacked package structures and associated interconnections make it difficult to assign general precepts or guidelines to predict the reliability of the solder joints as a function of mechanical shock. Computational modeling can take into account the constitutive properties of the PoP and PoPoP materials sets and solder alloys as well as the geometric configuration of the PWA and shock load properties. Modeling provides the optimized approach towards predicting solder joint reliability when being confronted by such a complicated, multivariate, three-dimensional problem.

13.3.3.2 Temperature Cycling

The attributes of stacked packages have garnered the attention of designers in the high-reliability systems community. Mechanical shock environments are a concern in some, but not necessarily a majority, of electronics for those applications. Moreover, many of the methods to mitigate the effects of mechanical shock discussed above already have a long history in military and satellite payload programs. However, high-reliability hardware is subjected to considerably wider temperature fluctuations, which places an emphasis on the impact of temperature cycling on long-term reliability.

Temperature Limits

The span of service temperatures is *typically* −40 to 85 °C [24]. Use conditions are somewhat more severe for automobile under hood environments (upper limit of 125 °C). Accelerated aging temperature minimum and maximum values cited in [24] are:

- 0 °C, 100 °C
- −25 °C, 100 °C
- −40 °C, 125 °C
- −55 °C, 125 °C
- −55 °C, 100 °C

The most-often used cycles are: [0 °C, 100 °C] and [−55 °C, 125 °C] for high-reliability electronics. Dwell times at the temperature limits range from 10 to 30 min and ramp rates must be less than, or equal to, 20 °C/min to avoid thermal shock conditions.

The general precept is that, the wider the temperature range designated as ΔT, the greater is the acceleration factor. Thus, fewer cycles are needed to bring about the failures required for a reliability analysis. However, the complex structures of stacked packages require that careful consideration be given to the accelerated aging test temperatures versus the T_g of the organic materials—molding compound, die attach, substrate (usually epoxy), underfill materials, encapsulants, and conformal coatings. *Unless the temperature cycles in service cause any of these materials to pass through their respective T_g values, the temperature cycle used in accelerated aging should not pass through T_g.*

Test Vehicle Construction

The second complication that arises in the thermal cycling of stacked packages concerns the daisy chain loops that are used to monitor for electrical opens per reference [24]. These components have a combination of first-level interconnections—one for PoP and two for PoPoP—and second-level interconnections, which are between the bottom package and the PCB. All signals must eventually pass through the second-level interconnections, which would "control" the occurrence of failures. It is for this reason that the temperature cycling of stack packages is also referred to as *board level reliability* (*BLR*) evaluations because of the de factor emphasis placed on second-level interconnections. In the event that one of the first-level solder joints fails prior to the second-level interconnections in that loop, that fact can only be learned, later on, by destructive analysis.

Materials Set for Computational Modeling

Numerous empirical studies have examined the reliability of PoP components under temperature cycling [28, 30]. The challenge faced with understanding and, more importantly, making use of those data is the multiple variables that can affect reliability. The PoP and PoPoP parameters are:

- Molding compounds: material type and thickness;
- Substrate: laminate material and Cu features;
- Solder joints: alloy composition, pitch, and ball size; and
- Surface finishes: compositions, which can affect solder joint fatigue.

There are also the non-PoP-specific parameters:

- Underfill: material type and geometry (some or all gaps);
- Encapsulants and conformal coatings: material properties and geometry;
- PCB under the package; laminate, thickness, and Cu loading; and
- Test environment: temperature limits, ramp rates, and hold times, which together must correlate to service conditions through the appropriate acceleration factor.

Even in the event that all of the temperature-dependent CTE, elastic modulus, and Poisson's ratio parameters are available, an underlying assumption is that they do not change significantly over time due to any intrinsic aging phenomenon.

Solder Alloy Fatigue Properties

The TMF properties of the Pb-free solders have been observed to change with isothermal aging over time. In several studies, the solder joints were aged at relatively high temperatures (125 °C) and durations of several months [34, 35]. As a result, the increased thickness of the IMC layer had a role in the failure mode. Zhao, et al. documented the effect of aging by SAC305 and SAC105 solders on the TMF of BGA solder joints, including lower temperatures of 25 °C [36]. Aging times of 12 months (25 °C) caused approximately a 20 % drop in characteristic lifetimes (2P Weibull). The loss of characteristic lifetime increased to approximately 40 % when aging was performed at 55 °C over the same 12 months.

Two caveats are important when considering the above effects vis-à-vis accelerated aging tests and use conditions. First of all, aging at 125 °C for time durations on the order of months is well outside the traditional qualification and acceptance testing regiments of high-reliability systems prior to being placed in the field where temperature cycling will take place. Secondly, it is a recommended practice to subject a PWA to an isothermal aging at 100 °C for 24 h to "stabilize" the solder microstructure prior to commencing with temperature cycling [24]. Nevertheless, the findings in [34–36] are not wholly unexpected when consideration is given to the point that solders are performing at a high homologous temperature, even when at room temperature. Changes are surely to take place to the solder microstructure, including recovery and recrystallization, which in turn are very likely to affect the mechanical properties, including TMF.

Effects of Alloy Composition and Underfill on Solder Joint Reliability: An Empirical Study

Vianco and co-workers have examined the effects of underfill materials on the TMF performance of PoP interconnections [8]. In that study, the first-level interconnections were SAC305. Given the continued use of Sn-Pb solder in the second-level assembly processes used for high-reliability electronics, the bottom joints were either SAC305 or mixed Sn-Pb (paste)/SAC305 (solder balls) that represent a *backwards compatibility* condition. Metallographic cross sections confirmed *full intermixing* between the Sn-Pb and SAC305 solder components. The 2P Weibull failure plot is repeated in Fig. 13.11 for the baseline test vehicles, that is, the PWAs that did not include underfill. The characteristic lifetimes, η, and slope, β, are included on the plot. The same failure mode was observed for both cases in Fig. 13.11, which was TMF crack propagation in the second-level interconnections. This crack behavior is shown by the optical micrographs in Fig. 13.12 (100 % SAC305 joints). The red arrows indicate the TMF crack. The similarity of failure modes was further substantiated by both data sets in Fig. 13.11 by having nearly the same β values. The mixed solder interconnections have a longer fatigue life than do the 100 % SAC305 solder joints. This observation assures designers that PoP technology is suitable when used in a backwards compatibility scenario.

The directions and magnitudes of warpage are illustrated by the quarter symmetry displacement contours in Fig. 13.13. The temperature profile is shown at the

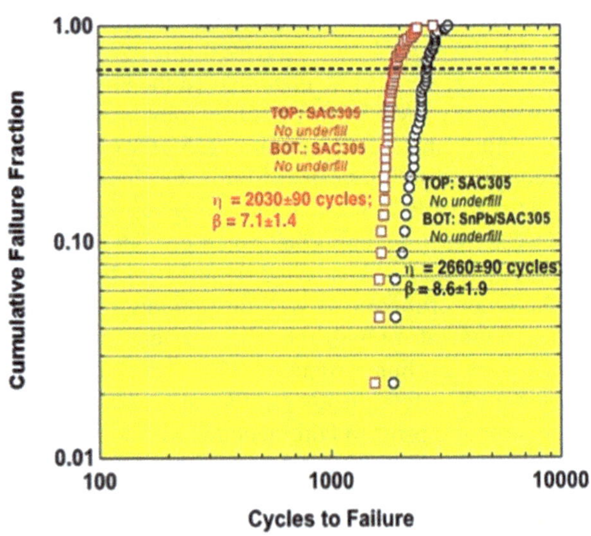

Fig. 13.11 Weibull failure plot repeats the earlier picture for two PoP assemblies [8]. Both assemblies used SAC305 alloy for the top solder joints. One test vehicle had 100 % SAC305 solder for the bottom joints and the other had backwards compatibility interconnections formed from SAC305 balls and Sn-Pb paste

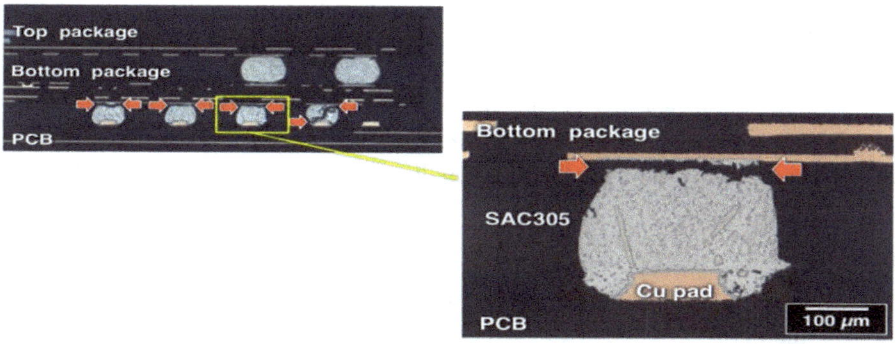

Fig. 13.12 Optical micrographs illustrate the predominant failure mode of both data sets in Fig. 13.11, which was TMF cracking in the second-level interconnections [8]. This particular case is that of the fully SAC305 interconnections

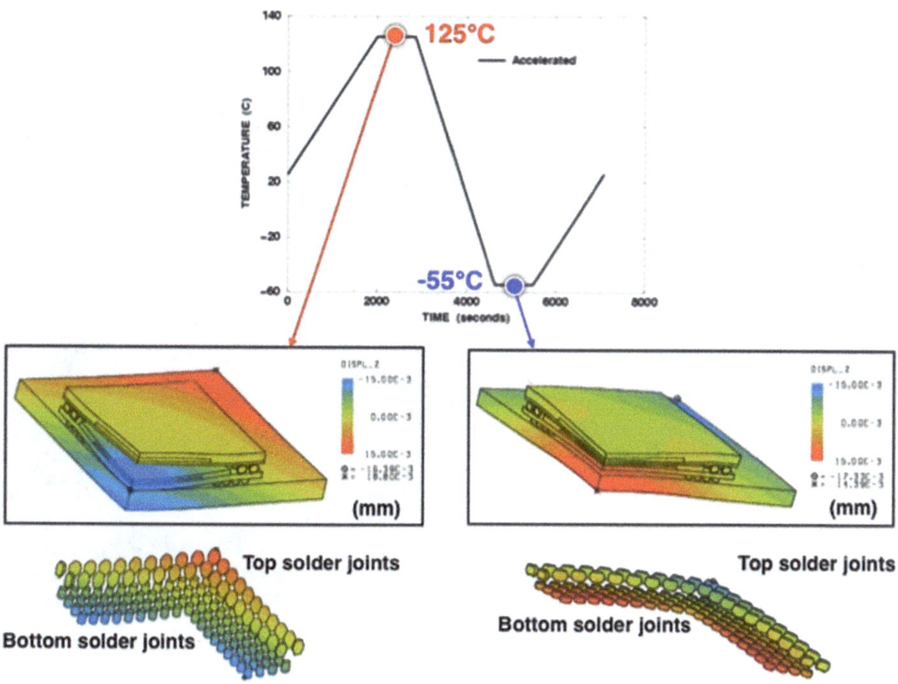

Fig. 13.13 The top graph is the accelerated aging temperature cycle to which the PoP components were exposed. The finite element displacement ("DISPL_Z") maps show the warpage of the package and surrounding printed circuit board at the thermal cycle temperature extremes of 125 °C and −55 °C (50× amplification). The bottom images display the model predictions of solder joints deformation (50× amplification) at the same temperature extremes

top of the figure. The displacement contours are shown in the middle of the figure for the two temperature extremes, −55 °C and 125 °C. These images have been magnified 50×. The "Displ.Z" scale refers to *displacement* of the contours in mm units relative to the zero position at each point. The difference between the highest and lowest points was approximately 30 μm according to the scales. The displacement difference across the PoP is 20–30 μm. The bottom displacement maps indicate the strain incurred by the solder joints. Note also that the locations of greatest displacement differ between the first- and second-level interconnections due to the complexity of the loading condition that is experienced by PoP solder joints.

Because the solder joints are in the solid state, the PoP warpage displacements are relatively small. Yet, they can generate significant stresses in *both* first- and second-level interconnections. This point is illustrated in Fig. 13.14, which shows optical micrographs of a first-level solder joint (SAC305). The tensile stresses were sufficiently high to cause catastrophic fracture at the Ni-Sn IMC/Ni-P interface (ENIG surface finish). Although observed infrequently, this failure mode illustrates the increased probability of a catastrophic failure mode caused by warpage and the loads it induces in the (solid) solder joints.

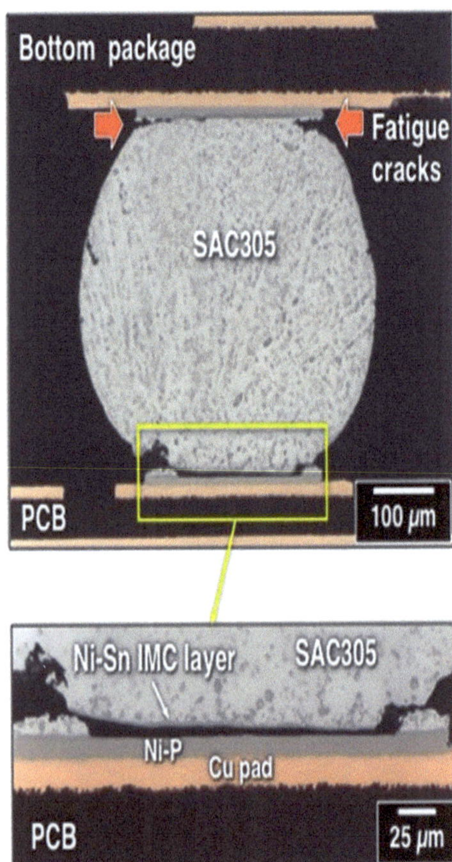

Fig. 13.14 Catastrophic failure of the SAC305 solder joint along the Ni-Sn IMC/Ni-P interface, which serves as a reminder of the complex loading configurations present in the stacked package structure, including tensile stresses of sufficient magnitude to have caused the failure in the (solid) solder joints

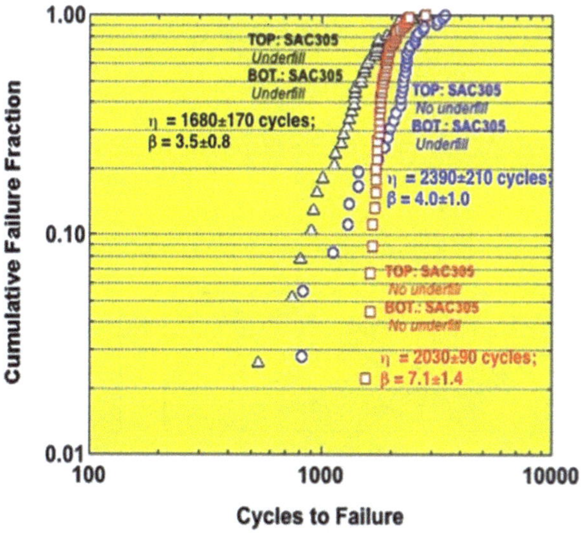

Fig. 13.15 Weibull failure plot for PoP assemblies that used 100 % SAC305 alloy for both the top and bottom (BOT) solder joints [8]. Also included in the plots are the 2P Weibull statistical parameters of characteristic lifetime, η, and slope, β. The conditions were defined by the presence of underfill: *red squares* no underfill at all; *blue circles* underfill only in the bottom (second-level assembly) solder joints; and *black triangles* underfill being present in both gaps

The effect of adding the underfill material is illustrated in Fig. 13.15. Here, both first- and second-level solder joints were SAC305. This analysis points to the necessity to fully understand the Weibull plot in light of the multiple factors affecting PoP solder joint TMF behavior. Although underfill in the bottom gap, only, resulted in the highest characteristic lifetime, failures started to occur at fewer cycles. The effect was further enhanced when underfill was placed in both gaps. These findings stemmed from the fact that underfill, whether present in the bottom gap, only, or in both gaps, decreased the slope (β) versus the case without underfill at all. A lower β value implies that there is more spread to the failure probability distribution function (PDF). The broader PDF results in more early failures as well as contributes to a greater overall uncertainty to the failure predictions.

The decrease in β value also suggests a change to the failure mode. This stipulation was confirmed when destructive physical analysis showed that the underfill caused more TMF deformation, and as such, an increased likelihood of failure, in the first-level interconnections. The transition of TMF deformation was more effective when underfill was present only in the bottom gap; it occurred to a lesser degree, when both gaps had underfill in them. The latter variant developed because the underfill reduced the overall warpage *between the packages*.

The above behavior is illustrated by the warpage displacement models in Fig. 13.16. Underfill was present in both gaps. Although the maximum displacement remained unchanged across the total package-plus-PCB configuration, the warpage in the PoP has been reduced considerably when compared to Fig. 13.13. In effect, the PoP component, as a whole, lost compliance because it was stiffened up, increasingly so by having the underfill in one gap, and then, in both gaps. Both rows of solder joints, but, more so, the first-level interconnections, experienced an increased degree of TMF due to the global CTE mismatch between the PoP and

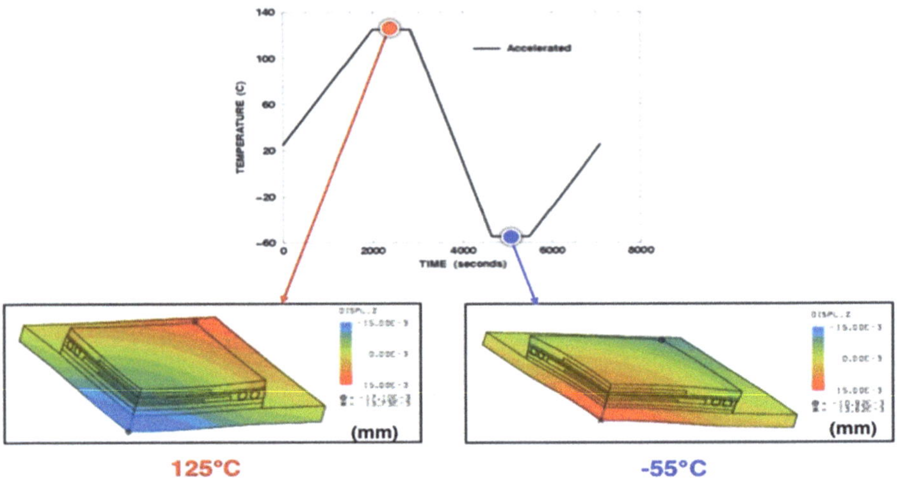

Fig. 13.16 The top graph is the accelerated aging temperature cycle. The finite element contour maps show the warpage ("DISPL_Z") of the package and surrounding printed circuit board at the thermal cycle temperature extremes of 125 °C and −55 °C (50× amplification). Underfill was present in both the top and bottom gaps

PCB as well as the local CTE mismatch between the packages, solder, and underfill. The increases of global and local CTE mismatch strains compensated, somewhat, for the decrease in warpage strains; the net effect was the change in Weibull distributions shown in Fig. 13.15.

The impact of the underfill on solder joint TMF can be observed by the dependence of cycles-to-failure as a function of the temperature-at-failure. The corresponding data are shown in Fig. 13.17. The data originated from test vehicles having 100 % SAC305 second-level solder joints. The symbol designations are: red circles, no underfill; blue circles, underfill in the bottom gap, only; and black circles, underfill in both gaps. When shear strains generated by CTE mismatch are principally responsible for TMF, the cycles-to-failure are randomly distributed across the temperature range. Or, as shown by the red circles in Fig. 13.17, the data have a slight skew towards the lower temperatures because there is less creep deformation to alleviate the residual stresses. However, both the blue circles (underfill in the bottom gap, only) and black circles (underfill in both gaps) are skewed to the *high temperature* end. High temperatures cause the underfill to expand, placing additional tensile stress into the solder that increases the likelihood of failure at high, rather than low, temperatures. Although not illustrated here, the effect was even more dramatic when the second-level interconnections had the Sn-Pb/SAC305 mixed metallurgy.

Finally, underfill played a significant role when used with the PoP test vehicles having the Sn-Pb/SAC305, second-level solder joints. This behavior is shown in Fig. 13.18. The mix-metallurgy joints experienced nearly an order-of-magnitude drop in characteristic lifetime, η. Similarly, the underfill caused the slope term, β, to

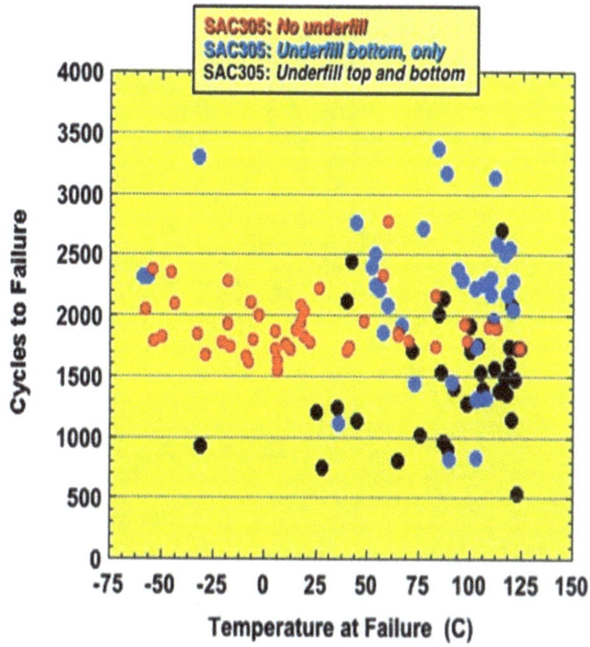

Fig. 13.17 Plot shows cycles-to-failure as a function of temperature-at-failure for the three cases of 100 % SAC305 solder in both first- and second-level interconnections: *red circles* no underfill; *blue circles* underfill only in the bottom gap; and *black circles* for the PoP components having underfill in both gaps

decrease to approximately one-half the value that was observed with no underfill in the gaps. While the data in Fig. 13.15 indicate that second-level interconnections having mixed metallurgy have better TMF life versus joints having 100 % SAC305 solder, this advantage is lost with the introduction of underfill into the gaps according to Fig. 13.18. Therefore, the use of underfills to mitigate against mechanical shock and vibration effect must necessarily consider TMF behavior of the interconnections when stacked package technology is introduced into high-reliability electronics systems.

13.4 Summary and Future Trends

13.4.1 Summary

The development of stacked packaging technology—materials, assembly processes, and reliability—has led to significant advances in the miniaturization and functionality of consumer electronics. Certainly, these advancements can provide a springboard for enabling PoP and PoPoP components in military, space, and satellite systems. However, the latter applications emphasize longer term reliability and, more so, under a significantly wider range of service environments. Therefore, computational modeling will continue to be a valuable tool for predicting the response of first- and second-level solder joints to the shock, vibration, *and*

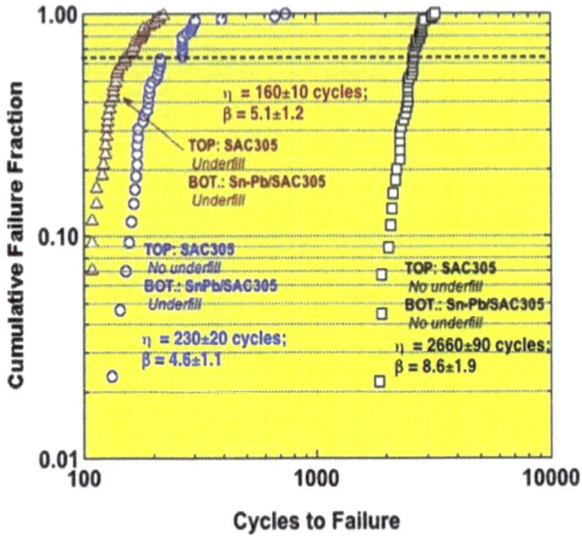

Fig. 13.18 Weibull failure plot for PoP assemblies that used 100 % SAC305 alloy for the top (first-level interconnections) and the Sn-Pb/SAC305 mixed solder metallurgy on the bottom (BOT) side [8]. Also included in the plots are the 2P Weibull statistical parameters of characteristic lifetime, η, and slope, β. The conditions were defined by the presence of underfill: *black squares* no underfill at all; *blue circles* underfill only in the bottom (second-level assembly) solder joints; and *dark-red triangles* underfill being present in both gaps

temperature cycling environments of those applications. Such models can readily account for not only the different use conditions, but also the complex construction (materials and geometries) of stacked packages. As the fidelity of those models improves further, fewer accelerated aging empirical studies will be required as their role transitions to simply model validation.

13.4.2 Future Trends

Stacked packages are considered to be a relatively mature technology for the consumer electronics sector. The quest for further miniaturization, especially product thickness, will place greater emphasis on *die stacking* in those product lines as opposed to increasing the number of packages stacked on top of one another.

Package-on-package and PoPoP components are being considered for military, space, and satellite systems. Much of the technology-base developed for stacked packages by the consumer electronics community can be adapted to their insertion into high-reliability systems. Of course, there are additional factors, including materials, qualification, and acceptance testing, as well as use environments, that are unique to these applications and which will need to be addressed by design and process engineers. The emphasis on reliability will continue to favor stacked packages over die stacking for now. However, improvements that are made towards screening for known-good die will support a transition from PoP and PoPoP components to die stacking package technologies in high reliability electronics.

Acknowledgments The author wishes to thank all of the individuals who have contributed to the projects that have allowed for the construction of this chapter: Jerry Rejent and Mark Grazier who performed all of the experimental work, the folks responsible for the microanalysis: Alice Kilgo, Amy Allen, Bonnie McKenzie, and Richard Grant; as well as the continued support by M. Rightly, R. Periasamy, and Dan Baker. Also, a "thank you" goes to Brian Wroblewski for his careful review of the manuscript. Sandia is a multiprogram laboratory operated by Sandia Corporation, a Lockheed Martin Company, for the United States Department of Energy's National Nuclear Security Administration under Contract No. DE-AC04-94AL85000.

The editors would like to thank Luhua Xu from Intel Corporation for his critical review of this Chapter.

References

1. R. Johnson, M. Strickland, D. Gerke, *3-D Packaging: A Technology Review* (NASA Marshall Space Flight Center, Huntsville, AL, 2005)
2. R. Priori, A. Burton, Development of a high reliability and large volume manufacturing assembly process for a stacked memory package, in *Proceedings of the IEEE Electronic Components and Technology Conference*, IEEE, New York, NY, June 2004, pp. 1142–1147
3. J. Zhao, L. Yuxiang, H. Zhenyu, M. Rong, Effects of package design on top PoP package warpage, in *Proceedings of the IEEE Electronic Components and Technology Conference*, IEEE, New York, NY, June 2008, pp. 1082–1088
4. M. Yim, R. Strode, A. Ravikumar, Y. Chan, Effects of material properties on PoP top package warpage behaviors, in *Proceedings of the IEEE Electronic Components and Technology Conference*, IEEE, New York, NY, June 2010, pp. 1071–1076
5. JESD22-B111, *Board Level Drop Test Method of Components for Handheld Electronic Products* (JEDEC, Arlington, VA, 2003)
6. JESD22-B110, *Mechanical Shock—Component and Sub-assembly* (JEDEC, Arlington, VA, 2013)
7. V. Wang, D. Maslyk, Analysis of the reliability of package-on-package devices manufactured using various underfill methods, in *Proceedings of the International Conference on Electronic Packaging Technology and High Density Packaging (ICEPT-HDP)*, IEEE, New York, NY, June 2008. doi: 978-1-4244-2740-6/08
8. P. Vianco, J. Grazier, J. Rejent, A. Kilgo, F. Verdi, C. Meola, Temperature cycling of Pb-free and mixed solder interconnections used on a package-on-package test vehicle, in *Proceedings of the Surface Mount Technology Association (SMTA) International Conference*, Edina, MN, Sept 2010, CD-ROM
9. A. Yoshida, J. Taniguchi, K. Murata, M. Kada, Y. Yamamoto, Y. Takagi, T. Notomi, A. Fujita, A study on package stacking process for package-on-package (PoP), in *Proceedings of the Electronic Components and Technology Conference*, IEEE, New York, NY, 2006
10. G. Padnos, PCB assembly system set-up for package-on-package (PoP) assembly. Global SMT and Pack, Mar 2010, pp. 13–17
11. J. Sjoberg, D. Geiger, T. Castello, D. Shangguan, Package-on-package (PoP) process development and reliability evaluation. Global SMT and Pack, Oct 2007, pp. 34–36
12. V. Srinivas, M. Osterman, R. Farrell, Reliability evaluation of one-pass and two-pass techniques of assembly for package on packages under torsion loading, in *Proceedings of the IPC APEX EXPO*, IEEE, New York, NY, 2009, CD-ROM

13. H. McCormick, I. Sterian, J. Chow, M. Berry, J. Trudell, R. Cortero, PoP (package on package): an EMS perspective on assembly, rework, and reliability, in *Proceedings of the SMTA Pan Pacific Conference*, Edina, MN, Feb 2009, pp. 178–183
14. V. Wang, D. Maslyk, Analysis of the reliability of package-on-package devices manufactured using various underfill methods, in *Proceedings of the International Conference on Electronic Packaging Technology and High Density Packaging*, IEEE, New York, NY, July 2008. doi: 10.1109/ECEPT.2008.4606944
15. E. Ibe, K. Loh, J. Luan, Y. Tee, Effect of unfilled underfill on board level reliability, in *Proceedings of the Electronic Components and Technology Conference*, IEEE, New York, NY, 2006. doi: 10.1109/ECTC.2006.1645687
16. M. Robins, Conformal coatings: to protect and serve. Electr. Proc. Prod. **39**(13), 5 (1999)
17. C. Tautscher, Conformal coatings: selection criteria. SMT **12**(7), 3 (1998)
18. L. Smith, M. Dreiza, A. Yoshida, Package on package (PoP) stacking and board level reliability, in *Proceedings of the Surface Mount Technology Association (SMTA) International Conference*, Edina, MN, Sept 2006, CD-ROM
19. M.-J. Yim, R. Strode, R. Adimula, C. Yoo, Effects of material properties on PoP top package warpage behaviors, in *Proceedings of the Electronic Components and Technology Conference*, IEEE, New York, NY, 2010. doi: 10.1109/ECTC.2010.5490834
20. K. Chiavone, Advanced analysis of package-on-package interconnect gaps during reflow assembly, in *Proceedings of the Surface Mount Technology Association (SMTA) International Conference*, Edina, MN, Sept 2015, pp. 786–791
21. P. Lall, K. Patel, V. Narayan, Process and material envelop for allowable package-on-package warpage. J. Surf. Mount Tech. **28**(3), 32–47 (2015)
22. *Rokot™ User's Guide EHB0003, Issue 5* (Euroket, Bremen, Germany, 2011), pp. 5-1–5-10
23. M. Berman, *Electronic Components for High-G Hardened Packaging, ARL-TR-3705* (U.S. Army Research Laboratory, Aberdeen, MD, 2006)
24. *Performance Test Methods and Qualification Requirements for Surface Mount Solder Attachments, IPC-9701* (Association Connecting Electronics Industries, Bannockburn, IL)
25. M. Dreiza, A. Yoshida, K. Ishibashi, T. Maeda, High density PoP (package-on-package) and package stacking development, in *Proceedings of the Electronic Components and Technology Conference*, IEEE, New York, NY, 2007
26. *Handbook for Robustness Validation of Automotive Electrical/Electronic Modules, Standard J1211_20121* (SAE International, Warrendale, PA, 2012)
27. *Road Vehicles—Environmental Conditions and Testing for Electrical and Electronic Equipment, ISO 16750-2:2012* (International Organization for Standardization, Geneva, Switzerland, 2012)
28. M. Dreiza, L. Smith, G. Dunn, N. Vijayaragavan, J. Werner, Package on package (PoP) stacking and board level reliability, results of joint industry study, in *Proceedings of the IMAPS Conference*, Research Triangle Park, NC, 2006
29. D. Shin, D. Lee, E. Ahn, T. Kim, T. Cho, Development of multi stack package with high drop reliability by experimental and numerical methods, in *Proceedings of the Electronic Components and Technology Conference*, IEEE, New York, NY, 2006, pp. 377–382
30. H. McCormick, J. Chow, M. Berry, J. Trudell, R. Brush, J. Qian, R. Cortero, Effect of assembly variations on package on package reliability in thermal cycling, in *Proceedings of the Surface Mount Technology Association (SMTA) International Conference*, Edina, MN, Sept 2009, CD-ROM
31. A. Daya, S. Ramkumar, Influence of alloy combination and PCB location on package on package (PoP) component assembly drop test, in *Proceedings of the Surface Mount Technology Association (SMTA) International Conference*, Edina, MN, Sept 2012, CD-ROM
32. Y. Ejiri, T. Noudou, T. Sakarai, Y. Arayama, Y. Tsubomatsu, K. Hasegawa, Influence of surface finishes and solder alloys on solder ball joint reliability, in *Proceedings of the Surface Mount Technology Association (SMTA) International Conference*, Edina, MN, Sept 2015, pp. 957–971

33. B. Roggeman, M. Meilunas, Assembly and reliability investigation of package on package, in *Proceedings of the IPC APEX*, 2009, CD-ROM
34. Z. Hai, J. Zhang, C. Shen, J. Evans, M. Bozack, Long term aging effects on reliability performance of lead-free solder joints, in *Proceedings of the Surface Mount Technology Association (SMTA) International Conference*, Edina, MN, Sept 2013, www.smta.org
35. J. Zhang, Z. Hai, S. Thirugnanasambandam, J. Evans, M. Bozack, Y. Zhang, J. Suhling, Thermal aging effect on the thermal cycling reliability of lead free fine pitch packages. IEEE Trans. Comp. Pack Manuf. Tech. **3**, 1348–1357 (2013)
36. C. Zhao, C. Shen, H. Zhou, J. Zhang, M. Bozack, J. Evans, Long term aging effects on the reliability of lead free solder joints in ball grid array packages with various pitch sizes and ball alignments, in *Proceedings of the Surface Mount Technology Association (SMTA) International Conference*, Edina, MN, Sept 2015, pp. 199–206

Chapter 14
Interconnect Quality and Reliability of 3D Packaging

Yaodong Wang, Yingxia Liu, Menglu Li, K.N. Tu, and Luhua Xu

14.1 Introduction

The challenges to quality and reliability for 3D IC packaging technology include but not limited to the followings. (1) New manufacturing processes that heavily involve through silicon via (TSV), microbumping, die stacking, multiple high temperature reflows; (2) New materials including new under bump metallurgies (UBM) with new Pb-free solder materials mainly based on intermetallic compounds, new underfill materials that flow into thin gap for microbumps; (3) Miniaturization in size, where solder joint is 5–10× smaller than C4 solder joints, which translates to a volume shrinkage by three orders of magnitude; (4) Thermal mechanical environment changes: Joule heating has to find new ways to dissipate and thermal gradient and thermal stress concentration are built up across layers; (5) Any new electromigration and thermomigration concerns should be studied, including failure in redistribution layers; (6) Reliability requirement changes due to new failure modes transition from desktop, laptop to mobile devices [1].

In this chapter, we discuss quality and reliability of interconnects in 2.5D IC and 3D IC packaging technology—microbump, TSV, UBM and copper interconnects, and we compare them to the quality and reliability concerns observed in the existing interconnects. We shall cover microstructure changes and failures driven by mechanical stressing, electromigration (EM), and thermomigration (TM). This way we can see how the transition, for example, from C-4 joints to microbumps

---

Recent job change to Apple inc

Y. Wang • Y. Liu • M. Li • K.N. Tu
Department of Materials Science and Engineering, University of California, Los Angeles (UCLA), Los Angeles, CA, USA

L. Xu (✉)
Intel Corporation, Chandler, AZ, USA
e-mail: luhua_xu@apple.com

Y. Li, D. Goyal (eds.), *3D Microelectronic Packaging*, Springer Series in Advanced Microelectronics 57, DOI 10.1007/978-3-319-44586-1_14

may affect the failure modes. On mechanical stressing, we emphasize the brittle nature as well as microvoid formation, especially Kirkendall void formation in microbumps. A string of voids in a brittle material can easily lead to fracture damage [2]. The interest in mechanical failures is because for mobile and wearable devices, the frequency of impact and dropping to the ground is high. On EM and TM in microbumps, we emphasize the enhanced failure mode due to Joule heating.

14.2 Quality Challenges for 3D Packaging

3D IC devices involve more complicated stacking processes where interconnect size becomes much smaller. Materials properties used in 3D IC influence the quality and reliability of electronic assemblies. For example, PCB substrate, silicon, microbump, copper layer and underfill have very different materials properties. Taking thermomechanical properties as example, Table 14.1 summarized the elastic modulus, coefficient of thermal expansion (CTE), and Poisson's ratio in room temperature. With more layers of stacking together in z-dimension, this becomes more challenging for 3D packaging processes to ensure both quality and reliability sound. One of the most obvious change in the continuous miniaturization of solder joint is the increase of volume fraction of intermetallic compounds (IMC) so that a microbump in 3D IC technology could be completely transformed into IMC. In other words, it is IMC instead of solder itself that plays a dominant role in determining microbump properties and reliabilities [3]. While IMC is electrically conducting, it is widely known to be mechanically brittle, see Table 14.2. Hence, the first concern is mechanical reliability. Besides its own brittle nature, its influence on low-k dielectrics on chip is also worth noticing. It is known that low-k dielectrics are susceptible to damage from chip-packaging interaction during thermal cycling. The presence of Cu-Sn IMC-based microbump has caused crack formation in the low-k dielectric layer [4]. This is one of the critical reasons that electronic industry depends on 2.5D IC for the reduction of chip-packaging

Table 14.1 Material properties used in 3D packaging

Materials	Young's modulus (GPa)	Poisson's ratio	CTE (ppm/°C)
PCB (FR4)	xy: 12.2–22, z:1.6–10	0.19–0.39	xy: 13–18, z:60
TSV/Pad (copper)	120–130	0.3–0.36	16.5–17
Solder (63Sn37Pb)	24.8–31.7	0.32–0.4	23.9–25
Solder (Sn3.8Ag0.7Cu)	44.4–58.0	0.35–0.4	23.5–26
Substrate (BT resin)	19–26	0.18–0.39	13–14
Solder mask	2.5–4.9	0.3–0.48	30–95
Silicon	106.9–169	0.22–0.35	2.3–2.7
Mold compound	12.5–16	0.25–0.35	20–25
Underfill	4.4–6	0.35–0.4	22–45
Die adhesive	3.45–7.4	0.27–0.35	45–100

Table 14.2 Properties of materials constituents of microbumps used in 3D IC technology [7]

Materials / Parameters	Cu_6Sn_5	Cu_3Sn	Ni_3Sn_4	Sn	Cu	Ni
Young's modulus (GPa)	116.89 ± 2.04 [8]	133.49 ± 4.44 [8]	141.12 ± 3.85 [8]	50	120	186 [9]
Indentation hardness (GPa)	6.35 ± 0.20 [8]	6.32 ± 0.15 [8]	6.31 ± 0.16 [8]	0.11 ± 0.05 [10]	1.7 ± 0.2 [10]	2.8 [11]
Yield strength (MPa)	2009 ± 63 [12]	1787 ± 108 [12]		35 ± 0.4	180 ± 9	950 [11]
Indentation fracture toughness (MPa $\sqrt{m}$)	2.73 ± 0.63 [13]	5.72 ± 0.86 [13]	3.88 ± 0.53 [13]	35–55 [14]	100–107 [14]	122 [15]
Poison's ratio	0.31	0.30	0.33	0.36	0.33	0.31
Coefficient of thermal expansion (ppm/K)	16.3 [16]	19.0 [16]	13.7 [16]	22.5 [9]	16.4 [16]	12.5 [9]
Electrical resistivity (μΩ cm)	17.5	8.8	28.5	11.5	1.71	6.99
Thermal conductivity (W/cm-K)	0.341	0.704	0.196	0.668	4	0.91
Melting point (°C)	415	676	795	232	1085	1455
Volume shrinkage (%)	5.0 % $6Cu + 5Sn = Cu_6Sn_5$ [17]	4.3 % $Cu_6Sn_5 + 9Cu = 5Cu_3Sn$ [17]	11.3 % $3Ni + 4Sn = Ni_3Sn_4$ [17]			
Crystal structure	Hexagonal (η) above 186 °C Monoclinic (η') below 186 °C	Orthorhombic (ε)	Monoclinic	Body-centered tetragonal	Face-centered cubic	Face-centered cubic

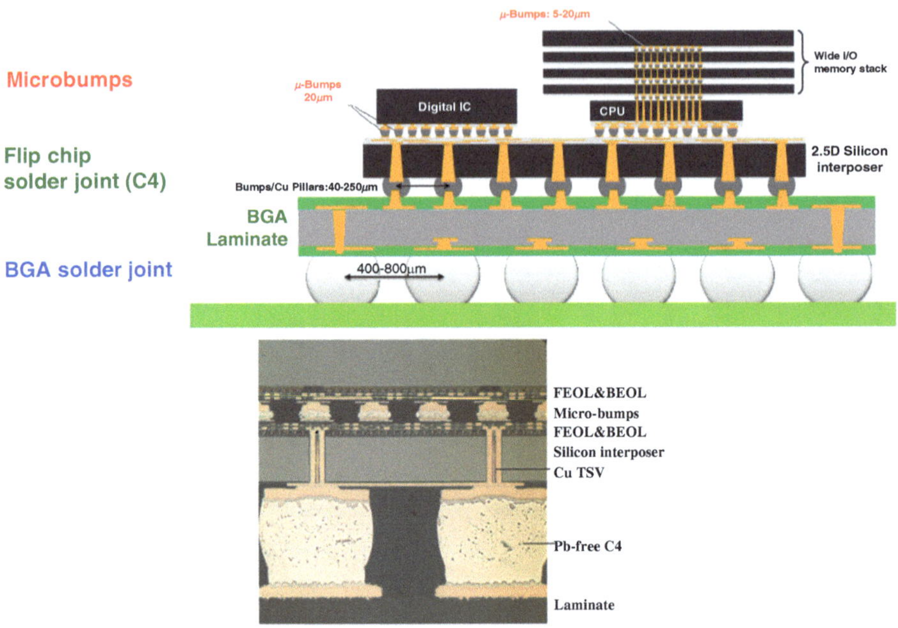

Fig. 14.1 Schematic diagram and optical image of a typical Chip-on-Wafer-on-Substrate (CoWoS) architecture using three-dimensional chip stacking with 2.5D silicon interposer ([5])

Fig. 14.2 A top view of 3D IC assembly [6]

interaction by using a thin layer of Si interposer, as shown in Fig. 14.1. The interposer is between the vertically stacked Si chips and the packaging substrate to greatly relieve thermomechanical stress in the Si chips. One of the prototypes of vertically stacked chips with Si interposer is shown in Fig. 14.2.

In a conventional flip chip solder joint, or Controlled-Collapse-Chip-Connection (C4) joint, the IC chips with pre-deposited solders are flipped over so that top surface of chip faces down and bonds with landing metal pads with a complex under bump metallurgy (UBM). One of the major constituents of UBM is a thin layer of Cu pad. In order to prevent the rapid Cu consumption by solder which may result in pancake-type electromigration (EM) void formation, typically a thin layer of Ni is deposited on top of Cu to change the metallurgy of UBM and reduce the Cu consumption rate. Hence only a small volume fraction of solder joint was converted to IMCs.

For the microbumps used in 3D IC technology, Thermal Compression Bonding (TCB) is used to meet the requirement for fine-pitch interconnect and high bonding yield, especially in wafer-level packaging [4]. Copper pillar, instead of Cu thin-film UBM, is used to control the fine gap between chip-to-chip stacking. This results in the utilization of low-volume and fine-cap Pb-free solder which is electroplated on Cu pillar before TCB process to form microbumps. For a small size microbump, its thickness is below 10 μm. Considering the large difference in terms of thickness and Young's modulus between Cu pillar (120 GPa) and Sn cap (50 GPa) of a microbump, see Table 14.2, Cu pillar may induce large thermomechanical stress at joint interfaces compared to the conventional flip chip solder joints. Hence, microbump is expected to have brittle fracture. In addition, the trend toward the miniaturization of electronic consumer products leads to the shrinkage of solder joint size, resulting in not only a high volume fraction but also a limited number of IMC grains within the microbump. Under a similar C-4 joint processing condition, i.e., about 240 °C and 1 min reflow time, a Sn-based microbump can be completely transformed into Cu-Sn or Ni-Sn IMC. Hence, microbump is no longer a solder joint. Instead, it is based on intermetallic compounds. For an IMC-based microbump, its mechanical reliabilities are far from understood because of the lack of field data.

For a chip about the size of our finger nail, ~1 cm^2, there are about a million of microbumps per chip, or 10^6 joints/cm^2, if we consider microbump diameter to be 5–10 μm. Therefore, how to study an individual microbump is quite challenging. In addition, for microbump diameter smaller than 5 μm, there is a strong likelihood that microbump properties are controlled by a few grains or even a single grain. For tens of thousands of IMC-based microbump interconnect, the microstructural anisotropy and texture effect may become even more important [1]. This in turn can influence microbump mechanical reliabilities such as early failure. Microstructure, in terms of grain size, morphology, texture, phase transformation, and phase stability, determines ultimately the properties, especially mechanical reliabilities of microbump. Therefore, an in-depth understanding of the properties and scalability of microbump for mass production requires a focused review.

14.3 Quality and Reliability of Microbumps

We present four types of microbumps that are widely used in the industry for 3D IC technology, in which the under-bump metallization (UBM) is in different metallurgical combinations. For each type of microbump, there are case studies and discussion of quality and reliability concerns. The aim is to understand the impact of different UBM or surface finishes on the mechanical reliability of microbump.

Microbump materials are typically Sn-based Pb-free solder, e.g., SnAgCu, SnAg, SnCu or SnAgCu-*X* where *X* represents the fourth elements. We simplified the discussion here by using "Sn" to represent solder and focus on the impact of UBM. No doubt the solder composition also plays a critical role to quality and reliability. With a fixed UBM layer, modulating solder composition and bonding temperature profile can significantly impact performance.

14.3.1 Type 1: Cu/Sn/Cu

14.3.1.1 Microstructure of Cu-Sn IMCs-Based Microbump

Cu/Sn/Cu is one of most widely used microbumps for chip-to-chip connection. When the joint height is less than 10 μm, under typical assembly condition at 240 °C for 1 min bonding time, all the microbumps can be converted completely into Cu-Sn IMCs. The kinetically favorable Cu_6Sn_5 is preferentially nucleated at the Cu/solder interface and continues to grow at assembly condition. Under further reflow or thermal annealing, Cu_6Sn_5 can be gradually transformed into the thermodynamically stable phase of Cu_3Sn. A typical microstructural evolution of Cu-Sn microbump in both as-received and reflowed condition is shown in Fig. 14.3.

14.3.1.2 Microstructural Characteristics of Cu_6Sn_5 in Microbump

Li et al. prepared Cu/Sn/Cu microbump samples by controlled compression of two polished copper surfaces sandwiching a tin foil in the middle with thickness 25 μm under different reflow temperature [19]. They monitored Cu_6Sn_5 microstructural evolution during continuous liquid state reaction where Sn is in the molten state. Shown in Fig. 14.4, Cu_6Sn_5 was formed at the Cu/Sn interface in scallop-type morphology. Growth and ripening of Cu_6Sn_5 on two opposing interfaces happened concurrently. Once the two opposing Cu_6Sn_5 grains came into contact with each other, the parallel grain boundaries surprisingly disappeared. No grain boundaries in lateral direction could be identified and only vertical columnar grain boundaries could be observed, Fig. 14.4c. The similar findings were also reported by Chuang [17] and Wang [20], Figs. 14.5 and 14.6. The columnar Cu_6Sn_5 was further consumed by Cu into more stable phase of Cu_3Sn in much finer columnar grain

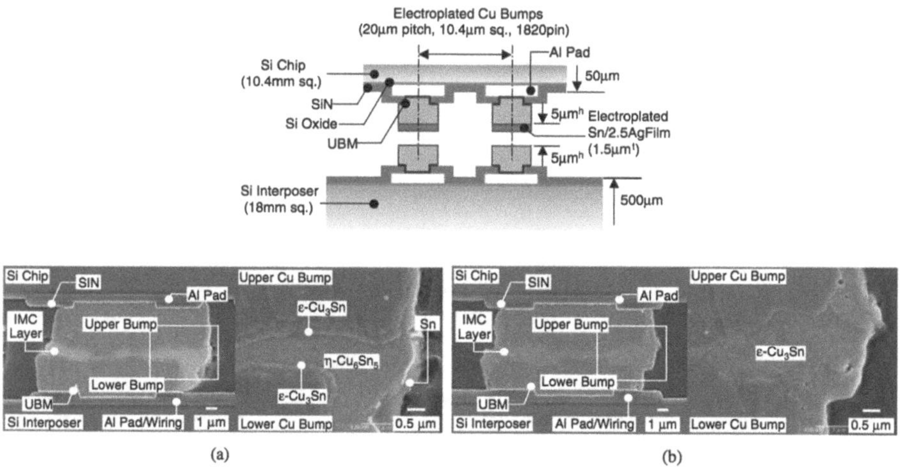

Fig. 14.3 Schematic diagram of Cu/Sn/Cu microbump-enabled chip-to-chip interconnection; SEM images of microstructural evolution of Cu-Sn microbump cross section (**a**) 240 °C, (**b**) 350 °C for 1-min bonding time [18]

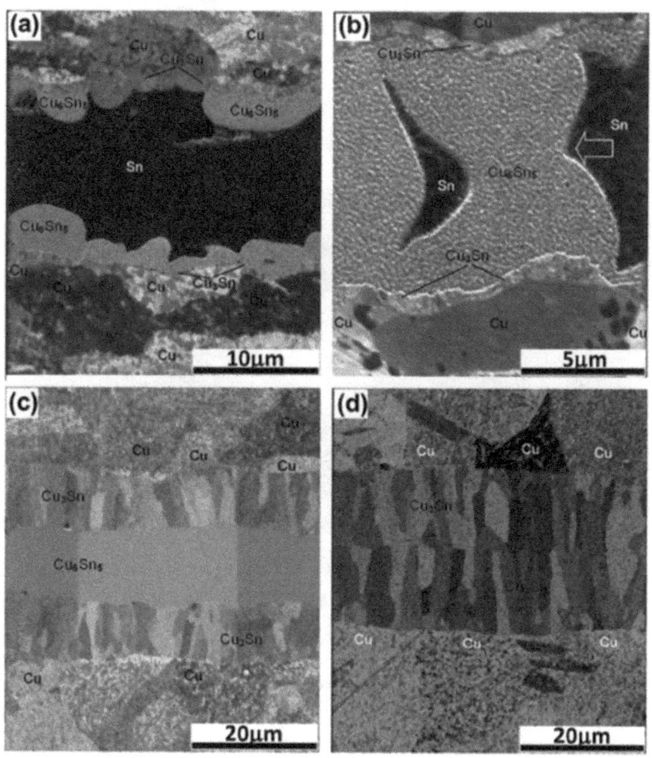

Fig. 14.4 FIB ion beam image of Cu/Sn/Cu micro-joint reflowed at: (**a**) 260 °C for 5 min; (**b**) 300 °C for 10 min; (**c**) 300 °C for 480 min; (**d**) 340 °C for 480 min [19]

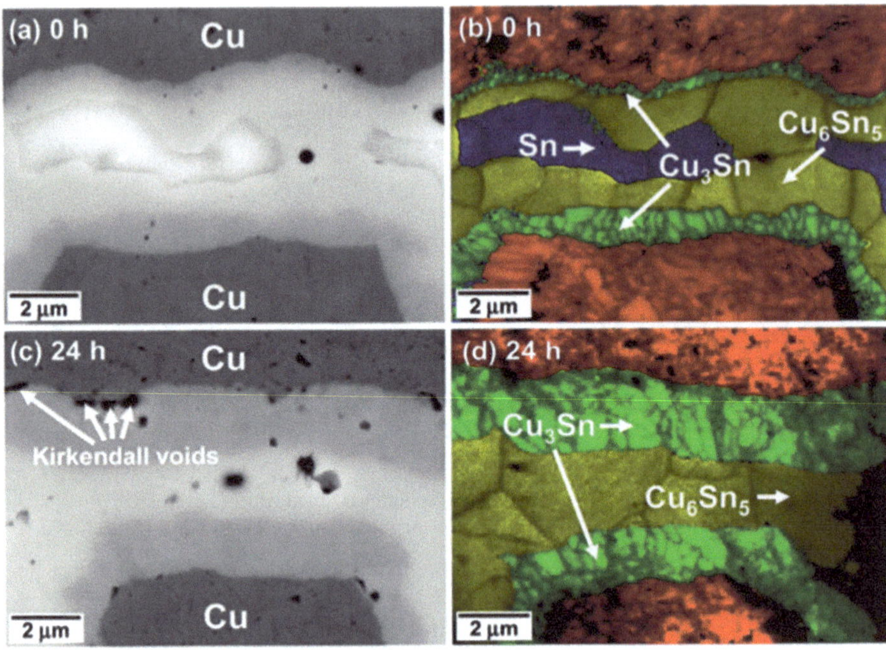

Fig. 14.5 BSE and EBSD phase mappings of Cu/Sn-Ag/Cu microbump in which IMCs microstructural evolution during isothermal annealing at 180 °C were characterized: (**a**, **b**) as-received, (**c**, **d**) 24-h annealing [20]

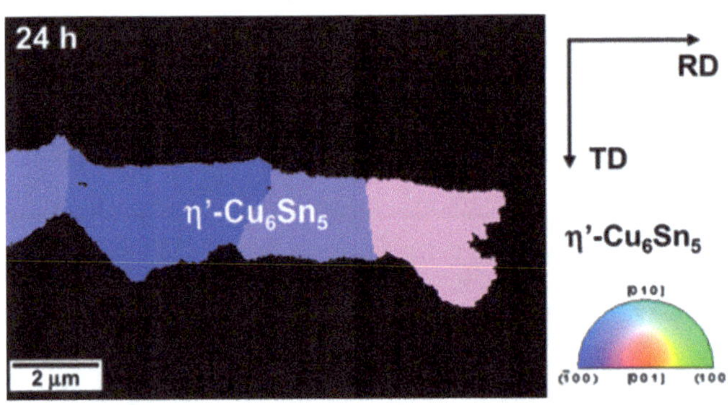

Fig. 14.6 Cu_6Sn_5-based microbump was indexed by low temperature η' - Cu_6Sn_5 phase and no lateral grain boundaries was identified in columnar-type Cu_6Sn_5-based microbumps [20]

structure. When the residual Cu_6Sn_5 was completely consumed, unlike Cu_6Sn_5 grains, two layers of impinging Cu_3Sn grains were left with both lateral and vertical grain boundaries. The underlying reasons that why opposing Cu_6Sn_5 grains tends to coarsen into a single grain with the disappearance of the lateral grain boundaries were not clearly understood. This remains a very puzzling question on columnar

Cu_6Sn_5 formation. However, the absence of lateral grain boundaries is certainly a benefit for microbump mechanical reliabilities because the horizontal grain boundaries across a microbump can degrade the microbump fracture strength by inducing crack formation and propagation.

14.3.1.3 Kirkendall Void and Porous Void Formation in Cu_3Sn

Cu_3Sn is known as the phase containing microvoid, i.e., Kirkendall void [21]. Kirkendall void is typically observed in Cu/Sn solder joints during a high temperature storage test or current stressing where Cu overconsumption leads to microvoid formation in the Cu_3Sn layer [21, 22]. Kirkendall void formation is due to unbalancing interdiffusion of Cu and Sn atomic flux where Cu diffuses much faster than Sn in the reverse direction [21, 23]. However, Kirkendall void can be effectively reduced by adding Co or Ni into Pb-free solder [24]. Besides, by controlling electroplating bath, or to have nanotwinned Cu as UBM can significantly reduce microvoid nucleation [25–28]. Kirkendall void can be a serious reliability issue. If it is not under control, Kirkendall voids tend to coarsen along the Cu_3Sn/Cu interface and induce micro-crack formation, as shown in Fig. 14.7

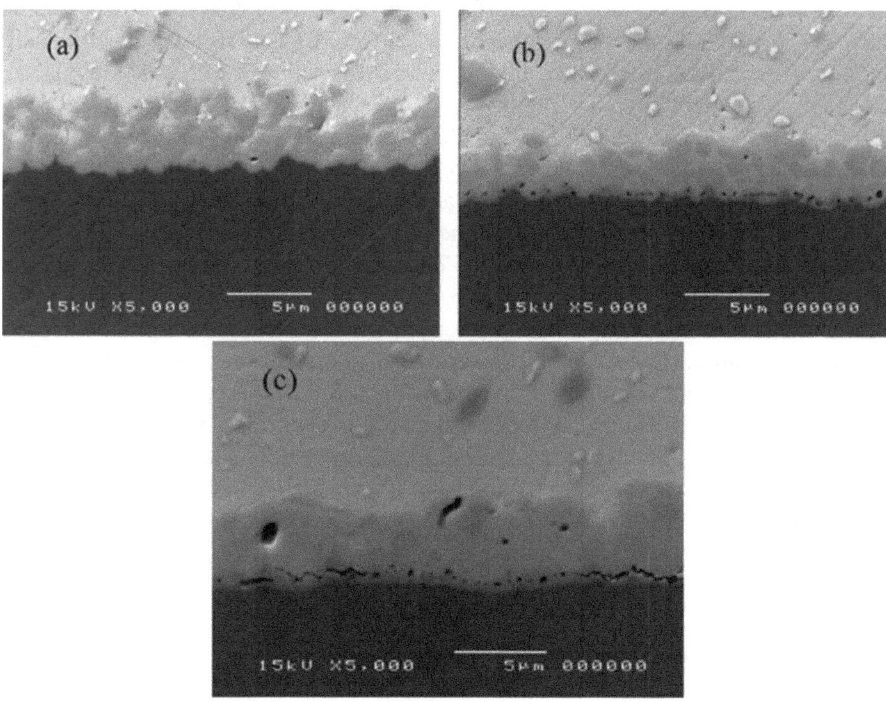

Fig. 14.7 SAC solder/copper pad cross-sectional interface after Ar^+ sputtering etch (**a**) as reflowed; (**b**) 500 cycles of thermal cycling that Kirkendall void formed and occupied about 10 % of interface; (**c**) 1000 cycles of thermal cycling at which Kirkendall void growth induced microcrack formation could be observed at the interface [28]

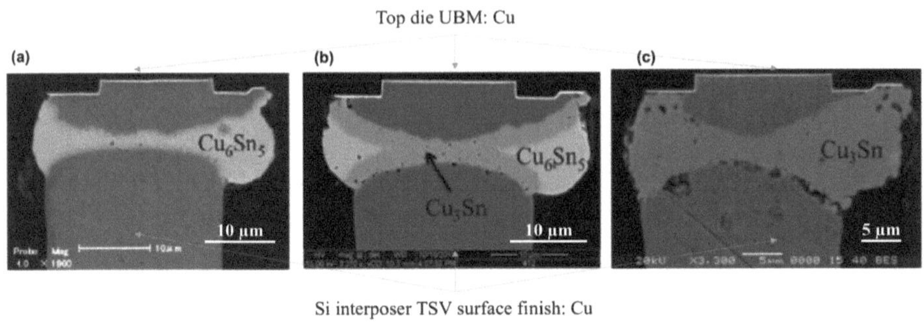

Fig. 14.8 Backscattered cross-sectional SEM images of Cu (top-die)-SnCu-Cu (TSV) microbump at different stages of thermal aging at 170 °C (**a**) as-received; (**b**) 83 h; (**c**) 1000 h [29]

[28]. No doubt, Kirkendall void is undesirable and deleterious to both mechanical and electrical performance of microbumps.

Wang et al. performed thermal annealing experiments on Cu-TSV/SnCu solder/Cu microbump at 170 °C. They found that Kirkendall void formation can induce crack formation at the interface between SnCu solder and Cu-TSV in the Si interposer after 1000-h annealing [29]. The Kirkendall void preferentially nucleated and grew on the TSV side of interface between Cu-TSV/microbump. This is attributed to different electrochemistry used in forming Cu UBM on the chip side and Cu-TSV in the interposer. In the latter, the Cu was filled in the via by electroplating where organic impurities such as S might be trapped inside and segregated to the end of Cu-TSV so that Kirkendall voids were preferentially nucleated at the Cu/Cu_3Sn interface because of heterogeneous nucleation. This was suggested to be the reason why Kirkendall void formation was only observed on the Cu-TSV interface instead of on the chip side (Fig. 14.8) [29].

Besides Kirkendall void formation, Chen et al. recently reported that there is another type of porous void formation which can happen inside Cu/Cu_3Sn/Cu microbump under aggressive thermal annealing or current stressing [30–32]. In porous microbump, Cu_3Sn can exist in two types of morphology: nonporous Cu_3Sn layer with Kirkendall void formation at the Cu/Cu_3Sn interface, and porous Cu_3Sn as middle layer sandwiched between its nonporous counterparts, Fig. 14.9. Wang et al. found that by conducting aggressive high temperature storage test at 210 °C for samples of Cu-wire/Sn/Cu-wire micro-joint in which Cu is of infinite amount. The mechanical strength of Cu wire degraded dramatically from 162 MPa from as-received state to 60 MPa after annealed for 1 week, Fig. 14.10.

14.3.1.4 Anisotropic Effect in Microbump

Suh et al. found that a very strong orientation relationship exists between Cu_6Sn_5 and Cu from the reaction between a pure (100) oriented single crystal Cu and eutectic SnPb. There are six types of preferred crystallographic orientation

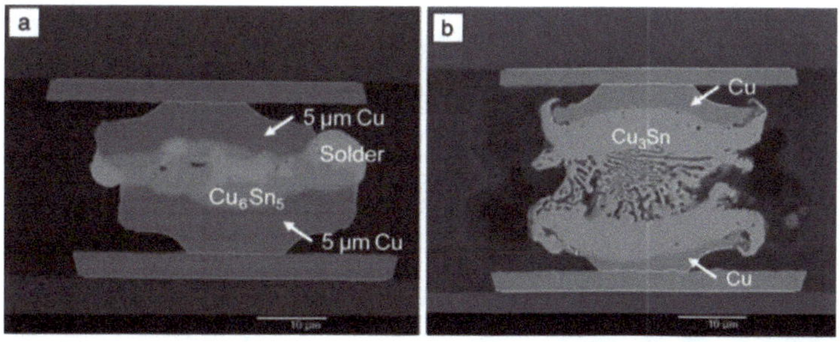

Fig. 14.9 (**a**) As-fabricated Cu/solder/Cu microbump with 8-µm-thick SnAg solder. (**b**) Porous Cu_3Sn is formed in the middle of microbump between two of nonporous Cu_3Sn layers after reflow at 260 °C for 12 h [32]

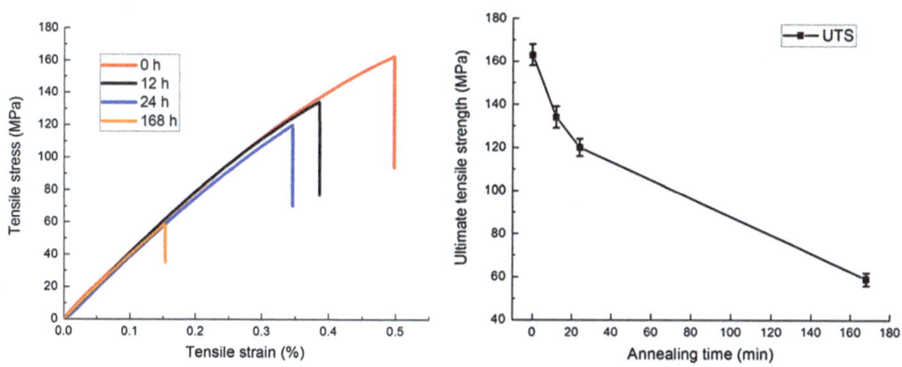

Fig. 14.10 (**a**) Tensile test curve of Cu/Sn(10-µm)/Cu micro-joint under aggressive high temperature storage at 210 °C up to 168 h. The adhesion strength degraded dramatically from 162 to 60 MPa [33]

relationship identified by Synchrotron X-ray microdiffraction [34, 35]. Figure 14.11a is an orientation mapping representing the angle between [101] direction of Cu and $[\bar{1}01]$ direction of Cu_6Sn_5. Figure 14.11b is a histogram orientation distribution corresponding to Fig. 14.11a. It indicates that most of the IMC grains have very strong orientation relationship between the [101] direction of Cu and $[\bar{1}01]$ direction of Cu_6Sn_5. A similar crystallographic orientation relationship also exists in the same kind of pure Cu and pure Sn diffusion couple, see Fig. 14.11c, d. The reason of preferred orientation relationship between [101] direction of Cu and $[\bar{1}01]$ of Cu_6Sn_5 is attributed to low misfit energy. The misfit between Cu and Cu_6Sn_5 is 0.24 %. The misfit direction lies on (001) plane of Cu.

$$(010)_{Cu_6Sn_5}//(001)_{Cu}\cdot\cdot[\bar{1}01]_{Cu_6Sn_5}//[101]_{Cu}$$

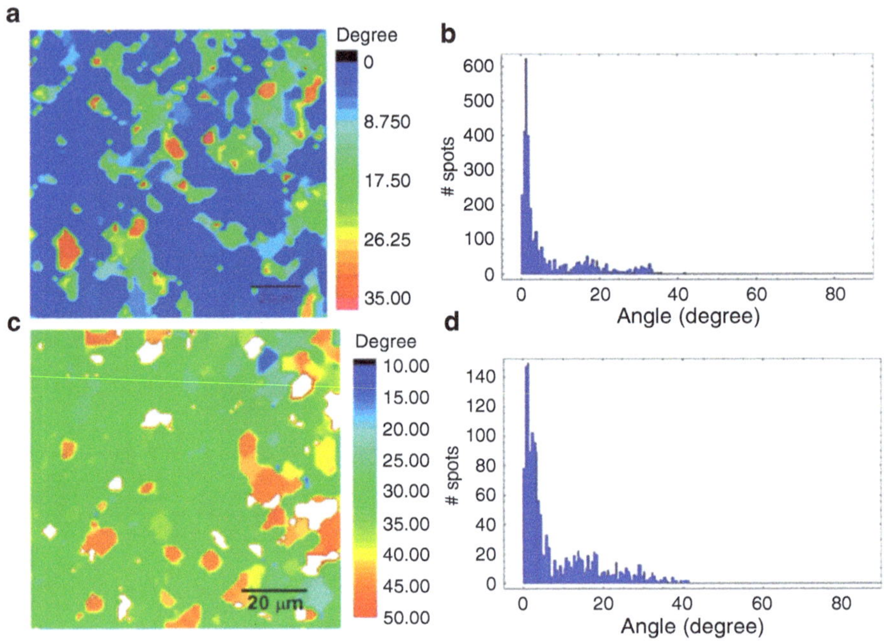

Fig. 14.11 Orientation mapping and histogram of (**a**), (**b**) Cu/SnPb and (**c**), (**d**) Cu/Sn sample with 4-min reflow time at 250 °C [35]

$$(343)_{Cu_6Sn_5}//(001)_{Cu} \cdot\cdot [\bar{1}01]_{Cu_6Sn_5}//[101]_{Cu}$$

$$(\bar{3}4\bar{3})_{Cu_6Sn_5}//(001)_{Cu} \cdot\cdot [\bar{1}01]_{Cu_6Sn_5}//[101]_{Cu}$$

$$(101)_{Cu_6Sn_5}//(001)_{Cu} \cdot\cdot [\bar{1}01]_{Cu_6Sn_5}//[101]_{Cu}$$

$$(141)_{Cu_6Sn_5}//(001)_{Cu} \cdot\cdot [\bar{1}01]_{Cu_6Sn_5}//[101]_{Cu}$$

$$(\bar{1}4\bar{1})_{Cu_6Sn_5}//(001)_{Cu} \cdot\cdot [\bar{1}01]_{Cu_6Sn_5}//[101]_{Cu}$$

An EBSD study indicates that the rooftop or prism-type Cu_6Sn_5 is elongated along two perpendicular $\langle 110 \rangle$ directions on the (001) plane of Cu, which also corresponds to low misfit directions of the Cu_6Sn_5. This study implies that if we can control the orientation of the UBM, i.e., (001) oriented Cu, the crystal orientation of Cu_6Sn_5 can also be controlled with preferred texture, resulting in unique grain morphology, i.e., rooftop, compared to randomly oriented scallop-type Cu_6Sn_5 grains. This orientation relationship might be potentially useful in controlling the texture of columnar grains-based microbumps, which is essentially dependent on the texture of Cu_6Sn_5 grains (Fig. 14.12).

This unique rooftop morphology was further studied by several groups [36, 37]. It was reported by Tian et al. that rooftop Cu_6Sn_5 morphology can be formed at elevated temperature, i.e., 300 °C while scallop-type Cu_6Sn_5 exists

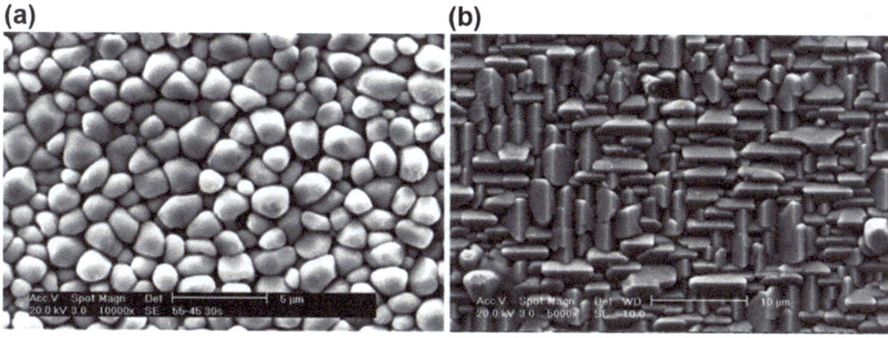

Fig. 14.12 Top view of exposed Cu_6Sn_5 grains after deep etching of eutectic SnPb solder joint reflowed at 200 °C for 30 s: (**a**) scallop-type Cu_6Sn_5 grains on polycrystalline Cu substrate; (**b**) rooftop-type Cu_6Sn_5 grains on (001) Cu [34]

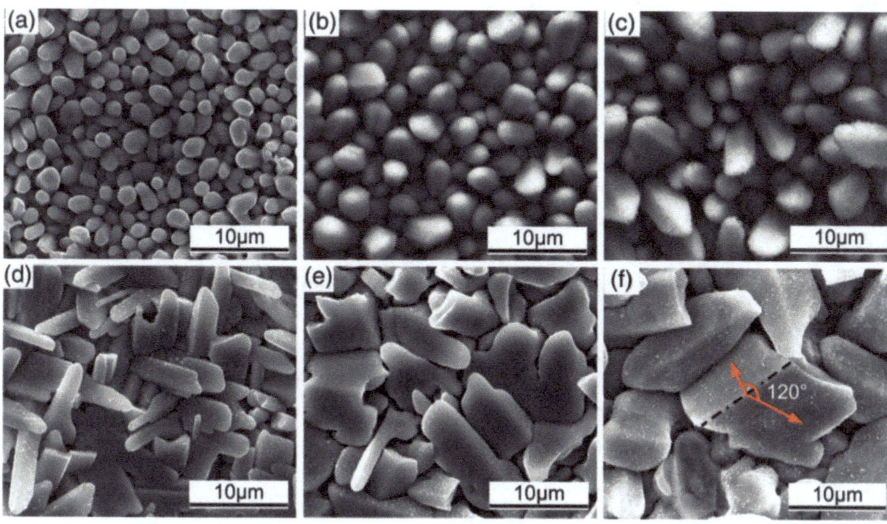

Fig. 14.13 Different interfacial morphology of Cu_6Sn_5 grains on polycrystalline Cu under different conditions: interfacial scallop-type morphology at 250 °C for (**a**) 10 s; (**b**) 60 s; (**c**) 300 s; and rooftop morphology at 300 °C for (**d**) 10 s; (**e**) 60 s; (**f**) 300 s [36]

during reflow at 250 °C. Both scallop-type and rooftop-type Cu_6Sn_5 have preferred [0001] direction parallel to the Cu surface normal (Fig. 14.13) [36].

In addition, the Cu_6Sn_5 grains of strong texture can also be obtained at typical reflow temperature, i.e., 240 °C for extended reflow time. Instead of using single crystal Cu, highly textured Cu_6Sn_5 grains can also form on polycrystalline Cu by reflowing at 240 °C at which η - Cu_6Sn_5 forms in [0001] direction which is aligned to the normal of polycrystalline Cu substrate. In order to induce strong preferred texture, a long-time reflow was conducted. As shown in Fig. 14.14, a clear texture

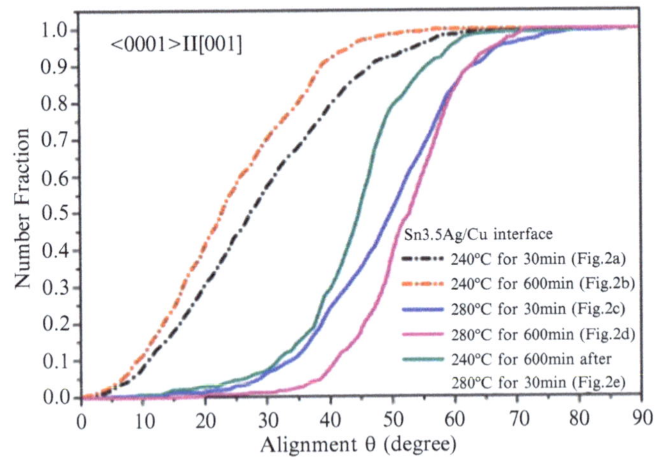

Fig. 14.14 Misorientation chart of alignment angle statistics of ⟨0001⟩-oriented Cu_6Sn_5 with polycrystalline Cu substrate toward [001] direction [38]

can be observed at 240 °C at which strongest orientation dependence of growth of Cu_6Sn_5 along the normal of the Cu substrate can be obtained, when it was reflowed at 240 °C for 600 min. At this condition, about 70 % of Cu_6Sn_5 grains have misorientation angle smaller than 30 °. Unlike the Cu_6Sn_5 texture formed on single crystal (100) Cu, the texture from polycrystalline Cu is a result of interfacial grain ripening.

The textured Cu_6Sn_5 was further evaluated from mechanical properties perspective. As shown in Fig. 14.15, Sn3.5Ag was used for reflow soldering for two conditions: 240 °C, 1 min, and 280 °C, 1 min for common polycrystalline Cu substrates. For comparison, a textured Cu_6Sn_5 substrate was used for reflow soldering at 280 °C for 1 min. Ball shear tests were conducted to evaluate IMC strength with preferred orientation and the average value was taken by 30 measurements and results were plotted in Fig. 14.15d. It indicates that the [0001]-oriented Cu_6Sn_5 formed at 280 °C for 1 min has highest joint shear strength [38]. This unique oriented growth of Cu_6Sn_5 is worth to be further studied and applied in a controlled growth of IMC-based microbump.

14.3.2 Type 2: Ni/Sn/Ni

For typical Ni/Sn/Ni microbump, Ni is used as diffusion barrier to retard IMC formation especially under reflow or thermal compression bonding. Hence, it could leave behind a sufficient amount of unreacted solder as mechanical buffer layer to retain the ductility of microbump. As shown in Fig. 14.16, for as-received Ni/Sn/Ni microbump, after bonding, Ni-Sn microbump was more or less transformed into Ni_3Sn_4 IMC at the center but with the remaining solder squeezed out on the

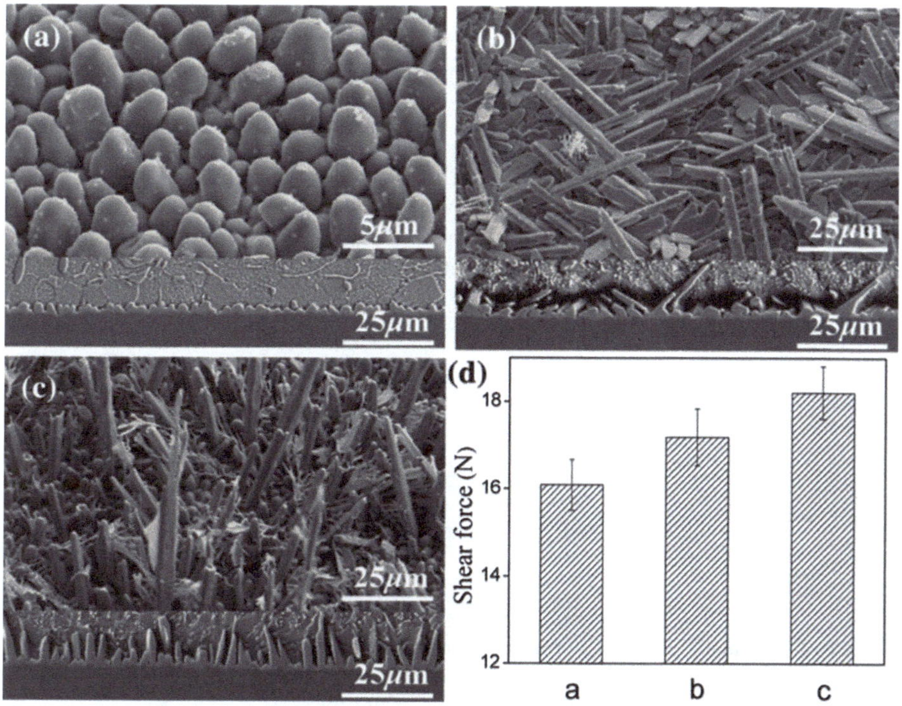

Fig. 14.15 Morphologies of Cu_6Sn_5 formed on Cu substrate at (**a**) 240 °C and (**b**) randomly-oriented Cu_6Sn_5 at 280 °C for 1 min, and (**c**) [0001]-oriented Cu_6Sn_5 at 280 °C for 1 min. The joints were cooled in air. (**d**) Shear strength of solder joint of (**a**), (**b**), and (**c**) condition [38]

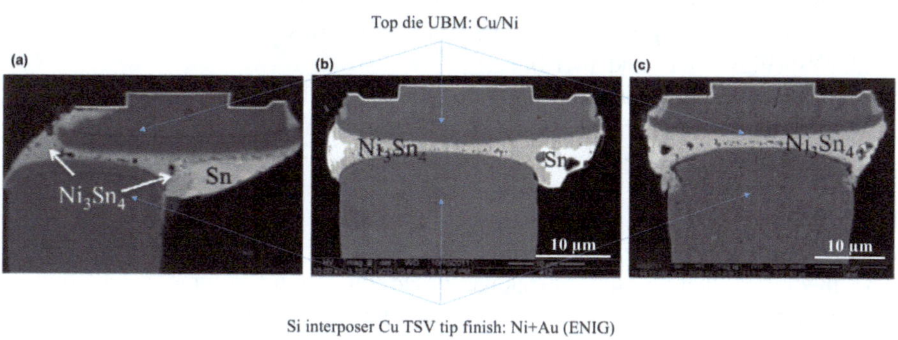

Fig. 14.16 BS-SEM images of cross sections of the Ni-Sn-Ni microbump under isothermal annealing at 170 °C (**a**) as-received; (**b**) 83 h; (**c**) 1000 h [29]

periphery of the microbump. The top and bottom layers of Ni_3Sn_4 were almost impinged into each other. Upon further thermal annealing, eventually the Ni/Sn/Ni microbump was fully converted into Ni_3Sn_4 IMC, in which an array of microvoids were trapped along the impinged Ni_3Sn_4 grain boundaries, which is susceptible to

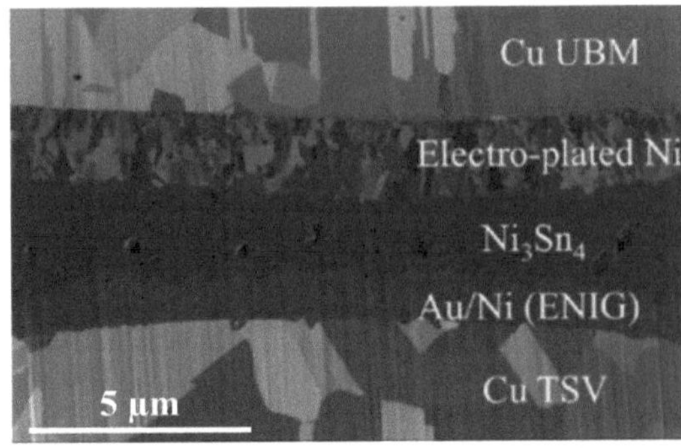

Fig. 14.17 FIB cross-sectional image of Ni-Sn-Ni microbump/TSV joint with microvoids trapped inside Ni_3Sn_4 IMC [29]

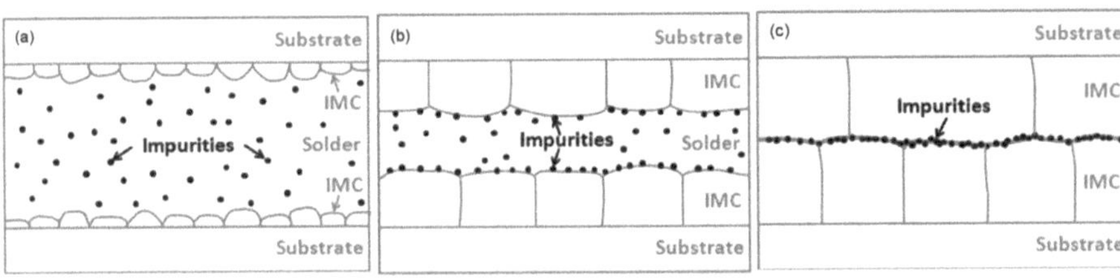

Fig. 14.18 Schematic diagrams showing the impurities concentration increases while residual solder is gradually transformed into IMC (**a**) As-reflowed condition; (**b**) two-third of solder consumed; (**c**) all of the solder completely converted into IMC [17]

grain boundaries embrittlement due to trapped impurities [29]. From the point of view of mechanical reliability, inter-grain fracture is expected to occur along this grain boundary which contains a large amount of microvoids and impurities. The detailed impurities migration to Ni_3Sn_4 grain boundaries was also given by Chuang et al. [39] (Figs. 14.17 and 14.18).

Besides impurities trapping, another type of mechanically weak link can be attributed to Ni/Sn chemical reaction in Ni-Sn-Ni microbump. The interfacial reaction of Ni and Sn is given by the following chemical reaction:

$$3Ni + 4Sn = Ni_3Sn_4$$

The theoretical net volume shrinkage based on atomic volume calculation is 11.3 %, see Table 14.2. After all of the remaining solder is completely transformed into Ni_3Sn_4, due to its brittle nature, the volume shrinkage can no longer be accommodated by solder thickness reduction. Instead, it results in void formation to dissipate the volume shrinkage. As shown in Fig. 14.19, void formation in Ni_3Sn_4-based microbump annealed at 180 °C for 240 h was identified. The cross section was polished by ion beam milling to avoid possible mechanical polishing induced micro-crack formation or polishing powder smearing effect. Figure 14.19 clearly demonstrates that the void was originated from Sn pockets during Ni/Sn IMC formation. As also indicated by Chen's Ph.D. thesis [40], for Ni-Sn-Ni

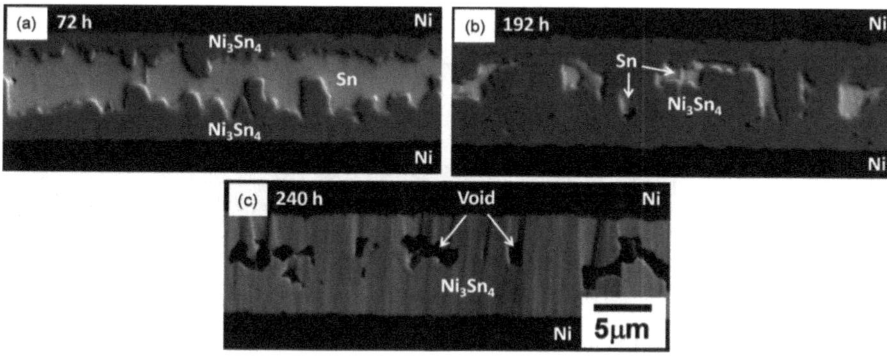

Fig. 14.19 Cross-sectional SEM images of Ni_3Sn_4 microstructural evolution in Ni/Sn (10 μm)/Ni micro-joint under isothermal annealing at 180 °C for (**a**) 72 h, (**b**) 192 h, and (**c**) 240 h [17]

microbump, Sn is faster diffusion species compared to Ni. The IMC growth during thermal annealing is also a flux-driven ripening process, which means larger grains will grow at the expense of smaller grains. The ripening effect tends to result in uneven interface due to opposing IMC impingement and hence result in forming Sn pockets in thermal annealing. Since Sn is faster diffusion species, the asymmetrical atomic flux results in vacancy accumulate inside Sn pocket. When the remaining Sn is completely consumed, the supersaturation of vacancy leads to void formation and therefore trapped inside the original position of Sn pocket. No doubt, the void trapped inside Ni_3Sn_4 microbump can adversely degrade IMC mechanical reliability. In terms of UBM used for Ni_3Sn_4 microbump, Ni ($E = 186$ GPa) itself is less mechanically compliant than Cu ($E = 120$ GPa), Table 14.1. It seems that Ni_3Sn_4 microbump, based on the above discussion, would have a more serious mechanical reliability concerns than Cu-based microbumps: the trapped impurities induced grain boundaries embrittlement and volume shrinkage resulted void or micro-crack formation. Traditionally, Ni was introduced to reduce IMC formation in order to address mechanical reliability concerns. Nevertheless, for the application of fine-pitch interconnects in 3D IC technology which requires joint thickness down to 5 μm, joint with full IMC formation is inevitable so that the utilization of Ni as UBM would not necessarily mitigate the full IMC formation and might actually result in serious mechanical reliability issues.

14.3.3 Type 3: Cu/Sn/Ni

The third type of metallurgical configuration of microbump is Cu/Sn/Ni. As shown in Fig. 14.20, Cu/Sn/Ni type of microbump was prepared to study IMC microstructural evolution and reliability [29]. The UBM on the top Si chip was composed of Cu stud while the TSV in the interposer was made of Cu coated with electroless-plated Ni as surface finish. Cross-sectional EDX analysis indicated that initial

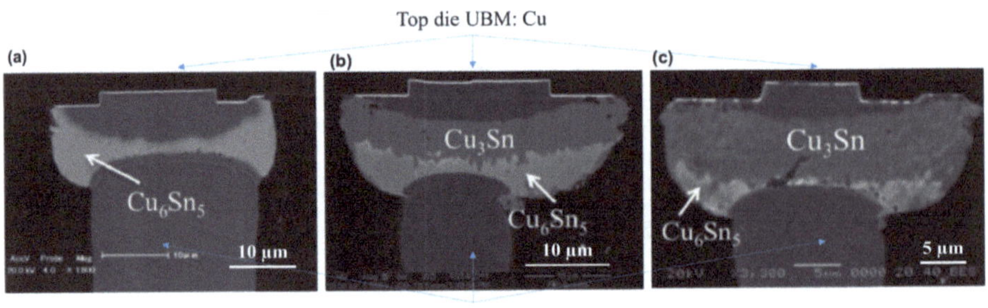

Fig. 14.20 Backscattered-SEM image of cross section of the Cu (top-die)/microbump/Ni (surface finish of Cu-TSV) isothermally annealed at 170 °C, (**a**) as-received; (**b**) 267 h; (**c**) 1000 h [29]

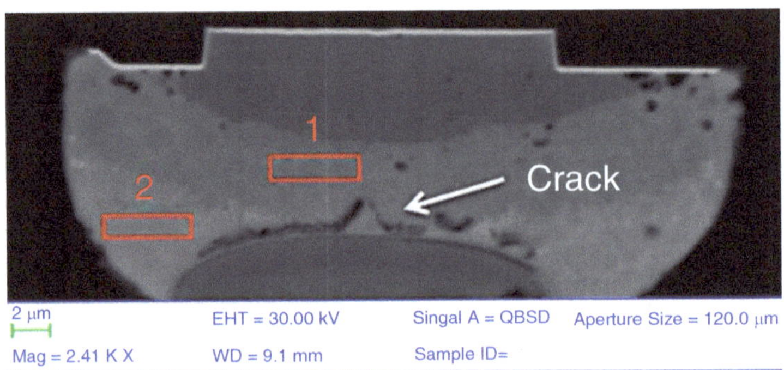

Fig. 14.21 A crack was identified near the interface between Cu_3Sn and Ni_3Sn_4 for Cu/Sn/Ni-type microbump [29]

thermal compression bonding has consumed all of the solder into IMC in the stoichiometric form of $(Cu,Ni)_6Sn_5$ with a limited amount of Ni dissolution. No $(Ni,Cu)_3Sn_4$ IMC was identified. The reason of preferential formation of Cu_6Sn_5 instead of Ni_3Sn_4 is due to the large volume difference between a limited amount of solder compared to the unlimited supply of Cu stud, which resulted in preferential formation of Cu_6Sn_5 instead of Ni_3Sn_4. Upon continuous thermal annealing, $(Cu,Ni)_3Sn_4$ layer grew thicker at the expense of Cu_6Sn_5. Since Ni was used as surface finish for diffusion barrier at the Cu-TSV side, the growth of Cu_3Sn was significantly restarted. This asymmetrical design of Cu-Sn-Ni microbump ultimately resulted in an asymmetrical IMC formation and excessive Cu UBM consumption at the top side of the stacked chips.

Crack formation was observed in Cu/Sn/Ni-type microbump, as shown in Fig. 14.21. The reason was attributed to thermomechanical stress, resulting from CTE mismatch of dissimilar IMC formation, for which CTE of Cu_3Sn and Cu_6Sn5, are respectively, 19.0 and 13.7 ppm/K, see Table 14.1. Besides, there might be another two reasons for the crack formation for Cu/Sn/Ni-type microbump. Once IMCs are formed inside the Cu/Sn/Ni microbump, Cu_6Sn_5 on the top die side was further transformed into Cu_3Sn, whereas on the bottom TSV side Sn diffusion is much faster than Ni diffusion. The unbalancing Sn and Ni diffusion flux results in vacancy supersaturation and void formation along the Ni surface finish/microbump interface, which ultimately leads to crack formation. The other possible reason is due to volume shrinkage difference between Cu-Sn and Ni-Sn IMC. As indicated in Table 14.1, volume shrinkage of Cu_6Sn_5, Cu_3Sn, and Ni_3Sn_4 are, respectively, 5 %, 4.3 %, and 11.3 %. The volume shrinkage could not be accommodated by IMCs deformation due its brittle nature. Hence, the mismatch of volume shrinkage might also result in crack formation.

14.3.4 Type 4: Cu/Ni/Sn/Ni/Cu

14.3.4.1 Typical Composition Parameters of Cu/Ni/Sn/Ni/Cu Microbumps

The fourth type of microbump is Cu/Ni/Sn/Ni/Cu. Two symmetrical Ni layers are used as diffusion barriers for two opposing Cu pillars joined by Sn cap during thermal compression bonding. For a typical symmetrical Cu/Ni/Sn/Ni/Cu microbump, its dimensional parameter is shown in Fig. 14.22. Sn2.5Ag was used as Pb-free solder. The microbump pitch is 30 μm. For each side of UBM, the thickness of Cu and Ni is, respectively, 5 and 3.5 μm. The passivation opening for Cu pillar is 12 μm. After assembly at 250 °C, IMC thickness from both sides is about 1 μm and the remaining solder thickness is 2.5 μm.

Under thermal cycling test between −55 and 125 °C, the failure rate of Cu/Ni/Sn2.5Ag/Ni/Cu microbump is 9.5 % for the standard 1000 cycle test. The failure rate only increased dramatically to 47.3 % at the same testing condition when the number of thermal cycles is beyond 1000, which exceeds generally accepted accelerated testing conditions. It indicates that Cu/Ni/Sn/Ni/Cu-type microbumps actually meet the mechanical reliability requirement (Fig. 14.23).

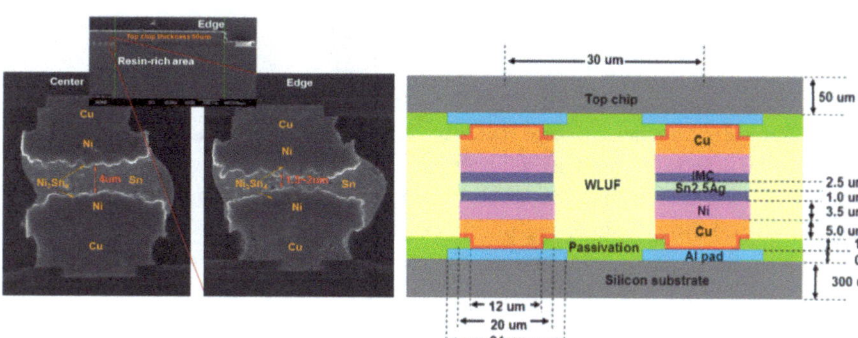

Fig. 14.22 Cross-sectional SEM images and schematic diagram of Cu/Ni/Sn/Ni/Cu microbump [41]

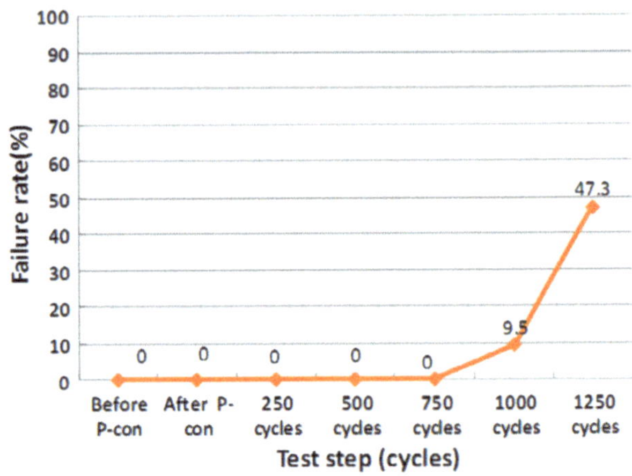

Fig. 14.23 Accumulative failure rate of Cu/Ni/Sn2.5Ag/Ni/Cu microbump during temperature cycling test between −55 and 125 °C [41]

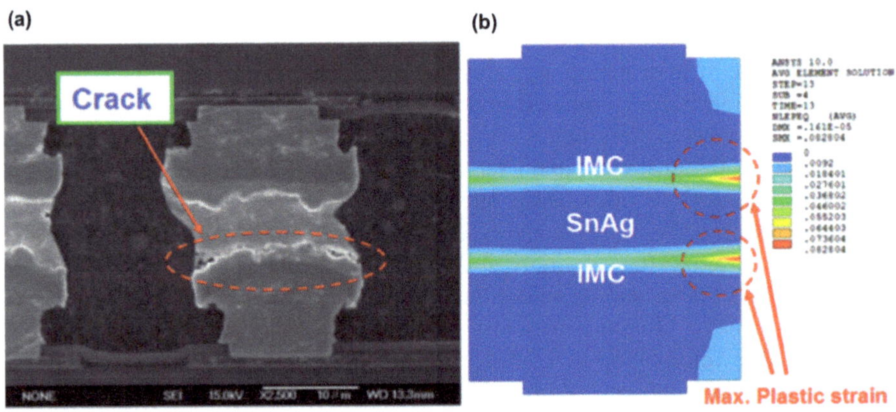

Fig. 14.24 Cross-sectional image of interfacial fracture of Cu/Ni/Sn2.5Ag/Ni/Cu microbump: (**a**) SEM image indicates that a crack propagates along IMC/solder interface; (**b**) finite element simulation result [41]

14.3.4.2 IMC/Solder Interfacial Crack Formation

Finite element simulation was performed on the Cu/Ni/Sn/Ni/Cu-type microbump with the geometry shown in Fig. 14.24b. The simulated result reveals that the maximum plastic strain exists at the corner of the interface between the IMC and remaining solder, which indicates that the micro-crack formation and rapid crack penetration through IMC layer are highly possible [41]. This postulation is indeed in accord with the experimental finding that crack formation was found to propagate along the IMC/solder interface, shown in Fig. 14.24a.

A similar finding of IMC/solder interfacial crack formation for the Cu/Ni/Sn/Ni/Cu-type microbump was also reported by Hsu et al., where crack formation and propagation was identified along the interface between the Ni_3Sn_4 and residual Sn2.5Ag-solder, as shown in Fig. 14.25 [42].

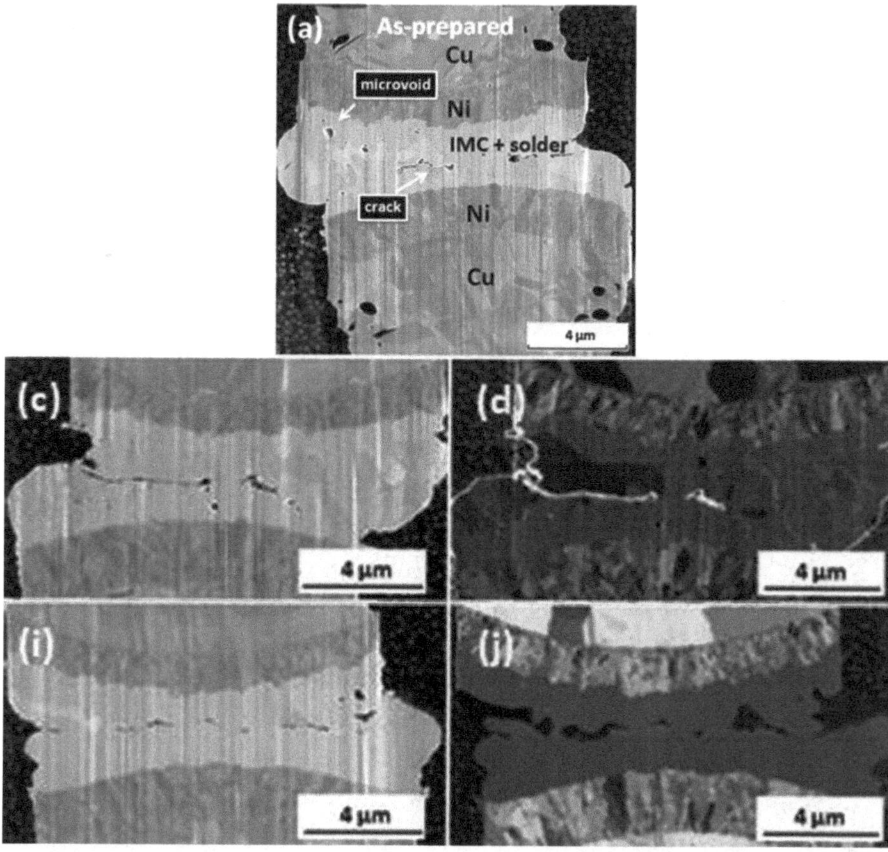

Fig. 14.25 Cu/Ni/Sn-2.5Ag/Ni/Cu microbump samples under solid state annealing at 150 °C for (**a**) as-received; (**c**) e-beam image of sample annealed for 25 h; (**d**) ion beam image of sample annealed for 25 h; (**i**) e-beam image of sample annealed for 250 h; (**j**) ion beam image of sample annealed for 250 h [42]

14.3.4.3 Ni as Effective Diffusion Barrier to Suppress Kirkendall Void Formation

For Cu/Ni/Sn/Ni/Cu microbump in which Ni is used as diffusion barrier, it has limited Kirkendall void formation. Shown in Fig. 14.26, three types of samples with one side of UBM fixed as Ni pad and the other side with different surface finishes of Cu pillar [25]. Sample A is 10-µm Cu with 10-µm SnAg. Sample B is 14-µm Cu with 8-µm SnAg. Sample C is 10-µm Cu with 2-µm Ni as diffusion barrier covered by 0.5-µm Cu with 10-µm SnAg. All of the three types of microbump samples were subjected to high temperature storage test at 150 °C for 500 and 1000 h. It clearly indicates that by depositing a diffusion barrier layer of Ni between Cu pillar and low volume of Sn-based solder, Kirkendall void formation is effectively eliminated. In

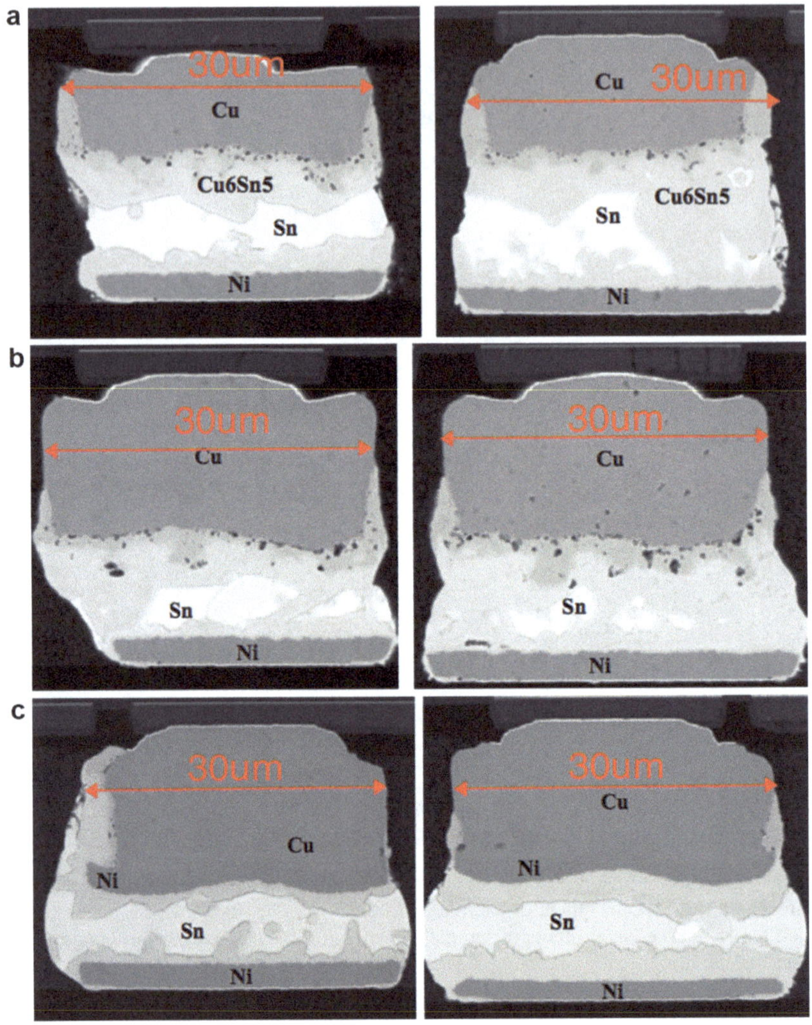

Fig. 14.26 (**a**) Sample A. 10-μm Cu/10-μm SnAg under thermal annealing. (1) 150 °C, 500 h; (2) 150 °C, 1000 h [25]. (**b**) Sample B. 14-μm Cu/8-μm SnAg under thermal annealing. (1) 150 °C, 500 h; (2) 150 °C, 1000 h [25]. (**c**) Sample C. 10-μm Cu/2-μm Ni/0.5-μm Cu/10-μm SnAg (1) 150 °C, 500 h; (2) 150 °C, 1000 h [25]

addition, under the same time of annealing, for microbump with Ni diffusion barrier, there is still about 50 % of microbump volume where the remaining solder was not converted to IMC yet. We expect that such joint has better mechanical properties. Therefore, in order to control Kirkendall void formation, besides adding Ni or Co into solder [24]; electroplating with nanotwin copper [26]; controlling electroplating bath chemistry [27], the addition of Ni between Cu and solder as diffusion barrier seems to be a promising approach. This is because the other three methods are more difficult in finding a practical and cost-effective route during operation and especially in control of electroplating bath chemistry for mass production.

14.3.5 Concluding Remarks

Among the four types of microbump discussed above, symmetrical Cu/Ni/Sn/Ni/Cu is one of the ideal and widely accepted metallurgical configurations of microbump in the industry for 2.5D/3D IC technology. It has superior mechanical reliability compared to other three metallurgical configurations: Cu/Sn/Cu; Ni/Sn/Ni; and Cu/Sn/Ni, in terms of Kirkendall void, porous void formation in Cu_3Sn, IMC volume contraction induced crack propagation, and asymmetrical impinging IMC interfacial cracking.

14.4 Field Performance Prediction of 3D Packaging

As compared to traditional desktop and laptop, the mobile devices tend to adopt more 3D technologies in order to reduce device size and weight. In the mean time, these devices are exposed to more stringent and dynamic environments due to handheld. However, the fundamentals to predict field performance are the same. One needs to understand (1) type of environmental/inherent stress that the devices experience and duration across its usage life (2) the failure modes that such stress can cause, (3) accelerated stress (run in lab) to duplicate such failure, and (4) acceleration model to interpret lab tests results to field performance.

Taking microbump open failure as an example, the cause of the failure is typically thermal cycling due to device power on/off and environmental temperature variation. Cyclic stress-strain will deform the solder material due to CTE mismatch in different electronic materials, thus inducing thermal fatigue failure. Thermal cycling (TC) or thermal shock (TS) tests can be used for accelerated testing of 3D packaging. JEDEC standard of temperature cycling provides different test conditions, where condition G (−40 to 125 °C), J (0–100 °C), K (0–125° C), L (−55 to 110 °C), and M (−40 to 150 °C) as shown in Fig. 14.27 can be used with frequency of 1 cycle/h and soak time of 15 min.

Two different approaches can be used to develop a life prediction model for microbump. One approach is to develop fatigue model by conducting displacement-controlled isothermal mechanical fatigue test using bulk solder material. The other is to develop fatigue model by combining actual reliability test data and FEA simulation results.

For the first method, Coffin-Manson strain-based model and Morrow's energy-based model have been reported:

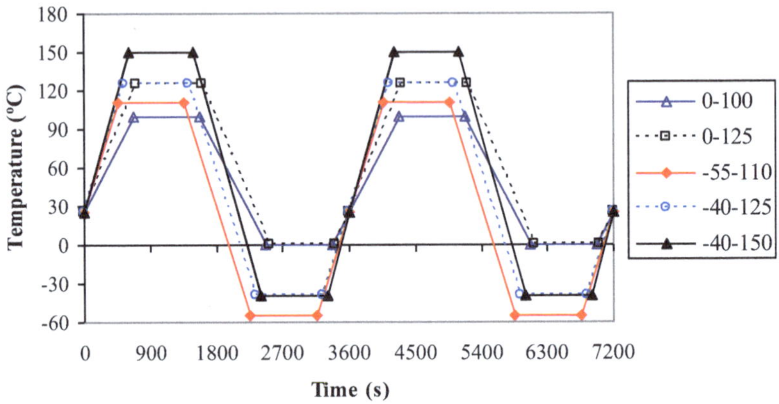

Fig. 14.27 Different thermal cycling profiles

$$N_f^m \Delta\varepsilon_{in} = C$$
$$N_f^n W_p = A$$

The fatigue ductility coefficient, C and A, and the fatigue exponent, m and n, for SAC bump can be obtained from strain-life and energy-life curves at different test conditions such as at 25, 75, and 125 °C with 0.1, 0.01, or 0.001 Hz, Temperature changes from –40 to 125 °C with 1 h per cycle during thermal cycling test. The highest temperatures of 125 °C and strain rate (low frequency at 0.001 Hz) in thermal cycling test have significant effect on solder joint fatigue failure. Hence, the isothermal test condition at temperature 125 °C with 0.001 Hz frequency was selected for further fatigue analysis. The m, n, C, and A are 0.853, 0.897, 9.2, 311.7 MPa (at 125 °C with 0.001 Hz), respectively reported by Pang et al. [24] (Fig. 14.28).

For the second method, the coefficient and exponent in fatigue model can be determined by combining numerical result and actual test data based on a damage mechanism. Schubert et al. proposed fatigue model for SAC bump solder determined by creep strain criteria and creep strain energy-based method based on FEA results and actual reliability test data (300 Ω as a failure criterion) from different assemblies such as FCOB with or without underfill and PBGA with Ni/Au board finish under different thermal cycling conditions:

$$N_f = 4.5{\varepsilon_{cr}}^{(-1.295)}$$
$$N_f = 345{W_{cr}}^{(-1.02)}$$

When elastic-plastic-creep constitutive model for solder was used in FEA simulation, the inelastic strain includes two part, plastic strain and creep strain. When creep part is dominant, fatigue life can be predicted by substituting creep

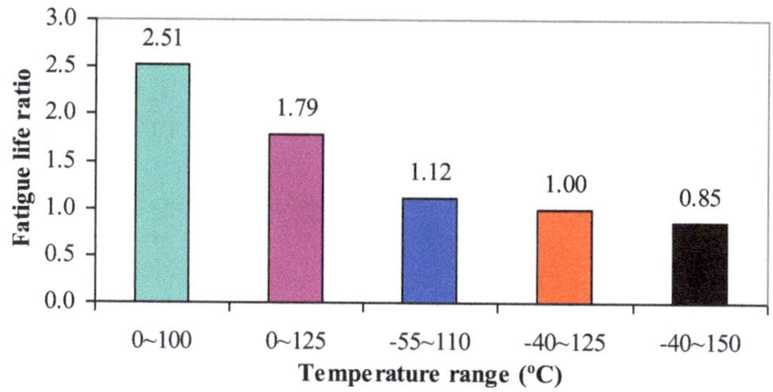

Fig. 14.28 Fatigue life ratio for different temperature ranges

strain or creep strain energy into above equations. When plastic part is dominant, fatigue life can be predicted by using plastic strain or plastic strain energy as fatigue damage parameter. Otherwise, fatigue life prediction also can be done using total inelastic parameter including creep and plastic parts as fatigue damage parameter. In addition, a combined creep-fatigue life prediction model for solder was proposed by Pang et al. [13, 29] as given below:

$$\frac{1}{N_f} = \frac{1}{N_p} + \frac{1}{N_c}$$

14.5 Electromigration Reliability for 3D IC Packaging

To connect the hundreds of millions or even billions transistors on a Si chip in VLSI circuit technology, multilayers of thin-film interconnect wires made of Al or Cu were used. Electromigration (EM) in interconnects has been one of the most crucial and persistent reliability issues in Si-based microelectronic technology. Al or Cu interconnects usually have a small cross-sectional area. Thus, a very high current density will be carried, typically 10^5–10^6 A/cm^2. Under such high current density, atomic diffusion and rearrangement are enhanced, leading to void formation at the cathode and extrusion in the anode. The cathode and anode are sites of atomic flux divergence, which has been recognized. The analysis of back stress accompanying EM has also attracted much attention. When flip chip technology introduced the C-4 solder joint, the effect of current crowding on EM was found to occur near the contact area of solder joints with a current density about 10^4 A/cm^2 [1–9]. As consumer electronic products are ubiquitous in the big data era, its dense packaging in using very small size microbumps in 2.5D/3D IC requires a careful thermal management. While EM damage is still affected by flux divergence, back-stress, and current crowding, Joule heating is now the new focus.

After introducing the basic concepts in electromigration very briefly below, we will show experimental findings of EM damages in interconnects and flip chip solder joints. Then we will discuss EM failures of 3D IC in using the vertical stacking of Si chips containing the new interconnect structure of microbumps and TSVs. Based on these new structures, we will present their EM behaviors. Especially, we show that system level EM study is becoming important. In the system, a surprise was the failure of the redistribution layer (RDL) between the C-4 solder joints and the microbumps. In the system level EM test, heat dissipation becomes a critical problem. Synergistic effect of Joule heating and EM will be discussed.

14.5.1 Introduction on Electromigration

Electromigration is the interaction between electron flow and atomic diffusion. The electrical force acting on a diffusion atom is expressed by Huntington as below [43]:

$$F_{\mathrm{em}} = Z^{*}eE = \left(Z_{\mathrm{el}}^{*} + Z_{\mathrm{wd}}^{*}\right)eE$$

where e is the charge of an electron and E is the electric field, and Z^{*} is the effective charge number of EM and it consists of Z_{el}^{*} and Z_{wd}^{*}. Z_{el}^{*} can be regarded as the nominal valence of the diffusing ion in the metal when the dynamic screening effect is ignored. $Z_{\mathrm{el}}^{*}eE$ is called the direct force (or static force) and it is acting in the direction opposing electron flow. Z_{wd}^{*} is an assumed charge number representing the effect of momentum exchange between electrons and the diffusing ions. $Z_{\mathrm{wd}}^{*}eE$ is called the electron wind force (or dynamic force), which is acting in the same direction as electron flow. In EM-enhanced atomic diffusion, atoms are found to move in the same direction as the applied electron flow direction. Thus, electron wind force is much bigger than the direct force.

14.5.1.1 Back Stress

Among all the experimental studies of EM in Al interconnects, a couple of them deserve mentioning. One is the addition of Cu to Al lines to retard EM, and the other is the use Al short stripes designed by I. A. Blech for a direct observation of EM and the finding of the effect of back stress. In the latter, short stripes of Al were deposited onto a titanium nitride (TiN) film and stressed at a high current density. As Al resistivity was much lower than that of the TiN layer, the short stripes would carry most of the current, which result in the transportation of Al atoms from the cathode to the anode. It leads to the formation of void at cathode and hillock at the anode, which become observable and in turn measurable for quantitative analysis of the driving force of EM. Furthermore, it was found that the longer the stripe, the more the depletion at the cathode side in EM. This was explained by the back stress below [43, 44].

The transport of Al atoms from the cathode to the anode will lead to compression in the latter and tension in the former. According to Nabarro-Herring model of creep where the equilibrium vacancy concentration is affected by the stress potential ($\sigma\Omega$), the tensile region has more vacancies and the compressive region has lesser vacancies than the unstressed region, so there is a vacancy gradient crossing the cathode to the anode. This gradient will induce an atomic flux of Al going from anode to cathode, which is opposite to the EM flux of Al. The vacancy concentration gradient depends on the length of Al stripe, and the shorter the stripe, the greater the gradient. At a certain short length defined as the critical length, the gradient is large enough to balance EM, so that no depletion at the cathode and no extrusion at the anode occur. At the steady state, the expression for the critical length is obtained as:

$$\Delta x = \frac{\Delta\sigma\Omega}{Z^* eE}$$

where Δx is the critical length, $\Delta\sigma\Omega$ is the stress potential difference, Z^* is effective charge number, e is electron charge, and E is electric field.

14.5.1.2 Statistical Analysis by Weibull Distribution Function in Reliability Study

In industrial manufacturing, it uses a statistical analysis for a large number of test samples in order to separate an early failure from the regular failure. The latter is acceptable because it follows the distribution and is expected, but the former is not because it is outside the distribution and is unexpected. Product assurance requires reliability tests of mean-time-to-failure (MTTF) besides the physical analysis of failure mode and failure mechanism. Knowing MTTF enables device engineers to predict the lifetime of devices. Here we will briefly introduce Weibull distribution and Black's MTTF equation. The Weibull distribution function is given below:

$$F(t) = 1 - \exp\left[-\left(\frac{t}{\eta}\right)^{\beta}\right] \tag{14.1}$$

where $F(t)$ is percentage or fraction of the failed sample as a function of time, η is the characteristic lifetime, and β is the shape factor or the slope of the Weibull plot. In Black's MTTF equation, there are three parameters: pre-factor A, current density power factor n, and activation energy E_a:

$$\mathrm{MTTF} = A(j^{-n})\exp\left(\frac{E_{\mathrm{a}}}{kT}\right) \quad (14.2)$$

With Black's equation, we can determine those parameters experimentally and obtain the statistical analysis. We need to conduct at the least the EM tests at two temperatures and two current densities. To find the activation energy and to determine n, we will not repeat them here, and the reader can use Google search or textbooks.

14.5.2 Experimental Studies of Electromigration in Al and Cu Interconnects

In the Al(Cu) interconnect, *W* via plugs are used to make vertical connections between different layers of Al lines, for example, the M2 or M1 metallization layers as shown in Fig. 14.29a. Since the atomic diffusion of *W* is much slower than Al, there will be flux divergence of Al atoms as well as vacancies located over the top of *W* via, and the divergence leads to hillock and void formation [45–48].

In comparison, the Cu dual-damascene structure is differently constructed [49, 50]. In Cu dual-damascene interconnects, the lower-level M2 trench and V1 that connects to the M1 level are simultaneously prepared before Cu electroplating,

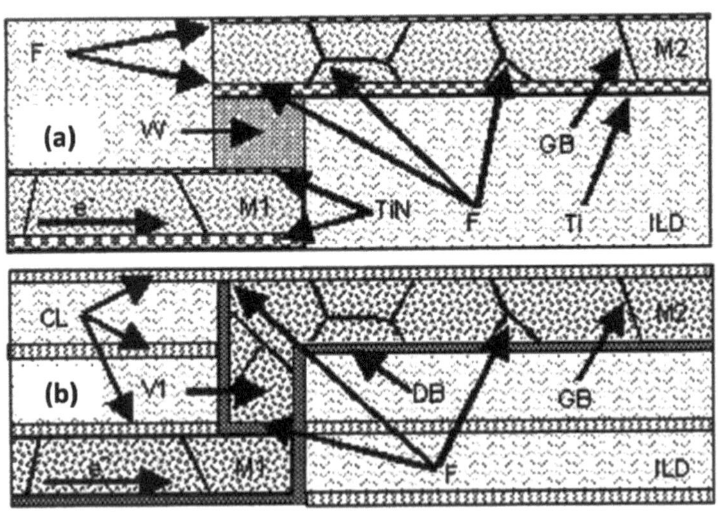

Fig. 14.29 (**a**) Idealized schematic of a two-level Al(Cu) or PVD Cu interconnect; (**b**) Idealized schematic of a two-level Cu interconnect. Where CL stands for capping layer, *DB* Ta barrier, e^- electron charge, *F* flux divergence region, *Ti* titanium layer, *TiN* titanium nitride layer, *V1* dual damascene via, *W* Tungsten plug, *GB* grain boundary, *ILD* interlevel dielectric, *M1* metal level 1, *M2* metal level 2

as shown in Fig. 14.29b. The M1 level is also separated from the V1 via by an electrically conducting diffusion barrier (DB) that also coats the inner walls of V1 and the side and bottom walls of the M2 trench. Consequently, the dual-damascene process places the flux divergence, not at the V1/M2 interface, which is the top of V1 as in Al(Cu), but at the M1/V1 interface at the bottom of V1. This via-bottom flux divergence is potentially problematic from a reliability standpoint because only a small amount of material depletion by EM will be sufficient for voiding damage to occur.

The dual-damascene interconnect is distinctly different from the previous Al (Cu) architecture in the other respect. It has been shown that EM in Cu interconnect occurs by surface diffusion at the device operation temperature of 100 °C. In processing the dual-damascene architecture, the top surface after electroplating must be chemical-mechanically polished and capped by a nitride layer before building of the next layer [51]. Because of the weak chemical bonding between the nitride and Cu, the interface can be regarded as a free surface of Cu diffusion and allows rapid surface diffusion of Cu resulting in EM reliability concern. As indicated in Fig. 14.29, a flux divergence exists at the cathode end of the M2 interconnect where Ta barrier and capping nitride layer intersect. And void would naturally form at this location.

14.5.3 Electromigration in Flip Chip Solder Joints

In C-4 flip chip solder joints, the current crowding usually occurs at the entrance and departure point of electron flow. This is because the cross-sectional area of thin Al line is much smaller compared to flip chip solder joints. Current crowding occurs at the entrance of the flip chip solder joint and tends to drive in Sn atoms in downward direction and hence vacancy comes back in reverse direction resulting in voids formation. After voids formation, current will move ahead of existing void, pushing Sn atoms down again and hence leading to the pancake type of void formation continuously along solder/UBM interface. As shown in Fig. 14.30, pancake type of void can be observed in bump No. 5, 3, and 1, which electron flow enters left corner of the solder joints [52–54].

14.5.4 System Level Electromigration Studies in 3D IC Packaging

Packaging technology is currently in transition from 2D ICs technology to 3D ICs to meet the requirements of consumer electronic products. Compared to flip chip technology in 2D IC, TSV and microbump are new. Besides, the dimension of microbumps in 3D ICs is shrunk by a factor of 10 and the manufacturing and

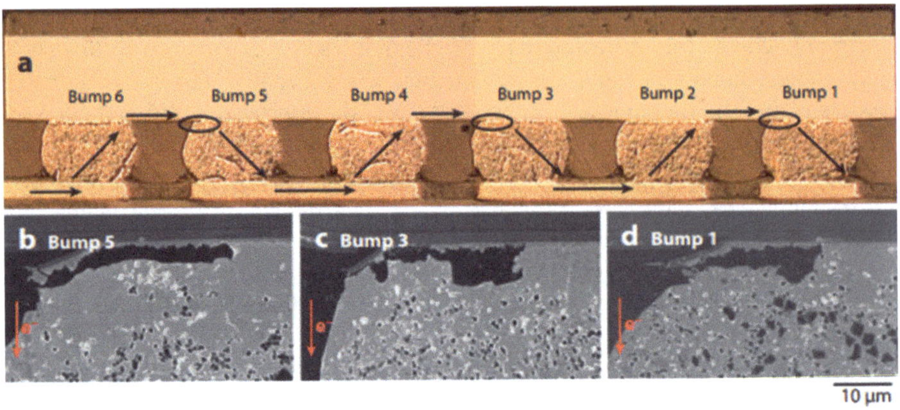

Fig. 14.30 (**a**) Optical microscope image to show current flow in daisy chain of solder joints. (**b**) SEM image to show package voids formation after electromigration [52]

structure of Cu-filled TSVs are also new. Based on these two reasons, the studies of electromigration failure in microbumps and TSVs have received much attention lately. In addition, system level electromigration study is new in 3D IC because heat dissipation becomes a problem. Synergistic effect of Joule heating and electromigration will be the trend of reliability failure in the near future.

14.5.4.1 Electromigration in Microbumps

Microbumps have a size reduction of ten times from that of a C-4 joint. Yet, the total volume shall shrink 1000 times and it affects EM behavior [29, 55–61]. It has been found that microbumps show a much better resistance to EM than traditional flip chip solder joints and no pancake type of void has been observed in microbumps. Figure 14.31 shows one of the EM testing results [20]. In the sample, microbumps adopted Ni as UBM. The testing condition of (a) and (b) was 4950 h at 8.8×10^4 A/cm^2 and 150 °C, and (c) and (d) was 2850 h at 1×10^5 A/cm^2 and 150 °C. There is no observable EM failure in microbumps.

The reason that microbump has robust EM resistance is partially because the percentage of intermetallic compound (IMC) in the microbump was very high. Furthermore, during EM, microbumps with remaining solder can be converted to total IMC microbumps at the early stage of electromigration, as we can see in Fig. 14.31, the entire bump has been changed into IMC after the EM test. Since IMC has a better resistance to EM, microbumps can survive at a higher temperature and higher current density during EM testing compared to C-4 solder joints.

Another possible explanation of superior EM resistance of microbump can be related to back stress [43]. As discussed in Sect. 14.5.1.1, at constant current density, when the length of Al stripe is smaller than the critical length, there is no EM damage. Due to the scaling trend, the height of the solder layer in microbumps

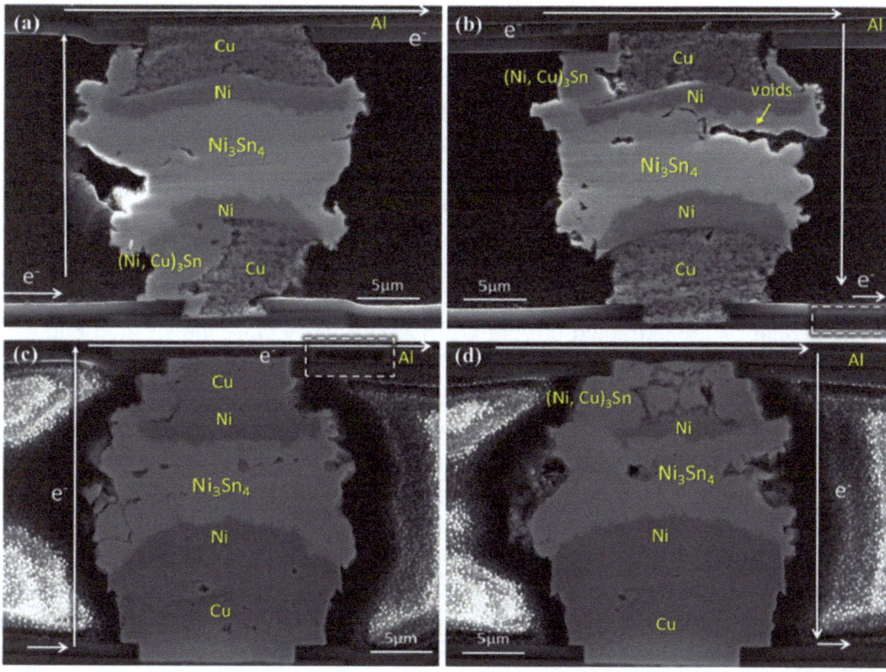

Fig. 14.31 Cross-sectional SEM images of microbump after current stressing of (**a**) and (**b**) 4950 h at 8×10^4 A/cm^2 and 150 °C, and (**c**) and (**d**) 2850 h at 1×10^5 A/cm^2 and 150 °C [55]

will shrink to less than 5 μm. This height is probably smaller than the critical length at the tested current density condition. However, typically it is a system that is under EM test. Usually when current density increased, Al lines, Cu redistribution layer or under ball metallization (UBM) tends to fail first. Actually, some studies even report that no failure of microbumps was observed before it was failed by Joule heating when the current density increased. Right now although many studies have been reported on electromigration failure related to microbumps, there is no direct observation of EM induced voids formation failure in microbumps.

14.5.4.2 Electromigration in TSVs

Early reliability studies of TSV started during the last decade with thermomechanical simulations, since the thermal expansion mismatch between Cu TSV and Si die is very large. Several studies evaluated, through finite element simulations, fatigue of Cu TSV, delamination at TSV interfaces, and Si cracking due to thermomechanical stress. These studies point out that the existence of the Cu TSV affects the stress distributions and the interconnection reliability [62–64]. For EM, most studies focus on matter flux diffusion modeling at TSV interfaces, but there is no sufficient emphasis on experimental results [65–67].

Frank, Moreau [68–70] and their group did EM experimental analysis of a high density Cu TSV-last technology of 2–4 μm diameters, and 15 μm of length. EM-induced voids were found in the adjacent metal level, right at TSV interface. Moreover, the expected lifetime benefit by increasing line thickness did not occur due to increasing dispersion of voiding mechanism. They also reported EM experimental reliability of Cu TSV-middle technology of 10 μm diameter and 80 μm thickness. EM revealed similar degradation mechanism and kinetics as those in TSV-last approach. Figure 14.32 shows the FIB cross-sectional image of the failure in both the thin and thick process on TSV-last approach.

For thin metal process, current density exponent n and activation energy E_A of Black's equation are extracted through MTTF data. Very close values are obtained for both configurations. $E_A = 0.9 \pm 0.1$ eV is the expected activation energy for surface diffusion of Cu on Cu damascene surfaces, ensuring by SiN capping. $n = 2 \pm 0.2$ is acceptable regarding the theoretical value of 2 to be expected. Yet it might be affected by local Joule heating in the TiN barrier as void becomes larger than the TSV section. As failure is revealed to be driven by void growth, any aging extrapolations should be made with a consistent but conservative value $n = 1$.

For thick metal process, activation energy and current exponent for downstream configuration are, respectively, of 0.8 ± 0.2 eV and $n = 2.4 \pm 0.3$, which are close to values of the thin metal level process, but it is less accurate due to the difference in lognormal standard deviation between MTTF distributions.

With the same $E_A = 0.9$ eV, and a conservative $n = 1$, the MTTF distributions of both thin and thick metallization processes are extrapolated at same operation conditions of current and temperature. At 50 % cumulative failure, the benefit of thick metallization process is about six times higher than thin metallization process. However, since lifetime is always extrapolated at 0.1 % cumulative failure, thicker process has at this cumulative failure no significant benefit (only 1.1 times thin process).

Another group from Samsung [71] also found the similar EM results. Electromigration-induced voids were found at Cu/SiN interface between TSV bottom and backside metal, but not at TSV itself due to the unexpectedly strong reliability of TSV.

14.5.5 System Level Weak-Link Failure in 2.5D Integrated Circuits

In system level EM test of 2.5D integrated circuits, a new failure mode due to synergistic effect of Joule heating and EM has been found [38]. In the test circuit, there are three levels of solder joints, two Si dies (one of them serves as interposer and has TSVs) and one polymer substrate. In addition, there are two redistribution layers, one between every two levels of the solder joints. It has been found that the redistribution layer (RDL) between the flip chip solder joints and microbumps is the

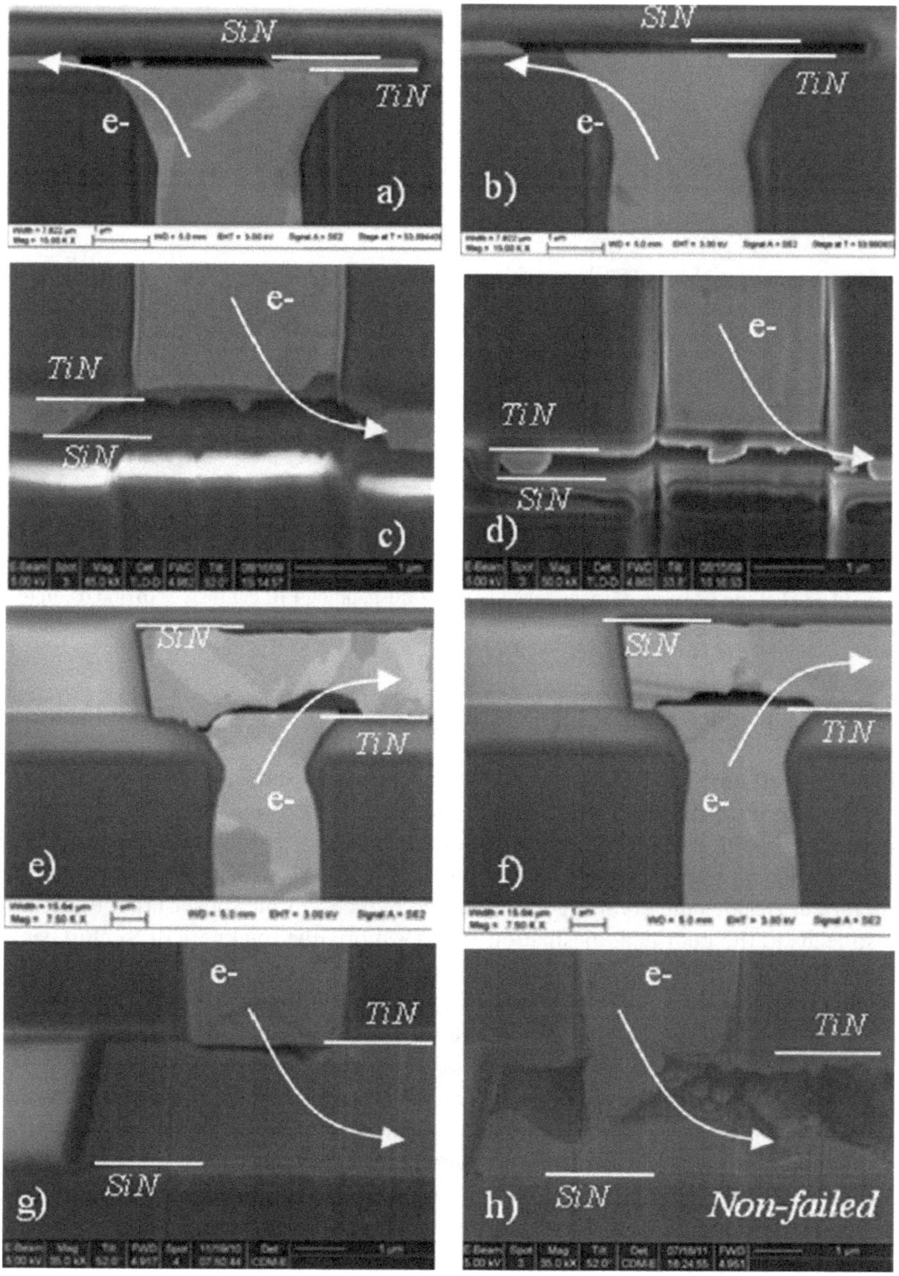

Fig. 14.32 Failure analyses of two samples for: (**a** and **b**) thin process upstream; (**c** and **d**) thin process downstream; (**e** and **f**) thick process upstream and (**g** and **h**) thick process downstream [68]

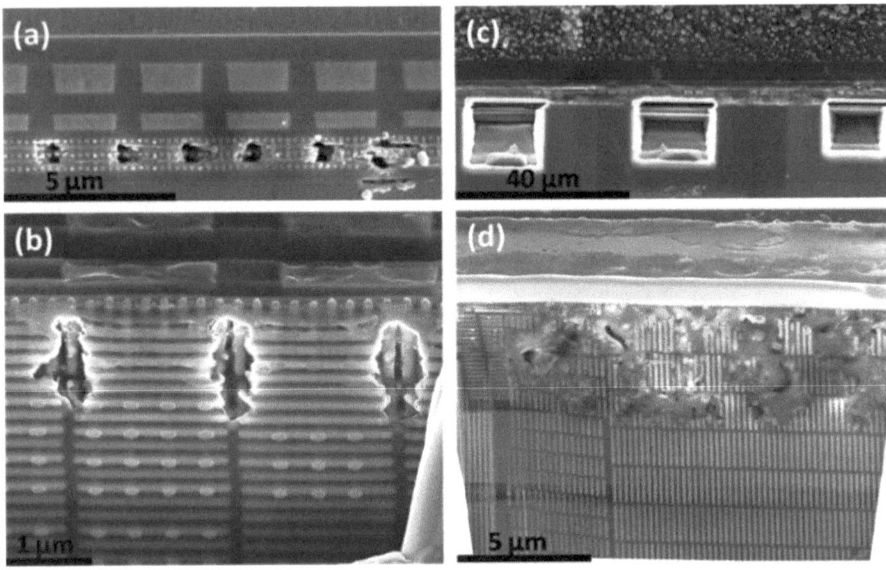

Fig. 14.33 Failure site images (**a** and **c**) and their corresponding FIB cut images (**b** and **d**). In (**c**), the failure site is really big, about 80 µm long. FIB was used to cut three holes at the beginning, the middle, and the end of the failure site [72]

weaklink and failed easily by burnout in EM test, as shown in Fig. 14.33. There are six periodic burnout holes in Fig. 14.33a. In Fig. 14.33d, the burnout area is quite big, about 80 µm. The failure is time-dependent with a sudden resistance increase.

The reason of this new failure mode to occur is due to the difficulty of thermal dissipation in the structure. The design of the network structure of Cu line in RDL is very compact when compared to previous Cu or Al structures studied in the literature. Besides the compact structure, the reduced Si interposer layer thickness also leads to the poor heat dissipation. What's more, the thickness of Si chip has a tendency of decreasing with the trend of smaller packaging size. Right now, in 3D IC, the stacked chip is about 50 µm, which is much less than the typical thickness of Si wafer at 200 µm. The heat conduction of Si interposer is much reduced and it becomes one of the reasons why heat dissipation is poor. These situations are very different from the previous EM studies carried out in Cu and Al lines reported in the literature. Joule heating generation and dissipation is a critical issue, leading to the time-dependent heat accumulation and temperature increase. These special situations result in the new failure mode, the time-dependent failure of burnout in EM test. The simulation results show that Joule heating has a positive feedback to EM in the redistribution layer and caused the thermal runaway failure [72].

This synergistic effect of Joule heating and EM on burnout is a new failure mode of reliability. Joule heating will be the critical issue as the density of transistor becomes increasingly higher while the form factor is reduced in mobile devices. This synergistic effect can couple with mechanical and other failure in the future. Clearly, it needs our attention and requires more in-depth understanding.

14.5.6 Concluding Remarks

Electromigration and Joule heating are of concern for 3D IC packaging. In microbumps, 100 % IMC can help increase MTTF at higher current density. TSV also has a higher resistance to EM failure. Nevertheless, EM failure can be transferred to RDL layer when it is poorly designed. In a complicated 3D IC system, Joule heating must be considered during the design stage to avoid current crowding not only in solder joints but also in redistribution metal layers. And EM failure always occurs in the weakest link of the system. By including EM design rules upfront in manufacturing may help predict or even prevent the weakest link failure in the system level happen.

14.6 Thermomigration in 3D IC Packaging

14.6.1 Introduction

For the thermal management in 3D IC, heat generation and dissipation among the stack of chips is the most critical reliability issue. When the Si chips are stacked vertically with the dielectric underfill in between, the heat generation from Joule heating of transistors in silicon chips is very large whereas the heat dissipation is very bad due to the poor thermal conductivity of the underfill materials. Meanwhile, heat dissipation depends on the temperature gradient according to Fourier's law. While a large thermal gradient is better for heat dissipation, it might lead to thermomigration (TM), which is another reliability concern in this review [73].

In flip chip solder joints, a temperature gradient of 1000 °C/cm is sufficient to induce TM according to many studies reported [74–76]. TM is a reliability concern because it will lead to phase redistribution and void formation, which will cause failure. For the eutectic Pb-Sn solder joint, Pb is found to be the dominant diffusing species in TM, and it will diffuse from the hot end to cold end [77]. TM for Pb-free solder joint is also well studied, and Sn is determined to move from the cold end to hot end with a positive heat of transport [78]. If we assume for microbumps, TM will also occur at the same temperature difference, i.e., 1 °C, across a microbump having a diameter of 10 μm, it will induce a serious thermomigration failure. It is hard to guarantee that we can control a temperature difference of 1 °C. Hence, the reliability problem associated with TM must be studied for microbumps. First, we shall discuss the fundamentals of TM. Then, we will discuss the observation of TM in flip chip solder joints. Finally, we will present the new challenges in TM for microbumps used in 3D IC packaging.

14.6.2 Fundamentals of Thermomigration

TM is the phenomenon of atomic diffusion driven by a temperature gradient. Classically, it is called the Soret effect. Mathematically, the driving force of TM can be expressed as

$$F = -\frac{Q^*}{T}\left(\frac{\partial T}{\partial x}\right)$$

where Q^* is the heat of transport, which is defined by the difference between heat carried by a moving atoms per mole and the heat of atoms per mole at the initial state (the hot end or the cold end), T is the working temperature, and $\frac{\partial T}{\partial x}$ is temperature gradient. The atomic flux due to the presence of TM has been expressed as

$$J = C\frac{D}{kT}\frac{Q^*}{T}\left(-\frac{\partial T}{\partial x}\right)$$

where C is the atomic concentration and D is the diffusivity of the diffusing atom at the working temperature [79].

14.6.2.1 Traditional TM Studies

To measure the Q^* and determine the dominant diffusing species in solder joints, insitu observation of TM has been conducted in traditional C-4 flip chip solder joints. TM in eutectic 63Sn37Pb flip-chip solder joints has been reported. Typically, TM is accompanied by EM when electric current is applied to the solder joints [80]. In order to decouple the electromigration effect, the neighboring unpowered solder joints, which are only subjected TM, are used to study TM effects. Figure 14.34 shows a typical arrangement of TM test. In this sample, there are 11 solder joints, but only a pair of them (number 6/7) are powered, and the rest of them are unpowered. The last bump (number 12) is at the as-received state for reference. The UBM thin films on the chip side were Al (~0.3 μm)/Ni(V) (~0.3 μm)/ Cu (~0.7 μm) deposited by sputtering. The bondpad metal layers on the substrate side were Ni (5 μm)/Au (0.05 μm) prepared by electroplating. The bump height between the UBM and the bond pad is 90 μm, and the solder bump material is eutectic SnPb [79].

Since the on-chip Al interconnects are the source of Joule heating and Si is a good thermal conductor, a temperature gradient was produced across all the solder bumps and caused TM in all of them. In this case, the Si chip side is the hot side while the FR4 substrate side is the cold side.

Figure 14.34 shows the SEM cross-section image after a DC current stressing of 0.95 A at 100 °C for 27 h. The average current density at the contact opening was

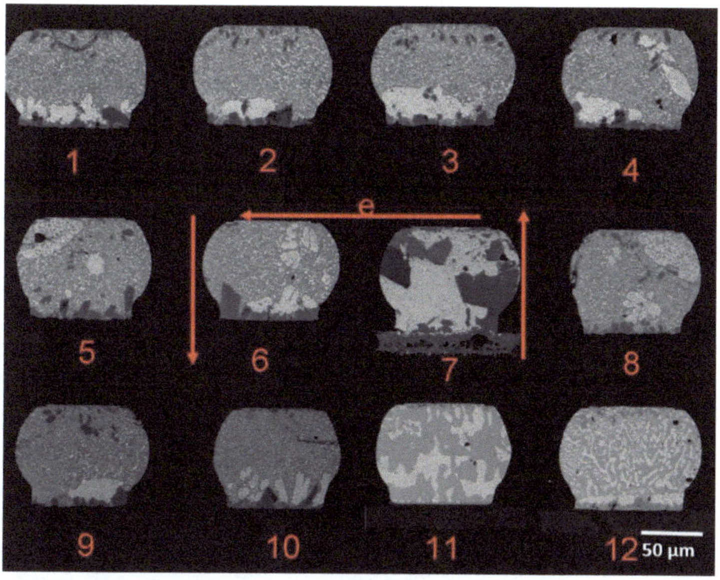

Fig. 14.34 SEM cross-section image after only one pair of them subjected to current stressing (Courtesy of Professor Ouyang Fanyi, National Tsing Hua University)

1.5×10^4 A/cm^2 [78]. The current flow direction is marked on the image. The lighter color in the SEM image represents Pb-rich phase and the darker color represents Sn-rich phase. Compared with the as-received sample, we can see that the Pb-rich phase has moved from chip side (hot side) to substrate side (cold side) in the unpowered bumps.

Due to environmental concerns, Pb-free solders are introduced to replace the eutectic SnPb solder. The TM effect in Pb-free SAC305 has also been studied for reliability concern [81–84]. For in-situ sample preparation, the sample was usually first polished down to the half of solder bumps, and then an array of markers was indented on one of the cross-sectioned solder joint surface to determine the direction and magnitude of atomic flux in TM [78]. Figure 14.34 shows the SEM cross-section of Pb-free solder bumps before and after TM at an AC current density of 1×10^4 A/cm^2 and at 100 °C for 800 h [81]. It is assumed that no EM occurs in AC, yet Joule heating remains. The temperature gradient across the solder bump was found to be around 2800 °C/cm. We can see that hillock formation occurs at the hot side. Also the Ag in solder was found to migrate to the cold side. The marker was found to move toward the substrate side (cold end), indicating that the dominant species in the TM has moved to the hot side. Based on the atomic flux measured through marker motion, the Sn transport heat is measured to be 1.36 kJ/mol in this study. The test structure similar to Fig. 14.34 is also designed for Pb-free solder bump to study TM, of which two bumps are under current stressing in order to generate Joule heating, while others are under TM [78]. Figure 14.35 shows the SEM cross-section of Pb-free unpowered solder bumps after TM at 150 °C. The

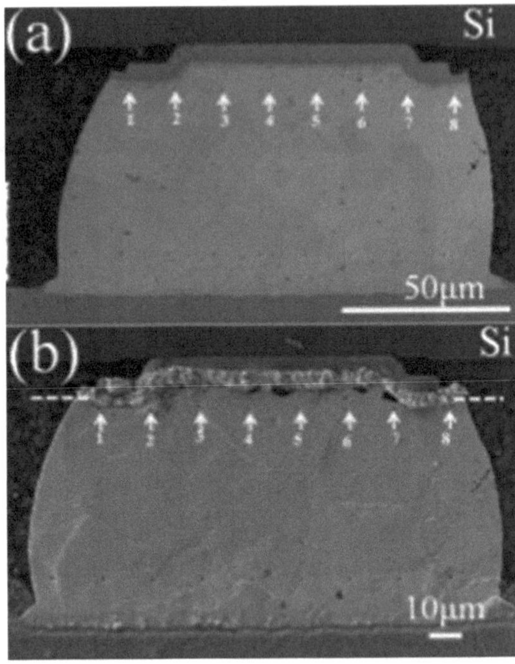

Fig. 14.35 Cross-section SEM images of the sample (**a**) before and (**b**) after thermomigration respectively at an AC current density of 1×10^4 A/cm^2 at 100 °C for 800 h [81]

temperature gradient is simulated to be around 250 °C/cm. We can see that the cold end has depletion while the hot end has protrusion. EDX data shows that Sn diffused to the hot end while Ag diffused to the cold end, which agrees with the TM study under AC stressing (Fig. 14.36).

14.6.3 Thermomigration Studies in 3D IC Packaging

In 3D IC packaging, chip-to-chip interconnections are enabled by through silicon via (TSV) and microbumps. The diameter of microbumps is usually below 20 μm, and is being scaled towards 5 μm. The solder joint is dominated by IMC due to rapid reaction with under-bump metallization (UBM) during reflow. In this section, we will discuss the TM issue in microbumps.

14.6.3.1 Thermomigration in Microbumps

The TM effect on microbumps reliability was first reported on the effect of asymmetrical IMC formation during reflow or solid state annealing with a temperature gradient [85]. Figure 14.37a, b displays the evolution of the microstructure of microbumps before and after subject to a temperature gradient of 7308 °C/cm at 145 °C, respectively [86]. Compared to the as-received samples shown in

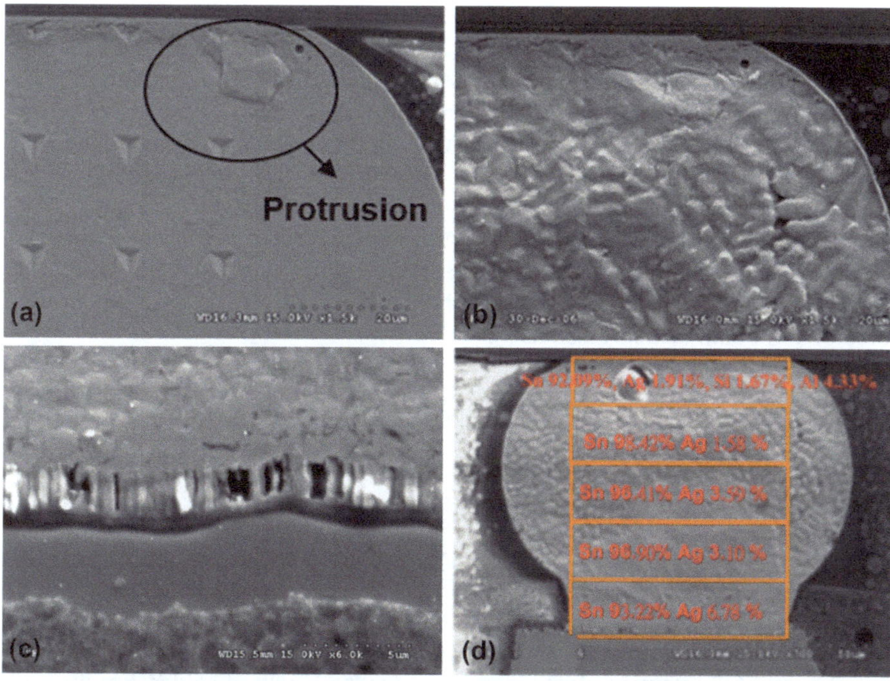

Fig. 14.36 SEM cross-section image of thermomigration bump at (**a**) 26 h; (**b**) chip side for 62 h; (**c**) substrate side for 62 h; (**d**) EDX data for the thermomigration bump after 62 h [78]

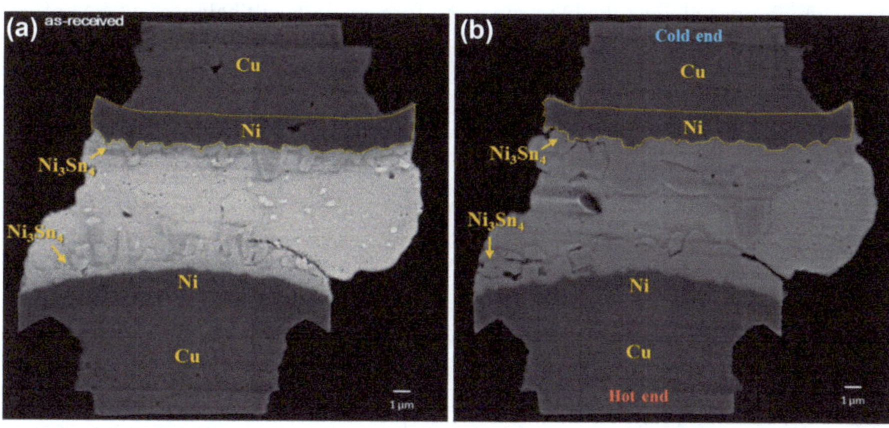

Fig. 14.37 Cross-sectional SEM images of a microbump (**a**) before and (**b**) after thermomigration of 7308 °C/cm at 145 °C

Fig. 14.34a, we see that Fig. 14.34b reveals a thicker IMC appeared in the cold end, which indicates that temperature gradient enhances the Ni diffusion interstitially inside Sn going from the hot end to cold end, and has caused more IMC formation at the cold end. What is interesting is that the TM effect on Sn is less significant for microbumps than that for flip chip solder joint [87], especially when the bump height is below 5 μm. One of possible reasons is that the back stress induced vacancy flux is in the opposite direction of TM-induced flux. In principle, when a solder joint is under a temperature gradient, TM induces mass transport of Sn from the cold to the hot end. The latter will be in compression and the former in tension. Based on the Nabarro-Herring model of equilibrium vacancy concentration in a stressed solid, the tensile region possesses more vacancies than the unstressed solid, whereas the compressive regions possess fewer vacancies, resulting in a stress-induced vacancy flux from hot to cold end. The driving force of TM is counteracted by this chemical-mechanical force, which is called back stress. Thus, the driving force of TM could be expressed as

$$F = -\frac{Q^*}{T}\left(\frac{\partial T}{\partial x}\right) - F_{\mathrm{BS}}$$

where

$$F_{\mathrm{BS}} = \frac{d\sigma\Omega}{dx} = \frac{(\sigma_1 - \sigma_2)\Omega}{d}$$

where F_{BS} is the back stress, σ_1 and σ_2 are the maximum hydrostatic stresses at each end of bump, Ω is atomic volume, and x or d is the height of solder. For flip chip solder joint, x is typically around 100 μm, while for microbumps, x is typically around 5 μm. The shrinkage of bump height will dramatically affect the back stress, and then affect the net effect of TM. In EM study, we note that there is also a back stress which goes against the electron wind force, and it will balance out EM effect when the metal stripe is under the critical length (Fig. 14.37).

14.6.3.2 Thermomigration in TSV

Through Silicon Via (TSV) is typically made of Cu, the TM of Cu is different from solder, because the latter is typically a binary alloy except pure Sn. So far, there is no literature reported regarding to the TM issue of TSV. However, TM will happen in pure metal. In [88], both the net diffusion of Cu and Au are studied under thermal gradient. It is found that Cu can diffuse from hot end to cold end when the temperature gradient is in the order of 1000 °C/cm [88] with a positive heat of transport. This means that for a TSV with 50 μm in height, 5 °C difference across TSV will trigger TM. However, TM reliability issue of pure metal interconnect, such as Cu interconnects, is not well reported. The reason is possibly related to back stress. Because device is usually working at 100 °C, this temperature is too low for creep to take place in Cu. When TM drives more Cu atoms from the hot side into the cold side (or the reverse direction depending on the heat of transport), the stress will build up at the end where atomic accumulation occurs, and then the stress gradient will produce an atomic flux of Cu against the TM.

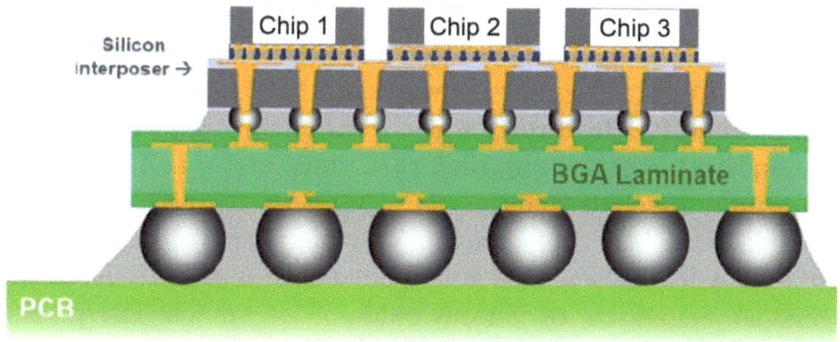

Fig. 14.38 2.5D multi-chip horizontal arrangement (adapted from Yole Développement)

Since TSV is connected to the redistribution layer at both ends, and the current crowding will occur at the interfaces, it is possible that the local Joule heating due to current crowding will create a temperature gradient that is high enough for TM to occur. Meanwhile, due to the high power density and poor heat dissipation in 3D IC, the working temperature around TSV might be high enough for creep to take place in Cu, in which case the back stress may not play a role.

14.6.3.3 Thermomigration Induced by Thermal Crosstalk

In 2.5D IC packaging design, a layer of Si interposer is used in between dies and substrate to mitigate the chip-packaging interaction. Nevertheless, this new design can also cause new reliability issue such as thermal crosstalk-induced thermomigration. Figure 14.38 shows that two Si chips are horizontally placed on one Si interposer. Since Si is a good heat conductor, if chip 1 is powered, the Joule heating can be transported to the bottom of chip 2. However, the top of chip 2 is not heated because the gap between chip 1 and chip 2 is filled with underfill materials, which is a poor thermal conductor. Therefore, chip 2 will undergo TM with the interposer side to be the hot side while chip 1 is powered. This phenomenon is called thermal crosstalk. The thermal cross-talk will be a reliability issue if we use Si interposer as the new substrate, and it deserves more research efforts.

14.6.4 Concluding Remarks

The TM effect of traditional interconnects has been summarized, including eutectic Pb-Sn solder joint and Pb-free flip chip solder joint. From the previous studies of flip chip solder joints, we can conclude that a temperature gradient of 1000 °C/cm will cause Pb to diffuse from hot end to cold end with Q^* to be −25.3 kJ/mol, and it causes Sn to diffuse from cold end to hot end with Q^* to be +1.36 kJ/mol. The materials driven away by thermal gradient will form void and cause reliability issue. The study of TM of new interconnect structures in 3D IC, such as microbumps and TSVs, is still ongoing. At present, microbump is found to be more resistant to TM because of its lower bump height (higher back stress), and stronger bonding in IMC. However, the asymmetrical IMC growth will occur inside microbumps under a temperature gradient. The study of TM in TSV is few. We can only conclude from precious study of pure Cu that 5 °C difference for a 10-μm-long TSV will induce TM. Finally, the use of Si interposer will cause new TM issue due to thermal crosstalk between the horizontal dies. It is a new reliability issue due to heterogeneous integration of various chips on a common interposer of good thermal conduction.

Acknowledgment The editors would like to thank Indranath Dutta from Washington State University and Tae-Kyu Lee from Portland State University for their critical review of this chapter.

References

1. K.N. Tu, Reliability challenges in 3D IC packaging technology. Microelectron. Reliab. **51**(3), 517–523 (2011)
2. Y. Wang, I.M. De Rosa, K.N. Tu, Size effect on ductile-to-brittle transition in Cu-solder-Cu micro-joints. In *2015 Proceedings of 65th Electronic components and Technology Conference*, San Diego, CA, 2015, pp. 632–639
3. S.F. Choudhury, L. Ladani, Grain growth orientation and anisotropy in Cu_6Sn_5 intermetallic: nanoindentation and electron backscatter diffraction analysis. J. Electron. Mater. **43**(4), 996–1004 (2014)
4. K. Sakuma, K. Sueoka, S. Kohara, K. Matsumoto, H. Noma, T. Aoki, Y. Oyama, H. Nishiwaki, P.S. Andry, C.K. Tsang, J.U. Knickerbocker, Y. Orii, IMC bonding for 3D interconnection. In *2010 Proceedings 60th Electronic Components and Technology Conference*, Las Vegas, NV, 2010, pp. 864–871
5. K. Sakuma, K. Tunga, B. Webb, An enhanced thermo-compression bonding process to address warpage in 3D integration of large die on organic substrates. In *Proceedings of Electronic Components and Technology Conference*, San Diego, CA, 2015, pp. 318–324
6. L. Li, P. Su, J. Xue, M. Brillhart, J. Lau, P.J. Tzeng, C.K. Lee, C.J. Zhan, M.J. Dai, H.C. Chien, S.T. Wu, Addressing bandwidth challenges in next generation high performance network systems with 3D IC integration. In *Proceedings of Electronic Components and Technology Conference*, San Diego, CA, 2012, pp. 1040–1046
7. C.C. Lee, P.J. Wang, J.S. Kim, Are intermetallics in solder joints really brittle? In *Proceedings of Electronic Components and Technology Conference*, Reno, NV, 2007, pp. 648–652
8. P.F. Yang, Y.S. Lai, S.R. Jian, J. Chen, R.S. Chen, Nanoindentation identifications of mechanical properties of Cu6Sn5, Cu3Sn, and Ni3Sn4 intermetallic compounds derived by diffusion couples. Mater. Sci. Eng. A **485**(1–2), 305–310 (2008)
9. L. Jiang, N. Chawla, Mechanical properties of Cu_6Sn_5 intermetallic by micropillar compression testing. Scr. Mater. **63**(5), 480–483 (2010)
10. R.R. Chromik, R.P. Vinci, S.L. Allen, M.R. Notis, Nanoindentation measurements on Cu–Sn and Ag–Sn intermetallics formed in Pb-free solder joints. J. Mater. Res. **18**(09), 2251–2261 (2003)
11. R. Schwaiger, B. Moser, M. Dao, N. Chollacoop, S. Suresh, Some critical experiments on the strain-rate sensitivity of nanocrystalline nickel. Acta Mater. **51**(17), 5159–5172 (2003)

12. X. Deng, N. Chawla, K.K. Chawla, M. Koopman, Deformation behavior of (Cu, Ag)-Sn intermetallics by nanoindentation. Acta Mater. **52**(14), 4291–4303 (2004)
13. G. Ghosh, Elastic properties, hardness, and indentation fracture toughness of intermetallics relevant to electronic packaging. J. Mater. Res. **19**(5), 1439–1454 (2004)
14. Granta Design Ltd, CES 2016 Selector materials selection software (2016)
15. R.A. Mirshams, C.H. Xiao, S.H. Whang, W.M. Yin, R-curve characterization of the fracture toughness of nanocrystalline nickel thin sheets. Mater. Sci. Eng. A **315**(1–2), 21–27 (2001)
16. T.-T. Luu, N. Hoivik, K. Wang, K.E. Aasmundtveit, A.-S.B. Vardøy, High-temperature mechanical integrity of Cu-Sn SLID wafer-level bonds. Metall. Mater. Trans. A **46**(11), 5266–5274 (2015)
17. H.Y. Chuang, T.L. Yang, M.S. Kuo, Y.J. Chen, J.J. Yu, C.C. Li, C.R. Kao, Critical concerns in soldering reactions arising from space confinement in 3-D IC packages. IEEE Trans. Dev. Mater. Reliab. **12**(2), 233–240 (2012)
18. K. Tanida, M. Umemoto, N. Tanaka, Y. Tomita, K. Takahashi, Micro Cu bump interconnection on 3D chip stacking technology. Jpn. J. Appl. Phys. **43**(4B), 2264–2270 (2004)
19. J.F. Li, P.a. Agyakwa, C.M. Johnson, Interfacial reaction in Cu/Sn/Cu system during the transient liquid phase soldering process. Acta Mater. **59**(3), 1198–1211 (2011)
20. S.J. Wang, L.H. Hsu, N.K. Wang, C.E. Ho, EBSD investigation of Cu-Sn IMC microstructural evolution in Cu/Sn-Ag/Cu microbumps during isothermal annealing. J. Electron. Mater. **43**(1), 219–228 (2014)
21. K. Zeng, R. Stierman, T.C. Chiu, D. Edwards, K. Ano, K.N. Tu, Kirkendall void formation in eutectic SnPb solder joints on bare Cu and its effect on joint reliability. J. Appl. Phys. **97**(2), 24508 (2005)
22. J. Yu, J.Y. Kim, Effects of residual S on Kirkendall void formation at Cu/Sn-3.5Ag solder joints. Acta Mater. **56**(19), 5514–5523 (2008)
23. K.-N. Tu, *Solder Joint Technology* (Springer, New York, 2007)
24. K. Nogita, T. Nishimura, Nickel-stabilized hexagonal (Cu, Ni)6Sn5 in Sn–Cu–Ni lead-free solder alloys. Scr. Mater. **59**(2), 191–194 (2008)
25. H. Zhang, E. Perfecto, V.L. Calero-DdelC, F. Pompeo, An effective method for full solder intermetallic compound formation and Kirkendall void control in Sn-base solder micro-joints. In *2015 I.E. 65th Electronic Components and Technology Conference*, San Diego, CA, 2015, pp. 1695–1700
26. W.-L. Chiu, C.-M. Liu, Y.-S. Haung, C. Chen, Formation of nearly void-free Cu_3Sn intermetallic joints using nanotwinned Cu metallization., Appl. Phys. Lett. **104**(17), 171902 (2014)
27. Y. Liu, J. Wang, L. Yin, P. Kondos, C. Parks, P. Borgesen, D.W. Henderson, E.J. Cotts, N. Dimitrov, Influence of plating parameters and solution chemistry on the voiding propensity at electroplated copper-solder interface: plating in acidic copper solution with and without polyethylene glycol. J. Appl. Electrochem. **38**(12), 1695–1705 (2008)
28. J.H.L. Pang, Effect of intermetallic and Kirkendall voids growth on board level drop Reliability for SnAgCu lead-free BGA solder joint. In *56th Electronic Components and Technology Conference 2006*, San Diego, CA, 2006, pp. 275–282
29. Y. Wang, S.-H. Chae, R. Dunne, Y. Takahashi, K. Mawatari, P. Steinmann, T. Bonifield, T. Jiang, J. Im, P.S. Ho, Effect of intermetallic formation on electromigration reliability of TSV-microbump joints in 3D interconnect. In *2012 I.E. 62nd Electronic Components and Technology Conference*, San Diego, CA, 2012, pp. 319–325
30. I. Panchenko, K. Croes, I. De Wolf, J. De Messemaeker, E. Beyne, K.J. Wolter, Degradation of Cu_6Sn_5 intermetallic compound by pore formation in solid-liquid interdiffusion Cu/Sn microbump interconnects. Microelectron. Eng. **117**, 26–34 (2014)
31. L. Mo, Z. Chen, F. Wu, C. Liu, Microstructural and mechanical analysis on Cu–Sn intermetallic micro-joints under isothermal condition. Intermetallics **66**, 13–21 (2015)
32. C. Chen, D. Yu, K. Chen, Vertical interconnects of microbumps in 3D integration. MRS Bull. **40**(March), 257–263 (2015)

33. Y. Wang, Mechanical reliabilities of porous type $Cu/Cu_3Sn/Cu$ micro-joints. Unpublished Manuscript (2016)
34. J.O. Suh, K.N. Tu, N. Tamura, Dramatic morphological change of scallop-type Cu_6Sn_5 formed on (001) single crystal copper in reaction between molten SnPb solder and Cu. Appl. Phys. Lett. **91**(5), 051907 (2007)
35. J.O. Suh, K.N. Tu, N. Tamura, Preferred orientation relationship between Cu_6Sn_5 scallop-type grains and Cu substrate in reactions between molten Sn-based solders and Cu. J. Appl. Phys. **102**(6), 063511 (2007)
36. Y. Tian, R. Zhang, C. Hang, L. Niu, C. Wang, Relationship between morphologies and orientations of Cu6Sn5 grains in Sn3.0Ag0.5Cu solder joints on different Cu pads. Mater. Charact. **88**(100), 58–68 (2014)
37. H.F. Zou, H.J. Yang, Z.F. Zhang, Morphologies, orientation relationships and evolution of Cu_6Sn_5 grains formed between molten Sn and Cu single crystals. Acta Mater. **56**, 2649–2662 (2008)
38. M. Li, M. Yang, J. Kim, Textured growth of Cu_6Sn_5 grains formed at a Sn3.5Ag/Cu interface. Mater. Lett. **66**(1), 135–137 (2012)
39. G. Hariharan, R. Chaware, I. Singh, J. Lin, L. Yip, K. Ng, S.Y. Pai, A comprehensive reliability study on a CoWoS 3D IC package. In *2015 I.E. 65th Electronic Components and Technology Conference (ECTC)*, San Diego, CA, 2015, pp. 573–577
40. Y.T. Chen, Chemical effect on diffusion in intermetallic compounds. Ph.D. thesis, University of California, Los Angeles, 2016
41. C.C. Lee, T.-F. Yang, C.-S. Wu, K.-S. Kao, R.-C. Cheng, T.-H. Chen, Reliability estimation and failure mode prediction for 3D chip stacking package with the application of wafer-level underfill. Microelectron. Eng. **107**, 107–113 (2013)
42. H.H. Hsu, S.-Y. Huang, T.-C. Chang, A.T. Wu, Nucleation and propagation of voids in microbumps for 3 dimensional integrated circuits. Appl. Phys. Lett. **99**(25), 251913 (2011)
43. H.B. Huntington, A.R. Grone, Current-induced marker motion in gold wires. J. Phys. Chem. Solids **20**(1), 76–87 (1961)
44. I.A. Blech, Electromigration in thin aluminum films on titanium nitride. J. Appl. Phys. **47**(4), 1203–1208 (1976)
45. C.K. Hu, K.P. Rodbell, T.D. Sullivan, K.Y. Lee, D.P. Bouldin, Electromigration and stress-induced voiding in fine Al and Al-alloy thin-film lines. IBM J. Res. Dev. **39**(4), 465–497 (1995)
46. C.K. Hu, M.B. Small, P.S. Ho, Electromigration in Al(Cu) two-level structures: effect of Cu and kinetics of damage formation. J. Appl. Phys. **74**(2), 969–978 (1993)
47. C.K. Hu, P.S. Ho, M.B. Small, Electromigration in two-level interconnect structures with Al alloy lines and W studs. J. Appl. Phys. **72**(1), 291–293 (1992)
48. E.T. Ogawa, K.D. Lee, V.A. Blaschke, P.S. Ho, Electromigration reliability issues in dual-damascene Cu interconnections. IEEE Trans. Reliab. **51**(4), 403–419 (2002)
49. Y. Morand, Copper metallization for advanced IC: requirements and technological solutions. Microelectron. Eng. **50**(1–4), 391–401 (2000)
50. H. Helneder, H. Korner, A. Mitchell, M. Schwerd, U. Seidel, Comparison of copper damascene and aluminum RIE metallization in BICMOS technology. Microelectron. Eng. **55**(1–4), 257–268 (2001)
51. C.K. Hu, R. Rosenberg, H. Rathore, D. Nguyen, B. Agarwala, Scaling effect in electromigration of on-chip Cu wiring. Interconnect technology 1999. In *IEEE International Conference*, San Francisco, CA, 1999, pp. 267–269
52. E.C.C. Yeh, W.J. Choi, K.N. Tu, P. Elenius, H. Balkan, Current-crowding-induced electromigration failure in flip chip solder joints. Appl. Phys. Lett. **80**(4), 580–582 (2002)
53. L. Zhang, S. Ou, J. Huang, K.N. Tu, S. Gee, L. Nguyen, Effect of current crowding on void propagation at the interface between intermetallic compound and solder in flip chip solder joints. Appl. Phys. Lett. **88**(1), 012106 (2006)

54. K.N. Tu, C.C. Yeh, C.Y. Liu, C. Chen, Effect of current crowding on vacancy diffusion and void formation in electromigration, Appl. Phys. Lett. **76**(8), 988 (2000)
55. F.Y. Ouyang, H. Hsu, Y.P. Su, T.C. Chang, Electromigration induced failure on lead-free micro bumps in three-dimensional integrated circuits packaging. J. Appl. Phys. **112**(2), 023505 (2012)
56. R. Labie, P. Limaye, K. Lee, C. Berry, E. Beyne, I. De Wolf. Reliability testing of Cu-Sn intermetallic micro-bump interconnections for 3D-device stacking. In *3rd Electronic System Integration Technology Conference (ESTC),* Berlin, 2010, pp. 1–5
57. R. Labie, W. Ruythooren, K. Baert, E. Beyne, B. Swinnen, Resistance to electromigration of purely intermetallic micro-bump interconnections for 3D-device stacking. In *2008 I.E. International Interconnect Technology Conference (IITC),* Burlingame, CA, 2008, pp. 19–21
58. Y.M. Lin, C.-J. Zhan, J.-Y. Juang, J.H. Lau, T.-H. Chen, R. Lo, M. Kao, T. Tian, K.-N. Tu, Electromigration in Ni/Sn intermetallic micro bump joint for 3D IC chip stacking. *In 2011 I.E. 61st Electronic Components and Technology Conference*, Lake Buena Vista, FL, 32011, pp. 351–357
59. C.C. Wei, C.H. Yu, C.H. Tung, R.Y. Huang, C.C. Hsieh, C.C. Chiu, H.Y. Hsiao, Y.W. Chang, C.K. Lin, Y.C. Liang, C. Chen, T.C. Yeh, L.C. Lin, D.C.H. Yu, Comparison of the electromigration behaviors between micro-bumps and C4 solder bumps. In *Proceedings Electronic Components and Technology Conference*, Lake Buena Vista, FL, 2011, pp. 706–710
60. H. You, Y. Lee, S. Lee, J. Kang, Reliability of 20 μm micro bump interconnects. In *2011 I.E. 61st Electronic Components and Technology Conference (ECTC)*, Lake Buena Vista, FL, 2011, pp. 608–611
61. S.Y. Huang, C.J. Zhan, Y.W. Huang, Y.M. Lin, C.W. Fan, S.C. Chung, K.S. Kao, J.Y. Chang, M.L. Wu, T.F. Yang, J.H. Lau, T.H. Chen, Effects of UBM structure/material on the reliability performance of 3D chip stacking with 30 μm-pitch solder micro bump interconnections. In *Proceedings Electronic Components and Technology Conference*, San Diego, CA, 2012, pp. 1287–1292
62. N. Tanaka, T. Sato, Y. Yamaji, T. Morifuji, M. Umemoto, K. Takahashi, Mechanical effects of copper through-vias in a 3D die-stacked module. In *Proceedings of 52nd Electronic Components and Technology Conference 2002,* no. 2, pp. 473–479
63. J. Zhang, M.O. Bloomfield, J.Q. Lu, R.J. Gutmann, T.S. Cale, Modeling thermal stresses in 3-D IC interwafer interconnects. IEEE Trans. Semicond. Manuf. **19**(4), 437–448 (2006)
64. K.H. Lu, S.K. Ryu, Q. Zhao, X. Zhang, J. Im, R. Huang, P.S. Ho, Thermal stress induced delamination of through silicon vias in 3-D interconnects. In *Proceedings Electronic Components and Technology Conference*, Las Vegas, NV, 2010, pp. 40–45
65. J. Pak, M. Pathak, S. K. Lim, and D. Z. Pan, Modeling of electromigration in through-silicon-via based 3D IC. In *2011 I.E. 61st Electronic Components and Technology Conference (ECTC)*, Lake Buena Vista, FL, 2011, pp. 1420–1427
66. Z. Chen, Z. Lv, X.F. Wang, Y. Liu, S. Liu, Modeling of electromigration of the through silicon via interconnects. In *2010 11th International Conference on electronic Packaging Technology & High Density Packaging (ICEPT-HDP),* Xi'an, 2010, pp. 1221–1225
67. Y.C. Tan, C.M. Tan, X.W. Zhang, T.C. Chai, D.Q. Yu, Electromigration performance of through silicon via (TSV)—a modeling approach. Microelectron. Reliab. **50**(9–11), 1336–1340 (2010)
68. T. Frank, S. Moreau, C. Chappaz, L. Arnaud, P. Leduc, A. Thuaire, L. Anghel, Electromigration behavior of 3D-IC TSV interconnects. Proc. Electron. Compon. Technol. Conf. **3**(1), 326–330 (2012)
69. T. Frank, S. Moreau, C. Chappaz, P. Leduc, L. Arnaud, A. Thuaire, E. Chery, F. Lorut, L. Anghel, G. Poupon, Reliability of TSV interconnects: electromigration, thermal cycling, and impact on above metal level dielectric. Microelectron. Reliab. **53**(1), 17–29 (2013)
70. S. Moreau, D. Bouchu, Reliability of dual damascene TSV for high density integration: the electromigration issue. IEEE Int. Reliab. Phys. Symp. Proc. **33**, 1–5 (2013)

71. H.J. Choi, S.M. Choi, M.S. Yeo, S.D. Cho, D.C. Baek, J. Park, An experimental study on the TSV reliability: electromigration (EM) and time dependant dielectric breakdown (TDDB). In *2012 I.E. International Interconnect Technology Conference*, San Jose, CA, 2012, pp. 4–6
72. Y. Liu, M. Li, D.W. Kim, S. Gu, K.N. Tu, Synergistic effect of electromigration and Joule heating on system level weak-link failure in 2.5D integrated circuits. J. Appl. Phys. **118** (13),135304 (2015)
73. K. Chen, K. Tu, Guest Editors, Materials challenges in three-dimensional integrated circuits. MRS Bull. **40**, 219–222 (2015)
74. A.T. Huang, A.M. Gusak, K.N. Tu, Y.S. Lai, Thermomigration in SnPb composite flip chip solder joints. Appl. Phys. Lett. **88**(14), 1–4 (2006)
75. A.T. Huang, K.N. Tu, Y.S. Lai, Effect of the combination of electromigration and thermomigration on phase migration and partial melting in flip chip composite SnPb solder joints. J. Appl. Phys. **100**(3) (2006)
76. D. Yang, Y.C. Chan, B.Y. Wu, M. Pecht, Electromigration and thermomigration behavior of flip chip solder joints in high current density packages. J. Mater. Res. **23**(9), 2333–2339 (2011)
77. F.Y. Ouyang, K.N. Tu, Y.S. Lai, A.M. Gusak, Effect of entropy production on microstructure change in eutectic SnPb flip chip solder joints by thermomigration. Appl. Phys. Lett. **89**(22), 26–29 (2006)
78. F.Y. Ouyang, C.L. Kao, In situ observation of thermomigration of Sn atoms to the hot end of 96.5Sn-3Ag-0.5Cu flip chip solder joints. J. Appl. Phys. **110**(12) (2011)
79. K. Tu, *Electronic Thin-Film Reliability* (Cambridge University Press, Cambridge, 2010)
80. H. Ye, C. Basaran, D. Hopkins, Thermomigration in Pb-Sn solder joints under joule heating during electric current stressing. Appl. Phys. Lett. **82**(7), 1045–1047 (2003)
81. H.Y. Hsiao, C. Chen, Thermomigration in Pb-free SnAg solder joint under alternating current stressing. Appl. Phys. Lett. **94**(9), 2007–2010 (2009)
82. H.Y. Chen, C. Chen, In-situ observation of the failure induced by thermomigration of interstitial Cu in Pb-free flip chip solder joints. In *2009 59th Electronic Components and Technology Conference*, San Diego, CA, 2009, pp. 319–324
83. X. Gu, K.C. Yung, Y.C. Chan, D. Yang, Thermomigration and electromigration in Sn8Zn3Bi solder joints. J. Mater. Sci. Mater. Electron. **22**(3), 217–222 (2011)
84. C. Chen, H.M. Tong, K.N. Tu, Electromigration and thermomigration in Pb-free flip-chip solder joints. Annu. Rev. Mater. Res. **40**(1), 531–555 (2010)
85. M.Y. Guo, C.K. Lin, C. Chen, K.N. Tu, Asymmetrical growth of Cu_6Sn_5 intermetallic compounds due to rapid thermomigration of Cu in molten SnAg solder joints. Intermetallics **29**, 155–158 (2012)
86. F.Y. Ouyang, W.C. Jhu, T.C. Chang, Thermal-gradient induced abnormal Ni_3Sn_4 interfacial growth at cold side in Sn2.5Ag alloys for three-dimensional integrated circuits. J. Alloys Compd. **580**(580), 114–119 (2013)
87. F.Y. Ouyang, W.C. Jhu, Comparison of thermomigration behaviors between Pb-free flip chip solder joints and microbumps in three dimensional integrated circuits: bump height effect. J. Appl. Phys. **113**(4), 043711 (2013)
88. C.J. Meechan, G.W. Lehman, Diffusion of Au and Cu in a temperature gradient. J. Appl. Phys. **33**(2), 634–641 (1962)

Chapter 15
Fault Isolation and Failure Analysis of 3D Packaging

Yan Li and Deepak Goyal

15.1 Introduction

3D microelectronic packaging is the industry trend to meet the ever-increasing market demand for increased performance, lower power, smaller footprint, lower cost, and integration of heterogeneous devices. Chapters 1 and 2 have discussed in detail the various types of 3D packages, having die-to-die 3D integration, package-to-package 3D integration, or heterogeneous 3D integration combining both package and die stacking.

Figure 15.1 illustrates the schematic of a System in a Package (SIP), a 3D package with a Package on Package (POP) configuration, where a field-programmable gate array (FPGA), in the form of a ball grid array (BGA) package, is mounted in tandem with a flip chip CPU die onto a common package substrate [1]. The resultant joint between the FPGA package and the substrate is termed mid-level interconnect (MLI). While the solder joints connecting the CPU to the substrate or the FPGA die to FPGA substrate are called first-level interconnects (FLI). With multiple levels of solder interconnects, Si dice, and packages integrated into one SIP package, numerous failures, for example, solder joint nonwet or cracks in FLI or MLI, substrate trace cracks in either FPGA substrates or SIP substrates, delamination in either the CPU or the FPGA die could happen in a single SIP package post reliability tests. Isolating the location of the failure and accessing the failing feature or interconnect for physical failure analysis in such complex package architectures is a significant challenge.

Figure 15.2 shows the schematic of a 3D package with stacked dice and TSV configuration in which multiple dice are stacked on top of each other with Through Silicon Via (TSV) and solder microbumps providing the electrical interconnection

Y. Li (✉) • D. Goyal
Intel Corporation, Chandler, AZ, USA
e-mail: yan.a.li@intel.com; Deepak.goyal@intel.com

Y. Li, D. Goyal (eds.), *3D Microelectronic Packaging*, Springer Series in Advanced Microelectronics 57, DOI 10.1007/978-3-319-44586-1_15

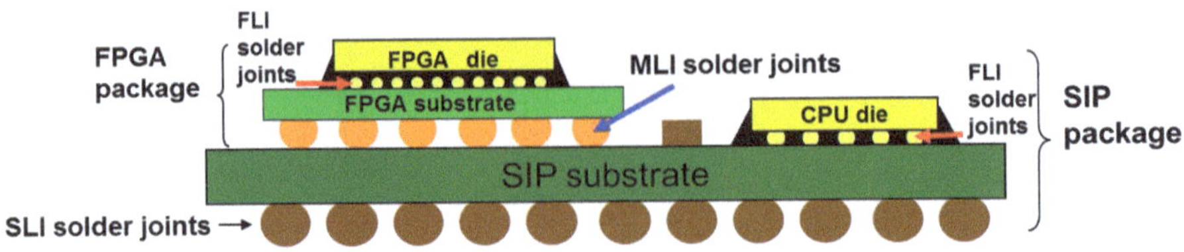

Fig. 15.1 Schematic of a SIP package with a packaging stacking configuration (adapted from Ref. [1])

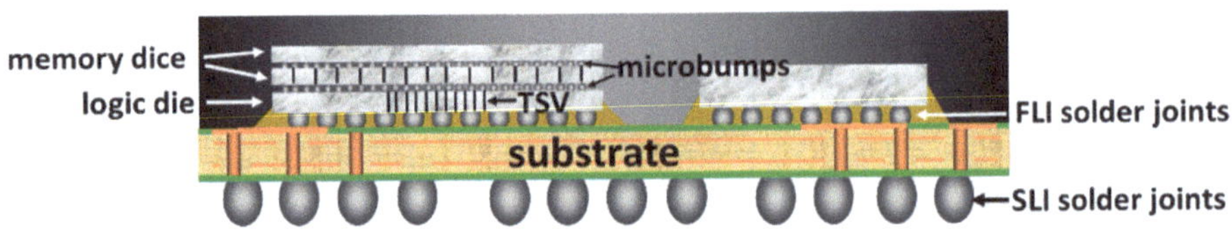

Fig. 15.2 Schematic of a 3D microelectronic package having a stacked die with TSV configuration (adapted from Ref. [3])

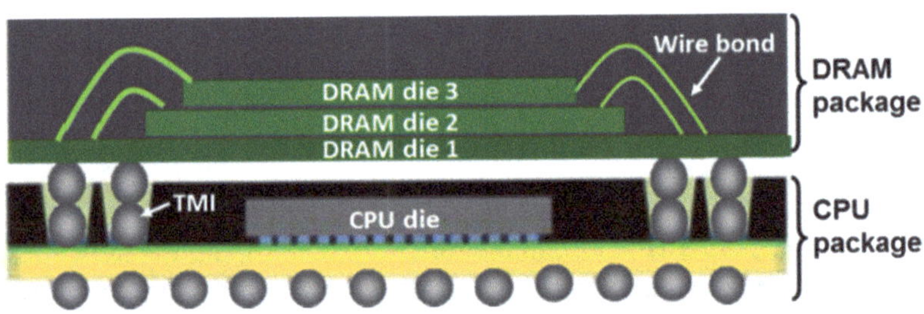

Fig. 15.3 Schematic of a 3D microelectronic package with a package stacking configuration demonstrating the integration of a wire bonded DRAM package onto a flip chip CPU package through TMI (adapted from Ref. [4])

between them [2]. Compared to large solder interconnects (100–500 μm in diameter) in a package stacking configuration, interconnects in a die stacking configuration are much smaller (5–25 μm in diameters), which results in additional fault isolation and failure analysis challenges arising from the extremely small size scale of both the feature and the potential defect [3].

Figure 15.3 demonstrates the integration of a wire bonded Dynamic Random Access Memory (DRAM) package onto a molded flip chip CPU package by Through Mold Interconnects (TMI) to form a 3D package with package stacking configuration [4]. Figure 15.4 displays a 3D package with a combination of both die and package stacking configurations. A couple of DRAM packages with four memory dice and one logic die piled on each other through TSVs and microbumps are assembled onto a common substrate along with a flip chip CPU die by MLIs. Manufacturing of these complicated packages involves multiple solder reflows, underfill (UF) curing, molding, and other assembly steps, thus diverse process

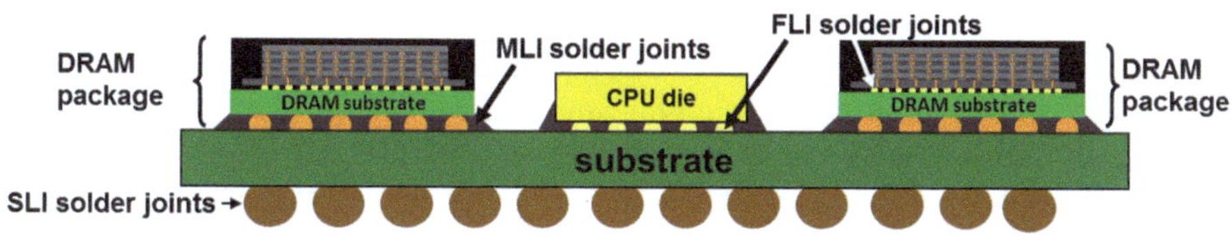

Fig. 15.4 Schematic of a 3D package with a combination of both die stacking and package stacking configurations (adapted from Ref. [4])

defects could exist in complex packages, and may result in many kinds of failures during reliability tests. Performing thorough root cause analysis of failures and providing solution paths by optimizing a specific assembly process are very challenging.

15.2 Fault Isolation and Failure Analysis Challenges for Advanced 3D Packages

3D packages are complex packages with multiple dice, substrates, and interconnects with different sizes, ranging from 5 to 500 μm, multiple failures typically exist in one unit post reliability tests. How to perform flawless failure analysis on each failure from the same unit is a big challenge. High resolution and nondestructive techniques, including fault isolation, imaging, and material analysis techniques are highly desired.

In addition to the challenge of analyzing multiple failures in one package, each electrical failure in a 3D package could come from various dice, assembly layers, or interconnects, precisely isolating every failure in such a complicated package is very challenging. For example, a short failure in a 3D package shown in Fig. 15.4 could come from second-level interconnects (SLI) solder joints, MLI solder joints between DRAM packages and the substrate, FLI solder joints between logic dice (the bottom die in DRAM packages) and DRAM substrates, and TSVs or microbumps between stacked dice of DRAM packages. Fault isolation techniques for both open and short failures, which could provide 3D information of defects are the key for 3D package failure analysis.

Once defects are isolated, physical failure analysis is needed to analyze defects and identify root causes. As some of the interconnects in 3D packages, for example TSVs, have large aspect ratio of 1:20, with small diameters (2–10 μm) and long lengths (~40–200 μm), artifact-free cross section technique with short TPT is critical for characterizing small defects in a relatively large cross sectional plane. Mechanical and conventional Focused Ion Beam (FIB) cross section techniques do not meet the demands. Alternative artifact free cross section techniques, which have a much higher milling rate while maintaining accurate cross section location control are more promising for 3D package failure analysis.

Complicated 3D packages could have subtle defects, such as plating defects in surface finish layers, contaminations at interfaces, impurities in interconnects, crystal size, and distribution abnormity in small interconnects. These subtle defects could result in failures, such as delamination, non wets, and cracks in interconnects post reliability tests. Innovative applications of material analysis techniques, which could provide elemental, chemical state, depth profile, and crystallographic orientation of materials around the failing area are crucial to the root cause investigation.

This chapter will discuss the advanced FI and FA techniques developed to address FI-FA challenges in 3D packages, including high-resolution nondestructive FI and FA methods, advanced sample preparation and material analysis techniques, efficient FI and FA flows, along with FA strategies and case studies to demonstrate their applications in the failure analysis of 3D microelectronic packages.

15.3 The Application of Nondestructive FI and FA Techniques to 3D Microelectronic Packages

15.3.1 Nondestructive Fault Isolation Techniques for Electrical Failures in 3D Microelectronic Packages

Nondestructive fault isolation techniques are the key to reveal multiple failures in a complex 3D package with shorter TPT and higher success rate. Detailed discussions of four important FI techniques for 3D packages, including time-domain reflectometry (TDR), Electro Optic Terahertz Pulse Reflectometry (EOTPR), Lock-In Thermography (LIT), and Scanning SQUID Microscopy (SSM) are provided in this section. Table 15.1 summarizes the resolution, capability, Pros, and Cons of each technique. Depending on package configurations, applications, TPT requirements, and budget limits, one or more FI techniques could be selected for the FI of electrical failures in 3D packages.

15.3.1.1 Time-Domain Reflectometry

For open and high resistance failures, TDR and EOTPR have been proven to be very effective in 3D packages [1]. TDR is a low cost and conventional nondestructive FI technique used to isolate electrical failures in packages. A step electrical pulse with 35–40 ps rise time is injected into the packaging interconnects, and the impedance variation along the circuit could be analyzed by interpreting the collected reflected signal from the package. The location of failures can be isolated to be either in the die or the package substrate by comparing the reflected waves from the failed unit with those from the good reference unit and the bare substrate [5]. The resolution of TDR depends on the rise time, TDR system bandwidth, and

Table 15.1 Pros and cons of nondestructive FI techniques for 3D packages

FI technique	Resolution[a] (μm)	Capability	Pros	Cons
TDR	~500	Open and short	Low cost	Relatively low resolution; need reference units; less sensitivity for short failures
EOTPR	~10	Open and short	High resolution in some applications	Need reference units; less resolution and sensitivity for some applications, like short failures
LIT	~5 (x and y) ~20 (z)	Short and high resistance open	Relatively short TPT; direct imaging of hot spots from defects	Have application limits, like multiple defects in the same short failure; not capable for open failures
SSM	~10 (x and y) ~40 (z)	Short and open	Providing current mapping	Data collection and interpretation may take time

Adapted from Refs. [1, 10–12, 14]
[a]Rough estimation in optimized condition

the material properties within packages. The resolution of TDR in time domain could be estimated as 1/10–1/5 of the TDR rise time [6]. The spatial resolution of TDR could be estimated by the following equation [7]:

$$\lambda_{\text{resolution}} = \frac{0.35}{\text{BW}_{\text{system}}} \cdot \frac{c}{2\sqrt{\varepsilon_{\text{eff}}}} \tag{15.1}$$

Where, $\text{BW}_{\text{system}}$ is the "overall system bandwidth," c is velocity of light in free space, and ε_{eff} is the effective dielectric constant of the package. It has been characterized that the spatial resolution of the TDR system with 35–40 ps rise time and 18–20 GHz $\text{BW}_{\text{system}}$ in conventional flip chip packages is about 500 μm [5].

By adding more reference spectra to define the locations of the multiple levels of solder interconnects in 3D packages, TDR has been demonstrated to be quite effective for detection of open failures in 3D packages [1]. As illustrated in Fig. 15.5a, in order to interpret and determine the failure location from TDR spectra obtained from a failed SIP package, reference TDR spectra are collected from the same BGA balls on a bare SPI substrate (without FPGA attached) and from a golden or passing reference unit with FPGA attached. The impedance raise in the spectrum obtained from the bare substrate corresponds to the location of the MLI joint where the electrical path terminates and hence can be used to compare with the spectra from the failure unit to determine if the failure is before or after the MLI joint. Figure 15.5b demonstrates how the location of the FLI joint is identified by comparing the blue spectrum from a SIP package without FPGA die reference unit with the green spectrum from a golden unit. Figure 15.6 shows a set of TDR spectra collected from the same SLI BGA ball of a failed unit, and reference units including a bare SIP substrate, a SIP package without FGA die, and a golden unit. The TDR spectra from the reference units are used to determine locations of the MLI and the

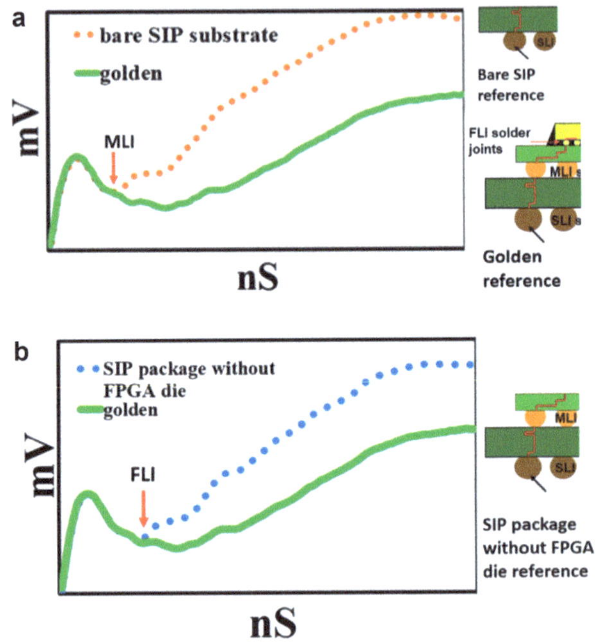

Fig. 15.5 (**a**) The location of the MLI can be defined by comparing the orange TDR spectrum from a bare SIP substrate with that from a golden reference unit collected from the same SLI BGA ball. (**b**) The location of the FLI can be identified by comparing the blue TDR spectrum from a SIP package without FPG die and that of a golden unit (adapted from Ref. [1])

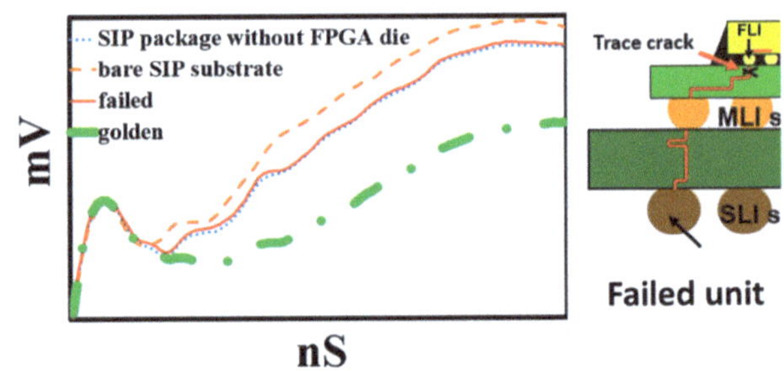

Fig. 15.6 Conventional TDR spectra from a failed unit, a golden unit, a bare SIP substrate, and a SIP package without FPGA die, showing the open failure is close to the FLI solder joint in the FPGA package of a SIP package (adapted from Ref. [1])

FLI. The overlapping of the blue spectrum from a SIP package without FPGA die with the red waveform taken from the failed unit suggesting that the failure is close to the FLI joint in the FPGA package. Based on the limited 500 μm resolution of TDR, the failure could be a crack in the trace routing very close to the FLI joint, the FLI solder joint open, or FPGA die defects close to the FLI joint.

15.3.1.2 Electro Optic Terahertz Pulse Reflectometry

EOTPR is the next generation of TDR, developed to further improve its resolution. Unlike TDR, EOTPR uses an ultrafast laser pulse to produce a 40 GHz–4 THz electrical pulse with a sharp peak, which is injected into the package under test

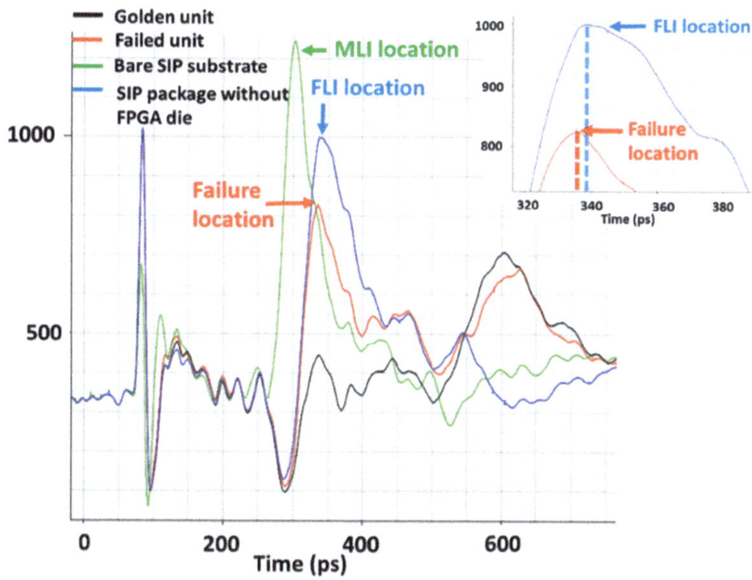

Fig. 15.7 EOTPR spectra from a failed unit, a golden unit, a bare SIP substrate, and a SIP package without FPGA die (adapted from Ref. [1])

[1]. Due to the much faster rise time of 5.7 ps and the THz range system bandwidth, the resolution of EOTPR could be about 10 μm, much higher than that of TDR [5]. Since EOTPR uses a pulse with a sharp peak as the input signal, open interconnects in packages show as peaks and shorting in interconnects as valleys in EOTPR spectra. As illustrated in Fig. 15.7, the same set of data shown in Fig. 15.6 was collected by EOTPR. By comparing with the blue spectrum defining the FLI location in FPGA package, the failure is isolated to be inside the FPGA substrate right before the FLI interconnect. The exact failure location can also be estimated by converting the time domain EOTPR raw data into the distance domain data through the phase velocity, ν_p, of the electromagnetic wave, which can be expressed as [8]:

$$v_p = \frac{c}{\sqrt{\varepsilon_{\text{eff}}}} \tag{15.2}$$

Where, ν_p, is the phase velocity, c is velocity of light in free space, and ε_{eff} is the effective dielectric constant of the package, which is different for different packages and can be estimated using known package structure locations in EOTPR waveform. As demonstrated in Fig. 15.8, the ε_{eff} can be estimated based on the MLI and FLI locations in the EOTPR time domain spectra and the actual distance between them, which is around 1.6 mm measured from the design file. Using the

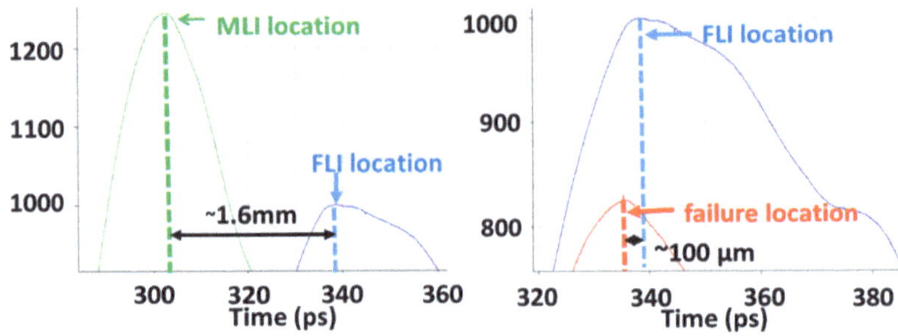

Fig. 15.8 The failure location can be estimated using known package structure locations in EOTPR spectra (adapted from Ref. [1])

estimated ε_{eff}, the distance between the failure location and the FLI can be calculated as about 100 µm.

By employing the same concept of adding more reference spectra to define the locations of multiple levels of interconnects, and comparing to the spectra obtained from the failing unit, both TDR and EOTPR could be applied to isolate open failures in 3D microelectronic packages with die stacking configuration [9]. For 3D packages with about six memory dice stacking on one another, reference units are created for TDR to define the locations of multiple levels of microbumps between the stacked dice [9]. The result demonstrates that locations of microbumps between the six stacked dice can be defined by TDR, which suggests the promising application of TDR in 3D packages with die stacking configuration.

15.3.1.3 Lock-in Thermography

The key to the precisely locating shorts and resistive open failures in 3D microelectronic packages is to obtain *z* dimensional information of defects in addition to the *x* and *y* location data. LIT is a very promising technique developed to locate the dissipated heat by resistive failures using the real-time graphical lock-in methodology, which is designed to detect small signal buried in random noise of higher magnitude [10]. The *z* dimensional locations of defects could be detected by LIT as the heat propagation is time depended. The time delay of the underlying thermal diffusion process determines the phase shift between the excitation signal and thermal response, which is related to the frequency of the stimulation signal (lock-in frequency) as well as thermal properties and thicknesses of different layers in the stack. The *z* position of hot spots within the package can be identified by their Lock-in frequency vs. phase shift curves which can be compared to reference units with defects at each relevant z-position. As illustrated in Fig. 15.9 [10, 11], it is demonstrated that LIT could differentiate hot spots in the stacked dice of a 3D package. With known application limitations and further development, LIT could be a very promising method for the nondestructive fault isolation of resistive failures in 3D packages [12].

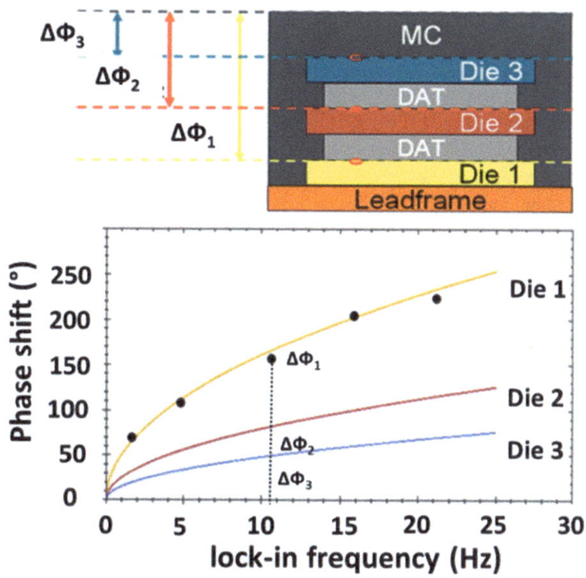

Fig. 15.9 Measurement results of Lock-in frequency vs. phase shift curves from reference units with short failures at die 1, die 2, and die 3 of a 3D packages with stacked die configuration showing in dots compared to a calibrated simulation model (lines) (adapted from Ref. [10, 11])

15.3.1.4 Scanning Superconducting Quantum Interference Device Microscopy

SSM is a generic name for FI techniques using a SQUID sensor to map the magnetic field produced by a current injected into the failing structure of the device under test (DUT) [13]. It is sometimes called Magnetic Field Imaging (MFI) or Magnetic Current Imaging (MCI) [14, 15]. For short and high resistance open failures, SQUID sensor typically operates at kHz frequency range. By increasing the bandwidth of SQUID electronics to RF range, SSM is capable for isolating open failures. The high frequency application of SSM is also called Space Domain Reflectometry (SDR) [14]. SSM has been applied as a nondestructive fault isolation technique for electrical failures in conventional packages [13–15]. Magnetic field generated by the input current inside the failed unit is detected and processed using a Fourier Transform inversion technique to obtain current density map of the sample. The *x* and *y* locations of the failure can be determined by comparing the current density map from the failed unit to a circuit diagram or a golden unit. The *z* dimension of the failure location can be estimated based on the fact that the distance between the minimum and the maximum magnetic field around a straight current path in the unit is twice the total distance from the SQUID sensor to the current path [15]. The distance from the sensor to the sample surface is fixed for each SSM scan; therefore, the distance from the sample surface to the current path could be estimated. Figure 15.10 illustrates how the *x*, *y*, and *z* locations of a short failure in a 3D package with stacked die configuration can be defined by SSM. As shown in Fig. 15.10a, *x* and *y* location of the short can be determined by the current analysis, while the *z* location of the defect is estimated to be 560 μm away from the Si

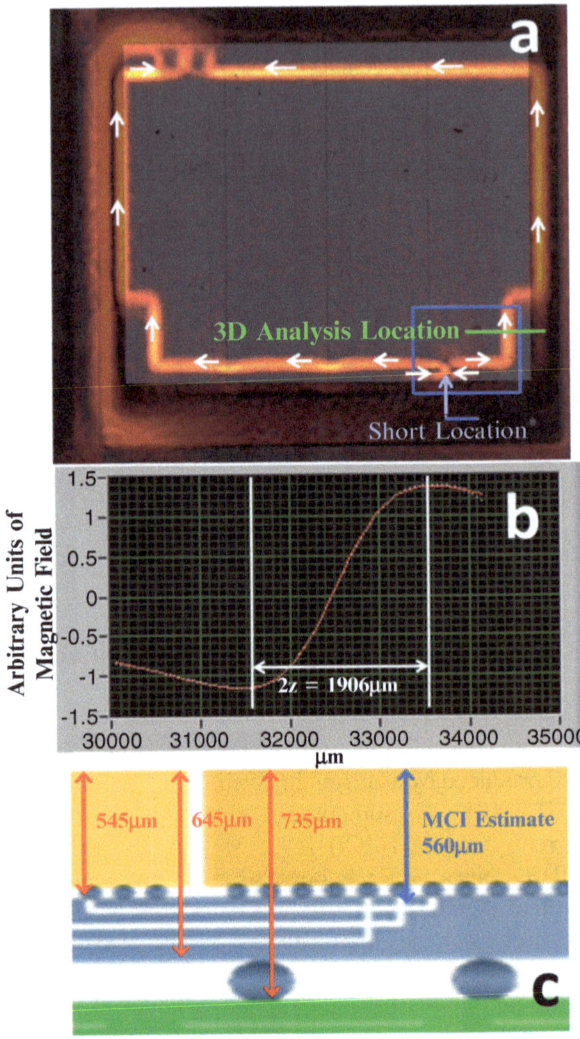

Fig. 15.10 (**a**) SSM analysis of a short failure in a 3D package with stacked die configuration. *x* and *y* location of the short can be determined by the current analysis. (**b, c**) *z* dimension analysis of the current path location shown in (**a**). The distance between the minimum and maximum magnetic field along the *green line* in (**a**) is measured as 1906 μm. The distance between the sample surface and the failure location is then estimated to be 560 μm, by subtracting the sensor to sample distance (393 μm) from the sensor to current path distance (953 μm, half of 1906 μm) (adapted from Ref. [15])

surface, by subtracting the sensor to sample distance (393 μm) from the sensor to current path distance (953 μm), which is half of 1906 μm, the distance between the minimum and the maximum magnetic field around the current path (Fig. 15.10b). By comparing to the design file, the failure is isolated to be in the microbump area between the two dice, shown in Fig. 15.10c [15].

An alternative method to acquire *z* information of the failure by SSM is to perform the analysis of the magnetic field signal by a different algorithm so that a 3D current path can be calculated instead of a 2D current path by the conventional method [16]. It has been demonstrated that the method is promising to map 3D current path in 3D packages with stacked die configuration [16].

SSM is a complementary technique to LIT, as it provides current mapping instead of locating the dissipated heat from short defects. For example, Defects causing short failures with extremely small resistance may not always get captured by LIT. High resistance routing in the failing structure, such as TSV, could also shadow the dissipated heat coming from short defects. The phase shifts obtained by LIT from multiple short defects nearby in the same failing structure could also be inaccurate. In these cases, SSM could be an alternative method to effectively identify 3D locations of defects.

15.3.2 High Resolution Non-destructive Imaging Techniques for 3D Microelectronic Packages

High-resolution nondestructive imaging techniques are crucial for defect screening and identification in complex 3D packages. Fundamental details along with application demonstrations of three commonly used nondestructive imaging techniques, Scanning Acoustic microscopy (SAM), 2D X-ray radiography, and 3D X-ray computed tomography (CT) are provided in this section. Table 15.2 summarizes the resolution, capability, Pros, and Cons of the three techniques. The nondestructive imaging techniques with high resolution, large field of view, and short TPT could be employed to directly detect failures without any fault isolation. They are also utilized as in-line monitoring tools for defect inspection. The high-resolution nondestructive imaging techniques with relatively smaller field of view and longer TPT could be applied after failure isolation to reveal defects causing the electrical failures.

Table 15.2 Pros and cons of high-resolution nondestructive imaging techniques for 3D packages

Nondestructive imaging technique	Resolution (μm)[a]	Capability	Pros	Cons
SAM	~1	Air gap (delamination, voids)	Short TPT; low cost	Not sensitive to vertical cracks
2D X-ray radiography	~5	Defects in interconnects and traces	Real time; low cost	Projective images; not for organic packing materials
3D X-ray CT	<1	Defects in interconnects and traces	3D imaging; higher resolution	Long TPT; high cost; small field of view; not for organic packing materials

Adapted from Refs. [3, 12, 30, 31, 37]

[a]Rough estimation in optimized condition, actual resolution depends on different equipment and set up

15.3.2.1 Scanning Acoustic Microscopy

SAM has been widely used to nondestructively detect voids in various packaging materials and delamination at multiple interfaces in packages [17–20]. As illustrated in Fig. 15.11, the acoustic transducer (piezoelectric element) excited by an extremely short electrical discharge transmits an ultrasonic pulse into a microelectronic package. Water is used as coupling medium to transmit the acoustic perturbation from the transducer to the package. Once the beam reaches any smooth interface, the reflection and transmission of acoustic wave from medium 1 to medium 2 are described by the Reflection coefficient "*R*" and Transmission coefficient "*T*," where

$$R(90^\circ) = \frac{Z_2 - Z_1}{Z_2 + Z_1} \tag{15.3}$$

$$T(90^\circ) = \frac{2Z_2}{Z_1 + Z_2} \tag{15.4}$$

Z_2 stands for the Acoustical Impedance of medium 2, while Z_1 means the acoustical impedance of medium 1. Acoustic Impedance Z is expressed as

$$Z = \rho c \tag{15.5}$$

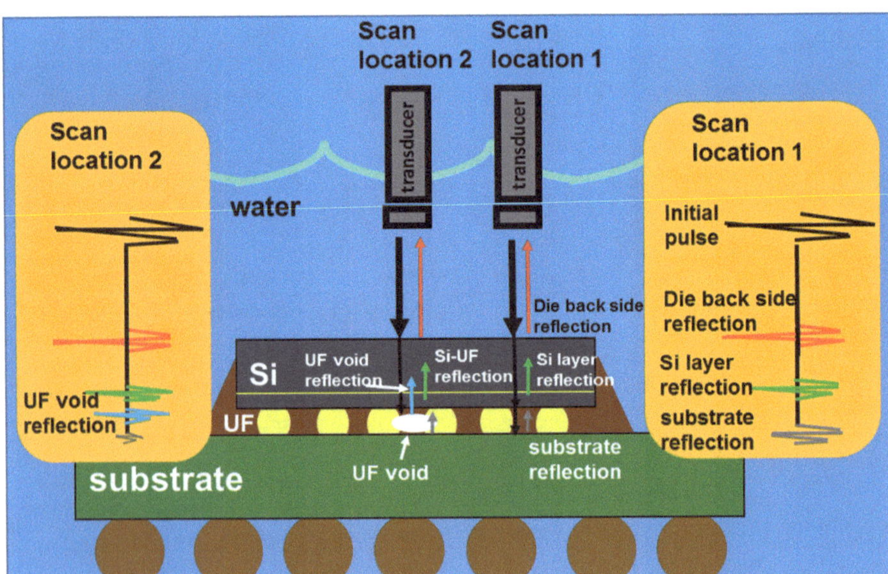

Fig. 15.11 Schematic of the SAM technique

"ρ" is the volumetric mass density of the medium, "c" is the acoustic wave velocity in the medium. For longitudinal acoustic waves traveling in a homogeneous and isotropic medium, c could be expressed as

$$c = \sqrt{\frac{E(1-\mu)}{\rho(1+\mu)(1-2\mu)}} \tag{15.6}$$

Where "E" stands for Young's modulus, "μ" is the Poisson's Coefficient. Thus, Acoustic Impedance is defined by density, Young's modulus, Poisson's Coefficient [21, 22] and is characteristic of materials.

From Table 15.3 [22], showing Acoustic Impedance of common materials in microelectronic packages, the Acoustic Impedance of air is about 0.389×10^{-3} [kg/m^{2}s] $\times 10^6$, very small comparing with other materials listed in the table. By Eqs. (15.3) and (15.4) with $Z_2 \approx 0$, the Reflection coefficient "R" and Transmission coefficient "T" from any medium to air are equal to -1, and 0, respectively, indicating almost 100 % reflection. Therefore, SAM is very sensitive to air gap defects in microelectronic packages, such as delaminations, blisters, cracks, and voids.

The SAM technique illustrated in Fig. 15.11 is also called Pulse-Echo method [23–25], which provides information associated with both intensity and time of flight of the ultrasonic beam. The same ultrasonic transducer that transmits the acoustic pulse is also used for receiving echoes occurring at boundaries of different acoustic impedance values. As illustrated in the scan location 2 of Fig. 15.11, the echo reflected by a UF void shown in blue has a unique time of flight compared with the die back side, thus contains the z dimensional information of the defect, and could be obtained by either C-scan SAM (CSAM) with Tomographic Acoustic Micro Imaging (TAMI) technique [26, 27] or B-scan SAM [28]. CSAM provides a plane view of the location and size of features of the test specimen. By setting a sequence of multiple gates of data acquisition adjusted to resolve the interfaces of interest, TAMI CSAM provides images from multiple depths within the sample simultaneously. B-scan SAM is virtual cross sectioning of the CSAM plane and could present a cross sectional view of defects [28]. As illustrated in Fig. 15.12, the UF void area shows up as "white contrast" in gate 7 to gate 9 of the TAMI CSAM images by employing a CSAM recipe with 12 gates regulated to analyze all the interfaces from the die backside to the substrate of the flip chip package shown in Fig. 15.11.

To image subtle defects in advanced 3D microelectronic packages, both lateral and axial resolution of the CSAM scan need to be improved. The resolution of CSAM scans is usually determined by CSAM transducers. As shown in Fig. 15.13, the characteristic features of CSAM transducers include frequency, shape of the cavity, focal length, spot size in water, and depth of field. Lateral or x–y dimensional resolution of CSAM images is related to transducer beam spot size, the smaller the better. While axial or Z dimensional resolution exposing the depth of delamination gaps is related to transducer frequency, the higher the better [28, 29].

Table 15.3 Acoustic impedance of common materials in microelectronic packages

Material	Air	Water	Silicon	Tin	Polyimide (Vespel SP-1)	Indium	Copper	Aluminum	Cast Iron	glass
Acoustic impedance [kg/m^{2}s] × 10^6	0.389 10^3	1.48	19.65	24.2	3.61	16.21	41.8	17.1	24.2	15.3

Adapted from Ref. [22]

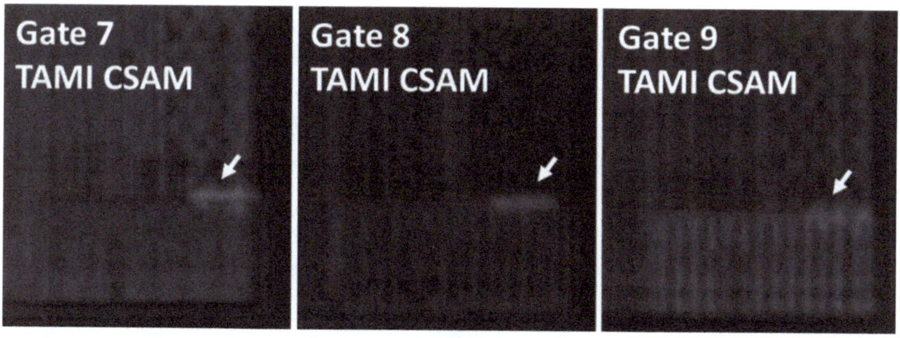

Fig. 15.12 TAMI CSAM scan showing a UF void in the FLI area, from gate 7 to gate 9

Fig. 15.13 Characteristic features of a CSAM transducer (adapted from Ref. [29])

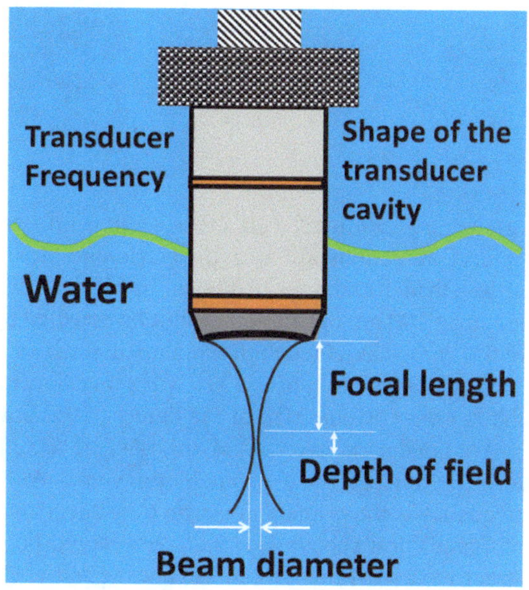

However, higher frequency acoustic waves have bigger attenuation in the medium, which may limit the nondestructive applications of high frequency transducers, especially for samples with thick Si. CSAM images of the same units performed with transducers having different frequency, spot size, focal length, and depth of field can be compared to optimize the CSAM recipe for a specific product. It is found that axial resolution sometimes is more critical for defect detection [29].

As indicated in Fig. 15.14a and b, the same sample is imaged by CSAM with transducers having different frequencies [29]. Figure 15.14a shows the image taken by a lower frequency transducer of 110 MHz, no obvious defect is observed. The image obtained with a higher frequency transducer of 230 MHz is shown in Fig. 15.14b. Three thin film defects (highlighted with white arrows) about 100 μm

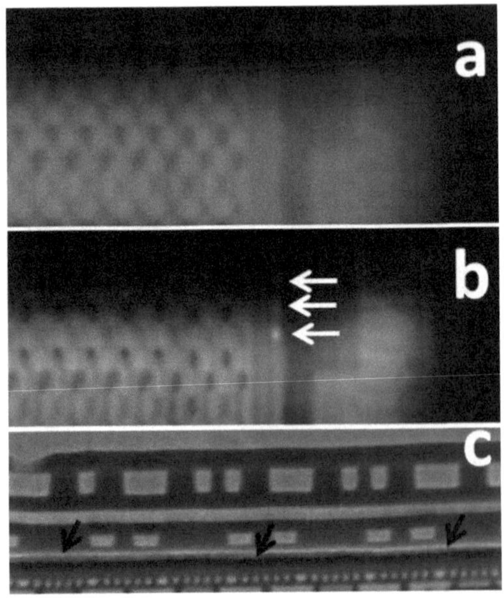

Fig. 15.14 (**a**) CSAM image taken by a 110 MHz transducer. (**b**) CSAM image of the same sample taken by a 230 MHz transducer. (**c**) FIB cross section on one of the defects in (**b**) showing subtle thin film delam with a delam gap depth of less than 100 nm (adapted from Ref. [29])

in size are identified. FIB cross section on one of the defects indicates that the defects are delaminations in dielectric thin films of the die, with a delam gap depth of less than 100 nm, as displayed in Fig. 15.14c. These 100 μm large defects with less than 100 nm delam gap cannot be detected by the 110 MHz transducer because of the limited axial resolution mainly defined by transducer frequency. Even though the lateral resolution of the 110 MHz transducer is about 20 μm, much smaller than the *x*–*y* dimensional size of the defect (~100 μm).

For nondestructive CSAM analysis, the focal length of CSAM transducers needs to be optimized for a given Si thickness. As the axial resolution or transducer frequency is the primary concern, transducers with high frequency and focal length designed for a specific die thickness would be desired for high-resolution nondestructive CSAM imaging of 3D packages.

GHz-SAM technique is developed to further improve both the lateral and axial resolution of SAM [30, 31]. As illustrated in Fig. 15.15, the frequency of the acoustic transducer is extended to the GHz domain, which ensures a much higher spatial resolution of ~1 μm, but also a smaller penetration depth. Instead of immersing the sample fully in water, only a drop of water is used to cover the sample surface. Figure 15.16 demonstrates the detection of TSV voids by GHz-SAM. Compared with the CSAM image using a 200 MHz transducer shown in Fig. 15.16a, the GHz-SAM image shown in Fig. 15.16b and c clearly reveal the TSVs with voids about 1 μm in diameter and 10 μm in depth, confirmed by the FIB cross section shown in Fig. 15.16d. In addition to spatial resolution improvement, GHz-SAM could also provide additional information, for example, defect-induced

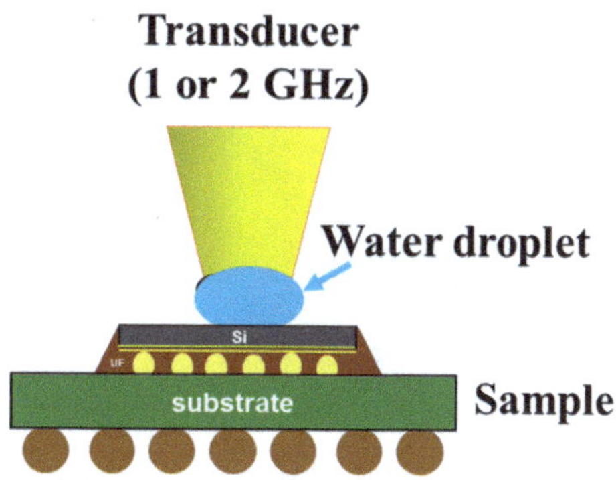

Fig. 15.15 Schematic of the GHz SAM technique (adapted from Ref. [31])

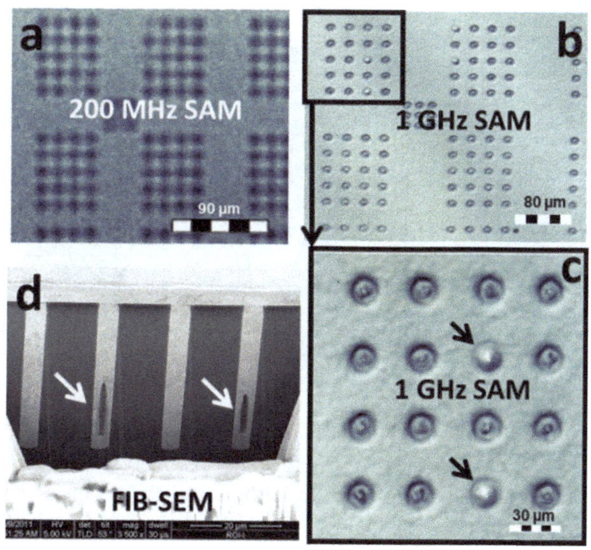

Fig. 15.16 CSAM images of TSVs taken by (**a**) 200 MHz transducer, (**b**, **c**) 1 GHz transducer. (**d**) FIB cross section on the TSVs having voids in GHz CSAM (adapted from Ref. [31])

stress around TSVs shown in Fig. 15.17, which is possibly due to reflection of surface waves generated by the GHz transducer [31].

To nondestructively detect defects embedded in stacked dice or complex packages, for instance, MLI UF voids, delamination in logic dice having a couple of memory dice stacked on the top, blisters in substrate layers, and defects in embedded Si chips of Printed Circuit Boards (PCB), acoustic transducers with higher penetration depth are desired. However, transducers with high penetration depth typically come with lower frequency and lower resolution. To improve the detectability of defects embedded deep in the 3D packages, Through Transmission SAM (TSAM) technique with newly designed transducers and improved signal-to-noise ratio is often employed.

As illustrated in Fig. 15.18, TSAM uses two transducers aligned facing each other from both sides of the specimen [23]. The emitter generates a pulse

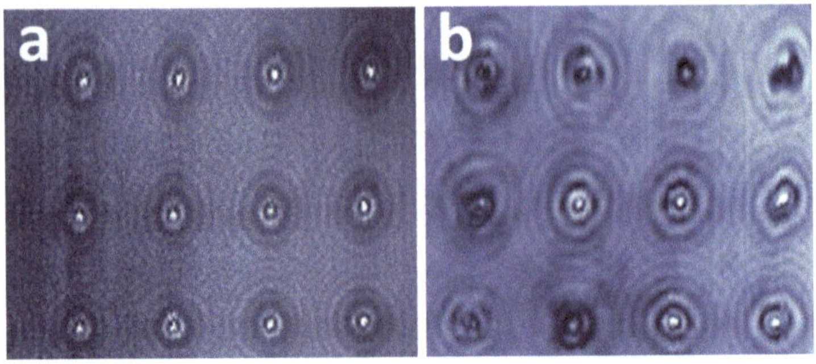

Fig. 15.17 (**a**) GHz CSAM image of TSVs without defects. (**b**) GHz CSAM image of TSVs with lots of defects showing fringes around the TSVs (adapted from Ref. [31])

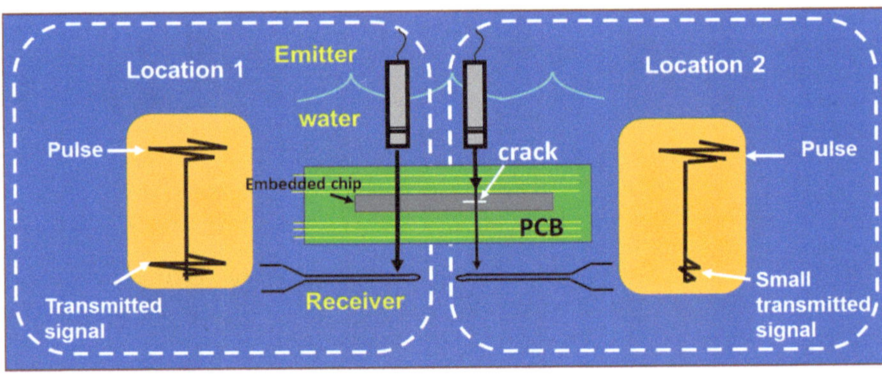

Fig. 15.18 Schematic of the TSAM technique

propagating through the specimen and the receiver transducer senses the intensity of the crossing wave at the rear side of the specimen. The beam energy attenuates as the pulse propagates through the specimen. At location 1 of Fig. 15.18, the acoustic wave goes through a clean path whereas at location 2, the propagation of the beam is interrupted by a crack or an air gap, which significantly attenuates the amplitude of the transmitted pulse, thus gives a "dark contrast" in the TSAM image.

Newly designed high power TSAM emission transducers [32] (directly focused without lenses) in the range of 20–50 MHz along with a focused 10 MHz reception transducer can improve the TSAM spatial resolution to 50 μm and have been used to study a PCB having embedded chips. Figure 15.19 shows a TSAM image taken from a test vehicle with embedded chips, defects around the four die corners of an embedded chip are disclosed by nondestructive TSAM technique.

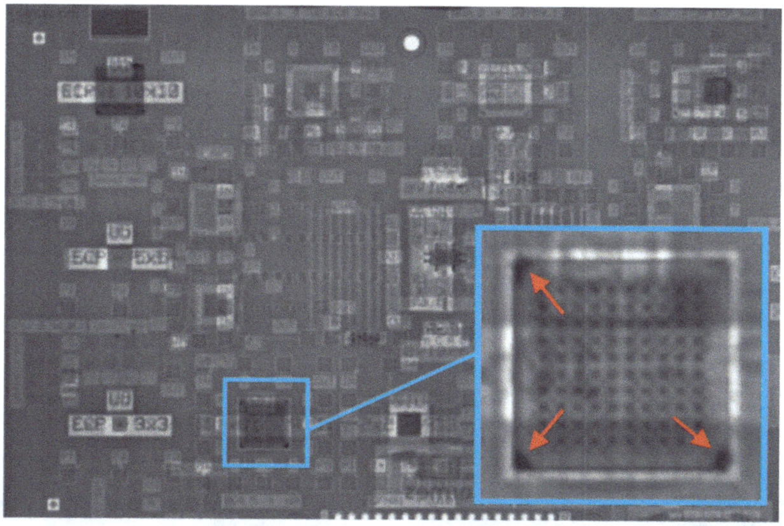

Fig. 15.19 TSAM of a PCB with embedded chips showing defects in one embedded chip (adapted from Ref. [32])

15.3.2.2 2D X-Ray Radiography

2D X-ray radiography has been widely used in the semiconductor industry to inspect solder joint defects in packages. Samples are put between the punctiform X-ray source and the detector, as illustrated in Fig. 15.20 [33]. The geometric magnification M is the ratio of the distance between the source and the detector (D), to the distance from the source to the sample (d). The combination of a very small X-ray beam spot size and the placement of the sample very close to the source is desired to get high-resolution 2D X-ray images [33]. Figure 15.20 also demonstrates that 2D X-ray radiography is a projection of a 3D object, thus information in depths and volume need to be obtained by observing the object from several different angles.

2D X-ray imaging technique has been demonstrated to be very efficient for the detection of solder void evolution post multiple reflows in SIP packages [34]. It is found that nondestructive real-time 2D X-ray imaging with optimized imaging conditions, such as sample tilt and rotation angles, could be used to scan the failing area and reveal package defects, for example, solder joint bridging, solder noncontact open, and nonwet. As illustrated in Figs. 15.21 and 15.22, correlation between 2D X-ray images and FLI solder joint defects can be established by cross section of the same solder joints post nondestructive 2D X-ray imaging. It has been proven that defective solder joints, such as partial nonwet, complete nonwet, noncontact open, and solder joint bridging could be easily analyzed via real-time 2D X-ray imaging with a very short TPT by comparing the image contrast and solder joint geometry shape with those from a golden solder joint [3].

Fig. 15.20 Schematic of a 2D X-ray radiography set up (adapted from Ref. [33])

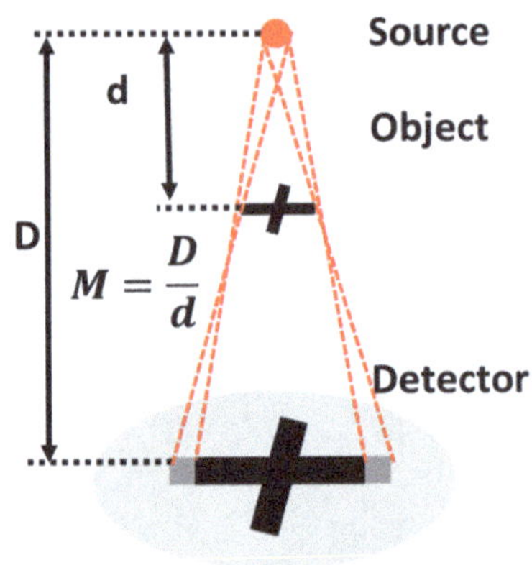

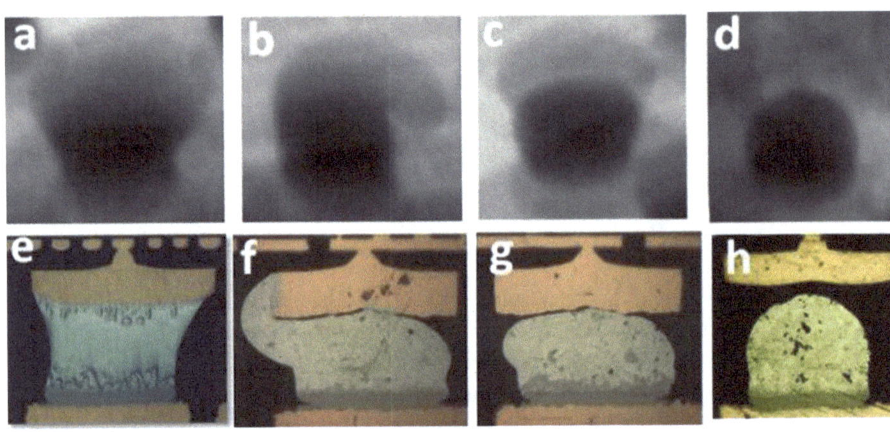

Fig. 15.21 2D X-ray images taken nondestructively and the corresponding optical images of FLI solder interconnects cross sectioned post-2D X-ray imaging: (**a**, **e**) normal or defect free, (**b**, **f**) partial nonwet, (**c**, **g**) complete nonwet, (**d**, **h**) noncontact open (adapted from Ref. [3])

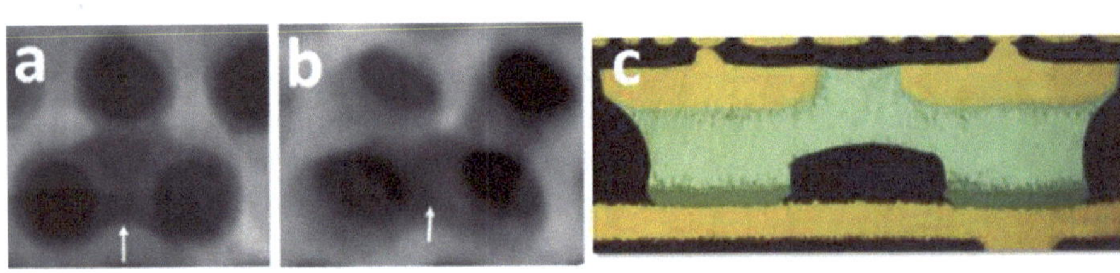

Fig. 15.22 2D X-ray images of FLI solder joint bridging taken nondestructively with a top-down view (**a**), a tilted view (**b**), and the corresponding optical image taken from the sample cross sectioned post nondestructive 2D X-ray imaging (**c**) (adapted from Ref. [3])

The high-resolution, nondestructive, and real-time 2D X-ray imaging enables in situ study of package failures happened during reliability tests, such as reflow, and provides direct observation of how the failures happens, thus reveals root causes of the failures [35].

However for defects in micrometer scale, such as substrate trace cracking, subtle solder extrusion between solder interconnects, microvoids in Cu vias, nondestructive 2D X-ray imaging may not be applicable, as the small contrast difference in a single 2D X-ray image is masked out by other layers or interconnects in packages [1].

15.3.2.3 3D X-Ray Computed Tomography (CT)

3D X-ray computed tomography (CT) has been demonstrated to be a very successful technique in detecting defects in micrometer or submicrometer scale [36, 37]. As illustrated in Fig. 15.23a, 3D X-ray computed CT (Zeiss Xradia system) relies on a basic operation principle of using an X-ray source to radiate the object at different tilt angles, with a rotating stage providing angular displacement in equally spaced angles, and a detector collecting the 2D X-ray images at each angle. All 2D images are mathematically superimposed and processed to obtain a 3D image of the sample, as shown schematically in Fig. 15.23b. Since the processed 3D X-ray CT data contain volumetric information of the sample under test, analysts can manipulate it to display virtual cross sections or slice views at any given location of the 3D data set. 3D X-ray CT images in Fig. 15.24 showing the substrate trace crack in a SIP package located by the EOTPR (Figs. 15.7 and 15.8) indicate that nondestructive FI techniques combined with high-resolution 3D X-ray imaging are very effective to detect interconnect and trace failures in 3D packages [1]. One of the key components of the 3D X-ray CT system used in these studies is an X-ray detector with proprietary X-ray optics and optical design that provides less dependency over magnification with respect to the source-to-sample distance [37]. The resolution is achieved in the submicron meter range without the need for placing the sample close to the source, thus allowing full sample rotation

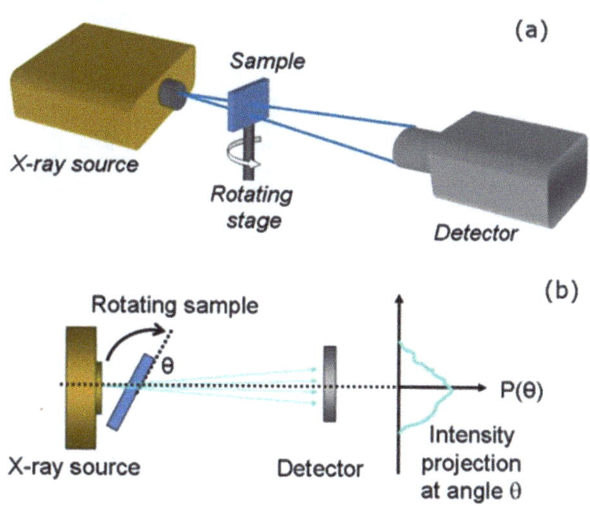

Fig. 15.23 (**a**) Schematic of the 3D X-ray CT setup consisting of X-ray source, rotating stage, and detector. (**b**) The intensity projection at each angle is superimposed and mathematically processed to generate the 3D image (adapted from Ref. [1])

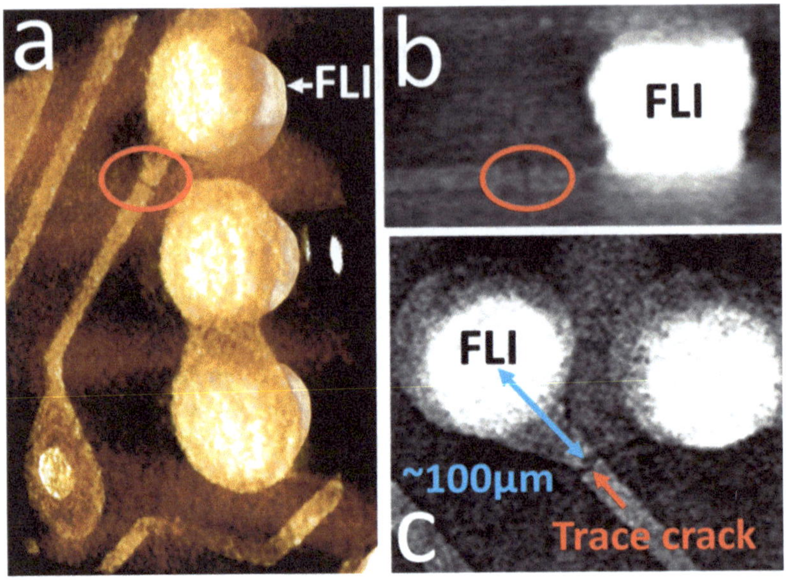

Fig. 15.24 (**a**) 3D view, (**b**) virtual Cross sectional view, and (**c**) virtual planar view of the 3D X-ray CT showing trace crack in a SIP package (adapted from Ref. [1])

Fig. 15.25 (**a**, **b**) Virtual X-sectional view of solder joints having partial nonwet and good solder joints, respectively [37]

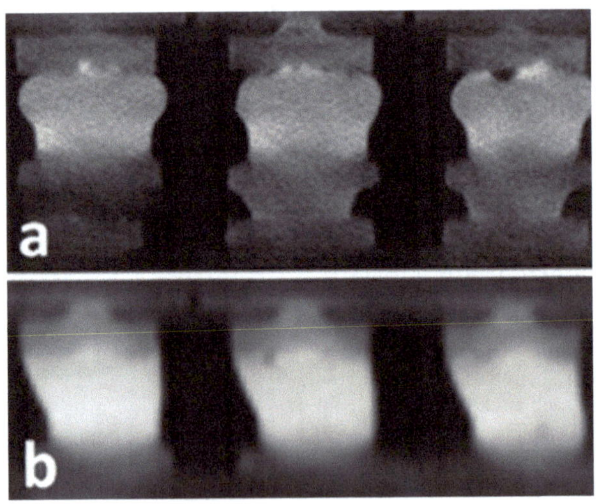

required for 3D X-ray imaging without sample–source collision. This is an important capability because it allows scanning intact samples, regardless of sample size.

3D X-ray CT techniques have found applications in the semiconductor package assembly process for monitoring micrometer size defects in 3D packages, such as partial nonwet, voids (<5 μm) in solder joints, and voids or cracking in Cu vias [37]. Figure 15.25a shows a virtual cross sectional 3D X-ray image of partial

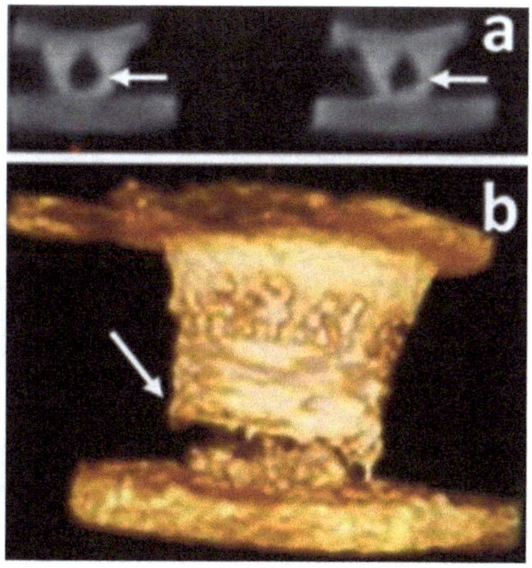

Fig. 15.26 (**a**) Virtual X-sectional view of voids in substrate Cu vias. (**b**) 3D view of cracks in Cu vias [37]

nonwet solder joints connecting between the die and the substrate. Compared with good solder joints shown in Fig. 15.25b, the partially nonwet solder joints have a pear shape and show small gaps between the die bumps and the bulk solder. Figure 15.26 demonstrates that 3D X-ray CT distinctly captures voids and cracks in small substrate Cu vias.

3D X-ray CT can provide high-resolution information of semiconductor packages nondestructively, which enables the progressive study of packages during reliability tests, such as reflow, temperature cycling, and consistent current flow at elevated temperatures [34, 37]. Void nucleation and growth in solder joints under current flow at elevated temperatures is one of the focused areas during the development of semiconductor packages having different design and form factors. Progressive 3D X-ray CT study can provide the location and size of voids at time zero and at intermediate readouts of electrical tests. As shown in Fig. 15.27a–d, voids in the solder joints start to nucleate and grow at the cathode during the electrical test due to the electron migration. The void growth kinetics can be obtained from progressive 3D X-ray CT images, which could be used for effective reliability risk assessment [37].

The current 3D X-ray CT techniques can provide high-resolution 3D information inside semiconductor packages nondestructively and have been applied in the field of failure analysis, process control, and kinetic study of electronic packages. However, the flux and brightness of lab scale X-ray sources are low, which results in relatively long exposure or image capture time, thus longer TPT for high-resolution imaging. On the other hand, X-ray beam with energy higher than 100 kV is usually used in semiconductor package imaging to cope with highly absorbing Cu or solder components. This sacrifices the resolution due to the beam spot size blooming at higher energy. Additionally, high energy X-ray imaging reduces the phase contrast and also makes the organic packaging materials “invisible.” These factors limit the applications of 3D X-ray CT to failure detection in

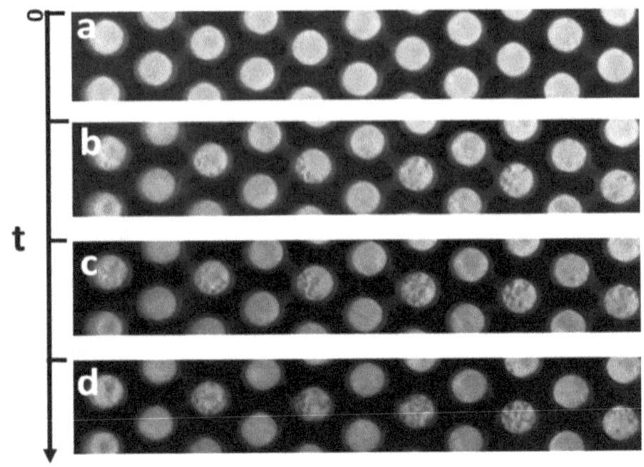

Fig. 15.27 (**a–d**) virtual lapping view of the same solder joints under consistent current flow with elevated temperature at time zero, and intermediate readouts, respectively [37]

materials other than metals with high Z numbers, for instance, cracks, voids, and delaminations in underfill, molding compound, solder resist, and other dielectric materials in electronic packages. The synchrotron 3D X-ray CT study of SIP packages reveals the possibility of expanding applications of 3D X-ray CT in microelectronics and also provides some suggestions to the development of next-generation lab-scale 3D X-ray CT systems [37, 38].

15.4 The Application of Sample Preparation and Material Analysis Techniques to 3D Microelectronic Packages

15.4.1 Sample Preparation Techniques

Root cause understanding of failures is critical for process improvement and optimization during product development, which typically needs destructive failure analysis, such as cross sectional data, to disclose the details of defects. As interconnects in 3D package could be as small as a few micrometers in diameter, it is very challenging to get good alignment as well as artifact-free finish using conventional mechanical cross sectional techniques. On the other hand, it is very time consuming to cross section TSVs with more than one hundred micrometers in length by the conventional FIB technique [39]. A couple of advanced cross section techniques, for example, plasma-FIB, femtosecond (fs) and nanosecond (ns) laser ablation, and broad-beam Argon ion milling, are developed in the industry to fill in the technical gaps. Table 15.4 shows the milling rate, spatial resolution, Pros, and Cons of each technique.

Table 15.4 Pros and Cons of sample preparation techniques for 3D packages

Cross section technique	Milling rate ($\mu m^3/s$)[a]	Spatial resolution (μm)	Pros	Cons
Laser ablation	~10^6	~10	High milling rate; short TPT	Thermal damage
Plasma FIB	~10^3	~0.1	High resolution; less milling artifacts; precise milling location and direction control	High cost
Ar ion milling	~10^4	~1	Low cost; less milling artifacts	Milling location and direction control is hard

Adapted from Refs. [39–41]
[a]Rough estimation in optimized condition, actual resolution depends on different equipment and set up

15.4.1.1 Nanosecond (ns) and Femtosecond (fs) Laser Ablation Techniques

As presented in Table 15.4, the laser ablation techniques have the fastest material removal rate, thus could provide a very short TPT. However, the relatively low spatial resolution and laser-induced thermal damage in the μm range heat affected zone may limit its application as an independent cross section technique [40]. Plasma FIB or ion milling is needed to clean up laser prepared samples. An integrated femtosecond (fs) laser ablation combined with conventional FIB and Scanning Electron Microscopy (SEM) is an alternative advanced cross sectional technique comparable with the plasma FIB and SEM method [40]. Compared with the conventional nanosecond laser ablation technique, fs laser ablation technique integrated with SEM and FIB has much better location control and very little laser-induced damage to the sample, while maintain a high material removal rate of up to 10^3 $\mu m^3/s$. It has been demonstrated that the system could perform artifact-free TSV cross sections with comparable TPT to the plasma FIB system [40].

15.4.1.2 Plasma Focused Ion Beam (FIB)

Commercially available Plasma-FIB system is equipped with an inductively coupled plasma (ICP) source providing a focused Xe beam in a wide beam current range from several pA up to 2 μA [39]. Compared with the conventional Ga ion beam-based FIB system with maximum beam current of ~65 nA, the TPT of sample preparation process by plasma-FIB system with similar sample finish could be 20–100 times faster.

Plasma FIB combined with SEM technique has been used to reveal subtle defects in a large cross sectional area, such as delamination and cracks in TSVs, microbumps, stacked dice with high quality, and fast TPT [39].

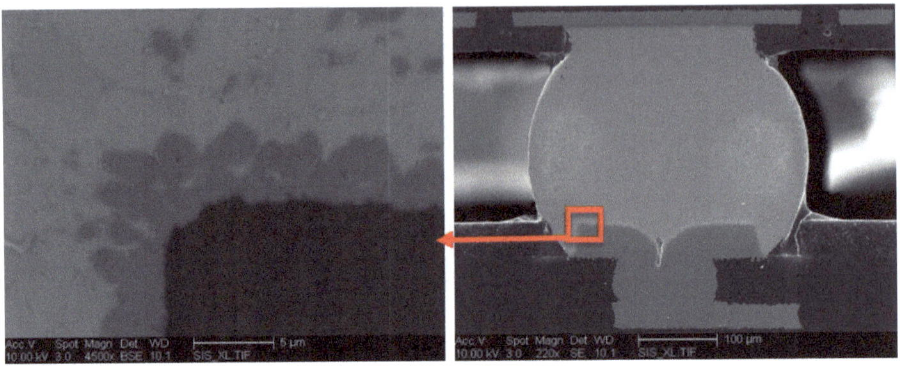

Fig. 15.28 SEM images of SLI solder interconnects cross sectioned by the broad-beam Argon ion milling technique (adapted from Ref. [3])

15.4.1.3 Broad-Beam Argon Ion Milling (EDX)

Unlike the advanced plasma-FIB and fs laser systems, broad-beam Argon ion milling is a low cost sample preparation technique for 3D packages. Units could be cross sectioned either by conventional mechanical technique or laser milling close to the area of interest before performing ion milling. It can produce about 1 mm^2 cross sectional area with artifact-free final finish within a couple of hours. Curtaining artifacts in FIB cross sections are minimized in the ion milling method because of the large area with homogenous ion dose and milling with varied incident angles. Figure 15.28 shows a cross sectional image of a SLI solder joint processed by the ion milling method, which reveals details of the intermetallic compound (IMC) between the solder joint and the substrate metal pad. The ion milling technique has been applied to perform TSV cross sections with artifact-free sample finish and fast TPT [41].

15.4.2 Material Analysis Techniques

Thorough material analysis around the defects is crucial for in-depth root cause investigation in order to find the solution paths during 3D packing technique development. Combined applications of material analysis techniques, such as Energy-dispersive X-ray spectroscopy (EDX), X-ray photoelectron spectroscopy (XPS), Time-of-Flight Secondary Ion Mass Spectrometry (TOF-SIMS), and Electron backscatter diffraction (EBSD) could provide elemental, chemical state, depth profile, and crystallographic orientation of materials around the area of interest. Table 15.5 lists capability, spatial resolution, Pros, and Cons of each technique. Case studies are provided in this section for each material analysis technique to demonstrate its application to 3D package FA.

Table 15.5 Pros and Cons of material analysis techniques for 3D packages

Material analysis technique	Material information provided	Spatial resolution[a]	Pros	Cons
EDX	Elemental analysis	SEM-EDX: ~2 μm TEM-EDX: 1–10 nm	Combined with electron microscopy	Less sensitive for light elements
XPS	Elemental and chemical state information; depth profile	5–30 μm	Analysis depth less than 2–5 nm; quantitative results	Hydrogen and helium not detectable
TOF-SIMS	Elemental and chemical state information; depth profile	~100 nm	Analysis depth less than 2–5 nm; all elements are detectable	Semiquantitative
EBSD	Crystallographic orientation	~10 nm	Visualization of phases, grain size and distribution	Sample preparation take time

Adapted from Refs. [42–44]
[a]Rough estimation in optimized condition, actual resolution depends on different equipment and set up

15.4.2.1 Energy-Dispersive X-ray Spectroscopy (EDX)

EDX is a material analysis technique used in conjunction with electron microscopy, including both SEM and Transmission electron microscopy (TEM), to produce spectra of X-ray counts vs. X-ray energy. When samples are bombarded by the electron beam, electrons ionize atoms in samples, and generates X-rays with characteristic energy unique to the ionized atoms. Depending on the energy and reaction volume with samples of the incident electron beam, elemental composition of the sample about 1–3 μm in depth could be obtained by SEM-EDX [42]. - TEM-EDX provides higher spatial resolution than SEM-EDX as the reaction volume of the high energy electron beam is much smaller in the ultrathin samples. However, TEM sample preparation is time consuming and only a small area could be inspected each time. With μm range spatial resolution, SEM-EDX could be employed as a nondestructive material analysis technique to quickly characterize materials on the sample surface.

Figure 15.29 shows an example of how SEM-EDX is employed to find out the root cause of a die crack. Some foreign materials (FM) shown in Fig. 15.29a are observed on the die back side impact location, which leads to a die front side crack. By comparing with the Si background EDX spectra shown in Fig. 15.29c, EDX analysis of the FM in Fig. 15.29b demonstrates that the FM is rich in aluminum, an element exist in the Thermal Interface Material (TIM) applied between the die back side and the heat sink. Follow-up SEM-EDX study on the heat sink confirmed the presence of aluminum in TIM. The root cause of die crack failure is suggested to be uneven loading stress on the die back side around the impact location, which could be solved by changing the heat sink loading process.

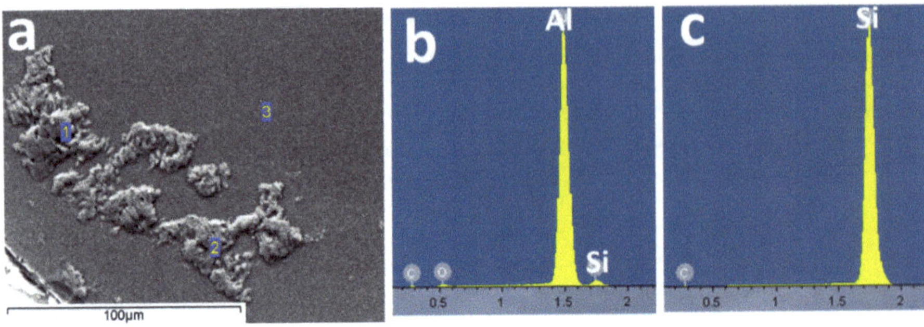

Fig. 15.29 (**a**) SEM image of the Foreign Material (FM) observed around the die back side impact point which lead to a front side die crack. (**b**) EDX spectra taken from the FM. (**c**) EDX spectra taken from the area without FM

15.4.2.2 X-ray Photoelectron Spectroscopy (XPS) and Time-of-Flight Secondary Ion Mass Spectrometry (TOF-SIMS)

XPS also known as Electron Spectroscopy for Chemical Analysis (ESCA) is usually accomplished by measuring the energy of photoelectrons emitted from the sample surface excited with monoenergetic Al kα X-rays. Quantitative elemental information along with chemical state could be determined with an average analysis depth of 5 nm [43].

Time-of-Flight Secondary Ion Mass Spectrometry (TOF-SIMS) provides elemental, chemical state, and molecular information from sample surfaces with about 1 nm analysis depth. It is achieved by using a time-of-flight analyzer to measure the exact mass of ions and clusters emitted from the sample surface excited with a finely focused ion beam. Thin film characterization with depth distribution information is obtained by combining TOF-SIMS measurements with sputtering [44].

In contrast to SEM-EDX which has a typical analysis depth of 1–3 μm, XPS and TOF-SIMS are surface analysis techniques with a typical analysis depth of less than 2–5 nm and are therefore better suited for the compositional analysis of thin layers. In addition, both techniques can be used to characterize molecular information in organic materials that would not be possible with SEM-EDX.

XPS or TOF-SIMS characterization of surface layers or thin film structures is critical for root cause understanding of failures in 3D packages related to impurity or contamination at surface and interface [45, 46], 47]. TOF-SIMS studies illustrated in Fig. 15.30 reveal the relation between Cu electroplating impurities and the Cu to Cu_3Sn IMC separation failure in lead-free solder interconnects post bake reliability tests [47]. As shown in Fig. 15.30a, b, the Cl concentration in the electroplated Cu layer with an electroplating current density of 10 mA cm^{-2} is about 200 times higher than that in the Cu layer electroplated at 0.5 mA cm^{-2}. One hundred and sixty-eight hours of Bake tests at 175 °C are applied on the solder joints assembled on Cu layers with two different electroplating current densities. Cross sectional SEM images of the two kinds of solder joints post bake with Cu

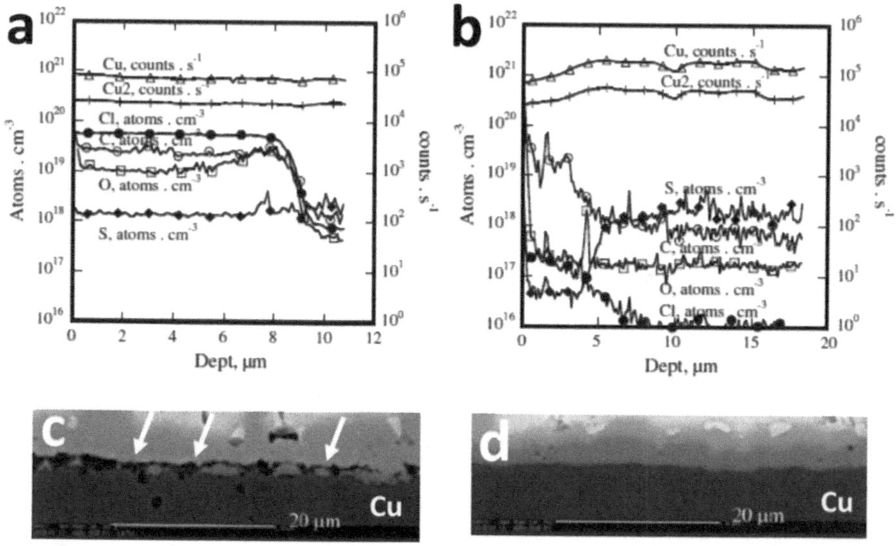

Fig. 15.30 TOF-SIMS analysis of electroplated Cu layers with an electroplating current density of 10 mA cm^{-2} (**a**), and 0.5 mA cm^{-2} (**b**). SEM images of solder joints assembled on the electroplated Cu layers with an electroplating current density of 10 mA cm^{-2} (**c**), and 0.5 mA cm^{-2} (**d**) post 168 h of bake at 175 °C (adapted from Ref. [47])

plating current density of 10 and 0.5 mA cm^{-2} are shown in Fig. 15.30c and d, respectively. Massive Cu to IMC separation is observed in the solder joint assembled on the Cu layer with more impurities (Fig. 15.30c), while the solder joint formed on the Cu layer with less impurities is free of separation. Impurity level in electroplated Cu layers is found to be a big modulator for the Cu to Cu_3Sn IMC separation failure in lead-free solder post bake reliability tests [47].

15.4.2.3 Electron Backscatter Diffraction (EBSD)

EBSD also known as backscatter Kikuchi diffraction (BKD) is a microstructural-crystallographic technique used to define the crystallographic orientation of materials. EBSD data is typically acquired from a cross sectioned unit sitting in SEM. It has been demonstrated that EBSD is critical for the root cause understanding of electromigration (EM)-induced open interconnects [48–50]. Electromigration (EM) in interconnects of 3D packages becomes a major reliability concern due to the much smaller interconnect dimension in advanced packages. EM can cause solder joint open failures due to Sn self-diffusion through a void nucleation and propagation mechanism. On the other hand, electron wind can also accelerate the dissolution of under bump metallization (UBM) on the cathode side and cause failures due to depletion of UBM. It has been reported that metallization dissolution is closely related to Sn grain orientations [48]. It is found that at given temperature

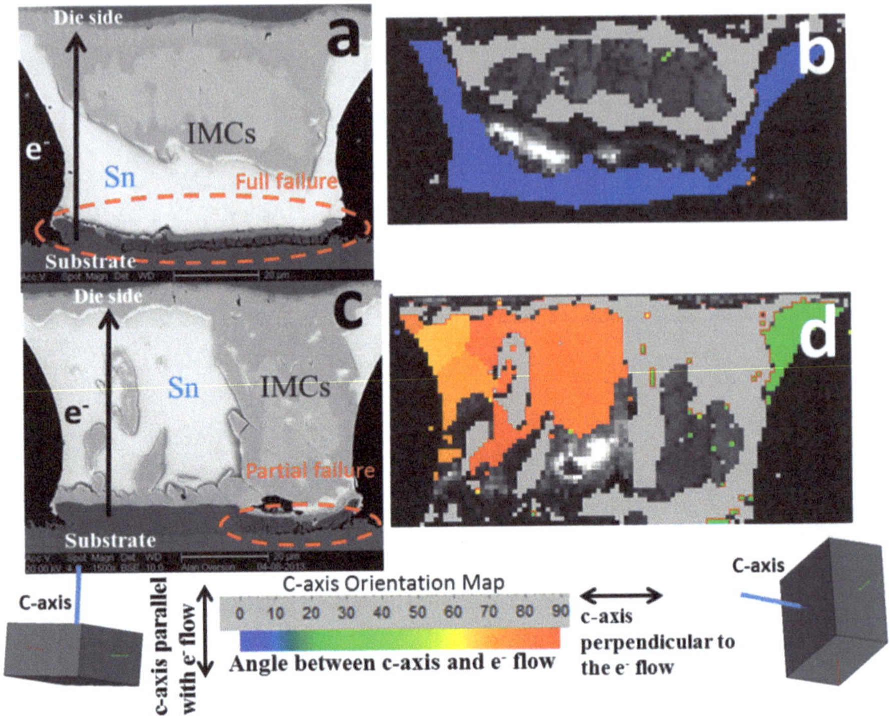

Fig. 15.31 (**a**, **b**) The SEM image and the corresponding EBSD mapping of a failed solder interconnect after EM reliability tests with fast Ni dissolution when the *c*-axis of the single Sn grain of the bulk Solder is in-line with the electron flow direction, respectively; (**c**, **d**) The SEM image and the corresponding EBSD mapping of a survived solder interconnect after EM reliability tests with slow Ni dissolution when the *c*-axis of some Sn grains in the bulk solder is perpendicular to the electron flow direction, respectively (adapted from Ref. [3])

and current density, the time to dissolve 2 μm Ni UBM can vary from 100 h (when Sn *c*-axis is in parallel to electron flow direction) to 1800 h (when *c*-axis is perpendicular to electron flow) [49]. EBSD is used to analyze the Sn grain orientation distributions to explore the relation between metal grain orientation and EM-induced open failures. As illustrated in Fig. 15.31, solder interconnects with *c*-axis of Sn crystals in-line with the electron flow direction have much faster Ni dissolution compared with those having *c*-axis perpendicular to the electron flow direction and could lead to open failures during the EM reliability tests [50].

TSV extrusion, which is shown to be related to the mechanical properties of TSV, has been a major concern for yield and reliability of 3D packages with stacked die and TSV configuration [51]. EBSD is employed to perform the Cu grain size distribution analysis in two types of TSVs with different reliability performance. It has been found that TSVs with smaller and more uniform Cu grains have a higher yield strength are more favorable for inhibiting extrusion [51].

15.5 Failure Analysis Strategies for 3D Packages

Good failure analysis strategies are critical for the efficient applications of multiple FI-FA techniques to achieve the ultimate goal of failure analysis: root cause finding and solution path suggestions.

15.5.1 Understanding the Package Assembly Process, Reliability Stress, and Failure Rate Distribution

Understanding the package assembly process, reliability stress, and failure density distribution is the first key step for flawless failure analysis service. 3D packages usually go through complicated assembly processes, for example, Si dicing, bonding processes for multiple levels of interconnects, underfill curing, and molding process, which involves chemical treatment, high temperature, and sometimes high pressure. Understanding the details of each step is essential to the root cause analysis of failures happened at End of Assembly Line (EOL).

Majority of failures happen post various kinds of reliability tests, which are well designed to mimic the use conditions of packages in an accelerated way. Standard reliability tests for packages include the following [52]:

1. Preconditioning (PC) Test including humidity soak and multiple reflow cycles to reproduce the shipping, storage, and assembly of packages to PCB.
2. Temperature Cycling (TC) Test with temperature cycles go from extremely low temperature to high temperature simulating cyclic "on" and "off" status of devices.
3. Baking Test at excessively high temperature to imitate the long "on" status of devices.
4. Unbiased Highly Accelerated Stress Test (UHAST) and Biased Highly Accelerated Stress Test (BHAST) with high temperature and relative humidity to replicate the harsh environment applications of devices.

Each reliability test initiates specific failure modes and mechanisms. For example, PC test can cause "popcorn" or delamination between package interfaces due to large pressure of vapors from adsorbed moisture or organic contaminations in packages, building up at elevated temperature during each reflow cycle [53]. Figure 15.32a and b shows the TSAM image and cross sectional optical image of big substrate blisters in a package with excessive moisture exposure before the PC test, respectively. Failures happen post-TC tests are typically associated with thermal mechanical stress, for instance, solder joint fatigue crack and substrate trace crack [54, 55]. Figure 15.32c presents a typical fatigue crack in solder joints postextended TC tests. Baking test can generate IMC aging or material degradation-related failures [56]. Figure 15.32e and f displays substrate solder resist layer crack post extended Bake tests, which lead to substrate trace cracks. UHAST failures

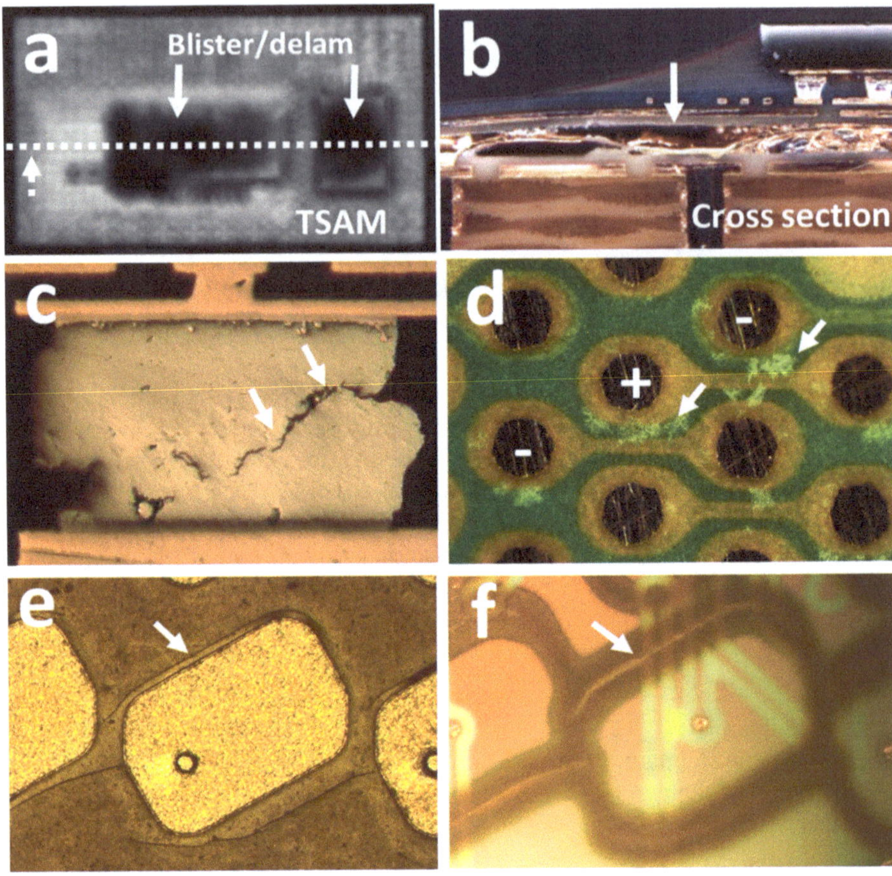

Fig. 15.32 (**a**, **b**) The TSAM and cross sectional optical image of big substrate blisters in a package post PC test due to "popcorn," respectively. (**c**) A typical fatigue crack in MLI solder joints post extended TC tests. (**d**) Cu dendrite between the adjacent Cu pads of FLI solder interconnects with different polarities post extended BHAST tests. (**e**, **f**) Substrate solder resist layer crack post extended Bake Tests, which leads to substrate trace cracks

are often related to corrosion [57], while BHAST failures also involve electrochemical reaction in addition to corrosion, as a bias is applied to the devices during the test [58]. Figure 15.32d shows electrochemical reaction induced Cu dendrite growth between the adjacent Cu pads with different polarities post extended BHAST tests.

Knowing the failure rate distribution in a reliability test is important to tell if the failure is defect driven, intrinsic, or wear out [52]. A well-known bathtub failure rate vs. time curve shown in Fig. 15.33 presents three different types of failures during reliability tests. Infant Mortality Failures are defect drive failure modes, having very high failure rates at early readouts of reliability tests. Intrinsic failures have a constant failure rate during reliability tests. Wear-out failures start to show

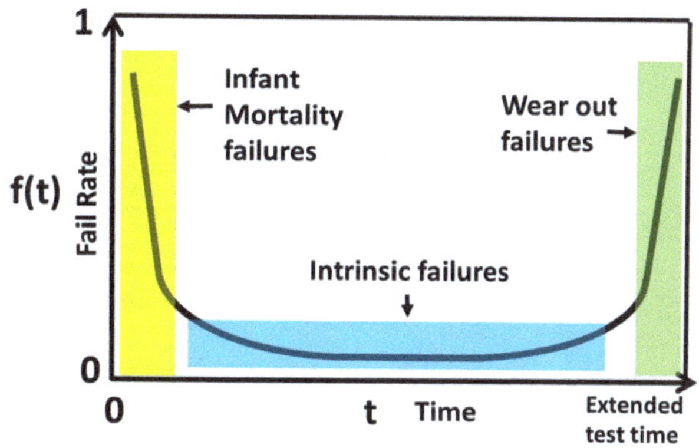

Fig. 15.33 Schematic of Bathtub Curve showing failure rate vs. reliability test time (adapted from Ref. [52])

up with high failure rates at extended test time. Different failures have distinct failure mechanism and root cause, thus need unique solution paths. Infant Mortality Failures are typically due to manufacturing defects and can be fixed by optimizing assembly process to reduce defects or performing a test to screen out defective units. Intrinsic failures are inherent to the design, component selection, and application of packages, whose solution paths are involving major changes to the assembly process. Wear-out failures typically happen at extended readout of reliability tests indicating the end of device life [52].

15.5.2 *Efficient FI-FA Flow to Identify Defects*

Efficient FI-FA flow is important for quickly identifying defects in complex 3D packages. An efficient FI-FA flow generally has the following features:

1. It is a systematic method containing multiple selected FI and FA techniques organized in an optimized and logical way, so that defects could be identified with a high success rate close to 100 %.
2. It has the shortest TPT possible. Short FA data turn is crucial for complex 3D packaging assembly, especially during the technology development stage.
3. It is FI-FA artifact free. Artifacts produced due to improper applications of FI-FA techniques are misleading and should be eliminated.
4. It is low cost. Cost is the primary concern in the competitive packaging industry. Extending capability of low cost FI-FA techniques to achieve comparable or better results is a major consideration during the building up of FI-FA flow.
5. It is flexible for adjustment according to each failure with different reliability test, electrical failure signature, and failure rate. Adjustable FI-FA flow guarantees high success rate with short TPT and low cost by eliminating nonnecessary FI-FA steps for specific failure modes.

Figure 15.34 displays an efficient FI-FA flow for electrical failures post a PC test. The failures are suspected to be Infant Mortality failures associated with

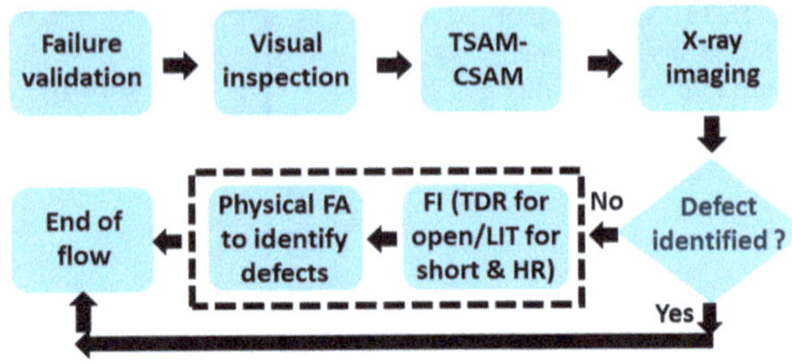

Fig. 15.34 Schematic of an efficient FI-FA flow for electrical failures post-PC test

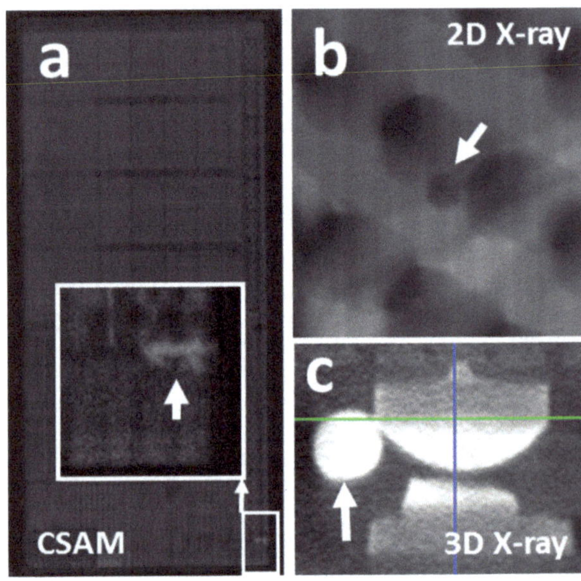

Fig. 15.35 (**a**) TAMI CSAM images showing a UF void in the failing area. (**b**) 2D X-ray image with titled view discloses the solder extrusion in the failing area having the UF void. (**c**) 3D X-ray CT around the solder extrusion revealed an open FLI interconnect

manufacturing defects as they happen post a quality stress test. Nondestructive imaging techniques, including both acoustic imaging and X-ray imaging are added to the FI-FA flow before "fault isolation" and "physical FA" to capture possible popcorn type of substrate blister, UF defects, die crack, nonwet solder joints, solder bridging, or any other gross defects. Using this efficient FI-FA flow, defects in more than 75 % of PC failures could be found by nondestructive acoustic and X-ray imaging, saving both time and cost by skipping nonnecessary analysis steps.

Figure 15.35 presents the FA result of a PC failure gone through the FI-FA flow shown in Fig. 15.34. TAMI CSAM image shown in Fig. 15.35a reveals an UF void covering a couple of FLI interconnects in the failing area. The 2D X-ray image illustrated in Fig. 15.35b discloses the solder extrusion in the failing area having the UF void. 3D X-ray CT shown in Fig. 15.35c demonstrates that the failure is due to an open FLI interconnect having liquid solder flowed into the UF void during reflow cycles of PC tests, a well-known failure mode in microelectronic packages because of UF defects [59].

15.5.3 In-Depth Failure Mechanism and Root Cause Understanding to Provide Solution Paths

The ultimate goal of failure analysis is providing solution paths. Identifying defects is just the first step of the in-depth failure mechanism and root cause analysis. Strategies of root cause analysis normally include the following:

1. Commonality check, including the investigation on commonalities of component, material, supplier, Si wafer, assembly time, tool, environment, etc. Root cause of failures related to process defects could be identified by a through commonality check.
2. Analysis of failure rate distribution and reliability tests. As mentioned earlier in this chapter, failure rate distribution can provide information about failure types, while reliability test analysis could help to determine if the failure is associated with moisture, thermal mechanical stress, corrosion, or electrochemical reaction.
3. Detailed physical analysis of defects, including artifact-free progressive cross section study to understand the 3D shape of defects, material characterization to understand the chemical composition of defects, and "crack tip" study to explore how defects are initiated and propagated.
4. Direct observation of the initiation and propagation of defects by nondestructive and in situ techniques, for example, 2D X-ray with hot stage [35], or progressive reliability test study combined with nondestructive imaging techniques [60]. Kinetic models could be developed along with through root cause understanding and solution paths by these studies.

As demonstrated in Fig. 15.36a, b [60], progressive reliability tests along with nondestructive 3D X-ray CT have been employed to study a new void migration mechanism in Cu vias with preexisting voids, where voids migrate and accumulated on the anode at a very low current density of 10^3–10^4 A/cm^2, quite different from a classical electromigration-induced void evolution with voids appearing on the cathode at much higher current density of 10^5–10^6 A/cm^2. Estimation based on Faraday's law of electrolysis indicates that it is reasonable to propose an electrochemical reaction-induced void migration. Cu from the anode may dissolve in voids filled with electroplating solution and has a potential to deposit on the cathode and hence the movement of voids take place toward the anode.

$$R_m \ \alpha \ j^n \ \exp\ (-E_a/kT)$$

Where R_m is the rate of mass transport, E_a is the activation energy, T is the temperature, j is the current density, k is Boltzmann's constant, and n is a variable current density exponent. As illustrated in Fig. 15.36c, d, E_a and n are estimated as 0.36 eV and 1.4, respectively, based on transported Cu volume calculated from 3D X-ray CT images on samples with different temperatures and current density levels. Both the simulation results and experimental data indicate that the void migration could happen at room temperature with relatively low current density, while no new

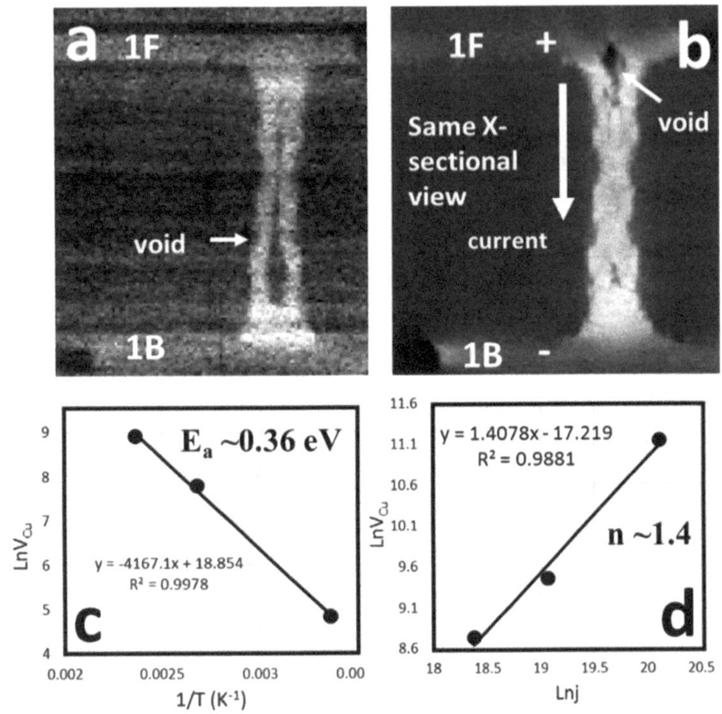

Fig. 15.36 (**a**) 3D X-ray CT virtual X-sectional image of a Cu via at time zero. (**b**) The same virtual X-sectional view of the Cu via after 1000 h of 3 A current flow at 150 °C revealed the void migration to anodes. (**c**, **d**) E_a and n in the Arrhenius Law used to simulate the void evolution are estimated as 0.36 eV and 1.4, respectively (adapted from Ref. [60])

void is created in the Cu vias not having any preexisting voids. Thus, void-free Cu vias are desired especially for packages designed to carry high currents [60].

15.6 Conclusions

Advanced fault isolation and failure analysis techniques for 3D microelectronic packages are reviewed along with FA strategies and efficient FI-FA flow design to demonstrate their applications. It has been illustrated that TDR is a low cost and effective fault isolation tool for locating open failures with spatial resolution of 500 μm. EOTPR could be used to precisely (~10 μm resolution) estimate the failure location with correct data interpretation. Both techniques have been successfully demonstrated for the detection of open and high resistance failures in 3D packages.

LIT and SSM are very promising nondestructive techniques that could provide fault isolation capabilities not just in the x–y plane but also through the depth of packages, which is critical for isolating short and leakage failures in 3D packages.

High-resolution nondestructive imaging techniques such as SAM, 2D X-ray imaging, and 3D X-ray CT could be applied to reveal defects in 3D packages either before or post the fault isolation, depending on different cases.

Sample preparation techniques, such as plasma-FIB, ultra-short pulse laser milling combined with conventional FIB and SEM, and broad-beam Argon ion milling are very efficient to provide artifact-free cross sections of interconnects in 3D packages. EDX, XPS, TOF SIMIS, and EBSD have been demonstrated to be very valuable failure analysis techniques for root cause understanding of failures by providing information about chemical composition, interface contamination, and crystal orientation of defects.

Understanding details of assembly process, various reliability tests, and failure rate distribution is the first key step of 3D package failure analysis. A systematic, artifact free, and efficient FI-FA flow well designed to save both time and cost is essential to quickly identify defects led to electrical failures. In-depth failure mechanism and root cause understanding to provide solution paths of failures is the ultimate goal of failure analysis. Direct observation of the initiation and propagation of defects by nondestructive and in situ techniques are proven to be effective for thorough root cause studies.

Acknowledgement The editors would like to thank Frank Altmann and Sebastian Brand from Fraunhofer-Institute for Microstructure of Materials and Systems IMWS in Germany for their critical review of this Chapter.

References

1. Y. Li, Y. Cai, M. Pacheco, R.C. Dias, D. Goyal, In *Proceedings from the 38th International Symposium for Testing and Failure Analysis (ISTFA, ASM International)*, Phoenix, 2012, p. 95
2. R. Agarwal, W. Zhang, P. Limaye, R. Labie, B. Dimcic, A. Phommahaxay, P. Soussan, In *Proceedings of Electronic Components and Technology Conference (ECTC)*, Las Vegas, 2010, p. 858
3. Y. Li, P.K.M. Srinath, D. Goyal, J. Electron. Mater. **45**(1), 116–124 (2016)
4. Zhiyong Ma and David G. Seiler, "Metrology and Diagnostic Techniques for Nanoelectronics", Pan Stanford Publishing, (2016), pp. 1089–1119. (To be published).
5. Y. Cai, Z. Wang, R.C. Dias, D. Goyal, In *Proceedings from the 36th International Symposium for Testing and Failure Analysis (ISTFA)*, 1 November 2010, ASM International, Materials Park, OH, p. 1309
6. D. Smolyansky, Printed Circuit Design, 20 (April 2002)
7. D. Abessolo-Bidzo, P. Poirier, P. Descamps, B. Domenges, In *Proceedings from the 12th International Symposium on the Physical and Failure Analysis of Integrated Circuits (IPFA)*, 2005, p. 318
8. S.H. Hall, G.W. Hal, J.A. McCall, *High-Speed Digital System Design: A Handbook of Interconnect Theory and Design Practices* (Wiley, New York, 2000)
9. K. Matsumoto, H. Otsuka, O. Horiuchi, Y.G. Han, W. Choi, H. Tomokage, Trans. Jpn. Inst. Electron. Packag. **6**(1), 57 (2013)
10. R. Schlangen, S. Motegi, T. Nagatomo, C. Schmidt, F. Altmann, H. Murakami, S. Hollingshead, J. West, In *Proceedings from the 37th International Symposium for Testing and Failure Analysis (ISTFA, ASM International)* 2011, p. 68

11. F. Naumann, F. Altmann, C. Grosse, R. Herold, In *Proceedings from the 40th International Symposium for Testing and Failure Analysis (ISTFA, ASM International)*, 2014, p. 130
12. M. Xie, T. Begala, K. Kijkanjanapaiboon, D. Goyal, Conference Proceedings from the 66th Electronic Components and Technology Conference. (ECTC), 755 (2016).
13. R. Dias, L. Skoglund, Z. Wang, D. Smith, In *Proceedings from the 27th International Symposium for Testing and Failure Analysis (ISTFA, ASM International)*, Santa Clara, 2001, p. 77
14. M. Xie, Z. Qian, M. Pacheco, Z. Wang, R. Dias, V. Talanov, In *Proceedings from the 38th International Symposium for Testing and Failure Analysis (ISTFA, ASM International)*, Phoenix, 2012, p. 11
15. J. Gaudestad, D. Nuez, P. Tan, In *Proceedings from the 40th International Symposium for Testing and Failure Analysis (ISTFA, ASM International)*, Houston, 2014, p. 43
16. A. Orozco, N.E. Gagliolo, C. Rowlett, E. Wong, A. Moghe, J. Gaudestad, V. Talanov, A. Jeffers, K. Torkashvan, F.C. Wellstood, S. Dobritz, M. Boettcher, A.B. Cawthorne, F. Infante, In *Proceedings from the 40th International Symposium for Testing and Failure Analysis (ISTFA, ASM International)*, Houston, 2014, p. 33
17. W. Lawton, J. Barrett, Microelectron. Reliab. **36**, 1803 (1996)
18. J.E. Semmens, Microelectron. Reliab. **40**, 1539 (2000)
19. D.A. Hutt, D.P. Webb, K.C. Hung, C.W. Tang, P.P. Conway, D.C. Whalley, Y.C. Chan, In *Proceedings from the 2000 IEEE/CPMT Int'l Electronics Manufacturing Technology Symposium*, Santa Clara, 2000, p. 191
20. J.E. Semmens, L.W. Kessler, Microelectron. Reliab. **42**, 1735 (2002)
21. B.T. Khuri-Yakub, Ultrasonics **31**(5), 361 (1993)
22. G.A.D. Briggs, O.V. Kolosov, *Acoustic Microscopy*, 2nd edn. (Oxford University Press, New York, 2009)
23. T.M. Moore, C.D. Hartfield, AIP Conf. Proc. **449**, 598 (1998)
24. A.J. Komrowski, L.A. Curiel, Q. Nguyen, D.J.D. Sullivan, L. Logan-Willams, In *Proceedings from the 30th International Symposium for Testing and Failure Analysis*, Worcester, 2004, p.1
25. L. Angrisani, L. Bechou, D. Dallet, P. Daponte, Y. Ousten, Measurement **31**, 77 (2002)
26. J. Sigmund, M. Kearney, Adv. Packag. July/August (1998).
27. S. Haque, G. Lu, J. Goings, J. Sigmund, Microelectron. Reliab. **40**, 465 (2000)
28. T.S. Leng, J.C.P. McKeon, H.S. Jang, In *Proceedings from the 32th International Symposium for Testing and Failure Analysis*, Austin, 2006, p. 480
29. Y. Li, L. Hu, G. Li, R. Dias, D. Goyal, In *Proceedings from the 39th International Symposium for Testing and Failure Analysis (ISTFA, ASM International)*, San Jose, 2013, pp. 468–470
30. S. Brand, A. Lapadatu, T. Djuric, P. Czurratis, J. Schischka, M. Petzold. J Micro. Nanolithgr. MEMS MOEMS **13**(1), 011207 (2014)
31. I. De Wolf, A. Khaled, M. Herms, M. Wagner, T. Djuric, P. Czurratis, S. Brand, In *Proceedings from the 41st International Symposium for Testing and Failure Analysis (ISTFA, ASM International)*, Portland, 2015, p. 119
32. J. Perraud, S. Enouz-Vedrenne, J. Clement, A. Grivon, In *Proceedings from the 38th International Symposium for Testing and Failure Analysis (ISTFA, ASM International)*, Phoenix, 2012, p. 332
33. M. Feser, J. Gelb, H. Chang, H. Cui, F. Duewer, S.H. Lau, A. Tkachuk, W. Yun, Meas. Sci. Technol. **19**, 094001 (2008)
34. Y. Li, J.S. Moore, B. Pathangey, R.C. Dias, D. Goyal, IEEE Trans. Dev. Mater. Reliab. **12**(2), 494 (2012)
35. Y. Li, R. Panat, B. Li, R. Mulligan, P.K. Muther Srinath, A. Raman, IEEE Trans. Dev. Mater. Reliab. **11**(1), 141 (2011)
36. M. Pacheco, D. Goyal, In *Proceedings of Electronic Components and Technology Conference*, p. 1263.

37. Y. Li, M. Pacheco, D. Goyal, J.W. Elmer, H.D. Barth, D. Parkinson, In *Conference Proceedings from the 64th Electronic Components and Technology Conference (ECTC)*, Lake Buena, 2014, p. 1457
38. J.W. Elmer, Y. Li, H.D. Barth, D.Y. Parkinson, M. Pacheco, D. Goyal, J. Electron. Mater. **3** (12), 4421–4427 (2014)
39. F. Altmann, J. Beyersdorfer, J. Schischka, M. Krause, G. Franz, L. Kwakman, In *Conference Proceedings from the 38th International Symposium for Testing and Failure Analysis (ISTFA, ASM International)*, Phoenix, 2012, p. 39
40. L. Kwakman, M. Straw, G. Coustillier, M. Sentis, J. Beyersdorfer, J. Schischka, F. Naumann, F. Altmann, In *Proceedings from the 39th International Symposium for Testing and Failure Analysis (ISTFA, ASM International)*, San Jose, 2013, p. 17
41. Meyer, G. Grimm, M. Hecker, M. Weisheit, E. Langer, In *Proceedings from the 38th International Symposium for Testing and Failure Analysis (ISTFA, ASM International)*, San Jose, p. 12 (2013)
42. H. Xu, H. Zhang, M. Xue, In *Conference Proceedings from the 15th Electronic Components and Technology Conference (ECTC)*, p. 403 (2013).
43. P. Van der Heide, *X-ray Photoelectron Spectroscopy: An introduction to Principles and Practices* (Wiley, Hoboken, 2011)
44. P. Van der Heide, *Secondary Ion Mass Spectrometry: An Introduction to Principles and Practices* (Wiley, New York, 2014)
45. F. Yang, S. Yiqiang, L.H. Sheng, F. Chao, In *Conference Proceedings from the 15th Electronic Packaging Technology Conference (EPTC)*, Singapore, 2013, p. 37
46. B. Pathangey, A. Proctor, Z. Wang, Z. Fu, R. Tanikella, IEEE Trans. Dev. Mater. Reliab. **7**(1), 11 (2007)
47. Y. Liu, J. Wang, L. Yin, P. Kondos, C. Parks, P. Borgesen, D.W. Henderson, E.J. Cotts, N. Dimitrov, J. Appl. Electrochem. **38**, 1695 (2008)
48. M.H. Lu, D.Y. Shih, P. Lauro, C. Goldsmith, D.W. Henderson, Appl. Phys. Lett. **92**, 211909 (2008)
49. Y. Wang, P.S. Ho, Appl. Phys. Lett. **103**, 121909 (2013)
50. P. Liu, A. Overson, D. Goyal, In *Conference Proceedings from the 65th Electronic Components and Technology Conference (ECTC)*, San Diego, 2015, p. 99
51. C. Wu, T. Jiang, J. Im, K.M. Liechti, R. Huang, P.S. Ho, In *Proceedings from the 21st International Symposium on the Physical and Failure Analysis of Integrated Circuits (IPFA)*, 2014, p. 312
52. R.R. Tummala, *Fundamentals of Microsystems Packaging* (McGraw-Hill, New York, 2001), pp. 878–923
53. S. Atkins, L. Teems, W. Rowe, P. Selby, R. Vaughters, Microelectron. Reliab. **38**, 773 (1998)
54. T. Akutsu, Q. Yu, Y. Nishimura, In *Conference Proceedings from the 12th Electronic Packaging Technology Conference (EPTC)*, Singapore, 2010, p. 550
55. M.M. Basit, M. Motalab, J.C. Suhling, Z. Hai, J. Evans, M.J. Bozack, P. Lall, In *Conference Proceedings from the 65th Electronic Components and Technology Conference (ECTC)*, San Diego, 2015, p. 106
56. D. Yang, Z. Cui, In *Conference Proceedings from the 2011 International Conference on Electronic Packaging Technology & High Density Packaging*, Shanghai, 2011, p. 1030
57. S.H. Kim, J.W. Park, S.J. Hong, J.T. Moon, In *Conference Proceedings from the 12th Electronic Packaging Technology Conference (EPTC)*, 2010, p. 545
58. K. Lee, S. Barbeau, F. Racicot, D. Powell, C. Arvin, T. Wassick, J. Ross, In *Conference Proceedings from the 63rd Electronic Components and Technology Conference (ECTC)*, 2013, p. 2138
59. X.L. Zhao, J.M. Chin, R.N. Master, In *Conference Proceedings of 12th IPFA*, Singapore, 2005, p. 258
60. Y. Li, L. Xu, P. Liu, B. Pathangey, M. Pacheco, M. Hossain, L. Hu, R. Dias, D. Goyal, In *Conference Proceedings from the 65th Electronic Components and Technology Conference (ECTC)*, San Diego, 2015, p. 432

Index

Y. Li, D. Goyal (eds.), *3D Microelectronic Packaging*, Springer Series in Advanced Microelectronics 57, DOI 10.1007/978-3-319-44586-1

FSC
www.fsc.org
MIX
Papier | Fördert gute Waldnutzung
FSC® C083411

Kolibri 360 GmbH
Ferdinand-Jühlke-Straße 7
99095 Erfurt, Deutschland
produktsicherheit@kolibri360.de